Risk Revealed

Explore the concept of risk through numerous examples and their statistical modeling, traveling from a historical perspective all the way to an up-to-date technical analysis.

Written with a wide readership in mind, this book begins with accounts of a selection of major historical disasters, such as the North Sea flood of 1953 and the L'Aquila earthquake. These tales serve to set the scene and to motivate the second part of the book, which describes the mathematical tools required to analyze these events, and how to use them. The focus is on the basic understanding of the mathematical modeling of risk and what types of questions the methods allow one to answer.

The text offers a bridge between the world of science and that of everyday experience. It is written to be accessible to readers with only a basic background in mathematics and statistics. Even the more technical discussions are interspersed with historical comments and plentiful examples.

PAUL EMBRECHTS is Emeritus Professor of Insurance Mathematics in the Department of Mathematics of ETH Zurich, Switzerland. He holds numerous distinctions and awards from universities and organizations worldwide. He co-authored the influential books *Modelling Extremal Events for Insurance and Finance* and *Quantitative Risk Management: Concepts, Techniques and Tools* and has published over 200 scientific papers in leading international scientific journals.

MARIUS HOFERT is Associate Professor in the Department of Statistics and Actuarial Science at The University of Hong Kong. He obtained his PhD in Mathematics from Ulm University in 2010. He then held a postdoctoral research position at RiskLab, ETH Zurich. Afterwards, he was Guest Professor in the Department of Mathematics at the Technische Universität München, Visiting Assistant Professor in the Department of Applied Mathematics at the University of Washington and Associate Professor in the Department of Statistics and Actuarial Science at the University of Waterloo. Marius' research interests are dependence modeling, computational statistics, data science and quantitative risk management. He has offered several courses, mini-courses, workshops, summer and winter schools in these areas, including courses on risk management at the Risk Management Institute at the National University of Singapore and at the 29th International Summer School of the Swiss Association of Actuaries. Marius also participates in the education of actuaries and risk managers by developing teaching material and software freely available on qrmtutorial.org.

VALÉRIE CHAVEZ-DEMOULIN is Professor of Statistics at the Faculty of Business and Economics, University of Lausanne (UniL). She is also co-founder and on the Executive and Scientific Board of the UniL research center ECCE (Expertise Center for Climate Extremes). Valérie holds a Master's degree in Mathematics from EPFL and a PhD in Mathematics (specializing in Statistics) from the same institution. She was a research fellow at the Department of Mathematics (D-MATH) at ETH Zurich and later an Invited Professor at D-MATH, ETH Zurich, for a sabbatical leave. Her domain of expertise is extreme value theory and in particular, the statistical modeling of univariate or multivariate extreme events in non-stationary or covariate-dependent contexts.

Risk Revealed

Cautionary Tales, Understanding and Communication

Paul Embrechts
ETH Zurich, Switzerland

Marius Hofert
The University of Hong Kong, Hong Kong

Valérie Chavez-Demoulin
University of Lausanne, Switzerland

CAMBRIDGE
UNIVERSITY PRESS

CAMBRIDGE
UNIVERSITY PRESS

Shaftesbury Road, Cambridge CB2 8EA, United Kingdom

One Liberty Plaza, 20th Floor, New York, NY 10006, USA

477 Williamstown Road, Port Melbourne, VIC 3207, Australia

314–321, 3rd Floor, Plot 3, Splendor Forum, Jasola District Centre, New Delhi – 110025, India

103 Penang Road, #05–06/07, Visioncrest Commercial, Singapore 238467

Cambridge University Press is part of Cambridge University Press & Assessment, a department of the University of Cambridge

We share the University's mission to contribute to society through the pursuit of education, learning, and research at the highest international levels of excellence.

www.cambridge.org
Information on this title: www.cambridge.org/9781009299800

DOI: 10.1017/9781009299794

First published 2024

Printed in the United Kingdom by CPI Group Ltd, Croydon CR0 4YY

A catalogue record for this publication is available from the British Library

Library of Congress Cataloging-in-Publication Data
Names: Embrechts, Paul, 1953- author. | Hofert, Marius, author. | Chavez-Demoulin, Valérie, 1969- author.
Title: Risk revealed : cautionary tales, understanding and communication / Paul Embrechts, Marius Hofert, Valérie Chavez-Demoulin.
Description: [Cambridge] : [Cambridge University Press], [2024] | Includes bibliographical references and index.
Identifiers: LCCN 2023040041 (print) | LCCN 2023040042 (ebook) | ISBN 9781009299800 (hardback) | ISBN 9781009299817 (paperback) | ISBN 9781009299794 (epub)
Subjects: LCSH: Risk assessment–Statistical methods.
Classification: LCC HM1101 .E44 2024 (print) | LCC HM1101 (ebook) | DDC 363.1/020727–dc23/eng/20231011
LC record available at https://lccn.loc.gov/2023040041
LC ebook record available at https://lccn.loc.gov/2023040042

ISBN 978-1-009-29980-0 Hardback
ISBN 978-1-009-29981-7 Paperback

To Emma, Olivia and Gerda

To Saisai

To Kiko and Camille

Contents

Preface

The desire to write grows with writing.
Erasmus of Rotterdam (1466–1536)

This book project was first announced during the farewell lecture of Paul Embrechts on May 31, 2018, on the occasion of his retirement as professor of mathematics from ETH Zurich; see Embrechts (2018, minute 44). The book's planned outline as well as its writing experienced a major disturbance after the coronavirus pandemic hit us all. We had only one longer stretch of joint collaboration, in Zurich in September and October 2019. From early 2020 onward, videoconferencing became our main method of communication. As one example of this shift in our *modus operandi* we would like to mention that we planned to spend some weeks in 2020 together at the famous Mathematisches Forschungsinstitut Oberwolfach in Germany's Black Forest, but sadly, it was not to be. The project got a major boost when Marius Hofert stayed at the Forschungsinstitut für Mathematik (FIM) of the Department of Mathematics (D-MATH) of ETH Zurich from March to August, 2021. This visit allowed us to communicate online in the same time zone and, occasionally, meet outdoors at Polyterrasse or Lindenhof. Our gratitude hence goes out to FIM for its financial and logistic support, also for the earlier visits in 2019 of Marius Hofert and Valérie Chavez-Demoulin to ETH Zurich. Our thanks go to Alessio Figalli and Andrea Waldburger. Likewise, we would like to thank the IT Support Group of D-MATH for helping us master the ever-growing size of the files being whizzed around electronically.

Throughout visits to ETH Zurich by Marius Hofert and Valérie Chavez-Demoulin, the RiskLab of D-MATH offered an academically stimulating environment; for this, we thank the current director, Patrick Cheridito. Galit Shoham made visiting RiskLab always a home-coming event. Her efficient and friendly support so often went well beyond the expected. As university teachers we are grateful that over many years all three of us have been able to teach generations of students on topics related to this book. Their questions so often improved our own understanding of the material taught. We are grateful to our home institutions for the stimulating academic environment they provide in which our academic ideas and dreams can grow into a concrete shape, in this case the book you are reading.

Paul Embrechts gratefully acknowledges the academic post-retirement hospitality and support from ETH Zurich's D-MATH, RiskLab and the ETH Risk Center. While working on the book, and in the isolation coming from the coronavirus pandemic, for him the Flemish radio program "Klara Blijf Verwonderd" provided a musical background, acting as a much-needed muse. During numerous discussions with his wife Gerda, going over the book's developing content was always met with a constructive, but also critical, response.

Marius Hofert would like to thank the Statistics and Actuarial Science departments at University of Waterloo and The University of Hong Kong for their support for travels abroad. Marius would also like to thank the data scientists at Rijkswaterstaat and Alessandro Amato of the Istituto Nazionale di Geofisica e Vulcanologia for providing even more relevant information about the data than we asked for, an unmatched first.

Valérie Chavez-Demoulin would like to thank HEC Lausanne and the members of the Department of Operations, with special thanks to her colleagues Marc-Olivier Boldi, Linda Mhalla, Maximilian Aigner, Juraj Bodik and Fabien Baeriswyl for stimulating discussions and the work-group atmosphere.

Of course, over the years we discussed with numerous people topics related to this book. Singling out some of these colleagues is near impossible. One name, however, that we do want to mention here specifically is that of our friend and mentor Hans Bühlmann. For many years he has been a constant source of important and relevant information on the topic of risk for all three of us, especially when insurance considerations enter the discussion. His insights on the topic of risk are second to none.

We are especially grateful to Saisai Zhang for ending our three-year-long struggle for finding a good title for the book in a brilliant flash of thought. We would like to thank Kiko Chavez for his excellent work on the illustrations, as well as Niels Hagenbuch, Matthias Kirchner and Linda Mhalla for proofreading our book.

We are indebted to numerous colleagues and friends, as well as anonymous reviewers, who commented on earlier versions of our manuscript or helped us in our understanding of the topic of risk. These include Guus Balkema, Bastian Bergmann, Roger Cooke, Anthony Davison, Rüdiger Frey, Claudia Klüppelberg, Marie Kratz, Hans Rudolf Künsch, Alexander J. McNeil, Thomas Mikosch, Philippe Naveau, David Raymont, Bozar Stojadinovic, Jonathan Tawn, Andreas Tsanakas, Olivier Wintenberger and Mario Wüthrich. We also thank the team at Cambridge University Press for their professional help in the production of this book, particularly our editors Clare Dennison, David Tranah and editorial assistant Anna Scriven.

Our deepest thanks go to our partners, Gerda, Saisai and Kiko, and our families. Even though it was not hard to convince them of the usefulness of this book project, every large-scale venture comes with considerable sacrifice and we are eternally grateful for their continuing support.

Introduction

So long as you write what you wish to write,
that is all that matters; and whether it matters
for ages or only for hours, nobody can say.

Virginia Woolf (1882–1941) in *A Room of One's Own*

Why we have written this book. For many years, we have taught courses on probability, statistics, insurance and financial mathematics and the modeling of extremes and quantitative risk management. Our students come from a wide range of disciplines. Beyond these, we have also offered courses and seminars to practitioners and regulators. Over the recent years, an increasing awareness of risk-related issues has become omnipresent in society. This no doubt is in part due to the continuous streaming of news bulletins through diverse electronic media. Transporting knowledge down from the academic ivory tower remains a difficult task that far too often is neglected by academics; see Figure 0.1 for a cartoon interpretation. This book offers a bridge between the world of scientific writing and that of everyday experience.

Figure 0.1 Coming down the academic ivory tower. Source: Enrico Chavez

We deliberately write *a* and not *the*, as our presentation is very much colored by our own scientific background as well as our personal experiences. Throughout, we address risks due to natural and man-made disasters, and we describe how such events and future risks can be understood, modeled and communicated. Much has been said and written about risk at the non-technical level, and more and more specialized literature concerning the modeling of risks is emerging. Our goal is to bridge this gap and provide a widely accessible treatment at a sufficiently technical level to appreciate the types of question mathematical modeling can answer. Clearly this is ambitious. We see our effort more as providing you, the reader, with a walking stick to discover the land of risk at your own pace. Occasionally we will encounter a more demanding path, but the experience of walking together is always more important than the ultimate goal of reaching a local summit. Of course, when we reach a local summit, it is worthwhile to look around and admire the view.

Whom we have written this book for. The book is written with a wide readership in mind. A unique feature is its mix of selected accounts of major historical disasters. Beyond these general presentations, we cover the understanding and mathematical modeling of risk described in an accessible way so that readers with only basic mathematics background (high school level, including, for example, the knowledge of limits, derivatives, integrals, elementary mathematical functions) should be able to follow our reasoning. Even the slightly more technical discussions are interspersed with historical comments and simple worked examples. Of course there exist several examples of researchers in the realm of risk who have contributed importantly to the public risk debate. The various chapters contain references to these personalities. One name that occurs often throughout the book is that of Sir David Spiegelhalter, the former Winton Professor of the Public Understanding of Risk in the Statistical Laboratory at the University of Cambridge. It is impossible to better the work done by David Spiegelhalter on risk communication and its understanding, aimed at a wider public. We do hope however that our book does justice to his efforts and offers complementary material on, as well as interpretations of, the multifaceted manifestation of risk.

Risk in its various disguises: a historical perspective. So far we have mentioned a couple of times the word *risk*, but which interpretation hides behind this word? Besides risk, we can equally well talk about chance, luck, or fate, say. For instance, we all have experienced "lucky escapes", or had good or bad luck in playing a game, or a lucky experience in daily life. But what is this elusive luck? In Roman and Greek mythology, Tyche (Greek) or Fortuna (Latin) personified the goddess of luck. Fortuna was much more important to the Romans than Tyche was to the Greeks, hence also the saying being "fortunate". We learn from Wikipedia (2020b) that "The Greek historian Polybius believed that when no cause can be discovered to events such as floods, droughts, frosts, or even in politics, then the cause of these events may be fairly attributed to Tyche." Especially for politicians this could be rather convenient. Whereas the goddess Fortuna mainly blessed citizens with beneficial events, Tyche embodied bad luck as well as good luck, and hence would address both sides of the risk medal. Etymologically, the word "risk" most likely derives from the ancient Greek word "$\rho\iota\zeta\alpha$" (rhiza). The word has many meanings, both literally as well as metaphorically. We concentrate on its translation to the root or branch of a tree – something that offers stability, something to hold onto. In this interpretation, we find the word back in Homer's account of Odysseus' passage through the

Strait of Messina between Scylla and Charybdis. The latter two are mythological sea monsters trying to kill Odysseus and his companions. A relevant detail in the story is that Charybdis lurked under a fig tree, the roots (hence rhiza) of which would assist Odysseus to pass. Later interpretations referred to Scylla as a rock and Charybdis as a whirlpool. In English, rhiza obtained a nautical translation as a hazard of sailing along rocky coasts. Nowadays, being "between Scylla and Charybdis" means being caught between two equally unpleasant alternatives; related to the advice "to choose the lesser of two evils". An excellent depiction of this interpretation is to be found in James Gillray's 1793 political cartoon "Britannia between Scylla and Charybdis"; see Figure 0.2. In the picture, on the summit of Scylla is a

Figure 0.2 "Britannia between Scylla and Charybdis", James Gillray, 1793. The three sharks with human heads (Richard Brinsley Sheridan, Charles James Fox and Joseph Priestley) closely pursue the boat. Source: Wikimedia Commons

large bonnet rouge with a tri-colored cockade, symbol of the 1789 French Revolution, and on the right the inverted royal crown in Charybdis' whirlpool symbolizing arbitrary power.

Everyday language is not precise when it comes to possible differences between the usage and the philosophical loading of words like risk, chance, fate, fortune, luck, to name a few. In this book, we will gradually translate risk into more technical, mathematical language; see, for example, Sections 8.4.2 and 8.4.5. When doing so, we will always be very well aware of the words spoken by Hamlet to Horatio in Shakespeare's play "The Tragedy of Hamlet" (Act I, Scene 5): "There are more things in heaven and earth, Horatio, Than are dreamt of in your philosophy." For a most readable and informative historical overview of risk, see Bernstein (1996). An excellent talk on the concept of luck is to be found in the entertaining

BBC 4 presentation Spiegelhalter (2019). In it, David Spiegelhalter ends his discourse with the following words:

> After listening to all these stories what would I say luck is. I heard that your real luck happens before you were born and after that it's a mixture of chance and your attitude that determines what happens to you. But even hardened gamblers often can't help feeling that they're in the grip of some temporary external force [Tyche?] of good or bad fortune. So if we think of chance as simply unavoidable unpredictability, then I agree with David Flusfeder [see Flusfeder (2018)] that luck is chance taken personally.

We will return to the interpretation of chance as "unavoidable unpredictability", such as the result of tossing a fair coin, in our discussion on probability theory in the more mathematical Chapter 8.

It is perhaps no coincidence that representatives of the animal kingdom are often used in idioms related to risk. You have surely heard of "the elephant in the room", "bear and bull markets" or "crying wolf". The latter epitomizes the problems faced by early warning systems, for example in tsunami or earthquake-prone regions, or false negative reporting in case of a pandemic. Together with Nassim Taleb's "black swan", see Taleb (2007), we could also have added a "gray rhino" (Wucker (2016)) as well as Didier Sornette's double metaphor "dragon kings" (Sornette (2009)). It may be our hunter–gatherer distant past that coded the animal link to risk into our genes.

Two examples from daily life. Before we delve a bit deeper into the structure of the book, we would like to recount two lighthearted examples where risk was involved. The stories below we learned from a close friend of ours, Agnes Herzberg, Professor Emeritus of Statistics at Queen's University, Kingston, Ontario, Canada. Since 1996, Agnes has organized international conferences on the topic of statistics, science and public policy. These annual conferences bring together a diverse mix of scientists, politicians, civil servants and journalists. The second conference in the series took place in 1997; see Herzberg and Krupka (1997) for its proceedings. The stories underscore two important aspects: (1) giving a precise and multipurpose definition of the concept of risk is difficult, if even possible, particularly as (2) any definition very much depends on personal attitude and historical perspective. Throughout the book, we will of course offer you some more technical 'translations'. For the moment, we leave you with the following quotes from Agnes' opening remarks to the above conference:

> [. . .] I returned to Kingston on a small plane from Toronto. I was in an aisle seat and just before take-off a woman arrived, wanting the window seat beside me. She was carrying many bags and a large cup of tea. I held the tea for her as she settled in her seat. Still holding the cup, I sat down, but did not realize that the lid was not secure. The movement caused the tea to spill everywhere. I began to read the book Risk by John Adams. [. . .] When I was asked "What is Risk?" I suggested it was perhaps risky to go on this airplane and go to Kingston. I returned to reading, only to be interrupted again and again for more examples. Finally, in exasperation, I replied, "Suppose you get on an airplane and someone gives you a cup of tea to hold and then it spills all over you; that is taking a risk!" No more was said.

The opening remarks end with:

> Professor Lewis Wolpert, a biologist and writer, used the following example [he presented at a] Royal Society meeting in London [on science, policy and risk]. His office is in the basement of a building at University College London. The hallways are made of brick and there are glass-covered signs stating that because of the fire hazard, no paper may be posted unless it is under glass. He then enters his office where every surface is covered with paper! Our attitudes toward risk are contradictory at best.

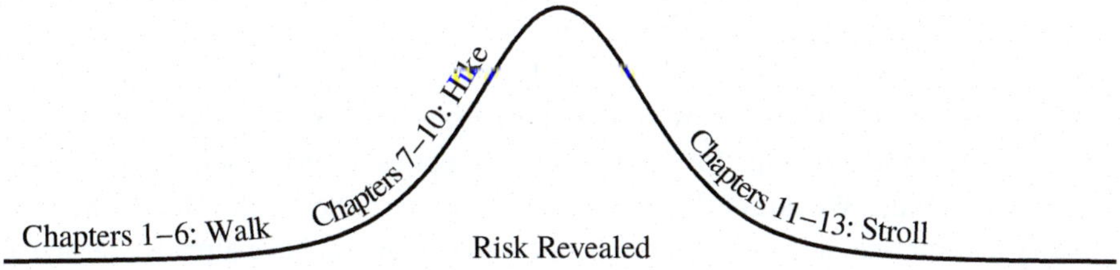

Figure 0.3 A walk, a hike and a stroll through the landscape of risk. Source: Authors

How the book is structured. We will take you for a guided tour through different terrains in the multifaceted world of risk. Our tour will consist of a *walk*, a *hike* and a *stroll*; see Figure 0.3. We start with several cautionary tales, which are presented in a non-technical, widely accessible way; hence this part of the book is just a *walk*. These tales also serve as motivation for the second part of the book, where we introduce you to some basic mathematical techniques from the realm of probability and statistics. The usefulness of these techniques is highlighted in several applications where we return to some of the cautionary tales. Our focus in the second part of the book is on obtaining a basic understanding of the mathematical modeling of risk and what types of questions mathematical methodology allows us to answer. This more technical *hike* requires a bit of stamina but only very rarely additional oxygen. We finish the book with a final, hopefully pleasant and relaxing *stroll*. Chapter 7 has a somewhat special character as it sits between the cautionary tales and the more mathematical, technical hike. Its aim is to offer you, the reader, a brief glimpse of the beauty of mathematics. As such, Chapter 7 acts as a bridge that we cross together before we start on the more serious hike.

A recurring theme. Throughout we ask questions of the *if* and *what if* variety. For example, on July 18, 2021, the catastrophic low-pressure system Bernd ravaged parts of Germany, Belgium, The Netherlands and Austria, causing several hundred deaths and a massive material loss. Also at that time, untold high temperatures, just short of 50 °C, scorched Northwestern USA and Canada. An *if* question corresponds to asking for the probability that a low-pressure zone like Bernd gets stuck for several days above a relatively small geographic area, as it did above the states of Rheinland-Pfalz and Nordrhein-Westfalen in Germany, inundating these areas with massive amounts of rain. The word "rain bomb" entered the journalistic vocabulary. Similarly, what is the probability of a high-pressure zone settling for an unusually long time above the west coast of Canada? One explanation that climate experts offer is the slowing down of the jet stream, forming local cusps and trapping highs and lows rather than moving them faster along its path. Putting a statistical estimate on such events, as well as proving climate change causality, is very difficult. It is precisely here that a *what if* scenario analysis becomes highly relevant. Skeptics may say that, before the event under discussion, the *if* event is deemed so rare that the *what if* stress scenario should not enter our risk radar. Our answer invariably is that this is not the point nor the question. We as individuals and society ought to consider the consequences of such a rare *if* event as much as possible beforehand. This is the essence of the *what if* thinking and related methodology like extreme value theory, a topic we will cover in Chapter 9. Skeptics may still object that this is all rather academic and quote that "after the event we always know better". This surely holds some truth and is related to the so-called hindsight bias corresponding to a "we always knew" attitude. The chapters to follow contain numerous examples through which we hope to convince you that there is hope for scientists and skeptics to meet each other and achieve a mutual understanding. We admit that we have set the bar rather high. The threats we face do not justify a lower one, however. We very much hope that you have already learned a first concrete lesson from our book: when faced with a discussion on risk, it does not suffice to ask the *if* question; we always need to ask the *what if* question, too. It will invariably broaden your understanding of the underlying topic.

Reader guidelines. From our brief discussion of its walk, hike and stroll structure, it should be clear that this book is not necessarily to be read in one go from cover to cover, though you are of course welcome to do so. Enjoy the material presented at your own pace. On occasion, the book's language may be deemed somewhat scholastic, in that even in the less technical parts we try to be precise on sources used and quotes included. However, we always try hard to keep the language used at a narrative level, and this despite the interspersed referencing. The references form an important part of the book. They are carefully chosen to offer you the possibility of delving deeper into a specific topic of interest. Also important to us is that the less technically versed among readers should not be frightened by the use of mathematical formulas in the more technical chapters. Even in those chapters you will always find some pedagogical and historically motivated examples to assist you in following the main exposé. The book is primarily oriented to individual readers, but it can certainly be used as a guiding text for a graduate course on the understanding and communication of risk. The various chapters can easily be structured into part self-study, part *ex cathedra* teaching as well as accompanying classroom projects.

A picture is worth a thousand words. In a somewhat different formulation, this famous quote goes back to the Norwegian writer and playwright Henrik Ibsen (1828–1906). If you flip through the pages of our book, you surely will notice the various graphs, figures and even cartoons. Those graphs and figures produced by us were made by Marius Hofert with the free software R (for statistical computing and graphics), LaTeX (a typesetting software) or, occasionally, Google Drawings (a web-based drawing software). When a graph has been produced by us, we have always taken great care to present it in its best, hence most informative, format. We have also made history come to life by inserting photographs of important risk-related events as well as of the personalities who played a key role in the stories told. A special feature is the several cartoons drawn by Enrico (Kiko) Chavez, the husband of Valérie Chavez-Demoulin. As a professional statistician, Kiko regularly provides cartoons for statistical publications such as *Significance* and the *Bulletin of the Swiss Statistical Society*. Like any cartoons, their main message is somewhat hidden in the presentation. Do spend some time looking at them in detail and discover the deeper insights that this contemplation typically will offer you on the topic under discussion.

Intelligibility versus correctness. When discussing the probabilistic modeling of extremes in Chapter 9, the name of Emil Julius Gumbel will frequently appear. He wrote an early classic on the field of extreme value theory; see Gumbel (1958). From that book's Summary section we quote the following headline: “A book should either have intelligibility or correctness. To combine the two is impossible.” With all respect for the greatness of Gumbel, on this issue we beg to differ. At least we have tried hard to achieve this combination. It is for you, the reader, to judge our efforts.

1

The 1953 Great Flood

[. . .] en in alle gewesten wordt de stem van het water met zijn eeuwige rampen gevreesd en gehoord.

Hendrik Marsman (1899–1940) in "Herinnering aan Holland"

[. . .] and in every county the water's warning of more catastrophes heard and heeded.

Translated by Michael Longley (from *Wavelengths: Various Translations*, Longley, M. and Morgan, J., Enitharmon, 2009)

1.1 The Night of January 31, 1953

On Saturday, January 31, 1953, around 18:15 local time, at the end of the evening news, every radio listener in The Netherlands heard the message below from the Dutch meteorological service (KNMI):

Boven het noordelijke en westelijke deel van de Noordzee woedt een zware storm tussen noordwest en noord. Het stormveld breidt zich verder over de noordelijke en oostelijke Noordzee uit. Verwacht mag worden dat de storm de hele nacht zal voortduren. Daarom werden vanmiddag om half zes de groepen Rotterdam, Willemstad en Bergen op Zoom gewaarschuwd voor gevaarlijk hoogwater.

Translated:

Over the northern and western parts of the North Sea, a strong gale rages from between northwest and north. The storm field is extending further north and east over the North Sea. It is expected that the storm will continue the whole night and given this fact, this afternoon at 17:30 the areas of Rotterdam, Willemstad and Bergen op Zoom have been warned of dangerously high water.

The experts of the KNMI, who had been monitoring the storm depression since Friday, became increasingly worried about potential consequences for the coastal areas of The Netherlands. This refers first to Zeeland, the southwest delta area which also borders on Belgium. A further area at high risk is the southwestern part of Holland, which contains the main shipping harbor Rotterdam. The northern part of the 20.5 km New Waterway canal linking the North Sea to Rotterdam's harbor is known as Hoek van Holland (Hook of Holland). Only two days before it had been a full moon, with the possible consequence of a so-called spring tide ("springvloed"). In combination with an expanding severe storm already reaching force 10 and more, and taking a direction straight to the coasts of Zeeland and Holland, expert expectations for an extremely high tide for the coming night increased.

Concerning the above radio message we would like to add a comment. During those ominous days, the KNMI meteorologist Klaas Rienk Postma (1913–2005) was on duty. For the above radio message at the time, there were only two official formulations available that were allowed to be wired as a warning, namely "flink hoogwater" (seriously high water) and the chosen "gevaarlijk hoogwater" (dangerously high water). Postma would have preferred to communicate the not allowed "zeer gevaarlijk hoogwater" (very dangerously high water).

Today, more than 60 years on, the population of a region at risk can (and does) get ample information on possible catastrophic weather conditions, through radio and television news, both screened almost continuously, and of course through the omnipresent internet and social media. Furthermore, today's storm predictions typically have a pre-event time lag of about 4 to 5 days with rather precise predictions of intensity, storm duration, storm path and geographic landing. KNMI did an excellent job in recognizing the threat but failed in getting its message sufficiently through to the people of Zeeland and Holland. In those days, risk communication was technologically underdeveloped.

In 1953 media warnings were extremely restricted; essentially only one warning was given, which was probably missed by many people. After 00:00 that night, no further warnings were transmitted to the population until 08:00 that Sunday morning, February 1. As every night, the radio program ended that Saturday midnight with the Wilhelmus, the Dutch National Anthem. The consequences were disastrous. People in Zeeland and Holland, who for centuries have been aware of the threat posed by living near the sea, went to bed trusting the existing dikes. That night the storm battered the coast with ferocious power. The ensuing seawater surge toppled the first dikes at around 02:00 Sunday morning, well before the expected high tide of 05:00. About one hour later, numerous other dikes along the coast of Zeeland and Holland started to collapse, and the resulting flooding was enormous; see Figure 1.1. Events

Figure 1.1 Left: The 1953 great flood in Zeeland and Holland. Right: The area flooded where we have additionally pointed out the location of Hoek van Holland. Sources: Herman Gerritsen (left) and Wikimedia Commons (right)

had taken their course. In The Netherlands alone, 1836 people died that night with most of the deaths occurring in Zeeland. Across the east coast of the UK, the coast of West Flanders (Belgium) and at sea, at least a further 500 casualties had to be added. In total, 9% of total Dutch farmland was flooded, 30 000 animals drowned and 47 300 buildings were damaged of which 10 000 were destroyed.

How severe was the 1953 flood from a historical perspective? Zeeland and Holland have experienced several other major storms in the past. By name, historically the most well-known ones are the so-called Saint Elisabeth flood of November 19, 1421 and the All-Saints flood of November 1, 1570; see Figure 1.2. Estimates for the number of fatalities of the latter storm stand above 20 000.

Figure 1.2 Left: Flood marks in the delta area of Zeeland. Right: The Saint Elisabeth flood of November 19, 1421, Master of the Saint Elisabeth panels. Sources: Herman Gerritsen (left) and Wikimedia Commons (right)

About a third of The Netherlands lies below sea level, with about two thirds being prone to flooding. The lowest point of The Netherlands, standing at minus 6.76 m, is Nieuwerkerk aan den IJssel (after a merger, now referred to under its new name Zuidplas). It is situated about 10 km northeast of Rotterdam. Under such extreme topographic conditions, an understanding and communication of (especially) flood risk is highly relevant and "living with risk" becomes very tangible. Historically, the protection of the country by coastal as well as river dikes has always been of eminent importance. Below we will give a brief discussion of the world famous Delta Works project, which finds its origin in the 1953 flood event. First however, let us define what we understand by the above "minus 6.76 m". For that, one clearly needs a proper gauge. In the case of The Netherlands, this is the NAP = "Normaal Amsterdams Peil", a well-defined North Sea average level, also referred to as the Dutch National Ordnance level; see Figure 1.3. NAP is also officially the zero level within the European Union. Storm surges and low/high geographical points are always recorded and communicated relative to the NAP. In the case of the 1953 flood, the highest water level reached at Hoek van Holland was 3.85 m above NAP whereas for Zeeland this was 4.55 m above NAP.

Many accounts of the 1953 flood are available and on several occasions, commemoration days have been organized. You should therefore have no problem in finding relevant sources, for example start by searching "Rijkswaterstaat, 1953" to find the website of Rijkswaterstaat, the Dutch Directorate-General for Public Works and Water Management. Before discussing in some more detail the Delta Works, we would like to add two stories, one is fiction, and the other one corresponds to reality. As already stated above, because of its geographic location, sea protection always has been a key aspect of daily life in The Netherlands, as also becomes clear from the country's name. Consequently, several books treat the topic in a direct or an indirect way. A well-known example of the latter is Dodge (1866). In this book, the author tells the famous story of how a boy saves his village from flooding by plugging a hole in a dike with his finger. Whereas this yields a wonderful, even metaphorical, story, during the

Figure 1.3 Left: NAP, the zero level of The Netherlands and the European Union in the NAP Visitors Centre of the City Hall of Amsterdam; Guus Balkema is pointing his finger at the official NAP zero-level marker and Paul Embrechts is climbing the stairs to find out how far the sea level stands above NAP in Vlissingen at that moment in time (January 20, 2006). Right: Dutch boy saving Haarlem from flooding by putting his finger in a crack of the dike, according to the story told in Dodge (1866). Sources: Authors (left) and Enrico Chavez (right)

1953 flood an event occurred not too far removed from the above story. Indeed, during the storm, some seamen were able to close a breach in an important dike in southwest Holland by first maneuvering and then grounding their boat, De Twee Gebroeders (The Two Brothers), in the already sizable gap. By doing so, they prevented a large, densely populated area near the city of Rotterdam from flooding, and as such no doubt saved many lives; see Figure 1.4. In memory of this act of courage, near the site of the event, a statue was erected with the title "Een dubbeltje op zijn kant" which in the Dutch language means a very rare event symbolized by a small coin ("een dubbeltje") being tossed and landing on its edge.

1.2 The Delta Works

It is important to realize that The Netherlands in the early fifties of the last century was struggling with logistic consequences and rebuilding efforts in the aftermath of World War II. This may also have contributed to the deterioration of some of the dikes along the west coast. In any case, very soon after the 1953 flood, the government took swift action. Already on February 18, 1953, the Delta Committee was set up by the Minister of Transport and Waterways in order to examine "which hydraulic engineering works should be undertaken in relation to those areas ravaged by the storm surge, (and) also to consider whether closure of the sea inlets should form one of these works." The Committee produced its first findings in May of the same year, together with a more specific report (standing at only seven pages) about one year after the catastrophic event, on February 27, 1954. This report led to the by now famous Delta Plan and the resulting Delta Works. It is important to understand that the Delta Plan was highly ambitious in proposing (of course in several stages) a fully new coastal protection as compared to "just strengthening and raising" the existing dikes. As

Figure 1.4 Plugging a hole in a dike near Nieuwerkerk aan den IJssel with the boat De Twee Gebroeders. Source: Historical Society of Nieuwerkerk aan den IJssel

stated in Gerritsen (2005): "Until the middle of the twentieth century, a dike was largely characterized by its height. Very little was known about the factors influencing its strength or failure mechanisms. [. . .] The dike height was simply the height of the highest recorded high water plus a safety level of approximately half a meter." Whereas a combination of statistical analyses based on historical data combined with the strengthening of individual (typically smaller) dikes still played a very important role, the Delta Plan contained in particular the construction of major new dams and barriers (13 in total) across the southwest delta area of The Netherlands; see Figure 1.5.

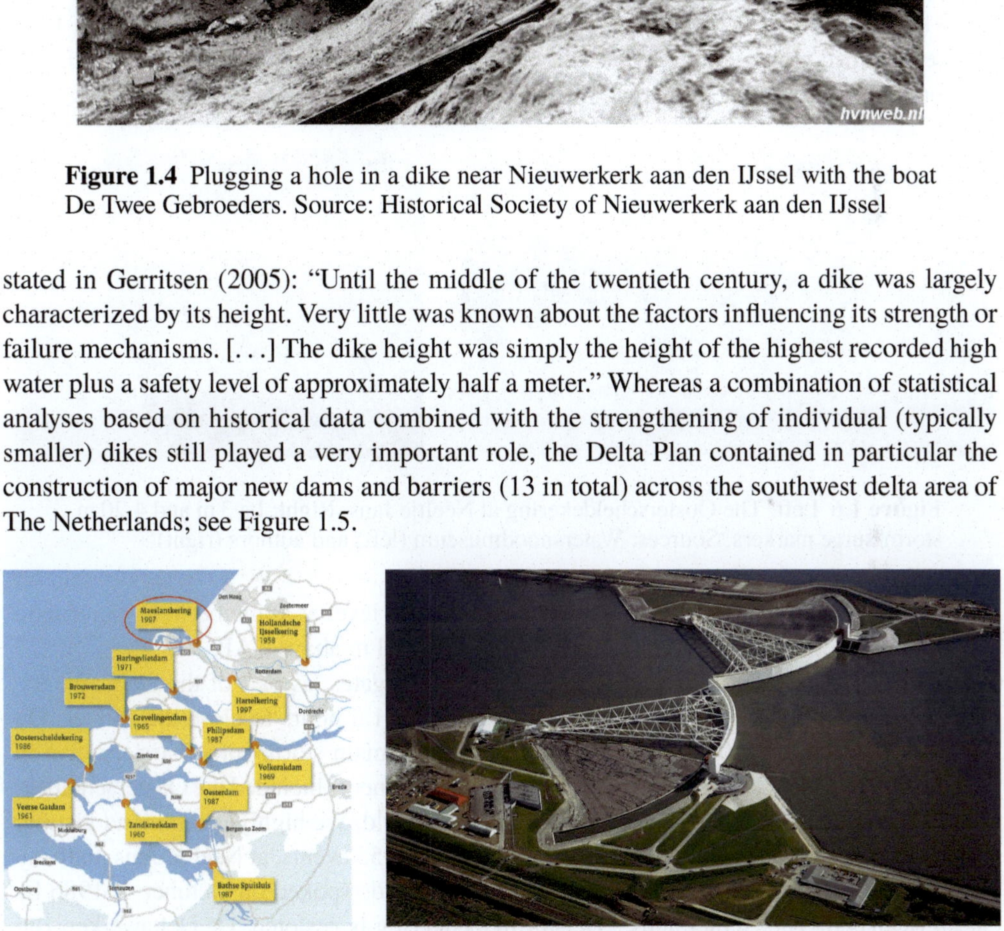

Figure 1.5 Left: The original 13 storm dikes and barriers from the Delta Works. Right: The Maeslantkering when closed. Sources: © Rijkswaterstaat (left) and Watersnoodmuseum (right)

Construction started with the "Stormvloedkering Hollandse IJssel" (Number 1, already inaugurated in 1958) and included, as main engineering highlights, the "Oosterscheldekering" (Number 9, 1986) and the "Maeslantkering" (Number 13, 1997). The latter consists of two moveable gates, each arm reaching in length the height of the Eiffel Tower; see Figure 1.5. They are designed to protect the important harbor of Rotterdam and its surroundings. The flood barrier at the Oosterschelde warrants a more detailed discussion; see Figure 1.6. The delta area of the Oosterschelde in Zeeland suffered the highest number of casualties during the flood. Original planning started in 1960, initially aiming at a fully solid dam closing off the delta area. This section of the Delta Works, however, took 26 years to complete, far more than any other section and at a considerably higher cost than that originally estimated. The main reason was that, early on, pressure from the local fishing industry as well as an increased environmental awareness caused a fundamental rethinking of the original plans. Prevention of the free in- and outflowing of seawater would destroy the rich delta fauna and flora and impair the livelihood of many people. By 1975, a solution was reached consisting of a 9 km barrier with 62 moveable steel gates that would close as soon as the sea level at a well-defined point along the coast reached 3 m above NAP. In 1953, at that particular location, the level stood at 4.20 m above NAP; see Figure 1.6. There is a further triggering

Figure 1.6 Left: The Oosterscheldekering at Neeltje Jans. Right: Its 3 m and 4.20 m storm-surge markers. Sources: Watersnoodmuseum (left) and authors (right)

event for the closure of the barrier, namely when the sea level at a line between Stavenisse and Wemeldinge, well within the Oosterschelde, reaches 1 m above NAP.

Since inauguration, individual moveable barriers and gates, as part of the Delta Works, have been closed on average once a year. It takes about 1.5 hours to close the sluices. On Wednesday, January 3, 2018, Rijkswaterstaat closed all main storm-surge barriers (five in total) on the same day. The American Society of Civil Engineers included the Oosterscheldekering in its list of The Seven Wonders of the Modern World. We highly recommend visiting its construction, which includes a museum reviving the 1953 events, at Neeltje Jans, halfway along the dam. Famous became the official opening words spoken on October 4, 1986 by Queen Beatrix of The Netherlands: "De stormvloedkering is gesloten. De Deltawerken zijn voltooid. Zeeland is veilig." (The flood barrier is closed. The Delta Works are completed. Zeeland is safe.) Of course, in a way the Delta Works will never be completed. Mechanical deterioration and corrosion, changes in the topography of the sea floor and possible adjustments because of climate change have to be monitored carefully. We will return to these issues in Section 9.5.2. Over centuries, the Dutch have learned to live with the sea both as a friend and as an enemy. As a "friend": the nearness to the sea contributed importantly to the Dutch Golden Age in the seventeenth century. Its struggle against the recurring storm events led to engineering knowledge now reaching out all over the world. The latter, the "enemy" part, is explicitly recognized in the title "Fighting the arch-enemy with mathematics" of the interesting paper by Laurens de Haan (1990).

Of course, the above brief description of the Delta Works cannot come close to doing justice to the multitude of scientific and engineering innovations that went into the (current) completion of the works. In the sections to follow, we shall touch upon some of the statistical, economic and political discussions which entered into the Delta Plan. The communication of risk will play an important role.

1.3 Inside the Delta Plan

Early on, Dutch scientists (mathematicians/statisticians, economists, engineers) paid attention to flood hazards. An early report concerning the north of The Netherlands is known under the name of the Lorentz Report, written in 1918. Hendrik Antoon Lorentz (1853–1928) was a mathematical physicist who received the Nobel Prize in Physics in 1902; his scientific work turned out to be crucial for Einstein's relativity theory. The 32 km long Afsluitdijk created the large fresh water IJsselmeer. It was built over the period 1927–1932. As reported in Kruizinga and Lewis (2018): "The Lorentz Committee's work combined state-of-the-art hydrographic modeling of the impact of a new dam on seawater flows with a historical analysis of previously recorded wind speeds and water levels. Based on these data, which the committee itself admitted [were] incomplete, it was suggested that the new Afsluitdijk should be raised by an additional meter to between 7.5 and 7.8 meters above NAP." This scientific approach, for the first time presented in the Lorentz Report, became the hallmark of Dutch dike constructions. The success of the Afsluitdijk was confirmed during the 1953 flood; though battered, the dike did not give way.

In June 1953, the mathematician and statistician David van Dantzig (1900–1959) accepted the invitation to contribute to the by now famous *Delta Report*; it contains several appendices, some of which we will highlight later on. The original Dutch version of the final report (Deltacommissie, 1961) was signed off by all committee members on December 10, 1960. For a version in English of Part 1, containing the main conclusions and recommendations, see Delta Committee (1962). The full report also contains a contribution by Jan Tinbergen, the Dutch economist who in 1969 was to be awarded the first Nobel Memorial Prize in Economic Sciences. He became one of the founding fathers of the field of econometrics. The key scientific paper, underlying van Dantzig's analysis for the Delta Committee, is van Dantzig (1956). His important input to Deltacommissie (1961) appeared, in Dutch, under the heading of "The contributions of the Mathematical Center (MC) on storm surges" as Part 3, Contributions II.1–II.5. It was worked out with several collaborators from the MC, in particular J. Hemelrijk, J. Kriens and H. A. Lauwerier. These scientific additions to the Delta Report, referred to as the *van Dantzig Report*, were officially published in September 1960, more than a year after the premature death of David van Dantzig on July 22, 1959. In his foreword to the van Dantzig Report, the then director of the Mathematical Centre, J. F. Koksma, wrote "His death, particularly in view of the current investigations, means a loss that cannot be estimated. There is so much the more reason for thankfulness, that he was allowed to see at least a great part of this work accomplished."

It is important to understand the challenges that these scientists encountered in influencing the final decision process; their recommendations first had to pass through the necessary political filters before the conclusions could find their way into policy papers ready for public discussion and final legal decision-taking. As several of the discussions and misconceptions from that time are still relevant today, below we recall some of the key ingredients of this process.

Tinbergen and van Dantzig reasoned that absolute safety with respect to flood protection cannot be reached. As a direct consequence, statistical safety measures enter, such as "what is the level of risk with which the population is willing to live?" Typical questions then become "Do we want safety corresponding to a 1 in 100, 1 in 1000 or 1 in 10 000 years flood event, say, and what does this mean? How does one communicate these numbers to politicians and the broader public?" Surely, the dike height to be constructed must be primarily a function of the lives saved but also of the economic value of the protected land and infrastructure. Further, how does one put a monetary value on "lives lost" in case of a disastrous flooding, and last but not least what are the resulting technical-engineering constraints. To put the potential economic losses into perspective, the material loss of the 1953 flood stood at about 10% of GDP. The van Dantzig Report contains an explicit mathematical formula for a necessary increase Δ_{height} of the existing dike height (see van Dantzig, 1956, p. 283; combine (12) and (14)):

$$\Delta_{\text{height}} = \frac{1}{\alpha}\log\left(\frac{100 p_0 V \alpha}{(\delta' - \beta)k} \times \frac{1 - e^{-(\delta'-\beta)T}}{1 - e^{-\delta' T}}\right). \tag{1.1}$$

The exact analytic formula is less important in the present discussion. We mainly want to highlight the various input parameters. The crucial output variable Δ_{height} stands for the difference between the new dike height and the current, pre-flood one, expressed in meters. It depends on various parameters (the positive p_0, V, α, k, δ', β, T). The constant p_0 denotes the probability that a high storm surge topples the current sea dike; it is determined through a parameter α which is estimated from daily high-tide observations. These parameters will become clear later in the book when we discuss extreme value theory. The crucial constant V stands for the value of the "goods" lost as a consequence of a dike breach. We refer to the paper for a discussion on deriving V on the basis of economic data.

The author also briefly discusses actuarial models for valuing "lives lost". The constant k denotes the cost of heightening the dike by one meter. The time horizon T is measured in centuries or fractions of a century. Over such long time horizons, the geological sinking of the land should be taken into account; here the constant β enters. Finally, δ' corresponds to an appropriate discount factor, accounting for the change of the value of money over time. For more specific discussions we again refer to the paper. It is to be hoped that the above summary of the various parameters entering into van Dantzig's formula will convince you that the formula stays close to the initial task set, that of "determining the optimal height of the dikes, taking account of the cost of dike-building, of the material losses when a dike-break occurs, and of the frequency distribution of different sea levels". David van Dantzig stressed very clearly the many shortcomings of the formula and methodological shortcuts he had to take. Personally, we find its derivation a true gem of applied mathematics and recommend that you go over the paper's content in full detail. Our main aim in presenting the formula explicitly was to highlight the power of mathematics in contributing to the solution of such an important problem as determining the necessary heights of protective sea dikes.

In van Dantzig (1956) we find only one specific, though in van Dantzig's words somewhat pessimistic, dike height (p. 284), for Hoek van Holland, of 6.73 m. The word "pessimistic" here takes into account uncertainty bounds for the various constants included in the formula. He personally advocated, as a compromise, a height of about 6 m. It is worthwhile to quote the precise wording used in the paper: "The combination of these extreme values for all constants, however, is rather pessimistic. Several reasonable combinations of values lead to the conclusion that roughly 6.00 m may be considered as a reasonable estimate of a sufficiently safe height". In the end, the Delta Committee went for 5 m above NAP, which corresponded at the time to a heightening of the existing dike by 1.15 m. Needless to say, van Dantzig was not particularly happy with this decision.

More recently, several papers have been published improving on or criticizing (parts of) the van Dantzig formula. With hindsight, some of these more recent criticisms are justified. However, the methodological, data-oriented, statistical as well as economic approach that was present in van Dantzig's work was innovative and of the utmost importance. It served (still serves) as a guiding light for generations of risk managers.

Below we recall part of the resulting political discourse surrounding the Delta Report, as its implications for wider risk management are highly relevant. We quote part of the decision process from Kruizinga and Lewis (2018) as it perfectly reflects the difficulties one encounters when communication of risk and rare events to politicians and a broader public becomes important.

> [...] the Delta Committee agreed that the report addressed to the Minister of Transport, Public Works and Water Management needed to be massaged in order to help him sell the necessity of spending about 1.5 to 2 billion guilders (roughly between 8 and 11.5 billion euros in 2018) on the new integrated system of dikes and sea defenses. It was also agreed that the report would omit any mention of methodological uncertainty. Furthermore, the report would not go into details as to the risks of future storms, as these would need to be expressed in the form of statistical probabilities. It was feared that the public would be confused by statements such as 'statistically once every 125 000 years'. The Delta Committee feared the public would misinterpret this statement and think that The Netherlands would be safe for the next 125 millennia rather than there being a 0.0008% chance of a storm of a certain magnitude occurring every year.

Important takeaways from the above quote are the need for a clear political communication, the suggested (even ordered) omission of uncertainty, and the possible misinterpretation of the statistical meaning of return periods by the public. The 1 in 125 000 years event mentioned in the above quote refers to the large economic and social loss potential for the "Randstad" (the built-up area around major cities like Rotterdam, Amsterdam and The Hague) in the case of a serious flood. In its conclusion, the Delta Committee settled for a 1 in 10 000 years safety measure resulting in a 5 m above NAP dike height at Hoek van Holland. The "1 in 10 000 years" became the so-called Dutch National Standard. This standard only applies to the more exposed areas like Hoek van Holland and indeed different safety requirements hold for different coastal areas and river basins, ranging from 1 in 250 to 1 in 10 000. We will continue this important discussion after we have analyzed historical sea-level data for Hoek van Holland in Section 9.5.2.

As already stated above, the Delta Works are never finished. And indeed, Veerman and Stive (2008) contains the findings of the new Delta Committee. Its mandate from the government was "to come up with recommendations on how to protect the Dutch coast and the low-lying hinterland against the consequences of climate change. The issue is how The Netherlands can be made climate proof over the very long term: safe against flooding, while still remaining an attractive place to live, to reside and work, for recreation and investment." One of its 12 recommendations was that all diked areas must be improved by a factor 10, hence the standard safety measure would move from 1 in 10 000 to 1 in 100 000. For a somewhat differentiated view on this recommendation, see Kind (2014). It is clear that climate change enters the equations in a fundamental way; the (new) Delta Committee concludes that

> [...] a regional sea level rise of 0.65 to 1.3 m by 2100, and of 2 to 4 m by 2200 should be taken into account. [...] These values present plausible upper limits based on the latest scientific insights.

This brings us to a very pivotal point in time. We started with a major flood catastrophe on January 31 to February 1, 1953, and discussed the way in which a country, in this case The Netherlands, faced the consequences and came up with a technical engineering solution in order to avoid such events in the future, and this with a (very) high degree of certainty. In the meantime, science has given us further knowledge on future climatological scenarios, which of course have a strong bearing on flood risk. Dutch society has to (and actually did) react to these threats and came up with protective recommendations for generations to come. The following quote from Veerman and Stive (2008) on the ever-increasing need for scientific advice is worth stating explicitly (we will come back to the work of the IPCC in Section 10.1):

> The Delta Committee sought scientific advice on a number of aspects, which form part of the present recommendations. In summary, these are the findings of a group of national and international experts, including those close to the IPCC (Intergovernmental Panel on Climate Change) and Dutch experts on flood protection and water management. This group of experts has supplemented the latest insights into climate scenarios, and come up with new estimates of extreme values.

We stress this statement as all too often scientific advice and expertise is frowned upon by (some) politicians worldwide, especially when it is related to climate change and its societal consequences. Besides possible reinforcements to the various existing dikes, most recently a further factor entered the discussion. As several of the moveable barriers are driven by IT systems, cyber risk becomes of great concern. In its 2019 Annual Report of the Delta Programme to the Dutch Parliament, the Delta Committee highlights this new type of risk. It strongly advises measures to be taken to make the computer systems of the various barriers sufficiently resilient against cyber attacks. Of course, the key question is whether in the end all the effort has paid off. In a way this question is unanswerable, especially as dike constructions are very long-term projects. What can be said is that, at the time of writing this book, The Netherlands have been safe behind their coastal defenses and that the various dikes have so far stood the test of time. An important question then naturally becomes: "What can other countries learn from the Dutch experience?" Of course on the engineering side a clear answer is available: "A lot!" Dutch engineers and firms are already exporting the technical knowledge obtained throughout the Delta Works to the rest of the world. Examples range from the building of levies in New Orleans to land reclamation in Singapore. However, as already stated above, one can never make a one-to-one translation from "what worked in The Netherlands" to "what would work in country X". This is perhaps less relevant for the technical engineering side, but it is surely true for the political socio-economic side. An excellent discussion on this, where X stands for the United States of America, is Iovenko (2018). From the latter paper we borrow the following statement: "The greatest lesson to be learned from the Dutch is perhaps less about engineering and more about mindset and culture." The author then continues by quoting from the book of Goodell (2017).

> It's easy just to talk about technological and engineering solutions, but a lot of the problems surrounding sea-level rise are legal and political. The Dutch have a legal and political system that is united around dealing with water issues; they've been doing it for a thousand years. Here in the US, it's not getting the right engineering ideas figuring out what technology or design ideas we're going to use. It's that our legal system and our political system are just not adapted to thinking about sea-level rise in any kind of holistic way.

At the end of this chapter, we would like to honor three scientists who contributed fundamentally to the success of the coastal protection in The Netherlands. Of course there are so many more names to be mentioned, but we chose Hendrik Antoon Lorentz, David van Dantzig and Jan Tinbergen; see Figure 1.7. They were able to step down from the pillars in their ivory tower, go beyond the typical academic thinking, and engage in truly interdisciplinary research. Their contributions have been fundamental towards making the Dutch population feel safe behind the dikes constructed. The world as a whole benefits from their original and courageous thinking.

Figure 1.7 Hendrik Antoon Lorentz (1853–1928) (left); David van Dantzig (1900–1959) (center); Jan Tinbergen (1903–1994) (right). Source: Wikimedia Commons

1.4 Lessons Learned

Throughout the book, we will return to the 1953 flood several times, as it is indeed a blueprint for lessons to be learned when dealing with extreme risks. We have learned how a country, The Netherlands in this case, reacted politically as well as scientifically to an existential environmental threat. The time scale, and hence the planning of underlying dike constructions, may run into hundreds of years and as such needs full societal support over a much longer time period than is normally found on political agendas. A key component underlying the Delta project is interdisciplinarity. In the case of the very important Oosterscheldekering, already environmental considerations were actively taken into account. We also learned how difficult the communication of an imminent risk to the population was in pre-social-media times. For the first time we met the notion of a risk measure, a return period, and the difficulty science faces in communicating statistical uncertainty in its estimation. In Chapters 8 and 9 we will review the key techniques from probability and statistics needed to address these issues. In particular, in Section 9.5.2 we present a detailed analysis of sea-level data at Hoek van Holland. By now, the technology underlying the engineering of coastal defenses has become a Dutch export product *par excellence*. An interesting exercise consists of comparing and contrasting different dike constructions worldwide; as examples we would like to mention New Orleans (USA), Pulau Tekong (Singapore), the Thames Barrier (UK) and the MOSE project (Venice), the first of which we will briefly meet later in the book.

2

The Space Shuttle Challenger Disaster

The future doesn't belong to the fainthearted;
it belongs to the brave.

President Ronald W. Reagan (1911–2004) in
"Address to the Nation, January 28, 1986"

2.1 The Morning of January 28, 1986

On January 28, 1986, at 11:38 (EST) at the Kennedy Space Center, Cape Canaveral, the space shuttle Challenger or OV-099 (Orbital Vehicle-099) took off for the 10th time; see Figure 2.1. Its mission was referred to as STS-51-L (STS standing for "Space Transportation System"),

Figure 2.1 Space Shuttle Challenger launch on January 28, 1986. Source: Wikimedia Commons

a reference that occurs in all official reports. This 25th launch within the STS took place amid very wide public interest: on board, one of the astronauts was Christa McAuliffe, a high school teacher chosen by the project "Teacher in Space". Once in orbit, she would address schoolchildren across the nation. Leading up to the launch, the American public became passionate about her story. While the whole world was watching, 73 seconds after the launch at 14 km above sea level, the shuttle suffered a fatal structural failure. The right rocket booster burst into flames, the Challenger was torn apart and disintegrated right before the eyes of the audience gathered at the space center and millions of television viewers around the world. The trauma at Cape Canaveral was immense. Burning debris rained down on the ocean and huge white clouds drew monstrous shapes in the late morning sky. Everyone witnessing this event live will always remember the shock of this historical tragedy, killing all seven crew members; see Figure 2.2. For the first time, the USA had lost human lives in a fully departed

Figure 2.2 Space shuttle Challenger crew (from left to right): Ellison Onizuka, Michael J. Smith, Christa McAuliffe, Dick Scobee, Gregory Jarvis, Ronald McNair and Judith Resnik. Source: Wikimedia Commons

rocket flight. In memory of the seven astronauts, on March 26, 1986, seven asteroids were named in their honor. Asteroid no. 3352 for Sharon Christa Corrigan McAuliffe is an Amor type asteroid that comes well inside the orbit of Mars in its closest approaches to the Sun.

2.2 The Rogers Commission

The "Presidential Rogers Commission" (the "Commission") chaired by the Attorney General and Secretary of State William P. Rogers was formed to analyze the possible causes of the accident. The commission included astronaut Neil Armstrong, from the first manned lunar landing, the first American woman in space, Sally Ride, and the Physics Nobel Laureate Richard P. Feynman, among others. By June 6, 1986, the commission produced a first detailed report (the "Report"); see Rogers (1986). It discussed technical, managerial as well as cultural causes. Later followed "Actions to Implement the Recommendations" (July 14, 1986) and "Implementation of the Recommendations" (June 1987). The conclusion from Chapter IV ("The Cause of the Accident") of the Report reads as follows:

In view of the findings, the Commission concluded that the cause of the Challenger accident was the failure of the pressure seal in the aft field joint of the right Solid Rocket Motor. The failure was due to a faulty design unacceptably sensitive to a number of factors. These factors were the effects of temperature, physical dimensions, the character of materials, the effects of reusability, processing, and the reaction of the joint to dynamic loading.

One has to realize that a complex spacecraft as the Challenger consists of thousands of highly critical hardware and software components, each of which can contribute to serious malfunctioning. There of course always remains the possibility of human error. The conclusion of the Report, however, puts the possible causal malfunctioning squarely in the camp of "the pressure seal in the aft field joint". The formulation that reached the general public was that of a malfunctioning of O-ring seals in the right solid rocket booster. Such rubber rings (nearly 11.6 m in diameter and 6.4 mm thick) were used as a seal for the joints in each solid rocket booster and should have prevented the gases produced by the burning solid propellant from leaving on the left side of the rocket booster ("blow-by") instead of at the aft end. O-ring seals play an important role in the immediate after-launch in order for the shuttle to go into orbit. In the Report, we read that

Just after liftoff at .678 seconds into the flight, photographic data show a strong puff of gray smoke was spurting from the vicinity of the aft field joint on the right Solid Rocket Booster. [. . .] The vaporized material streaming from the joint indicated there was not complete sealing action within the joint [Figure 2.3].

Further, the Report says that

Boosters were increasing their thrust when the first flickering flame appeared on the right Solid Rocket Booster in the area of the aft field joint. This first very small flame was detected on image enhanced film at 58.788 seconds into the flight [Figure 2.3].

The Challenger was on course for disaster.

It is clear that every manned space mission operates under time constraints. In this case, after several delays, no doubt mounting public expectations emerged. Numerous documents in the public domain comment on this cultural aspect. We will refrain from a discussion and concentrate on the O-rings in the right solid rocket booster. Clearly, for Challenger the O-rings did not provide the necessary sealing under the extreme conditions that persisted at launch. It was concluded that a major contributing factor to the seal degradation was the prevailing exceptionally cold weather that fatal morning. The night before, ice had accumulated all over the launch pad. The official temperature at launch time ranged from 36 °F (2.2 °C) from a weather tower 1000 feet away from the launch pad to an estimated 28 °F (−2.2 °C) near the failed solid rocket booster. The decision to launch Challenger on January 28, 1986, was made the day before during a three-hour teleconference between NASA's launch facility (Marshall Space Flight Center) and the company (Morton Thiokol) which manufactured the O-ring seals. During the teleconference, the motor engineers of the company expressed their reservations about the scheduled launch the next day, arguing that the weather conditions were unfavorable owing to the expected exceptionally low temperatures. They clearly recommended a no-go. The problem of the degrading effect of low temperature

Figure 2.3 Dark smoke escaping from the right solid rocket booster at 0.678 s into the mission (left) and a white plume on the right solid rocket booster at 58.788 s (right). Source: Wikimedia Commons

on O-ring performance was known. However, anxious to keep their schedule, NASA managers did not communicate this information to their superiors and decided to maintain the launch. The Report gives a very detailed and sobering account of these events.

As a member of the Commission, Richard Feynman played a crucial role in exposing the O-ring vulnerability given the ambient temperature conditions. Very much in line with Feynman's personality, early on in the commission's work he wanted to speak to the "workers on the shop floor", those engineers who were closest to the actual manufacturing and testing process. This investigative attitude towards risk is often also referred to as "talk and listen to the guys in the boiler room". A brilliant example of the efficient communication of risk for a general public is his famous C-clamp demonstration during the February 11, 1986 hearings of the Commission; see Figure 2.4.

He put a piece of O-ring rubber in a C-clamp, squeezed it a bit and submerged it into a glass of ice water. He then addressed Lawrence B. Mulloy, the project manager of solid rocket boosters at the Marshall Space Flight Center of NASA:

I took this stuff that I got out of your seal [O-ring] and I put it in ice water, and I discovered that when you put some pressure on it for a while and then undo it, it doesn't stretch back. It stays the same dimension. In other words, for a few seconds at least and more seconds than that, there is no resilience in this particular material when it is at a temperature of 32 degrees [Fahrenheit].

Figure 2.4 Richard Feynman dipping the C-clamp holding a piece of O-ring in ice water during the February 11, 1986 hearings of the Commission. Source: Getty Images

He then concluded his little experiment with the famous words: "I believe that has some significance for our problem." For the whole world to see, and in a fully transparent way, Richard Feynman brought the Challenger disaster to its simplest causal core. You can consult Feynman and Leighton (1985) and Feynman and Leighton (1988) for very readable accounts of Richard Feynman as a scientist and as a person. Richard Feynman was not only a famous physicist but also a highly interesting person. The above two references are full of relevant quotes. Here is one of them, which has a bearing on the topics treated in this book:

> The scientist has a lot of experience with ignorance and doubt and uncertainty, and this experience is of very great importance, I think. When a scientist doesn't know the answer to a problem, he is ignorant. When he has a hunch as to what the result is, he is uncertain. And when he is pretty damn sure of what the result is going to be, he is still in some doubt. We have found it of paramount importance that in order to progress, we must recognize our ignorance and leave room for doubt. Scientific knowledge is a body of statements of varying degrees of certainty – some most unsure, some nearly sure, but none absolutely certain. Now, we scientists are used to this, and we take it for granted that it is perfectly consistent to be unsure, that it is possible to live and not know. But I don't know whether everyone realizes this is true. Our freedom to doubt was born out of a struggle against authority in the early days of science. It was a very deep and strong struggle: permit us to question – to doubt – to not be sure. I think that it is important that we do not forget this struggle and thus perhaps lose what we have gained.

2.3 A Statistical Analysis of O-Ring Failure Data

By now, it is clear that the extremely low temperature the night before and the morning of the launch played a very important role in the malfunctioning of the O-ring sealing of the sections of the solid rocket boosters. Leading up to the launch, the engineers of Thiokol warned very explicitly. Already in a memo of July 1985 to his superiors, Roger Boisjoly (1938–2012), an engineer at Thiokol, voiced serious concerns about O-ring vulnerability. These warnings remained stuck at the so-called Level III and never made it up to the relevant decision levels II and I within NASA. A further telling point is that at an early morning meeting on launch-day, Jerry Mason (a Senior Vice President at Thiokol) told Robert Lund

(Thiokol's Vice President of Engineering) "to take off his engineering hat and put on his management hat". At the time, Mason was the ranking executive at the company's Utah installation. In the words of Robert Lund:

From this point on, management formulated the points to base their decision on. I felt personally that management was under a lot of pressure to launch and that they made a very tough decision, but I didn't agree with it. We couldn't prove absolutely that that motor wouldn't work.

NASA appeared to be requiring a contractor, in this case Thiokol, to prove that it was not safe to launch, rather than proving that it was safe. During these crucial hours before launch, the availability of historical data on past O-ring damage as a function of launch temperature became relevant. Except in the case of Columbia STS-4 on June 27, 1982, NASA recovered the solid rocket boosters of past space shuttle flights from the ocean; those of STS-4 could not be recovered due to parachute malfunctions. Hence, experts could, and indeed did, examine post-launch O-ring degradation. Of course, one should carefully define "degradation" as well as give a precise definition of "temperature" at launch, that is, the O-ring (or joint) temperature as opposed to bulk temperature of the propellant or ambient temperature in the air at a specific point near the launch pad. Throughout the discussions the days before the launch, Thiokol often stated that a launch should not take place with a joint temperature below 53 °F (12 °C), the coldest launch experienced so far, corresponding to Discovery shuttle STS-51-C on January 24, 1985. The flight of STS-51-C was delayed by one day because of freezing weather. In the case of STS-51-L, the Challenger, the predicted joint temperature was of the order of 30 °F (−1.1 °C), which moved the mission very much outside the range of the available historical data. The actual value was 31 °F (−0.6 °C).

It is worthwhile reading the Commission's statement on a possible statistical analysis, especially as it captures the key underlying low-temperature issue; see the discussion on Temperature Effects on pp. 146–147 in the Report:

The record of the fateful series of NASA and Thiokol meetings, telephone conferences, notes, and facsimile transmissions on January 27th, the night before the launch of flight 51-L, shows that only limited consideration was given to the past history of O-ring damage in terms of temperature. The managers compared as a function of temperature the flights for which thermal distress of O-rings had been observed, not the frequency of occurrence based on all flights [...]. In such a comparison, there is nothing irregular in the distribution of O-ring "distress" over the spectrum of joint temperatures at launch between 53 degrees Fahrenheit and 75 degrees Fahrenheit. When the entire history of flight experience is considered, including "normal" flights with no erosion or blow-by, the comparison is substantially different (Figure 7).

Figure 2.5 shows the number of O-ring failures that had occurred before the ultimate launch (with mission numbers added for each mission with at least one O-ring failure). This corresponds to "Figure 7" referred to in the above quote; for better visibility, equal observations are stacked. To assess the effect of joint temperature on the number of O-ring failures, NASA managers excluded the flights where no failures happened (the unlabeled points in Figure 2.5). Without these points, there is indeed "nothing irregular in the distribution of O-ring "distress" over the spectrum of joint temperatures at launch between 53 degrees Fahrenheit and 75 degrees Fahrenheit". However, flights with no O-ring failure typically happen at higher temperature; this information is statistically highly relevant and indeed makes "the comparison [...] substantially different". Leaving out the launches with zero O-ring failures is a typical example of *selection bias* in statistics. As reported in Hand (2020a), ignoring the data with no O-ring failures is a very simple example of so-called *dark data*. The author writes:

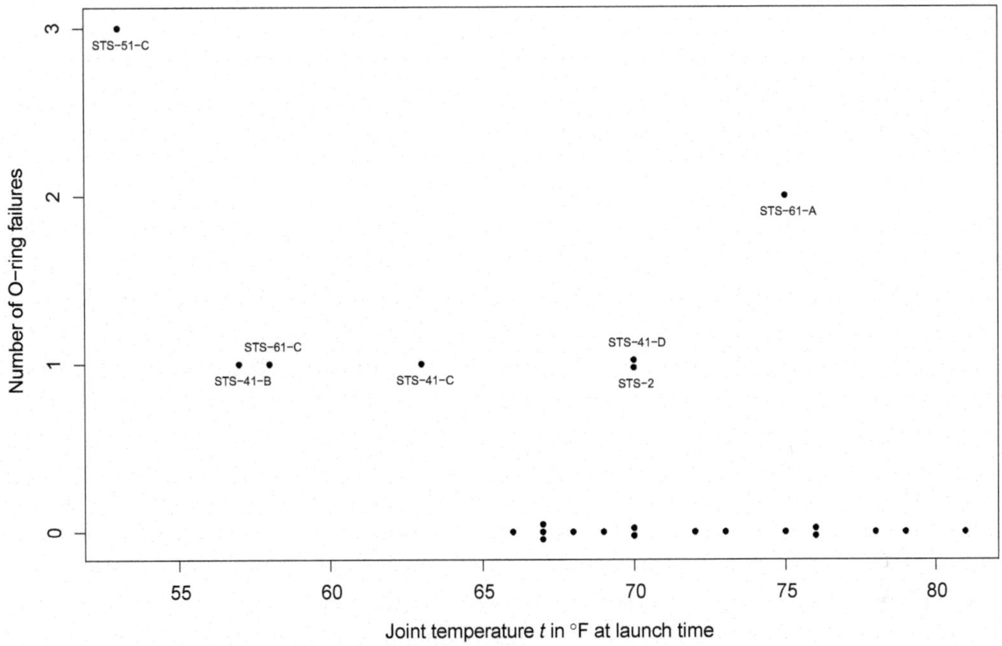

Figure 2.5 Number of O-ring failures versus joint temperature (the labels indicate those missions with at least one O-ring failure). Source: Authors

> Dark data are data you do not have. They might be data you thought you have, or hoped to have, or wished you had. But they are data you don't have. You might be aware that you do not have them, or you might be unaware. In any case, as a result of these dark data, the data missing from your view of the world, you are at risk of misunderstanding, of drawing incorrect conclusions about the way the world works, of making poor predictions, of getting things wrong. Just as in the Challenger example. [. . .] sometimes data are darkened deliberately.

For a textbook discussion on dark data, see Hand (2020b).

By fully concentrating on the O-rings, in no way do we disregard the broader managerial issues that contributed to the catastrophe. In that sense, "O-ring failure" should not be put equal to "disaster". It was however a very important, even causal contributing technical factor, once the decision to go ahead with the launch was taken. Because of the more statistical slant of our book, we want to look into the O-ring issue in a bit more detail. We will therefore perform a rather standard statistical analysis of the O-ring failures versus temperature data with the aim of explaining the probability of at least one O-ring failure as a function of joint temperature at launch time. One of the first papers addressing a statistical analysis of the Challenger data was Dalal et al. (1989). Our analysis is conducted with the statistical software R, and we use the dataset `Challeng` of the R package `alr4`, which contains the (calculated) joint temperature at launch time and the number of failures of O-rings of the earlier space shuttle launches. There were 23 sets of solid rocket boosters recovered from the previous 24 launches; 16 were classified as no O-ring failure, five as one O-ring failure and two as two O-ring failures, so 16 non-failure cases and seven cases with at least one failure. We use standard terminology and notation from probability theory and statistics, most of which will be introduced in Chapter 8 (probabilities, means). We recommend that the reader should try to follow the gist of the technical argument. We take up the non-technical discussion with the findings in Section 2.4 below.

We utilize a *generalized linear model* of the form

$$g(p(t)) = \beta_0 + \beta_1 t, \tag{2.1}$$

where t is the (calculated) joint temperature in °F at launch time, $p(t)$ is the probability of at least one O-ring failure at temperature t, and β_0, β_1 are parameters of the model which are estimated from the 23 O-ring incidents from former space shuttle launches. As the name suggests, the *link function* g links $p(t)$ to the linear right-hand side of (2.1). With β_0, β_1 estimated by $\hat{\beta}_0, \hat{\beta}_1$ from the 23 O-ring incidents (which can be accomplished with statistical software), we can compute an estimate $\hat{p}(t)$ of the failure probability $p(t)$ for a given temperature t via

$$\hat{p}(t) = g^{-1}(\hat{\beta}_0 + \hat{\beta}_1 t), \tag{2.2}$$

where g^{-1} denotes the inverse of the link function. In particular, for $t = 31$ we can obtain an estimate $\hat{p}(31)$ of the failure probability $p(31)$ for the Challenger mission at launch time.

When modeling probabilities, a commonly used link function is the *logit* link function, $\operatorname{logit}(p) = \log\big(p/(1-p)\big)$, $p \in (0, 1)$, with corresponding inverse $\operatorname{logit}^{-1}(q) = e^q/(e^q + 1) = 1/(1 + e^{-q})$, $q \in (-\infty, \infty)$. A generalized linear model with logit link function is known as a *logistic regression model*, and, by (2.2), we have that

$$\hat{p}(t) = \frac{1}{1 + e^{-(\hat{\beta}_0 + \hat{\beta}_1 t)}} \tag{2.3}$$

in this case.

We obtain the estimates $\hat{\beta}_0 = 15.0429$ and $\hat{\beta}_1 = -0.2322$ of β_0 and β_1. By (2.3), we thus have $\hat{p}(31) = 0.9996$, an estimated probability of failure of at least one O-ring larger than 99.9% for a Challenger launch at temperature $t = 31$. Figure 2.6 shows the predicted failure probability $\hat{p}(t)$ as a function of t (solid black line), the smallest t being $t = 31$. The red and green dots indicate the data to which the logistic regression model was fitted, with the seven red dots corresponding to past failures of at least one O-ring and the 16 green dots corresponding to no failures; as before, equal observations are stacked for better visibility. From here we immediately see that the temperature at launch time lies far below the available past data. As a consequence, the corresponding estimated probability $\hat{p}(t)$ of O-ring failure comes with considerable uncertainty, which translates into wide 95% "confidence intervals" at each t, indicated by the gray area. The important statistical notion of confidence interval will be made precise in Section 8.6.3. For this example we are fine with the mathematically sloppy interpretation that the 95% *confidence interval* contains the unknown parameter $p(t)$ with a probability of 0.95. At 31 °F (−0.6 °C), $\hat{p}(31)$ is nearly 1 and the lower endpoint of the 95% confidence interval is close to 0.5, which translates to a high risk of failure.

A more thorough and detailed statistical analysis is provided in Dalal et al. (1989). Besides the temperature t, the authors also utilized the leak-check pressure (the air pressure behind the primary O-ring as a measurement of its seal) as a variable to predict or explain the probability of O-ring failure. Such variables are known as *covariates*, and it turns out that the pressure at launch time is indeed a significant covariate in predicting or explaining the probability of O-ring failure.

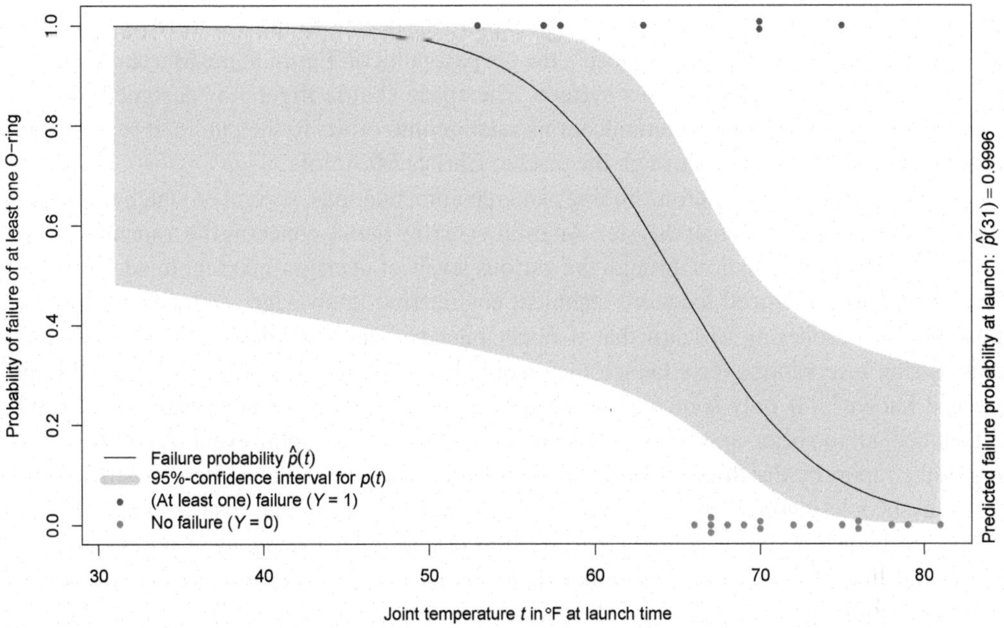

Figure 2.6 Predicted probability $\hat{p}(t)$ (solid line) of failure of at least one O-ring as a function of joint temperature t in °F at launch time; see (2.3). The logistic regression model underlying this plot was fitted to 16 non-failure cases (green dots) and seven cases with at least one O-ring failure (red dots). For each t, the gray region indicates a 95% confidence interval for $p(t)$. At the far left (for $t = 31$), we see the predicted probability $\hat{p}(31)$ of failure and the 95% confidence interval for $p(31)$ at launch time, January 28, 1986. Source: Authors

We will leave out all the extra engineering issues underlying the functioning of O-ring seals and refer to the Report for these issues. We finish this brief analysis of the launch data with two related findings from the Report:

- Finding 6 (p. 149):

  A careful analysis of the flight history of O-ring performance would have revealed the correlation of O-ring damage and low temperature. Neither NASA nor Thiokol carried out such an analysis; consequently, they were unprepared to properly evaluate the risks of launching the 51-L mission in conditions more extreme than they had encountered before.

- Finding 4 (p. 162):

  Little or no trend analysis was performed on O-ring erosion and blow-by problems.

2.4 The Aftermath

Besides the technical issues leading to the disaster, NASA and the Challenger mission no doubt experienced public, political and commercial pressure. The last two surely played a role in the higher echelons of the organization. As we already mentioned, the fact that a schoolteacher was on board raised considerable expectations across the country. The name Challenger comes from Her Majesty's Ship "HMS Challenger", a Royal Navy steam-assisted corvette launched in 1858 with as mission to explore the oceans from 1872 to 1876. This ship also gave its name to "Challenger Deep", the deepest point on Earth, in the Marianas Trench off Guam, 10 916 m below the sea surface. The space shuttle flight was designed for test purposes and to place a telecommunications satellite into orbit. To the public, it represented space discovery personified through the teacher Christa McAuliffe.

The Report contains numerous findings and recommendations, several of which transcend the specifics of the Challenger disaster. An ever recurring theme concerns the transmission of mission-relevant information through the various levels of decision-making. In addition, the Report mentions the need for more technical engineering knowledge at higher managerial positions. It is sobering to learn that if upper management had known about the serious engineering reservations for a launch they would have aborted. Unfortunately, the "if only we had known", "if only someone would have told us", "we were not aware" or "I can't remember" all too often appear as lines of defense after a catastrophic event. A careful reading of the report by the Rogers Commission, together with the various interviews that were given by people involved in the decision-making clearly shows that disaster management in the pre-launch period was below par. All too often higher management hides behind a horizontal line of communication rather than the much more relevant vertical, top-down, communication and the even more relevant bottom-up communication. Here Richard Feynman's approach of "talking to the guys in the boiler room" is crucial, and this is true well beyond the space shuttle program. In the case of the Dutch dike disaster of 1953, the most relevant shortcomings of disaster risk management in the immediate aftermath of the storm were to be learned from talking to the people who lived through the agonizing moments between life and death, sitting out the flood. An excellent book in this respect (in Dutch) is by Slager (2003). Another example we find in the 2007–2009 financial crisis, where the software programmers of the over-complicated financial credit products knew very well that the hype was entirely based on extremely shaky economic and technical assumptions; see Section 3.3.

A main theme of our book, and indeed very much lying at the basis of disaster risk management, is the proper understanding of extremes. NBC News space analyst James Oberg, who spent 22 years at NASA's Johnson Space Center as a Mission Control operator and an orbital designer, summarized his thoughts as follows (January 25, 2011):

> The disaster need never have happened if managers and workers had clung to known principles of safely operating on the edge of extreme hazards – nothing was learned by the disaster that hadn't already been learned, and then forgotten.

On September 29, 1988, NASA successfully launched STS-26 as the first STS mission after the Challenger STS-51-L disaster. However, on February 1, 2003, disaster struck once more. During the atmospheric re-entry phase of STS-107 Columbia, the 113th mission within the Space Shuttle program, the shuttle disintegrated, killing all seven crew members on board. The source of the failure was a piece of heat-resisting foam that broke off during launch and damaged the thermal protection system on the leading edge of the orbiter's left wing. Atlantis STS-135 flew the last Space Shuttle mission on July 8, 2011. After 135 missions, the STS ended.

2.5 Lessons Learned

The Challenger disaster happened "while the whole world was watching". As a consequence, it brought vulnerabilities underlying technological endeavors into everyone's living room. *The* key macro lesson learned, no doubt, is that in the case of technological risk, it is crucial to keep lines of communication concerning key technical project components open throughout the entire process. In this case, this concerned the non-resilience of the O-rings at low launch temperature. The critical information on potential O-ring vulnerability failed to reach the right company level at the most critical moment. At a more micro-level, we learned through a standard statistical fitting procedure that it is always crucial to consider *all* relevant data; in this case, also the data that did not involve to O-ring failure. In this context we met the concept of *dark data*. The famous press conference experiment by Physics Nobel Prize winner Richard Feynman has become a classroom example on how to communicate risk to a broader public. Feynman also stressed the importance of "talking to the guys in the boiler room", meaning that it is always crucial to listen to critical information voiced at company levels further away (or, indeed, down) from the level of managerial decision-making.

3

The 2007–2008 Financial Crisis

These heroes of finance are like beads on a string;
when one slips off, all the rest follow.

Henrik Ibsen (1828–1906)

This chapter is written for novices to the world of international finance as well as specialists. We want to make it clear however that we mainly aim for the former group. One should be able to 'enjoy' this chapter without having to enter into too many technical details; indeed the world of finance is full of technical jargon that quickly shrouds the broader underlying issues. We have put the word 'enjoy' in quotation marks as the story to be told is not a particularly nice one. The chapter contains numerous quotes in order to give a voice to the protagonists involved at the time. In contrast to technical books, broad general-purpose books on risk often neglect financial risk, or the risk inherent in financial markets. Nevertheless, it is a most important type of risk, which affects us all, and does so over longer periods of time. Think for instance of the 1929 great depression, the various market crashes, and indeed the 2007–2008 financial crisis. We very much hope that through our relatively brief and possibly somewhat biased discussion, we will be able to communicate the main aspects underlying the financial crisis and consequently contribute towards a better understanding when it comes to discussing the question "Can this happen again?" The description "possibly somewhat biased" reflects the fact that several parts of the story are reported in the way that we authors, working as mathematicians in the field of quantitative risk management, experienced them. In light of the first sentence of this chapter, we had to leave out some more technical aspects and events, but there is plenty of literature on the topic out there to satisfy each reader's hunger for more.

3.1 Introduction

In the present chapter, the 2007–2008 (or indeed 2007–2009) global financial crisis will alternatively be referred to as the *financial crisis*, the *subprime crisis* or simply the *crisis*. Numerous reports, articles, books (journalistic, academic, governmental, regulatory, etc.), as well as movies and plays have described the events unfolding around the crisis in every way possible. We mainly concentrate on aspects of the story as they fit into the overall concept of the book. At the same time, the choices made are colored by our personal experiences. In doing so, we hope to add relevant details to the story that are perhaps less well known.

It must have been around 2006, when, during a meeting with bankers, we were asked the question: "Value-at-risk throughout the global banking industry is down, but where is all the credit risk hiding?" Value-at-risk (VaR) is a regulatory risk measure (see Section 8.4.2), which throughout the financial industry was touted as a gauge for market safety, just like the dike height in the case of the Dutch coastal defenses discussed in Chapter 1. Its implementation in the early 1990s revolutionized risk management for larger international banks; see, for example, the book Jorion (2006), whose 624 pages give an indication that VaR calculations became a major, though non-trivial, daily task for risk-management functions throughout the industry. Like the calculation of a sufficiently high level of sea dikes to withstand a 1 in 10 000 years flood event, the calculation of the necessary regulatory capital to limit a bank's financial losses in the case of (rare but severe) distress became one, maybe *the* key, radar indicator for financial safety and soundness. Whereas the Dutch dikes have functioned perfectly well ever since their construction, the financial VaR-based dikes have crumbled on several occasions. This chapter gives a brief outline of how, worldwide, the financial dikes started collapsing around 2007. The structural failures of these "dikes", however, go back much further in time.

3.2 A Timeline of the Crisis

In late 2007 and early 2008, all over the western world the crisis started knocking on people's doors. On September 13, 2007, the BBC headlined a troubling story related to Northern Rock, a bank in the northeast of England, "Northern Rock gets bank bail out". And, further, "The Bank of England has agreed to give emergency financial support to Northern Rock, one of the UK's largest mortgage lenders [. . .]". In the following days, this led to a run on the bank, the first on a British bank since the case of Overend, Gurney and Company in 1866. People worldwide were shocked by images of customers queuing up outside bank premises to withdraw their savings; see Figure 3.1.

Figure 3.1 A run on the bank, customers queuing to make withdrawals. Source: Enrico Chavez

In a 10-year revisit of the event, the then Governor of the Bank of England, Marvin King, made it clear in an interview with the BBC on September 11, 2017, that "My advice was very clear – we should not reveal publicly the fact we were going to lend to Northern Rock." King obviously feared panic reactions. The rescue action by the Bank of England, however, leaked early and indeed market panic set in. Financial markets have always reacted nervously to rumors, whether well-founded or not; see also Figure 3.7 below. In the case of Northern Rock, the government's communication around the unfolding events did not calm the public's mood. On February 22, 2008, Northern Rock went into state ownership. The developing crisis had shown its ugly face for the first, but not for the last time. The shadow of the 1930s great depression loomed not only over Wall Street but also over Main Street; the latter idiom typically refers to the world outside finance.

We would like to recall three important dates. The first, March 16, 2008, concerns Bear Stearns, an 85-year-old investment bank. Over the period of a couple of days, its share price dropped from close to 100 US dollars (USD) to 2 USD; see Figure 3.2. On March 16, 2008, it was sold to J. P. Morgan Chase for 10 USD a share. The fire-sale-like low price masked the

Figure 3.2 Bear Stearns share price over the period from January 2004 to March 2008. Source: Authors

fact that this buy-out was not a friendly takeover but a necessary move to avoid bankruptcy and ignite a potential domino effect of institutions linked through products or business with Bear Stearns. A second important date, perhaps the most important, is September 11, 2008. On that day, the global financial services firm Lehman Brothers, founded in 1847, filed for bankruptcy protection, the so-called "Chapter 11". The evening news and the next-day newspapers showed Lehman bankers moving out their office belongings in carton boxes; see Figure 3.3. On September 15, 2008, Lehman was declared bankrupt. With Lehman, the fourth largest US investment bank, a crucial domino stone had fallen, some would say had been allowed to fall. More were to follow. A particularly visible case was an insurance giant, the American International Group (AIG). On September 16, 2008, our third important date,

Figure 3.3 Lehman Brothers banker leaves the building after the bank filed for bankruptcy. Source: Enrico Chavez

the US government authorized the Federal Reserve Board to make a 85 billion USD loan to AIG; more financial aid was to follow. At the time, AIG was deemed "too big to fail". It had operated for so long at the center of the world's financial web, with so many counterparties, that its collapse would be felt in every corner of the financial globe. The Federal Reserve Board, with the backing of the Treasury Department, had taken control of what had been one of the most successful private enterprises ever; we discuss this case in somewhat more detail in Section 3.4.

Of course, the seeds for this crisis were sown much earlier through the hands of politicians, bankers, rating agencies, traders and, we must stress, investors in the search for (excessive) yield. As so often, an unbridled growth of stock markets and a much-relaxed screening for credit quality helped to create a market for financial products of unprecedented complexity as well as astronomic volume. It is fair to say that the main thrust of the crisis came from a very aggressive Anglo-American business model for international investment banking. The latter was conveniently fueled by an overheated US housing market. The banking sector as a whole and no doubt most of its employees worldwide acted responsibly. However, some trading units (often bank-internal hedge funds) managed to grow relatively small but well-shielded kingdoms. The latter were typically populated by an all-male, young, aggressive and vastly over-paid group of bankers or traders. Greed played an important role, as the trading horizon often was the next bonus round. Many lived and worked to the adage "Heads the bank wins, tails you lose". This had to go wrong, and indeed, it did. The highly skewed system of incentives and the extreme potential for its misuse contributed to a situation where relatively few people created such havoc that it brought the global financial system to the edge of the abyss.

Many warned early on, but few anticipated the size of the financial earthquake and the harm it would do. Commemorating "ten years after Lehman", Christine Lagarde, at the time the head of the International Monetary Fund (IMF), was quoted as "I have said many times, if it had been Lehman Sisters rather than Lehman Brothers, the world might well look a lot different today." She further stressed that the male domination of banking could lead to another financial crisis. The article van Staveren (2014) discusses the need for gender diversification in banking and uses the "Lehman Sisters" idiom. No doubt gender diversification throughout the banking system is very much needed.

In his book, originally published in Spanish in 1998, Eduardo Galeano (2001) re-coined the phrase "profits are privatized, losses are socialized". Its origins go back at least to Andrew Jackson, the seventh president of the United States (in office from 1829 to 1837), who used the phrase on the occasion of the closing of the Second Bank of the United States in Philadelphia. This view on capitalism is folklore within so-called *lemon socialism*; see Figure 3.4 for a corresponding cartoon. Lemon socialism typically refers to government

Figure 3.4 During the financial crisis, the fate of several financial institutions turned sour, leading to widespread government intervention (lemon socialism). Source: Enrico Chavez

intervention for (almost) failing or "sour" companies, hence "lemons". The terminology also appears as "socialism for the rich, capitalism for the poor". This saying certainly turned out to be true for the financial crisis. The ensuing financial relief initiatives by governments worldwide saved the global financial system from a rude crash landing. The Occupy Wall Street movement from 2011 voiced concern about the consequences of this kind of lemon socialism, which further highlighted growing socio-economic inequalities. "Why save the bankers" became a political battle cry; it resonates loudly even today.

3.3 How Did we Get There?

Rather than appearing in a Japanese martial arts movie, during the financial crisis a *ninja* referred to a client whose credit risk was of dubious quality: *n*o *i*ncome, *n*o *j*ob or *a*ssets. Ninja-targeted loans and mortgages started flooding the American housing market. Such loans are also referred to as subprimes and gave the crisis its alternative name, *subprime crisis*. These loans originated during the housing bubble of 2003–2007 and became notorious during the crisis. Under no circumstances would a ninja client have gotten a loan from a bank that used a serious credit-quality screening system. However, then came out one of the magic wands of Wall Street: *securitization* (*Verbriefung* in German, *titrisation* in French). Take a big product kettle, throw in a number of such subprime (ninja) loans, together with some mediocre ones and sufficiently many good quality ones, and stir this financial soup with your wand; perhaps "potion" would better reflect the content of the kettle. After sufficient stirring, it is time to give it a financial engineering flavor. Repackage the payments coming out of this loan portfolio (kettle-)construction into tranches called equity, mezzanine and (super-)senior, ordered according to decreasing riskiness. We will come back to this construction. Now we need a sufficiently catchy name: "collateralized debt obligations" (CDOs) sounds good. The underlying interest payments on, and the repayments of, the loans are indeed collateral to (that is, depend on) these financially engineered investment products.

At this point, you may want to pause briefly and watch the brilliantly funny BBC sketch "Bird & Fortune – silly money – investment bankers"; see Bird and Fortune (2007).

So now we have packaged perhaps thousands of different loans into a CDO. We have sliced and diced them into tranches of different credit quality. The *equity tranche* is the most risky, attracting the highest risk premium. The *super-senior tranche* is the safest, with the *mezzanine tranche* sitting somewhere in between. In order to achieve this ordering we bring on the *waterfall principle*. Payment delays or defaults from the loans to the CDO will first affect the, by design, most risky equity tranche; this tranche hence bears the first portfolio losses. Then comes the mezzanine one, followed by the 'super safe' super-senior tranche. Regarding the latter tranche, for the situation to arise where it would have to bear losses and hence depreciate in value, the CDO's underlying loan portfolio, say, would need to encounter a very high percentage of defaulting loans. The latter was considered highly remote and hence of very low probability. Note that we put 'super safe' in quotes; we will indeed find out shortly that this quality label turned out to be 'not so safe at all'.

The 'super safe' was the sales pitch. Before CDO tranches can be sold to customers, like your pension fund, they need an official credit rating. Enter the next protagonists: rating agencies such as Moody's, Standard & Poor's (S&P) and Fitch, to name the three most important ones. Investment banks setting up CDO investment products would instruct and hence pay for a rating agency to deliver a credit rating; notice the lack of independence here. Once CDO senior tranches got a top rating, for example "AAA" in the case of S&P, your pension fund and financial institutions all over the world could load them up on their balance sheets – and they did. Not just that, they gobbled them up. This was just the start! Payment streams coming out of many CDOs were themselves repackaged into a new CDO structure called *CDO-squared*, and even one step further up the financial engineering ladder gave us *CDO-cubed*. The latter is a so-called triple derivative, a derivative of a derivative of a derivative; such a product was also referred to in the industry as a "derivative on steroids".

With Wall Street now on the loose, there was no stopping. Whereas CDOs still had regular debt cash products such as loans and mortgages as their input, *synthetic CDOs* would move away from that premise and include as input non-cash derivatives such as credit default swaps (see below), options and other financially engineered contracts. In McLean and Nocera (2010, Chapter 17), the authors refer to the market introduction of synthetic CDOs as follows:

> [The year] 2006 saw the creation of one of the most unnatural and destructive financial products that the world has ever seen, the synthetic CDO, it turned the keg of dynamite into the financial equivalent of a nuclear bomb.

Regarding such products, nobody fully understood their hidden risks and therefore did not know how to correctly price nor hedge them, nor was there an economic rationale for their introduction besides the dubious promise of "an extra yield to investors". When Wall Street talks about "extra yield", always look carefully for the small print. On several occasions in this chapter we will meet the word "hedge", as in "hedge fund". In financial jargon, to *hedge* refers to a financial construction aimed at reducing risk, just as hedges are planted around a field for protection and cover. Here we would like to make a link to the story of Richard Feynman and the Challenger disaster in Chapter 2. In Section 2.2 we mentioned that Feynman always wanted to talk to the "workers on the shop floor". In any serious risk situation, talk to the people at the heart of the business, not just to those sitting at the top of the hierarchical pyramid. In the case of the financial crisis, these shop floor workers certainly included the scores of computer programmers who had to write the computer code for all these new products. You may want to listen to the short interview with Michael Osinski (2018). For a while in the 1990s he and his wife were writing computer software for these ever more complicated financial products. Their software helped to turn mortgages into securities for Lehman Brothers; in the interview, he refers to these collateralized products as "You take chicken and you put it into the grinder, out comes sirloin." He retired years before the 2008 crash and now farms oysters on Long Island. At the run-up to the crisis and during its early stages, software development could hardly keep up with the speed and volume that investment banks were putting on show. These software developers knew that the boiler's pressure was getting out of control.

We would now like to add some sobering quotes. Enter Fabrice Tourre, a former trader of Goldman Sachs, who, in 2014, was found liable for defrauding investors in soured mortgage deals of the above type. He, no doubt, was in the eye of the synthetic CDO hurricane. The *Financial Times* of January 29, 2007, quotes Tourre as follows: "Well, what if we created a 'thing', which has no purpose, which is absolutely conceptual and highly theoretical and which nobody knows how to price?" Further, in the *Financial Times* of June 13, 2007: "I've managed to sell a few Abacus bonds [a particular synthetic CDO] to widows and orphans that I ran into at the airport, apparently these Belgians adore synthetic ABS CDO2 [CDO-squared]." The acronym ABS stands for asset backed securities, indeed like most of these CDO-like products. The word "sickening" comes to mind. Around that time, David J. Hand, a former President of the Royal Statistical Society stated that "On Wall Street, gain-maximisation often leads to constrained optimisation problems moving ever closer to the ethical boundary." The above examples clearly show that in this quote "ever closer to" can safely be replaced by "well beyond".

The short discussion above should already give you some idea of how parts of the investment-banking system had become fully out of control by the early 2000s. If you have had enough financial jargon for the time being, you may want to jump to Section 3.4. If, however, you still have the stamina to dig a little bit deeper, you next have to become aware of a phenomenally successful class of financial products, the *credit default swaps* (CDSs); above, we briefly mentioned CDSs as possible inputs to synthetic CDOs. They were created in the early 1990s and appeared on the market in their current form in 1994. The original introduction of CDSs was (and still is) economically convincing as well as natural from a market point of view. The very basics of a so-called *covered* CDS involves three parties, A, B and C; see the left-hand side of Figure 3.5. Suppose by way of example that company A

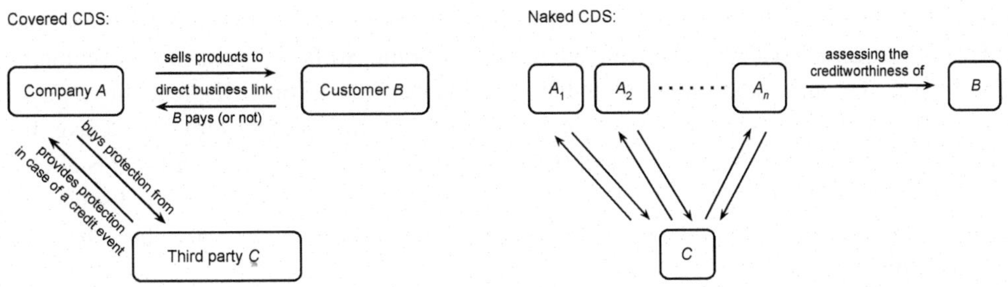

Figure 3.5 Schematic diagram of a covered CDS (left) and a naked CDS (right). Source: Authors

sells its products to a *direct* customer B, hence the deal is "covered"; the deal is covered by a direct and clear business link. In every such transaction, A does run the risk, however remote, of not being paid by B for its delivered goods. This "non-payment" is referred to as a *credit event* or *credit default*, which can become more involved as partial payment or delayed payment, say. This is where the third party, C, steps onto the scene. By paying C a premium, A buys protection from C for non-payment by B. Hence, the CDS *swaps* the non-payment risk for delivered goods from A to C. In the case of a covered CDS this may sound somewhat akin to buying insurance. The premium paid by A to C is calculated on the basis of fairness, with a small extra loading to be added to a fair price, indeed as in the calculation of an insurance premium. Corporation or customer B can also be a country, and then we are talking about a *sovereign* CDS rather than a *corporate* CDS. A credit event for a country could be the default on, or a restructuring of, its government debt; famous examples here are Russia in 1998 and Greece in 2015. The economic rationale for a covered CDS, where one producing company A has a direct business link to its client, company B, is clear. So far so good. Now let us move away from this direct business link through the introduction of a *naked* CDS. For this, consider for instance n further companies (call them investors or, if you prefer, gamblers) $A_1, A_2, \ldots, A_n$ with no immediate business links to B (hence the word "naked"); see the right-hand side of Figure 3.5. They each may have their own assessment of the creditworthiness of B and buy from C a CDS accordingly. These investors are gambling on the financial health of B. This is where the direct comparison with insurance stops; a naked CDS is akin to you buying fire insurance on your neighbor's house. Through this multiplication of actors (gamblers) involved, the CDS market went well beyond its original intention of providing protection for those directly involved and what was sold were pure investment products. On the other hand, a well-functioning CDS market can be viewed as a thermometer for a company's or country's credit health. The latter surely applies for those who claim that "the market always knows best". There are, no doubt, two sides to the CDS coin. Here the market volume of such products enters, and given that these volumes are large (we will see that they were astronomical), where do these products end up? Or, reformulated, who bears the risk when things do not go according to plan?

The above number *n* of naked positions could be very large indeed. Further, structured finance went well beyond CDSs as input to CDOs, but we will leave that path and without delay turn to the *value* and *volume* of the resulting markets. Below we quote some truly astronomical numbers. We (have to) forego a precise, and indeed non-trivial, definition of "value"; just absorb the numbers as presented. If after reading this paragraph your reaction is "This is just unbelievable" then we have made our point. So fasten your money seatbelt. In June 2006, the outstanding nominal value of CDSs was 30 trillion USD, reaching a peak above 60 trillion USD in the second half of 2007. Let us put this number into perspective. In 2006, the world's gross domestic product (GDP) was 58 trillion USD; the US GDP is about 14.5 trillion USD. We will come back to such bewilderingly large numbers in a discussion on long versus short scales in Section 6.6. For the moment, it suffices to know that a trillion is a 1 followed by 12 zeros!

We do realize that nominal values neglect important realities surrounding such derivatives markets, for example *netting*, aimed at aggregating positions to arrive at a net obligation. In these highly interconnected markets, one has to avoid the double counting of off-setting positions. However, and this is a crucial "however", the necessary market clearing needed to arrive at such netted positions is only good insofar as markets function in times of distress. Here market *liquidity* plays a crucial role, that is, the ability to instantly buy or sell at a given price. We therefore find it relevant also to keep an eye on the nominal values of such highly complicated financial products. They surely yield an indication of market developments galloping away from economic reality; we will revisit this statement when in Section 3.4 we discuss the Swiss regulator's report on the demise of UBS, an international Swiss bank.

Returning to "the necessary market clearing" mentioned above, most, if not all, of the more complicated products mentioned so far are of the so-called *over-the-counter* (OTC) type. They are traded via a broker and dealer as opposed to being transacted on a fully regulated exchange. There is no central clearing house to safeguard and settle the resulting payments, or for that matter non-payments. Throughout the financial crisis, the nominal value of OTC markets ballooned to about 680 trillion USD by June 2008. By the end of June 2020, it stood at around 600 trillion USD with a gross market value of 15.5 trillion USD. By the end of 2008, the latter was somewhat over 30 trillion USD. The publications by the Bank for International Settlements (BIS) yield useful information on the topic; see, for instance, BIS (2020). We do hope that you agree with our choice of the term "astronomical" before.

And yet, at the time, these developments were looked at favorably. In an article in the *Financial Times* of December 28, 2002, John Kay quotes famous economist and former Chair of the Federal Reserve Alan Greenspan on this brave new world of financial engineering:

> The use of a growing array of derivatives and the related application of more sophisticated methods for measuring and managing risk are key factors underpinning the enhanced resilience of our largest financial institutions. As a result, not only have individual financial institutions become less vulnerable to shocks from underlying risk factors, but also the financial system as a whole has become more stable.

Figure 3.6 Alice in Wonderland "You're nothing but a pack of cards." Source: Sir John Tenniel, Public Domain Review

Further, in its April 2006 Global Financial Stability Report, the International Monetary Fund (IMF), even at that time, still expressed the following view on these new credit instruments:

> There is growing recognition that the dispersion of credit risk by banks to a broader and more diverse group of investors, rather than warehousing such risk on their balance sheets, has helped make the banking and overall financial system more resilient. [...] The improved resilience may be seen in fewer bank failures and more consistent credit provision. Consequently the commercial banks may be less vulnerable today to credit or economic shocks.

In the latter quote, we encounter another bit of banking jargon, *warehousing*. In the context of CDO transactions, warehousing is the accumulation and custodianship (keeping within the bank) of bonds or loans that will become securitized through a CDO transaction, hence that enter as input into a tranching construction, as discussed above. Proper securitization would have implied a truly widespread distribution of risk. That this did not happen is in part because, contrary to the basic principles of securitization, several larger banks filled up their balance sheets (hence the word "warehousing") with super-senior tranches, which were perceived as riskless. In doing so, they constructed a world-wide network of interconnected financial products and interdependencies that would soon light the fuse of the financial powder keg.

3.4 The House of Cards Comes Tumbling Down

Towards the end of *Alice's Adventures in Wonderland* (Carroll, 1865) Alice faces the Queen of Hearts in an unpleasant court case. As the situation becomes more and more unreal, Alice suddenly cries out "You're nothing but a pack of cards," upon which the whole set-up falls apart in a rain of falling cards. Alice had exposed Wonderland as an illusion; see Figure 3.6.

In a similar way, the subprime hype in the end turned out to be an illusion; however, a very costly one. Early on in the crisis, the Swiss bank UBS ran into serious problems. Between 2007 and 2009, it had to write down 50 billion USD owing to its exposure to subprime and related markets. We quote from the highly interesting and informative report (Swiss Federal Banking Commission, 2008):

In addition, UBS had to take write-downs of about USD 21.7bn on investments in Super Senior CDOs. This was because the CDO Desk had not only securitized CDOs and sold such CDOs to investors, but had retained the Super Senior CDOs it structured as a long-term investment strategy [...] The bank's investment in these Super Senior CDOs increased in the first half year of 2007 and amounted to USD 50bn prior to the onset of the subprime crisis. The Super Senior CDOs retained by UBS were the greatest single source of loss for the bank.

The above points are further highlighted in Acharya et al. (2010). The authors discuss the business model followed by so-called large complex financial institutions (LCFIs) in the run-up to the crisis:

These LCFIs ignored their own business model of securitization and chose *not* to transfer credit risk to other investors. Instead they employed securitization to manufacture and retain tail-risk that was systemic in nature and inadequately capitalized. [...] Starting in 2006, however, the CDO group at UBS noticed that their risk-management systems treated AAA securities as essentially riskless, even though they yielded a premium (the proverbial free lunch). So they decided to hold onto them rather than sell them. After holding less than $5 billion of these securities in February 2006, the CDO desk was warehousing a staggering $50 billion of them by September 2007. [...] In a similar fashion, by the late summer of 2007, Citigroup had accumulated over $55 billion of AAA-rated CDOs. [...] it was not a true originate-and-distribute model that was at work but rather the new banking model of "originate-distribute-and-hold", incurring massive systemic tail-risks that brought the financial sector down.

In view of a comment made in Section 3.3 concerning keeping an eye on the nominal values of perceived safe investments, the following quote from the Swiss regulators' report is telling:

In assessing the risks of Super Senior CDOs, too much weight was attributed to the securities' AAA-rating, which was consequently reflected in the risk measurement metrics. Because these positions were viewed as being VaR neutral [that is they did not need extra regulatory capital set against them], they were not reflected in key internal risk reports and thus disappeared from the radar screen. A ceiling on their nominal volumes [see our earlier comments], which would have counterbalanced this weakness, did not exist for most of the business lines.

On October 16, 2007, the Swiss Government and the Swiss National Bank announced a 60 billion USD rescue plan. UBS became the most exposed foreign bank on the US real estate funds and derivatives market. A group of about 250 investment bankers and traders had brought down a bank with close to 80 000 employees worldwide.

We will concentrate our further discussion on the American International Group (AIG). A key question we have not answered yet is "Who wrote protection for all these CDOs and CDSs?" In other words, who acted as company C in the example involving companies A, B and C above? Enter AIG Financial Products (AIG-FP), a subsidiary of American International Group, an American multinational finance and insurance corporation. AIG-FP was founded on January 27, 1987, as a division focused on complex derivatives trades that took advantage of AIG's AAA credit rating. AIG-FP grew to about 400 employees by early 2000. AIG, which at its peak was one of the largest publicly traded companies in the world with core business rooted firmly in insurance, wanted to branch out into the world of finance. We do mention this "branching out" as other insurance companies at the time also walked that route, albeit with less grandeur. Before too long, these companies would find out that such moves could turn sour. AIG-FP would write CDSs insuring CDOs against default. For a while, it cashed risk premiums for a business that was perceived as virtually risk-free. Had Wall Street finally managed to achieve the medieval alchemist's ultimate dream of turning iron into gold? The equivalent in financial jargon being that of a free lunch.

This alchemist's dream, however, evaporated at the same time as the US housing market started to tumble. It reached a peak mid-2006. Suddenly all these perceived riskless investments turned out to carry considerable risk. Their apocalyptic volume, widely spread interconnectedness, as well as incredibly complicated structure, exposed several Achilles heels in the whole edifice. One after the other, financial institutions started to fall, needing government rescue packages, or, as in the case of Lehman Brothers, defaulted. The various customers were the real victims of reckless institutions involved in the various transactions. This was very much reflected in the many legal cases brought against the participating banks in the aftermath of the crisis. Not only did the volume of complicated financial products balloon, so also did the fines levied on those banks, especially through the US Department of Justice. The Economist of August 18, 2016, reports that legal settlements since 2009 amount to 219 billion USD, with the Bank of America topping the list with 70 billion USD. At the time, this constituted 50% of its market capitalization, where the latter is calculated by multiplying the price of one stock by the total number of stocks held by all the shareholders of the company.

Let us return to AIG's involvement through AIG-FP, and recall that the latter sold protection on senior tranches of CDOs where the underlying portfolio consisted of loans, debt securities, asset-backed securities, mortgage-backed securities, . . . , hence the lot! Some of the public statements at the time from within AIG-FP are worth revisiting, if only as a public warning for potential similar future developments. We mainly base the quotes below on the lengthy 291-page report of a Congressional Oversight Panel chaired by Elizabeth Warren; see Congressional Oversight Panel (2010). For an excellent summary of the story, we can recommend a three-part series by the Washington Post staff writers Brady Dennis and Robert O'Harrow Jr. that appeared on three consecutive days in December, 2008, under the suggestive titles "The beautiful machine", "A crack in the system" and "Downgrades and downfall" (O'Harrow and Dennis, 2008a,b,c). In the quotes summarized below, the above sources are referred to as COP, and WP1, WP2 and WP3, respectively.

Howard B. Sosin, a co-founder of AIG-FP in 1987, who stayed with the company until 1993, early on stated, "You know, we're not going to do trades that we can't correctly model, value, provide hedges for and account for." (WP1) One should indeed handle the ever-increasing complexity of financial products in this way. If you cannot correctly model or hedge these products, do not sell them. The interpretation of "correctly" is, however, debatable. According to WP3:

> Financial Products [AIG-FP] had built itself on data, analysis and a culture of healthy skepticism. Even as the firm grew to about 400 in 2005 from 13 employees in 1987, it sought to maintain its discipline. At Financial Products, God had always been in the details, and the details were always rooted in the math.

We will come back to the expression "rooted in the math" in the next section. We should note that AIG-FP later moved on from its original careful set-up. Of course, also the markets evolved. During the period from 1965 to 2005, the powerful and feared CEO of AIG was Maurice Raymond "Hank" Greenberg. He guarded the AAA rating of AIG jealously, as the following quote makes clear in no uncertain terms: "You guys up at AIG-FP ever do anything to my AAA rating, and I'm coming after you with a pitchfork." Schiff (2005) describes in more detail Greenberg's dramatic exit from AIG. For a while, AIG-FP delivered the goods. The news from the front was one of extremely remote risk, and yet an excellent return. Quotes from these glory days include "The models suggested that the risk [about credit default swaps] was so remote that the fees were almost free money. Just put it on your books and enjoy." (WP2) This was said by the President at the time of AIG-FP. And further, according to WP3, "Financial Products executives said the swaps contracts were like catastrophe insurance for events that would never happen." The latter quote yields an important view on the market situation at the time. It is fair to say that, in the end, those banks involved in these CDO-CDS-like constructions as well as their warehousing were long (in the sense of having bought) a massive catastrophe bond waiting to be triggered.

However, according to COP, p. 27, "AIG-FP continued to assume through the beginning of 2008 that the credit risk from its CDS portfolio was virtually non-existent given the super-senior credit ratings of the reference securities." This clearly shows the important role played by the rating agencies. Joseph Cassano, the head of AIG-FP at the time, made one of the most quoted statements, and this was during the company's second quarter 2007 earnings call with analysts and investors; see COP, p. 27:

> It is hard for us, without being flippant, to even see a scenario within any kind of realm or reason that would see us losing $1 in any of those transactions.

And further:

> AIG's then-CEO, Martin Sullivan, asserted in an investor presentation in December of 2007 that because AIG's CDS business is "carefully underwritten and structured with very high attachment points to the multiples of expected losses, we believe the probability that it will sustain an economic loss is close to zero."

Soon after, the tide turned and the cards started falling down. According to COP, p. 19, "AIG's downfall stemmed in large part from its CDS on multi-sector CDOs, which exposed the firm to the vaporization of value in the subprime mortgage market." On p. 21: "Federal Reserve Chairman Bernanke characterized AIG-FP as a 'hedge fund attached to a large and stable insurance company.'" And on p. 29:

> While market conditions remained similarly illiquid, ratings downgrades on the reference securities and valuation losses by market participants helped establish two of the three primary triggers for collateral payments, making it more difficult for AIG to continue to hide behind its models. [. . .] Subsequent downgrades of AIG's credit rating in turn precipitated additional collateral calls.

As a consequence, the above proverbial "catastrophe bond" was triggered, and losses mounted as counterparties started demanding collateral payments; soon liquidity (cash) dried up.

As we have already seen above and indeed in the previous paragraphs (COP, p. 41):

> Credit rating agencies played an exceptionally important role in AIG's collapse and rescue. Credit rating downgrades were a factor in AIG's problems [. . .]. Large insurance companies in general are dependent on a sound credit rating that permits them to access the bond markets cheaply. [. . .] Although AIG profited for many years from its AAA credit rating, it also became particularly vulnerable to the negative consequences of ratings downgrades.

According to WP3, "The diversification was a myth – if the housing market went bust, the subprimes would collapse, like a house of cards." Here we encounter a second magic wand of Wall Street: *diversification*! When the word "diversification" is used in a financial investment context, especially in new and intricate markets, always ask, "What do you exactly mean by diversification?" and do not let yourself be sidelined too easily. You may know the standard guideline to achieve diversification: "Do not put all your eggs in one basket."

American International Group (AIG), a company of around 100 000 employees, brought to its knees by a small subsidiary of 400 employees, is an example of a failure of risk management, both at the division and the group level. AIG almost went bankrupt because it ran out of cash. At 9 pm on Tuesday, September 16, 2008, the Federal Reserve Board, with the support of the US Department of the Treasury, announced that it had authorized the establishment of an 85 billion USD rescue fund for AIG (COP, p. 57). Clearly, AIG miserably failed the *what if* test: *What* happens to our portfolios *if* we lose our top rating due to a downgrade?

This comment concerns the important difference between the *if* and *what if* approach to risk management. The probability of a downgrade of AIG (*if*) was indeed considered remote, however, the real question was *what if*. The difference between *if* and *what if* thinking and questioning is highly relevant in any risk management situation. You can do the mental exercise in the case of the Dutch dike constructions in Chapter 1, or the destruction of parts of the nuclear power plants at Fukushima in the event of the 2011 Tōhoku earthquake and tsunami; see Chapter 4 for a discussion.

Perhaps the ultimate summary of AIG's case remains the following; see WP1:

> At the end, though, the story of Financial Products [AIG-FP] is not about math and financial formulas. It is a parable about people who thought they could outwit competitors and market forces alike, and who behaved as though they were uniquely positioned to sidestep the disasters that had destroyed so many financial dreams before them.

3.5 Blame the Mathematicians

Soon after the main events of the crisis unfolded, the finger-pointing started. Whereas it should be clear from the previous sections that many were to blame, mathematicians and financial engineers (typically referred to as "quants") became convenient scapegoats. We saw "mathematics" come up occasionally in some of the quotes in the previous section. However, no other publication caught the public's interest as much as Salmon (2009). Its title alone, "Recipe for disaster: the formula that killed Wall Street", invites further reading to learn more about the "villains" responsible for the world's resulting misery. The author, Felix Salmon, received the 2010 Excellence in Statistical Reporting Award of the American Statistical Association with as quotation "We reprint his article, first published as the cover story of *Wired* magazine, because it brilliantly conveys complex statistical concepts to non-specialists." For the occasion, the article was reprinted in the ASA–RSS journal *Significance* (Salmon, 2012). But even more than the paper, one of its formulas traveled the world, the infamous Gaussian copula formula (see below for an explanation of the symbols)

$$\mathbb{P}(T_A < 1, T_B < 1) = \Phi_2\big(\Phi^{-1}(F_A(1)), \Phi^{-1}(F_B(1)), \gamma\big). \tag{3.1}$$

Gaussian copula cartoons, T-shirts and even an example of someone carrying a Gaussian copula tattoo on the arm surfaced. The caption of the figure showing the Gaussian copula formula in the original article reads "Here is what killed your 401(k) [your pension fund savings]".

In an article (Jones, 2009) from the April 24, 2009 *Financial Times*, we read:

> In the autumn of 1987, the man who would become the world's most influential actuary landed in Canada on a flight from China. [. . .] And if he could apply the broken hearts maths to broken companies, he'd have a way of mathematically modelling the effect that one company's default would have on the chance of default for others. [. . .] [David] Li, it seemed, had found the final piece of a risk management jigsaw that banks had been slowly piecing together since quants arrived on Wall Street.

The article starts with a photograph of the American country singer Johnny Cash (1932–2003) with his wife June Carter (1929–2003). Johnny cash died (September 12) only four months after June Carter (May 15). In the epidemiological literature this fits within the context of the "broken heart syndrome", and the "broken hearts" in the above quote refers to that. The broken heart syndrome is a temporary heart condition that is often brought on by stressful situations and extreme emotions, for instance the death of a spouse, especially at a later stage of life.

As an actuary, with a PhD from the University of Waterloo, David Li was well aware of actuarial models for the survival times of coupled lives. From coupled lives of spouses to coupled times until default of companies is only a relatively small step. This is exactly where the Gaussian copula model from Li (2000, Equation 12) enters. Let us read together through (3.1) without entering into the precise mathematical details. In doing so, we hope to slightly demystify the mathematical symbols. On the left-hand side of the equation, we see the joint probability ("$\mathbb{P}$"; we will also use this notation in Chapter 8) that two companies A and B both default within one year (so that their default times T_A and T_B satisfy $T_A < 1$ and $T_B < 1$). On the right-hand side of the equation, we see the probability $F_A(1)$ that $T_A < 1$ and the probability $F_B(1)$ that $T_B < 1$ (so these are the individual probabilities that A, respectively B, default before year end). The remaining symbols (the $\Phi_2(\Phi^{-1}(\cdot), \Phi^{-1}(\cdot), \gamma)$) are the Gaussian copula model; the symbol Φ^{-1} will be covered in Chapter 8. We see that the model "couples" (hence the name "copula") the probability of a joint default of A and B with the probabilities of the individual defaults of A and B. The Gaussian copula therefore controls the dependence between the individual defaults and the joint default. We will not cover Φ_2 in this book, but you can see at the end of the formula that Φ_2 depends on γ. This parameter γ controls the range of possible dependencies between the default times T_A and T_B of A and B in the Gaussian copula model. About γ, Salmon (2012) writes:

> The all-powerful correlation parameter, which reduces correlation [the dependence between the two default times] to a single constant – something that should be highly improbable, if not impossible. This is the magic number that made Li's copula function irresistible.

The magic is that if we only have models for the individual or separate times until the default of A and B and believe in the Gaussian copula model, then we have an estimate for the probability of a joint default of A and B. There is no real restriction in considering only two companies. Needless to say that the crux lies in the above "and believe in the Gaussian copula model". This is exactly where all discussions start.

Rather soon, the rating agencies Moody's and S&P started using the formula, whereas on the various trading floors there was much more skepticism. For a readable discussion on the latter, see Pollack (2012). From the start, David Li always stressed the underlying model assumptions necessary before one can use the formula. It surely is interesting to hear his views on the important problems that credit markets face; see Puccetti and Scherer (2018). Pollack (2012) ends with the lines:

> The crisis was caused not by 'model dopes', but by creative, resourceful, well-informed and reflexive actors quite consciously exploiting the role of models in governance. [...] Where does all of that leave the Gaussian copula then? Looking a bit less like the killer it was made out to be, we hope.

Without doubt, the killer-formula-of-Wall-Street will keep inspiring stories. The Gaussian copula even made it into a tattoo on the forearm of Jared Elms, a Creative Director/Writer; see Moore (2011). Elms saw the tattoo as a "eulogy for humankind's unchecked greed".

3.6 Early Warnings

In the previous sections we have already encountered some early warnings concerning the financial markets developing in the 1990s and early 2000s. As we have seen, the products that mainly ignited the crisis typically belonged to credit markets, that is, loans, mortgages, etc. We also saw the danger caused by the two Wall Street magic wands securitization and diversification, especially when these concepts are badly misused. Our preferred early warning came very early, in 1992; Joseph Stiglitz, recipient of the 2001 Nobel Memorial Prize in Economic Sciences, wrote, in a very clear way (Stiglitz, 1992):

> I went on to explain how securitization can give rise to perverse incentives [...]. [...] [Has] the growth in securitization been a result of more efficient transactions technologies, or an unfounded reduction in concern about the importance of screening loan applications? [...] [We] should at least entertain the possibility that it is the latter rather than the former.

Stiglitz further stresses that in the past, banks had demonstrated an ignorance of two very basic aspects of risk: the importance of correlation (interpreted in the wider sense as interdependence, like domino stones one falling after the other), and the possibility of a substantial price correction (think of the burst of the housing bubble leading to the crisis). Taken together, these warnings are to the point and reflect rather severely on the industry.

For mathematicians, to reach the higher echelons of international political establishment and financial regulation is not easy. Famous economists have an easier access. The Jackson Hole Economic Symposium is an annual symposium, sponsored by the Federal Reserve Bank of Kansas City since 1978, and held in Jackson Hole, Wyoming, since 1981. Every year, the symposium focuses on an important economic issue faced by world economies. Participants include prominent central bankers and finance ministers, as well as academic luminaries and leading financial market players from around the world. The overall theme of the 2005 conference was "The Greenspan Era: Lessons for the Future"; you may recall the Greenspan quote in Section 3.3. On Saturday, August 27, 2005, Raghuram G. Rajan, at the time of the conference Economic Counsellor and Director of the IMF's Research Department, took the stage and in front of the prime of international banking delivered his speech "Has financial development made the world riskier?" His very clear answer was "Yes!". This did not make him very popular at the conference, especially since Mr. Greenspan gave the closing remarks at the conference. If you are interested in early warnings about the crisis not far around the corner you should read the speech, it is chillingly to the point; see Rajan (2005a). The talk was based on Rajan (2005b); see also Rajan (2010) for an updated view on the world of finance. The first two quotes concern the liquidity provision to, and procyclicality of, the financial system:

> But perhaps the most important concern is whether banks will be able to provide liquidity to financial markets so that if the 'tail' [the extreme, *what if*] risk does materialize, financial positions can be unwound and losses allocated so as to minimize the consequences to the real economy.

> While it is hard to be categorical about anything as complex as the modern financial system, it's possible that these developments are creating more financial-sector induced procyclicality [one negative event causing a next] than in the past. They may also create a greater (albeit still small) probability of a catastrophic meltdown.

Through prudential regulation, excessive risk taking should be curtailed:

> We want to ward off excessively risky short term investment strategies. Rather than limiting or constraining compensation, incentive regulation might simply require long-term investment of a portion of top investment managers' compensation in the claims issued by the investment that is being managed. In other words, you have to invest say 10 percent of your pay in the assets you manage, and it stays invested till one year after you quit. In other words, this is a form of 'own' capital regulation. It also has countercyclical properties – the more returns you make, the higher your salary, and the more you have invested in your strategy. [Nassim Taleb (2018) refers to this as 'Skin in the game'; see also Section 3.2.]

The latter quote brings us to prudential supervision and regulation. In this context, a name we surely want you to remember is that of Brooksley E. Born, a lawyer, who in the 1990s, during Bill Clinton's presidency, was Chair of the Commodities Futures Trading Commission (CFTC). She warned that unregulated derivatives trading posed a risk to the nation's financial stability and wanted more transparency of this dark market. Most unfortunately, her warnings not only fell on deaf ears, but the CFTC was eventually stripped of its power to regulate securities. The old-boys-network had pulled its strings. Shortly after the latter decision, on June 1, 1999, she resigned as chairperson. In 2009 she received the JFK Profiles in Courage Award.

An excellent article, partly an interview, on Brooksley Born is Schmitt (2009). We find this article an absolute must-read, especially for all aspiring young women facing difficulties trying to make a stand in an occasionally male-dominated professional world. We would like to add a 1997 quote, of the *if* versus *what if* type, from Schmitt (2009). It relates to the efforts of Born in trying, through regulation, to reign in excessive risk taking in the ballooning OTC derivatives market:

> What is more, all the growth had taken place at a time of economic prosperity. Some people were beginning to ask what would happen if the market suffered a major reversal.

The reversal came in September 1998 when the hedge fund Long-Term Capital Management (LTCM) blew up.

We have already encountered several lessons that should have been learned coming out of the 2007–2008 financial crisis. As mathematicians, we would first like to come back to comments made and feelings felt throughout the industry, as well as politics, when it comes to the role of mathematics in the context of the crisis. Once more, we base our reporting on several quotes. We first recall a sharp reaction by Michel Rocard, former Prime Minister of France (1988–1991), published in *Le Monde*, which reads as follows (Rocard, 2008):

Or, ce qui frappe, c'est le silence de la science. Les grands économistes se taisent. [. . .] La vérité, c'est que planquer des créances pourries parmi d'autres, grâce à la titrisation, comme l'ont fait les banques, c'est du vol. [. . .] On reste trop révérencieux à l'égard de l'industrie de la finance et de l'industrie intellectuelle de la science financière. Des professeurs de maths enseignent à leurs étudiants comment faire des coups boursiers. Ce qu'ils font relève, sans qu'ils le sachent, du crime contre l'humanité.

Translated:

What hurts most is that science remained silent. The famous economists did not raise their voice. [. . .] The simple truth is that hiding rotten investments among others using securitization is an act akin to stealing by the banks. [. . .] We have become far too admiring of the high world of finance and the intellectual industry of the science of finance. Professors of mathematics were teaching their students how to make money on the stock market. Without being aware, in doing so, they were committing a crime against humanity.

The final line in particular caused consternation among our French colleagues. This is a formidable accusation: a crime against humanity! The famous French mathematician Jean-Pierre Kahane (1926–2017) wrote an excellent reply, which *Le Monde* unfortunately did not publish (it was reprinted much later as Kahane, 2018):

Les mathématiciens n'ont pas le pouvoir de créer la demande sociale dans les domaines où elle répondrait aux besoins réels de l'humanité présente et à venir, mais ils peuvent aider à la faire s'exprimer.

Translated:

Mathematicians don't have the power to create social demands in fields where such demands respond to the real current and future needs of mankind; they can however help in clearly formulating such needs.

Two quotes on the topic of the role of mathematicians and mathematics in finance with which we personally very much agree are by Chris Rogers (University of Cambridge) and Steven Shreve (Carnegie Mellon University). Both are highly respected mathematicians who contributed extensively to the field of mathematical finance. In a personal communication Chris Rogers wrote:

The problem is not that mathematics was used by the banking industry, the problem was that it was abused by the banking industry. Quants were instructed to build models which fitted the market prices. Now if the market prices were way out of line, the calibrated models would just faithfully reproduce those wacky values, and the bad prices get reinforced by an overlay of scientific respectability! [. . .] The standard models which were used for a long time before being rightfully discredited by (some) academics and the more thoughtful practitioners were from the start a complete fudge; so you had garbage prices being underpinned by garbage modelling.

And by Steven Shreve (2008):

The quants know better than anyone how their models can fail. For banks, the only way to avoid a repetition of the current crisis is to measure and control all their risks, including the risk that their models give incorrect results. On the other hand, the surest way to repeat this disaster is to trust the models blindly while taking large-scale advantage of situations where they seem to provide trading strategies that would yield results too good to be true. Because this bridge [connecting lenders to borrowers via mortgage-back securities] will be rebuilt, the way out of our present dilemma is not to blame the quants. We must instead hire good ones and listen to them.

3.7 Lessons Learned

At the time of the crisis, several events presented in this chapter had been considered as highly unlikely or even impossible to happen. On several occasions, we stressed the difference between an *if* versus *what if* consideration of risk. From a mathematical point of view, in later chapters we shall address this crucial difference through the introduction of extreme value theory and a discussion of models that allow for "life beyond the bell curve". You have learned how markets in financial products ballooned beyond what was economically reasonable in both volume as well as complexity. We stressed that keeping an eye on the notional volume of outstanding and warehoused derivatives, especially in OTC markets, is advisable. We also demystified the concepts of securitization and diversification and warned that a naive use of these important concepts may easily lead to nonsense. Hidden in the chapter, Feynman's "listen to the guys in the boiler room" from Chapter 2 resonates. Partly based on personal experience, we have highlighted the role of mathematics and very much stressed that, before blaming mathematicians, one must make sure to hire good ones and listen to them. Political friendships often blocked much needed actions that were based on the early warnings raised by the supervisory authorities. A sad summary of the crisis is that "as long as the music plays, we (the investment bankers) keep on dancing."

In this chapter, we have mainly highlighted the link between financial markets on the one hand and on the other hand financial engineering in general and, more particularly, mathematics. These markets are however also driven by psychology, rumors and sentiment, leading to herding; see Figure 3.7.

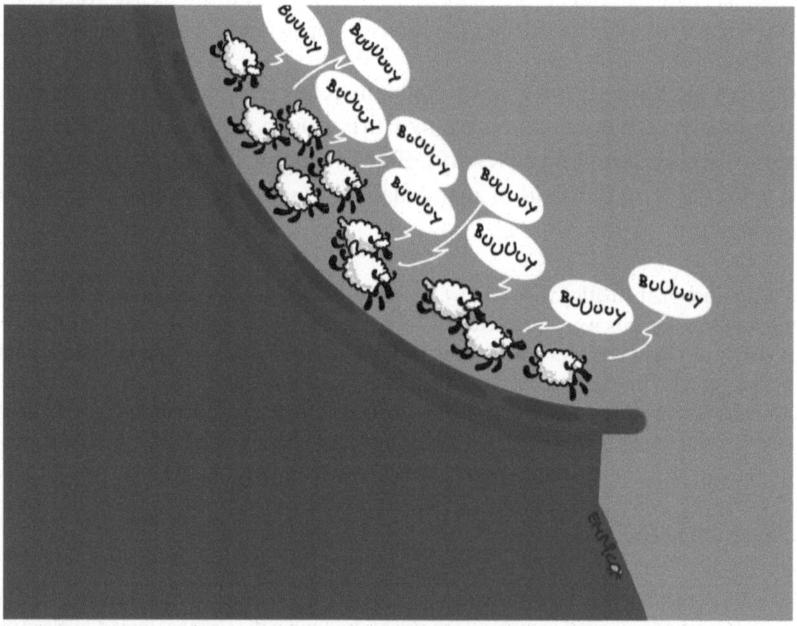

Figure 3.7 An old trading adage: "Buy the rumor, sell the fact." Source: Enrico Chavez

4

Earthquakes and Tsunamis

Whether it is a tsunami, or whether it is a hurricane, whether it's an earthquake – when we see these great fatal and natural acts, men and women of every ethnic persuasion come together and they just want to help.

Martin Luther King Jr. (1929–1968)

4.1 Measuring the Magnitude of an Earthquake

The word tsunami is of Japanese origin consisting of "tsu", meaning port or harbor, and "nami", meaning wave. Though not really depicting a tsunami but rather a large offshore wave ("okinami"), "The great wave off Kanagawa", an iconic woodblock print by the Japanese artist Katsushika Hokusai (1760–1849), has become the artistic personification of a tsunami; see Figure 4.1. Tsunamis can be triggered by landslides, volcanic eruptions, meteorite impact

Figure 4.1 "The great wave off Kanagawa" by Katsushika Hokusai (1760–1849). Source: Authors

or earthquakes. In the sections to follow we will discuss two examples of tsunamis triggered by earthquakes and one example triggered by a volcanic eruption. Non-seismic sources contribute to the generation of about one fifth of all tsunamis globally, and there are several examples of such tsunamis causing devastation.

Earthquake reporting changed its measurement units from the local magnitude scale M_L (also known as the Gutenberg–Richter scale, or simply the Richter scale, named after Beno Gutenberg (1889–1960) and Charles Francis Richter (1900–1985)), dating back to a series of papers published around 1935, to the moment magnitude scale M_W introduced in the 1970s. Here "*L*" stands for "local", and "*W*" for "wave(form)s". Whereas both magnitude scales are rather similar until about magnitude 6, the latter is supposed to give a more reliable earthquake-size estimate for large earthquakes, above $8\,M_W$, say. Public reporting often still refers to "the Richter scale" though the numbers typically correspond to the moment magnitude scale. It is fair to say that, when it comes to the public communication of the size of an earthquake, changing between the two scales does not really pose a problem, as for most public discourses on the subject the differences are not very big. What is important however is that they are logarithmic scales. A logarithmic function mathematically, and in a convenient way also graphically, transforms an exponential function into a linear one, which is much easier to grasp and compare; see Figure 6.2 in Chapter 6. The same happens with magnitude scales for earthquakes; in order to focus our attention on the main ideas, we refer to the logarithmic Richter scale. No doubt we have all seen on television the erratic up and down movement of the recording needle of a seismograph measuring the amplitude *A* of an earthquake. One first transforms *A* logarithmically into $\log_{10}(A)$ and further calibrates the latter through a function of the distance to the epicenter. The main consequence is that the (physical) magnitude of an earthquake grows as powers of 10. As an example, an increase of 3.0 on the Richter scale leads to a magnitude increase of $10^{3.0} = 1000$. A one-point change in magnitude, however, corresponds to about a 32-fold increase in the energy released. Hence in the case of a 3.0 increase on the Richter scale, the amount of energy released corresponds to a $32^{3.0} = 32\,768$-fold increase. The strongest earthquake that has actually been measured was the 9.5 Valdivia earthquake off the coast of Chile on May 22, 1960; see Figure 4.2. This raises the interesting question whether a magnitude 10 earthquake is at all possible

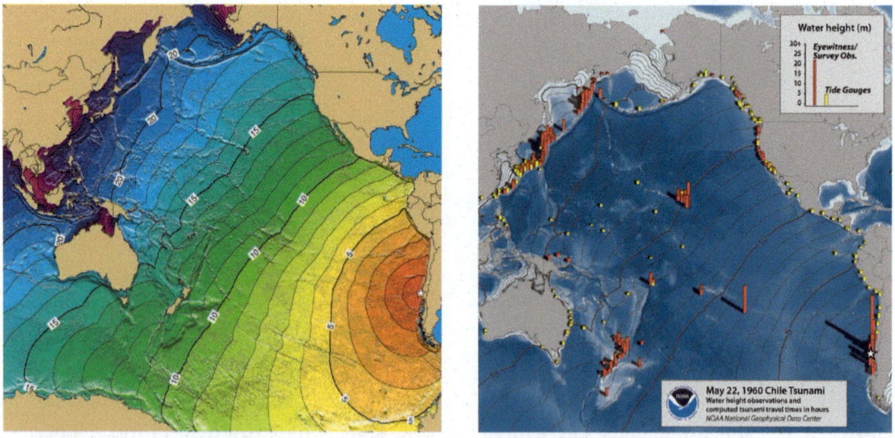

Figure 4.2 Left: Chilean earthquake of May 22, 1960 with tsunami travel times in hours. Right: Tsunami coastal heights in meters. Source: National Centers for Environmental Information

(*if*-question) and what the consequences would or could be (*what if*). In Section 4.2, we will briefly discuss how earthquakes typically occur through the subduction of tectonic plates, that is, the sliding movement of one of Earth's plates under another. From this it follows that the length of the fault line along which this subduction takes place plays an important role concerning the total release of energy. The 9.5 Chilean earthquake corresponded to more than 1500 km of such a zone. Earthquake scientists claim that for a magnitude-10 earthquake, the subduction zone would need to be at least the circumference of the Earth, that is, about 40 000 km. Such fault lines do not exist. One can perhaps speculate on some extreme "pasting together" of different fault lines or on a truly massive subduction event; both of these scenarios seem however purely academic. This leaves the possibility of an asteroid impact like the one that is thought to have caused the extinction of the dinosaurs. Here a *what if* discussion presumably needs to take into account the extinction of human civilization. For more details on maximal-magnitude earthquakes, see Matsuzawa (2014).

4.2 The 2004 Boxing Day Tsunami

On Boxing Day (December 26), 2004, a tsunami with wave heights in the range of 10–30 m hit the coastlines of Indonesia, Sri Lanka, India and Thailand, all situated around the Bay of Bengal. With more than 220 000 casualties, this natural disaster was one of the worst in recorded history. It was triggered by a massive undersea earthquake (also referred to as a seaquake) of magnitude 9.1–9.3 with epicenter in the Indian Ocean, 250 km southeast of the Indonesian city of Banda Aceh; see Figure 4.3.

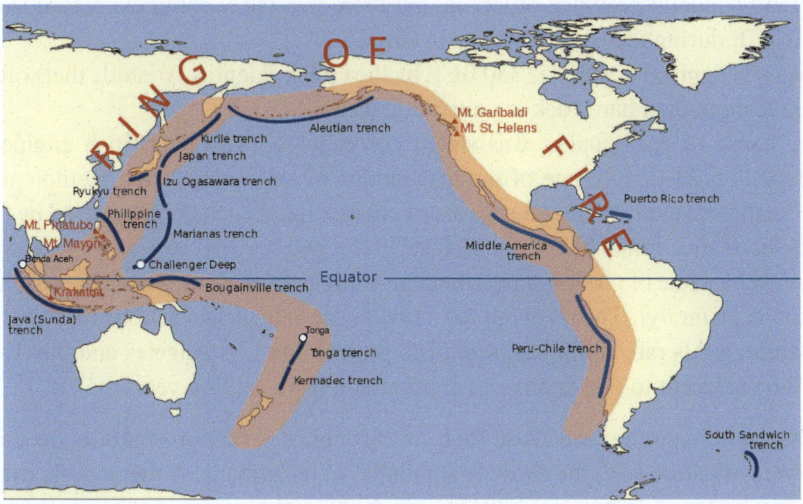

Figure 4.3 The epicenter of the Boxing Day earthquake (near Banda Aceh, Indonesia; see on the left) along the Ring of Fire. We have also highlighted Tonga (lower left center); see Section 4.4. Source: Wikimedia Commons

The Earth's crust consists of seven major techtonic plates (for example the Indo-Australian and the Eurasian plates), 15 minor plates (for example the Indian and the Burma plates) and a large number of micro plates. The location of the epicenter was at a point along the separation line between the major Indo-Australian and Eurasian plates. In particular, the earthquake was triggered by a millennia-long build-up of energy where the minor Indian plate was being pushed under the minor Burma plate. These about 100 km thick tectonic plates move in different directions with speeds in the range of 1–10 cm/year. Where the plates "rub against each other" or one "slides under another", referred to as subduction, there is an enormous built-up of potential energy, which is eventually released through earthquakes. The fault line underlying the Boxing Day earthquake lies along the so-called Ring of Fire (Figure 4.3), a geographic region with high tectonic activity surrounding the Pacific Ocean; about 90% of the world's earthquakes occur there. The *hypocenter* of an earthquake is its point of origin, typically located at one of the many fault lines of the Earth's crust; the *epicenter* corresponds to the projection of the hypocenter onto the Earth's or the sea's surface. Excellent pedagogic information on earthquakes can be found on the website of the United States Geological Survey (USGS). It contains some interesting "Cool facts (CF) about earthquakes"; here are some examples, see USGS (2021a):

(CF1) There are about 500 000 detectable earthquakes each year of which 100 000 can be felt and 100 cause damage.
(CF2) The world's deadliest recorded earthquake occurred in 1556 in central China. It struck a region where most people lived in caves carved from soft rock. These dwellings collapsed during the earthquake, killing an estimated 830 000 people. In 1976 another deadly earthquake struck in Tangshan, China, where more than 250 000 people were killed.
(CF3) The earliest recorded evidence of an earthquake has been traced back to 1831 BCE in the Shandong province of China, but there is a fairly complete record starting in 780 BCE during the Zhou Dynasty in China.
(CF4) It was recognized as early as 350 BCE by the Greek scientist Aristotle that soft ground shakes more than hard rock in an earthquake.
(CF5) The cause of earthquakes was stated correctly in 1760 by British engineer John Michell (1724–1793), one of the first fathers of seismology, in a memoir where he wrote that earthquakes and the waves of energy that they make are caused by "shifting masses of rock miles below the surface".
(CF6) The average rate of motion across the San Andreas Fault Zone during the past 3 million years is 56 mm/yr. This is about the same as the rate at which your fingernails grow. Assuming this rate continues, scientists project that Los Angeles and San Francisco will be adjacent to one another in approximately 15 million years.

The fact that so little motion can cause such catastrophic destruction contrasts the enormous hidden powers of nature with the extreme smallness of us humans. A memorable example in this sense can be found in van Son (2015), which indicates a vanishingly small Earth with an arrow in our galaxy, the Milky Way.

Though scientists have studied the topic in depth for a long time, it is fair to say that, prior to 2004, people in general were not particularly aware of the physical nature of a tsunami, or of the various ways in which it can manifest itself. The total energy residing in a tsunami can be enormous. Recall that energy is the ability to do work; it can appear in various forms such as heat, light, electricity, gravitation, etc. It is also important to recall that energy cannot be created nor destroyed and hence can only be converted from one form into another. This important property is referred to as the *conservation of energy*. Energy often manifests itself both as potential and kinetic energy, the former corresponding to the position of a body with mass, the latter to its velocity. Einstein's famous formula $E = mc^2$ relates energy E to mass m and the speed of light c. It just shows how much energy is stored at the sub-atomic level of matter. The power of the atomic bomb is a sad testament to that enormous amount of stored energy. You further learned in high school that water is an incompressible fluid; this fact plays an important role and relates to the eponymous principle of Daniel Bernoulli (1700–1782). For an incompressible fluid, the principle states that the total energy (the sum of the kinetic, potential and pressure energy) at any point in the fluid is constant. The principle applies to air and other gases; this will play an important role in Section 4.4.

An immediate translation yields that, as the speed of a fluid flow increases, the pressure decreases and vice versa. The principle explains why an aircraft can achieve lift by the shape of its wings, which are curved on the top versus flat on the bottom. Because of this difference in geometric shape, air has to travel a longer distance, hence faster, over the top of a wing (resulting in less pressure) than below (higher pressure). The pressure difference provides the necessary lift and pushes the aircraft up. The same principle explains why roofs are blown off houses in a storm. The fast flowing air creates a pressure difference between the air outside a house and the air inside. It is eventually the inside air that starts pushing on the roof. An easy experiment involves a hair dryer and a ping pong ball. Blow the air upwards and place the ball in the air stream, it will nicely stay there oscillating a bit. Why? Bernoulli's principle!

In the case of the Boxing Day, or Sumatra, earthquake, the amount of potential energy released at the Earth's surface and hence available to power the ensuing tsunami in the form of kinetic energy was equivalent to 1500 times the energy released by the Hiroshima atomic bomb. Though this release of energy is mind-bogglingly large, it still is less than the energy released by the largest nuclear weapon ever tested. On October 30, 1961, the former Soviet Union detonated the Tsar Bomba, a hydrogen aerial bomb, near Novaya Zemlya in the Arctic Ocean. The explosion had an energy equivalent of about 57 megatons of TNT. This corresponds to more than 1570 times the combined power of the atomic bombs dropped on Hiroshima and Nagasaki. A further comparison underlines the devastating power of a nuclear weapon like the Tsar Bomba. It corresponded to more than ten times the total ordnance exploded in World War II and was originally planned to be twice as powerful. This just goes to show how far the human race has walked on the path of potential self-destruction.

The potential energy residing at the hypocenter of the Boxing Day earthquake, at a depth of about 30 km below mean sea level, was several orders of magnitude higher than at the epicenter. The energy released, however, acted on the full sea-floor-to-sea-surface water column. This initiated a radiating energy transfer through the seawater very much like the concentric waves one observes when one throws a stone in a pond. As a consequence, waves radiated in all directions away from the epicenter. When it is reported that tsunamis travel at 700 km/h, as indeed in this case, it is important to understand that this is the speed of kinetic energy propagation through water particles (not the velocity of the water itself). This propagation is a consequence of the vibration of water particles. The above speed v can, approximately, be calculated through the formula $v = \sqrt{gh}$, where $g = 9.81\,\text{m/s}^2$ is the gravitational acceleration constant and h is the height in meters of the sea surface above the ocean floor. The depth of the sea bed below the epicenter was about $h = 4.4\,\text{km}$. From this, with the above formula, we can approximate the tsunami's speed (by converting to m/s and then to km/h) as $v = \sqrt{4400\,\text{m} \times 9.81\,\text{m/s}^2} \times (3600\,\text{s/h})/(1000\,\text{m/km}) \approx 748\,\text{km/h}$. The water particles themselves travel at a much lower speed of around 70 km/h; this is the reason why observers on a ship far from the coast do not, or hardly, notice the passing of a tsunami. Indeed, far out at sea, the sea surface may exhibit waves with only around 0.5 m amplitude but with a very long wavelength (crest-to-crest distance) of more than 100 km. This is also the reason why a tsunami can travel long distances at open sea without losing much of its energy; once more, Bernoulli's principle applies here. The height of a tsunami wave very much depends on the depth of the ocean floor near the coast; see Figure 4.4. It

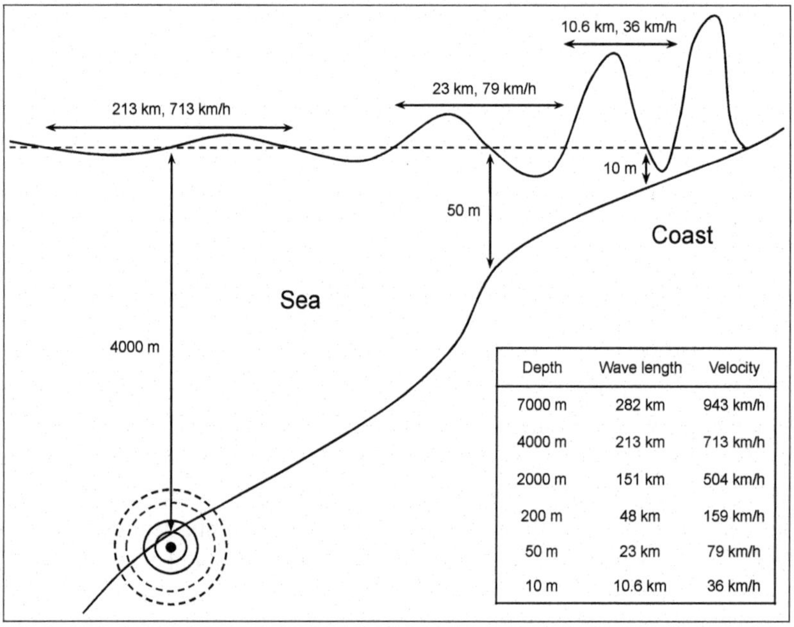

Depth	Wave length	Velocity
7000 m	282 km	943 km/h
4000 m	213 km	713 km/h
2000 m	151 km	504 km/h
200 m	48 km	159 km/h
50 m	23 km	79 km/h
10 m	10.6 km	36 km/h

Figure 4.4 Sketch of the formation and evolution of an earthquake-driven tsunami. Source: Authors (based on a figure by North-Eastern Atlantic and Mediterranean Tsunami Information Centre (NEAMTIC))

is in the shallow waters that a tsunami unfolds its destructive power. While slowing down in energy propagation speed due to the topography of the seafloor near the coastline, the enormous available energy is converted into wave height and surge. This is referred to as wave shoaling. Another phenomenon often encountered is that of receding sea water before a coastal tsunami, the reason being that a wave consists of crests and troughs. If a trough arrives first, this has unfortunately led to people rushing to the receding water's edge looking for stranded fish or exposed colorful corals. Doing so, they ran into the arms of the deadly tsunami waves. Friends of ours escaped death at a Thai beach, as a local companion, upon noticing the receding water, immediately urged them to run inland. We recognized another friend while we were watching the evening news on television some days after the event. Only dressed in shorts, T-shirt and sandals, he and his wife explained to the journalist that they survived the tsunami because they were indeed out at sea on a boat. A more detailed modeling of the propagation of tsunami waves needs more work, of course; for instance the change in depth and coastline structure comes in. But in the end, it all comes down to Daniel Bernoulli's famous equation. After the tsunami, many authors tried to explain the detailed underlying physics to a more general public. We recommend Margaritondo (2005), from whom we quote:

> The tsunami that devastated southern Asia on 26 December 2004 attracted the attention of a broad and diversified public to this extreme wave phenomenon. Many Internet sites can be found explaining the basic features that make it so destructive and distinguish it from other types of water waves: its causes, the extremely long wavelengths and the behaviour when it reaches a coastal region. However, I could not find a source presenting in simple terms the physical origin of these characteristics and in particular of the dispersion relation that underlies the last one. [...] It should be noted that a simple but correct way to comprehend tsunamis is important beyond mere scientific curiosity. A large portion of the casualties in south Asia was caused by trivial misunderstandings of the physical mechanism. For example, the lack of knowledge about large time intervals between different phases of the phenomenon prevented many people from taking simple life-saving precautions – such as leaving the beaches and getting to high ground after the first anomalous water retreat from the shore. Hopefully, a more widespread knowledge of the nature of tsunamis could mitigate their human impact.

Since this disaster, a lot of effort worldwide has gone into the creation of early warning systems for tsunamis. Because of the jet-plane speed with which the tsunami energy waves radiate away from the earthquake's epicenter, time is of the utmost importance. Upon receiving a warning, people are advised to move as quickly as possible to higher ground or inland. Coastal populations in areas prone to tsunamis have to fully understand the nature of a tsunami and be well informed about the risks involved as well as the necessary precautions and actions to be taken. Going forward, false-positive tsunami warnings do, however, pose a considerable communication challenge for the authorities involved. Early prediction of the occurrence of tsunamis as well as the estimation of their potential impact are notoriously difficult. The same holds true for the estimation of their return periods, which for catastrophic tsunamis can range from several hundred to a couple of thousand years. By the same token, multiple tsunamis can take place over a much shorter time span. The following two sections underscore this with full force.

4.3 The 2011 Tōhoku Earthquake and Tsunami

On March 11, 2011, a massive earthquake not only hit nearby Japan, but it would also rock the world. The Tōhoku 9.0–9.1 earthquake, also referred to as the Great Sendai earthquake, had its epicenter 130 km east of the coast of Sendai, northeast Japan. It triggered a tsunami, killing more than 15 800 people. The earthquake was the strongest ever recorded in the history of earthquake-prone Japan; see Figure 4.5. An important consequence of this event was the meltdown of three nuclear reactors at the Fukushima Daiichi nuclear power plant (NPP). Toxic radioactive material was released into the environment and thousands of people were forced to evacuate their homes and businesses. This event had serious repercussions worldwide. With the Three Mile Island (March 28, 1979) and Chernobyl (April 26, 1986) nuclear accidents still very much in people's minds, several countries, especially in Europe, decided to phase out their nuclear energy programs.

Whether this was a good decision or whether it was too rushed, given the environmental impact of continuing to rely on oil and coal in the short term, is a question on which society

Figure 4.5 The location of the earthquake's epicenter together with the NPPs at Fukushima and Onagawa (left) and Yanosuke Hirai (1902–1986), the civil engineer who oversaw the construction of the Onagawa nuclear power plant (right). Source: Wikimedia Commons

has to reflect. From a risk management perspective, one would need to trade off the risk of instantaneous rare events of high impact (nuclear accidents) against the steady, long-term, climate effects of burning fossil fuels like coal and oil. Owing to the complexity of these two "systems" it is fair to say that which source of energy will prevail in the relatively short-term future largely depends on the technical, economic, geopolitical as well as societal assumptions and projections made. For instance, the geopolitical relevance has been made abundantly clear through the war between Russia and Ukraine. Needless to say, both types of power generation are doomed to fail in the long run but will continue to be used for now (partly because of the respective lobby groups). Sornette et al. (2019) offers additional food for thought on this topic. We come back to these developments in Chapter 10.

A considerable amount of research has gone into coastal defense structures against storm surges in general and tsunamis in particular; the latter especially when nuclear power plants are at risk. We will use the Fukushima event in order to exemplify the important difference, from a risk management perspective, between an *if* discussion and a *what if* one. Whereas our short discussion cannot do full justice to the various important issues involved, we do give some key references that should enable you to obtain a broader view of this disaster. From the four NPPs that were affected along a 230 km stretch of the Sendai coast, we will mainly concentrate on the reactor meltdowns and hydrogen explosions, followed by radioactive contamination of the surrounding area, experienced at Fukushima Daiichi. Further, much can be learned from the fact that some reactors nearby did not go into a meltdown. For instance, both the NPPs at Onagawa (three reactors) and Fukushima Daiichi (six reactors) experienced 13 m maximum tsunami wave heights, but had entirely different fates. They were both reached by the tsunami's main waves after approximately 40–45 min; this stresses once more the speed at which the tsunami energy waves travel at open sea and the relatively short time span between earthquake instance and tsunami impact. At Daiichi, three reactors (especially reactor 1) experienced (partial) meltdowns, whereas at Onagawa none did. Comparing and contrasting the different fates between the Fukushima Daiichi and Onagawa NPPs is highly enlightening. At a distance of 70 km, Onagawa's position was the larger town closest to the epicenter of the earthquake; for Fukushima the distance was 180 km. Thus Onagawa experienced the heaviest shocks. Two major facts made the difference between the fates of the two NPPs: First, the protective seawall at Onagawa was 14 m tall, the one at Fukushima only 5.8 m. And second, despite flooding at Onagawa, its emergency power lines were much better protected and able to function. After the earthquake, the seawall at Onagawa was raised to 17 m. For a detailed study on why the Onagawa NPP survived the tsunami, see Ibrion et al. (2020). Although for any disaster it is difficult to pick single heros, in the case of Onagawa we would like to make an exception. The civil engineer Yanosuke Hirai (1902–1986), see Figure 4.5, was responsible for overseeing the construction of the NPP. He did this in a most dedicated and careful way. He showed remarkable foresight, scientific stubbornness and a tenacity that resulted in a much less vulnerable final NPP design and the construction of a sufficiently high protective seawall. A comparison with David van Dantzig, whom we met in Section 1.3, is called for.

TEPCO stands for the Tokyo Electric Power Company; it holds the final responsibility for the functioning and the running of the Fukushima NPPs. These responsibilities include the establishment of proper safeguards against adverse seismic events, including tsunamis. On April 24, 2012, it released an official company report in order to counter numerous comments made in the public domain. In this document, the company states:

> Although TEPCO believed that nuclear power plant safeguards had been secured per the standard set down by the Society of Civil Engineers, much consideration was given to applying the latest knowledge and research to the power plant design and operations. In addition, close attention was paid to the latest research trends dealing with earthquakes and tsunamis for the purpose of conducting in-house investigations.

On the basis of a 2006 in-house probabilistic tsunami hazard analysis, it was concluded that a tsunami wave exceeding 10 m corresponded to a 100 000 to 1 000 000 year event. The company added the following, somewhat nebulous, statement to this estimate:

> However, this analysis was conducted for the purpose of confirming adaptability and improving methods that at the time were still in development. Thus, TEPCO did not interpret this figure to literally mean the actual frequency of tsunamis that could strike the nuclear power plant in Fukushima.

In TEPCO (2012a), the company refers to an internal 2008 report and comments on trial calculations leading to tsunami heights in the range of 8.4–10.2 m. The report gives an estimate for a maximum tsunami flood height of 15.7 m "on the south side of the premises for major buildings of Units 1–4 at [. . .] Fukushima Daiichi". The analysis referred to the Sanriku, also called Jōgan, earthquake of 869 CE with an estimated magnitude of 8.4 which may have been as high as 9.0 on the Richter scale, hence coming into the range of the Tōhoku earthquake. The latter estimate was in part based on sand transport 4 km inland by the tsunami. In light of the above discussion, the following statement from TEPCO (2012a) is worthwhile repeating:

> The tsunami that occurred in this disaster was of a scale that vastly exceeded pre-disaster assumptions. The main reason was an enormous earthquake with a magnitude of 9.0, a size that could not be envisaged from

the history of earthquakes in Japan that stretches back for several hundred years, erupted as an earthquake with a wide epicentral area that interlocked several regions.

And, from TEPCO (2012b):

Consequently, based on the accumulated knowledge at the time, the Tohoku–Chihou–Taiheiyo–Oki earthquake could not have been foreseen and TEPCO also considers the height (scale) of the tsunami that followed this earthquake to have been unpredictable.

This sounds like a Black Swan, in the terminology of Nassim Taleb (2007). In light of so much uncertainty, the precautionary principle, which we will discuss in more detail in the context of the coronavirus pandemic in Chapter 6 (see Section 6.4 in particular), could have offered useful guidance, even if that would have led to considerable extra costs to reduce plant vulnerability in light of TEPCO's perception of an "unpredictable event". In any case, these extra costs would have constituted only a very small fraction of the total cost already incurred by the "cleaning up" after the nuclear disaster. In our discussions above, the obvious *if* event is the occurrence of a catastrophic earthquake followed by a tsunami. The *what if* event was a major tsunami wave going over the protective seawall followed by a flooding of the reactor sites, in particular of the installation rooms of the emergency diesel generators (EDGs) needed, in particular, for the operation of back-up cooling systems. The potential strength of the earthquake and the maximal tsunami wave heights were estimated too conservatively. Hence the necessary extreme event modeling failed. A sufficiently severe *what if* stress testing of the EDG installation measures was not performed; here we notice a glaring difference from the Onagawa site. Of course a *what if* analysis might not have fully prevented the damage to the reactors but it certainly would have saved time during which engineers could have repaired critical components. One would then no doubt have been able to address the loss of reactor-cooling capacity in a much more timely manner, if indeed the EDGs had broken down.

Synolakis and Kânoğlu (2015), in their very critical publication, enter into more detail:

The first elevation wave of the 2011 tsunami arrived at Fukushima Dai-ichi about 40 min after the earthquake. [...] The cooling was lost in Unit 1 after several hours, in Unit 2 after about 71 h and in Unit 3 in about 36 h, the three units in operation. The stage was set for the most costly NPP accident in history, this far, and the largest loss of life and civil disruption in Japan since World War II. What was unprecedented was the almost simultaneous loss of both onsite AC [alternating current] power (EDGs) and DC [direct current] power, an event that simply had not been imagined [it simply did not enter the *what if* discussion]. Yet, what doomed the Fukushima Daiichi was the elevation of the EDGs. The plant personnel were in such dire need of emergency power that they even connected car batteries just to get instrument readings.

The fact that TEPCO used over-conservative scenarios, in contradiction to their own, earlier, internal studies from 2008, led the authors of the above paper to voice the suspicion of criminal negligence. The rather strong wording used is based on a comparison made with a court case related to the 2009 L'Aquila earthquake in Italy, as we will discuss in Chapter 5. This comparison will become clearer from the excellent and very detailed official government report (Kurokawa et al., 2012). In National Research Council (of the USA) (2014) it is explained how a proper probabilistic risk assessment (PRA) would have been highly relevant in the case of the NPP at Fukushima Daiichi. Interestingly, a PRA typically asks three questions:

(1) What can go wrong?

(2) How likely is it to happen (the *if* question)?
(3) What are the consequences if it does happen (the *what if* question)?

A further quote from Synolakis and Kânoğlu (2015) is definitely relevant for the Fukushima disaster. It holds some lessons that should be understood much more widely in disaster management beyond tsunami risk, as it very much relates to educational and interdisciplinarity issues:

> It is clear to us that what is missing in contemporary hazard assessments are regulatory guidelines for the training of the scientists and engineers who work on estimating the maximum probable tsunami. [...] Hazard studies need to be considered in fairly broad context, and include the author's personal experiences in the interpretation of the results. How can someone who has not walked on flooded lands, compared the inundation between adjacent beaches, wondered why a single structure was left standing while all others were flattened, debated whether a debris mark on a surviving tree was from the tsunami or carried by the wind, and listened to sometimes widely different eyewitness accounts for the same location, how can he interpret a historic report, or understand what real difference a change in grid resolution can make, or ultimately put his flooding prediction in the proper context.

A further interesting quote from the same reference, again relevant for many applications, is:

> The six-unit Fukushima Dai-ichi was commissioned in 1966 and is operated by TEPCO. Its initial application was based on the 3.122 m measurement local tsunami height (above Onahama Peil (O.P.) reference sea level) observed during the 1960 Great Chilean tsunami at or close to Fukushima. [Incidentally, "peil" is the Dutch word for level or gauge; we have already encountered it in NAP in Section 1.1]. The flooding estimate was provided to the nearest millimeter, underscoring a false sense of accuracy in the safety assessment that is impossible even today.

The last sentence contains an important point that was also observed during the coronavirus pandemic (Chapter 6) where the ubiquitous reproduction number R_0 was occasionally quoted to two decimal places as if one could measure this quantity reliably at that accuracy. In all these discussions, underlying model uncertainty plays a major role and needs careful communication. Statistics allows one to quantify aspects of model uncertainty through the construction of confidence intervals, a topic we will return to later. In most of the above studies, confidence intervals were not reported, giving indeed a false sense of understanding. This communication of model uncertainty becomes especially relevant when one is considering rare, extreme events. Not doing so may even be considered a warning sign of a lack of good modeling practice.

4.4 The Hunga Tonga – Hunga Ha'apai Volcanic Eruption

On January 15, 2022 at 17:14 local time, the submarine volcano Hunga Tonga–Hunga Ha'apai erupted, sending shockwaves and tsunamis all around the world; most fortunately, the number of casualties remained very low. The Kingdom of Tonga is a Polynesian group of islands situated about 1800 km northeast of New Zealand's North Island. Every year, on January 1, Tonga makes the news as the first country to celebrate the arrival of the New Year. The volcano, which lies on the Pacific "Ring of Fire" (see Figure 4.3), links two uninhabited twin islands, Hunga Tonga and Hunga Ha'apai. The energy released from this particularly violent eruption was reported by NASA's Goddard Space Flight Center as 10 megatons of TNT equivalent. One megaton is 1000 kiloton, or 1 000 000 tons. It was accompanied by several hundred thousand lightning bolts, and volcanic ash was ejected about 55 km into the air, reaching the mesosphere, a layer between 50–85 km of the Earth's atmosphere. It was considered a 1 in 1000 years event; see Figure 4.6. What made this

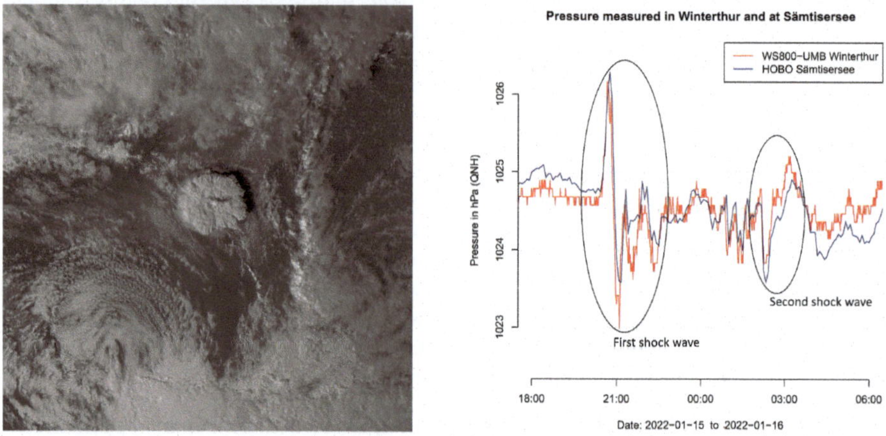

Figure 4.6 Left: A satellite picture taken at the time of the eruption. Right: Air pressure fluctuations from the Tonga eruption recorded by the Zurich University of Applied Sciences (ZHAW) at two locations (red and blue) in Switzerland. Sources: Wikimedia Commons/Japan Meteorological Agency (left) and Curdin Spirig, Research Unit Meteorology, Environment & Aviation, Centre for Aviation, ZHAW (right)

event extra special, however, was the phenomenon of atmospheric shockwaves on top of the seawater-based tsunami waves, with a resulting second type of tsunami. The main reason for this is that the Tonga eruption concerned a 2 km high volcano having only its summit, of about 100 m, above sea level. Bernoulli's principle, as discussed in Section 4.2, applies to energy propagation through atmospheric waves as well as in water. An interesting fact is that so-called meteotsunamis were created through the interaction on the seawater below by atmospherically expanding circular shockwaves, on which these latter acted as a kind of energy pump. These tsunamis did arrive earlier than the predictions based on standard tsunami models for seaquake-driven events. For instance, smaller meteotsunamis started arriving in Japan about three hours before the main tsunami waves. Tiny tsunami waves, just 10 cm tall, were detected around the same time in the Caribbean Sea, which is in an entirely different ocean basin. So what was going on?

First of all, the reason for the difference in the arrival times between the different types of tsunami waves stems from the difference in the speeds of energy transmission through air (which was almost supersonic, close to 1200 km/h) versus through seawater (about 700 km/h, as discussed in Section 4.2). Moreover, atmospheric shockwaves are not confined to traveling over sea and hence can reach areas of the world normally not reachable by a tsunami, which travels over sea. For a particular point on Earth, circular shockwaves typically cause two noticeable air pressure fluctuations. In the case of Switzerland, the first wave, with fluctuations around 2.5 hPa (1 hectopascal = 1 millibar), arrived about 15 hours after the eruption. As the waves traveled along expanding concentric circles emerging from the point of eruption, at the antipodal (that is, diametrically opposite) point to Tonga, in southern Algeria, the shockwaves converged and rebounded, leading to the second noticeable fluctuation in air pressure, about six hours later in the case of Switzerland. These relatively small drops in air pressure may have caused headaches for people in very distant, even antipodal, locations. The reason is that the pressure in a person's sinuses, which are filled with air, becomes higher than the atmospheric air pressure. A similar event that we may experience is the well-known "ear popping" in the middle ear when sitting on a descending airplane. These rather extraordinary events around the Tonga eruption are commented upon in Andrews (2022). We quote:

> [Mark] Boslough [a physicist at the University of New Mexico, Albuquerque] believes shock wave tsunamis may have been triggered by even stronger eruptive explosions, such as the 1883 outburst at Indonesia's Krakatau volcano, or the most explosive phases of Yellowstone's megaeruption 2.1 million years ago. "And although considerably rare on human time scales", he says, "a volcanic explosion mighty enough could potentially create a big tsunami in all ocean basins."
>
> Powerful, speedy shock waves are also generated by the midair self-destruction of meteors. Boslough has long suspected that these can make their own potentially devastating tsunamis, but the shock wave from Tonga's eruption has broadened his perspective. "What didn't occur to me", he says, "is that these shock waves can make tsunamis on the opposite end of the planet. I was thinking out of the box, but not [far] enough outside of the box, I think."

The above need for sufficient "outside of the box" thinking is akin to the "*if* versus *what if*" distinction which we encounter on several occasions in our book. The Tonga eruption brings two risk-related aspects to the forefront. First, we are living on a sphere (flattened at the poles). As a consequence, (extreme) events are related through this geographic constraint. Certainly the modeling of environmentally relevant (extreme) events has to take a global approach; indeed, we all share the same atmosphere. Second, it is always good to put ourselves, as well as our planet, into the vast perspective of our solar system (van Son, 2015) or even the universe. It pays to be humble in the face of nature's potentially destructive power.

4.5 Lessons Learned

Using three recent examples, in this chapter we highlighted the potentially devastating effect of tsunamis. As part of our discussion of the 2004 Boxing Day tsunami, we explained the basic physics underlying the propagation of tsunami waves and their destructive power upon their arrival at coastal areas. The Bernoulli principle of the conservation of energy in flows of incompressible fluids plays a pivotal role here. A better understanding of the physical processes also leads to a more consistent way of educating and preparing people living near the coast about what to do "when the next tsunami strikes". In the case of the 2011 Tōhoku earthquake, it was not only the loss of numerous lives but also the ensuing nuclear disaster at Fukushima that brought tsunami risk to the forefront of any risk radar. The latter event will have lasting consequences for energy portfolios across the world. Fukushima gave a decisive push for a green energy transition, which became symbolized by the German word *Energiewende*. Through a comparison of the tsunami impacts on Fukushima and Onagawa we were once more reminded of the crucial importance of the difference between an *if* versus *what if* attitude towards rare-event risk. The final government reports on the earthquake's consequences for the nuclear reactors at Fukushima paint an unflattering picture of those responsible higher up in the managerial and political echelons. This latter point will be taken up in a dramatic way in the next chapter, on the 2009 L'Aquila earthquake in Italy. Finally, the Tonga volcanic eruption once more highlights Hamlet's "There are more things in heaven and earth, Horatio, Than are dreamt of in your philosophy" quoted in the Introduction to our book. The geometric constraints of our planet have important implications on the modeling of environmental risk.

5

The L'Aquila Trial and the Public Communication of Risk

The single biggest problem in communication
is the illusion that it has taken place.

George Bernard Shaw (1856–1950)

5.1 The L'Aquila Earthquake

On April 6, 2009, at 03:32:34 local time, a 6.1 magnitude earthquake hit the old medieval town of L'Aquila in the region of Abruzzo of central Italy, 100 km northeast of Rome, killing 308 people. At 6.1, this was a medium-power seismic event, with, however, a disproportionally large human impact. The older part of the thirteenth-century city was very badly damaged; see Figure 5.1. There are several reasons why this particular earthquake caused a lot of

Figure 5.1 The earthquake damage to the Palazzo del Governo (Prefettura) in L'Aquila. Source: Wikimedia Commons

casualties and extensive damage to property. One reason is the complicated fault lines, which make that part of Italy very prone to earthquakes. Another reason is that, prior to the actual

6.1 magnitude earthquake, many smaller tremors, about 900 during the six months before the main shock, gradually deteriorated the fabric of the older buildings.

A week before the main earthquake, on March 30, 2009, a larger shock hit the region (see below). Numerous shocks plagued the area in the months to follow. These shocks caused further damage to buildings. Owing to these aftershocks, the dome of the Anime Sante Basilica in L'Aquila, already heavily damaged by the main shock, almost entirely collapsed. A particular problem was that older buildings are much less robust with respect to lateral ground movements, so-called side-to-side shear forces. In earthquake-prone areas, such as parts of central Italy, the continuous reinforcement efforts of older (not just very old) buildings is an important government task. Modena et al. (2011) discussed the seismic vulnerability of historic buildings and ways to remedy their structural weaknesses.

An important reason for including L'Aquila in this chapter is the discussion that took place between the local population, earthquake experts and government, in the weeks leading up to the disaster. As earthquakes are so very common in that part of the world, on the basis of centuries of unfortunate experience (Figure 5.2) popular lore surfaced concerning the

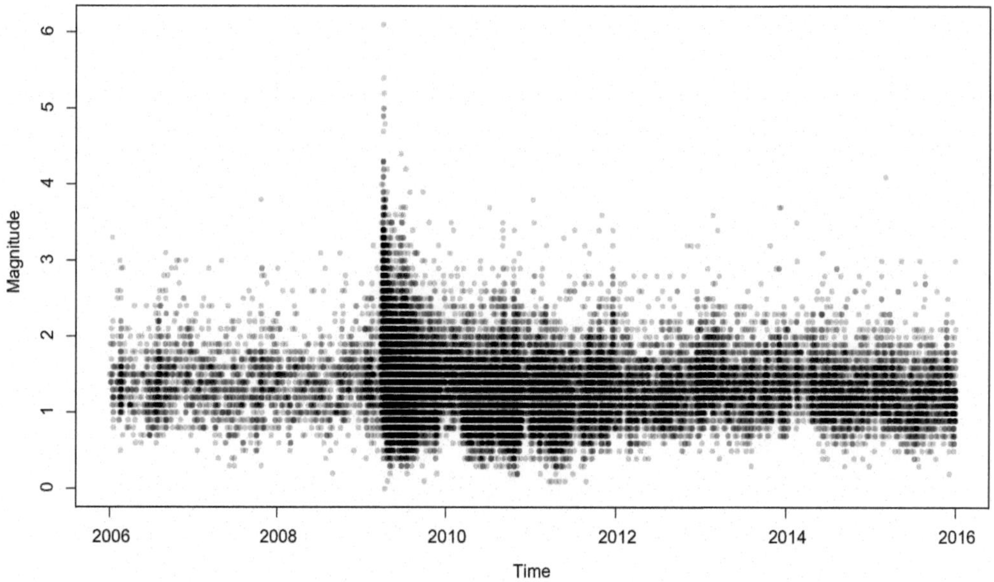

Figure 5.2 Seismic shock data in a 50 km radius around L'Aquila in central Italy, showing the magnitude peaking on April 6, 2009; the plot shows white horizontal gaps owing to magnitude data being recorded only to the first decimal. The data were obtained from the Istituto Nazionale di Geofisica e Vulcanologia (see Istituto Nazionale di Geofisica e Vulcanologia, 2021). In Section 9.6.2 we will come back to the story behind obtaining these data. Source: Authors

prediction of an imminent earthquake. The basis for these predictions ranged from local folklore on special animal behavior to a measured increase in subsoil emissions of radon gas made by a local amateur seismologist. In this case, the latter played an important role leading up to several controversies between local people and government authorities. For the moment we leave out a discussion on the "understanding" of earthquake risk and turn our attention to its "communication". The subtitle of our book, *Cautionary Tales, Understanding and Communication*, indeed played a crucial role in the weeks leading up to the disaster and the tragedy that happened during that fateful night as well as the aftermath. In Section 9.6 we will briefly discuss the statistical modeling of earthquake risk, in particular the case of the L'Aquila earthquake.

A main issue turned out to be whether the population had been sufficiently warned and this, especially in light of the many precursory shocks and incidental damage to several buildings. Several people even decided to sleep out on the street to avoid being caught inside collapsing houses in the event of a larger earthquake. Because of the foreshocks on March 30, on March 31 from 18:30 until 19:30 the local government organized a meeting of experts under the auspices of the Italian Department of Civil Protection (DCP); it was held at the Abruzzo Region Headquarters in L'Aquila and was immediately followed by a press conference. The committee consisted of six scientific experts and one former government official. The meeting's topic was announced as

> [. . .] to carefully analyze the scientific and civil protection issues related to the seismic sequence occurring in L'Aquila Province over the last four months and which culminated in the 4.0 [later increased to 4.1] earthquake on 30 March 2009.

The outcome of that fateful press conference became the basis for a court trial of the seven members of the committee, resulting, initially, in six year jail sentences. The court case led to a worldwide outcry among scientists; see the next section for quotes by several national academies. It goes without saying that the communication of a possible earthquake risk prior to March 31 was all but perfect. The precise "who said what and when and to whom" we leave for the courts to decide. If interested, you may consult Imperiale and Vanclay (2019). Upon appeal, on November 20, 2015, the Italian Supreme Court exonerated the six scientists on the committee. The senior public servant of the DCP who took part in the press conference was however confirmed guilty of inappropriate public communication and given a two-year (suspended) jail sentence; see Cartlidge (2015). The head of the DCP at the time of the earthquake, who did not take part in the March 31 meeting but communicated with town officials beforehand, was cleared of manslaughter on September 30, 2016; see Cartlidge (2016). This brought to an end, at least at the level of the courts, a highly controversial discussion on guilt and innocence, science communication and public risk assessment in the face of a natural disaster. Comparing and contrasting the social versus the scientific dimension of disaster communication, we quote from Imperiale and Vanclay (2019):

> Evidently, risk was understood in the legal process as likelihood of an earthquake, not the interplay between the hazard itself, hazard exposure, the extent of vulnerability and resilience, and likely negative social consequences. [. . .] Over-reliance on the techno-scientific approach demonstrates there is still a lack of understanding about how social vulnerability, risk and impacts are theoretically and practically related and about how science can contribute to enhancing local sustainable transformation. This lack of understanding of the social dimensions of disaster results in disaster risks being narrowly defined in regulatory frameworks and in inadequate procedures for managing disaster risk, conducting proper risk assessments and pursuing sustainable transformation. [. . .] While the trial established that the only responsibility the scientists had was to refer to the best available scientific knowledge, now, ten years after the L'Aquila disaster, it is high time to consider the questions: does the best available scientific knowledge concerning DRR [disaster risk reduction] only relate to seismological analysis of physical hazards? Should there be an interdisciplinary risk assessment protocol the MRC [Italian Major Risks Committee] should follow to consider the multiple dimensions of disaster risk?

The latter part of the quote we find highly relevant well beyond this particular case and this especially in light of the coronavirus pandemic discussed in Chapter 6. From a risk-communication point of view, Hasian et al. (2014) addressed the various ways in which the different participants in the L'Acquila dispute argued, or claimed to have the right to argue, in favor or against a trial:

> In this article, we argue that the controversies surrounding this case need not be viewed as binary disputes that pit supporters of the trial against those who viewed the proceedings as an attack on science. [. . .] We contend that in focusing attention on how the journalists, prosecutors, and judges understood or misunderstood the consensus of Italian and international scientific opinion regarding objective truth as it relates to seismic forecasting, non-rhetorical approaches often miss the ideological dimensions of this dispute and fail to investigate the motives underlying claims about who has the right to speak authoritatively about earthquake forecasting or the uncertainty surrounding earthquake prediction. This, in turn, leads to truncated ways of thinking about risk communication.

One could change "earthquake risk" to "pandemic risk"; the altered quote would then fit well into Chapter 6 on the coronavirus pandemic.

Figure 5.1 shows a badly damaged (former) local government building, the Palazzo del Governo, in L'Aquila. We can look at it in two different ways. First, the picture clearly underscores the seismic vulnerability of historic buildings. At least as important, and indeed not without a sense of historic irony, through the crumbling masonry around its entrance with "Palazzo del Governo" chiseled in big letters, the picture also epitomizes the political failures in addressing the run-up to this earthquake. As always, one hopes that the necessary lessons coming out of this disaster have been learned.

5.2 Some International Reactions

We first want to correct a possible misconception concerning the actual trial. The defendants were not sentenced for their inability to predict the earthquake but for having provided the local population with inaccurate, incomplete and contradictory information.

Needless to say, there was widespread reaction to the court case. Examples include official statements made by several academic institutions, for instance by Ralph J. Cicerone (1943–2016), former president of the US National Academy of Sciences, and Sir Paul M. Nurse, former president of the Royal Society (UK) from October 25, 2012. They highlighted the difficult task that scientists face when dealing with risk communication and uncertainty. Their cautionary words below very much reflect a statement made by Richard Feynman, quoted in Section 2.2 on the Challenger disaster:

> Much as society and governments would like science to provide simple, clear-cut answers to the problems that we face, it is not always possible. Scientists can, however, gather all the available evidence and offer an analysis of the evidence in light of what they do know. The sensible course is to turn to expert scientists who can provide evidence and advice to the best of their knowledge. They will sometimes be wrong, but we must not allow the desire for perfection to be the enemy of [the] good. That is why we must protest the verdict in Italy. If it becomes a precedent in law, it could lead to a situation in which scientists will be afraid to give expert opinion for fear of prosecution or reprisal. Much government policy and many societal choices rely on good scientific advice and so we must cultivate an environment that allows scientists to contribute what they reasonably can, without being held responsible for forecasts or judgments that they cannot make with confidence.

A joint Statement by the German National Academy of Sciences Leopoldina and the French Académie des Sciences of November 12, 2012, going in the same direction but further, called for an independent review of the events leading to the trial:

> The German National Academy of Sciences Leopoldina and the French Académie des Sciences therefore expressly supported the Accademia Nazionale dei Lincei, the Italian National Academy of Sciences, in its endeavors to set up an independent expert commission of geologists and legal experts. The role of this commission would be to examine the scientific and legal aspects of the L'Aquila verdict. Scientific research is substantially motivated by the aim of providing greater protection against natural disasters. In the case of uncontrollable events such as cyclones, earthquakes and volcanic eruptions, scientific forecasting methods are becoming increasingly important. Scientists and representatives of state institutions must work together with mutual trust in order to inform the public responsibly, and on the basis of reliable data, about possible risks. In their risk forecasts, scientists assess the probabilities of future events. Probability-based statements are *per se* fraught with uncertainty. At all times, scientists must communicate this fundamental fact as clearly as possible.

Whereas the overall assessment is somewhat different, both statements do agree on the necessity for science to get involved. Scientists have to provide clear facts; however, they cannot shy away from pointing out issues of uncertainty, especially when it comes to earthquake prediction. There is no disgrace in saying "we don't know" or "we don't know for sure". It is interesting to note that the trial took place in the country that was home to one of the most famous court cases against science and scientific evidence: the case against Galileo.

5.3 Galileo's Dialogue

Galileo Galilei was perhaps the first real science communicator. On the basis of telescope observations he became convinced of *heliocentrism* (the Earth revolves around the Sun) as opposed to *geocentrism* (the other way around). In order to communicate his findings to a wider public, he introduced his new ideas through a series of dialogues between two philosophers (Salviati and Sagredo) and a layman (Simplico), published in Italian in 1632; see Figure 5.3. For an excellent English translation, see Drake (1967). The book became an instant success. The rather lengthy discussion runs over four days with arguments going back and forth. Salviati explains and defends the observation-based views of Galileo. On the other hand, Simplico is a staunch believer of the geocentric views of Aristotle and Ptolemy; they both are represented in Raphael's School of Athens in Figure 9.1. An interesting protagonist is Sagredo, who embodies the initial disbeliever in heliocentrism but who gradually becomes more and more convinced by Salviati's arguments. Even today, the *Dialogue* contains a wealth of interesting material concerning the struggle science faces when it comes to the communication of its findings. A case at hand is no doubt to be found in the coronavirus pandemic, discussed in Chapter 6. In this respect, the final "thank you" words from Salviati to his discussion partners Simplico and Sagrado are worth recalling. First, Salviati thanks Sagredo for having become more and more convinced of heliocentrism. Sagredo has shown that he listened carefully to the scientific evidence provided and gradually changed his view away from initial disbelief. The final words addressed to Simplico are more subtle. Recall that, throughout the whole four-day discourse, Simplico remains a staunch disbeliever of heliocentrism. Salviati first thanks him for his ingenuous questioning of the arguments put forward. He further states that he became fond of Simplico's unwavering and forceful defense

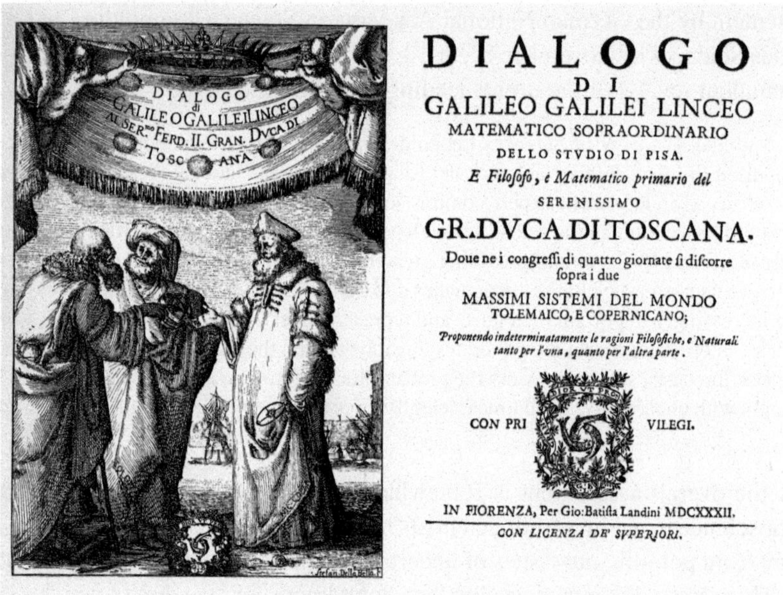

Figure 5.3 *Dialogo sopra i due massimi sistemi del mondo* (translated: "Dialogue on the two chief systems of the world"), in which Galileo Galilei presents a debate on heliocentrism between two experts and a reader, who, in the course of the narration, seeks explanations, starting from simple observations on everyday life. Source: Wikimedia Commons

of the worldviews of "his master" Aristotle. Recalling the various heated discussions, Salviati finally thanks both for their cooperation, which surely led to a deeper understanding of the underlying matter.

The above-mentioned translation of the *Dialogue* contains a foreword by Albert Einstein (1879–1955), who very early on faced disbelief concerning his theory of relativity. Einstein first heralds the economic and political importance of the *Dialogue*. Galileo showed both intelligence and courage in standing up for his evidence-based beliefs. In the *Dialogue*, these are presented in a truly passionate way. Einstein's closing words are worth repeating verbatim: "[Galileo's] unusual literary gift enables him to address the educated men of his age in such clear and impressive language as to overcome the anthropocentric and mythical thinking of his contemporaries and to lead them back to an objective and causal attitude toward the cosmos, an attitude which had become lost to humanity with the decline of Greek culture."

In a letter Galileo wrote to Johannes Kepler (1571–1630) on August 19, 1610, he laments that the disbelievers in heliocentrism even refuse to look for themselves through a telescope, to see a miniature solar system around Jupiter. The *Dialogue* contains, at the start of the second day, a wonderful example of how staunch believers in a cause refuse to change opinion even if proof of the contrary stares them in the face. Simplico and Sagrado muse over their last night feelings after a first full day of discussions. Simplico starts with "I find [yesterday's material] to contain many beautiful considerations which are novel and forceful. Still, I am much more impressed by the authority of so many great authors [like Aristotle and Ptolemy]." This statement Sagrado immediately counters with a wonderful story he once experienced in the house of a famous doctor in Venice. During that visit, the question under discussion was whether nerves originate in the heart, as Aristotle and followers from his Peripatetic school claimed, or in the brain, as claimed by followers of the Greek physician Galen of Pergamon (129 to c. 216 CE). In order to settle the matter, an anatomical dissection was performed that clearly showed that nerves originate from the brain, run through the spine, end up all over the body with only a single thread arriving at the heart. Matter settled one would say! Not, however, for one doubting attendant, referred to as a peripatetic philosopher, who was invited to participate in order to become convinced by the scientific evidence produced right in front of him. After the dissection, he was asked whether the evidence provided finally swayed his beliefs. His answer is rather remarkable. We quote from Drake (1967): "You have made me see this matter so plainly and palpably that if Aristotle's text were not contrary to it, stating clearly that the nerves originate in the heart, I should be forced to admit it to be true." On hearing this story, Simplico remarks "Sir, I want you to know that this dispute as to the source of the nerves is by no means as settled and decided as perhaps some people like to think." Similarities between then and today should be obvious. In 1633 the *Dialogue* was placed on the *Index Librorum Prohibitorum* (List of Prohibited Books) by the Roman Curia, and was not removed from that list until over 200 years later, in 1835. We will encounter another example of this kind of public communication through a spoken discourse in Chapter 13, where we discuss Joseph de la Vega's *Confusión de Confusiones* written in 1688, in which the author describes the Stock Exchange in Amsterdam in the seventeenth century. In this case the author uses as participants of the discourse a merchant, a philosopher and a trader.

5.4 Science Communication

In 1633, under pressure from the Inquisition, Galileo recanted (had to recant) his theory of heliocentrism. Towards the end of the process he is supposed to have muttered to the inquisitors "Eppur si muove" ("And yet it [Earth] moves"). Whether or not Galileo said these words is, today, not known. An excellent paper tracing the historical truth behind the term "supposed" is Livio (2020), the closing lines of which reflect excellently on science communication in modern times:

> Even if Galileo never spoke those words, they have some relevance for our current troubled times, when even provable facts are under attack by science deniers. Galileo's legendary intellectual defiance – "in spite of what you believe, these are the facts" – becomes more important than ever.

At the time, the Inquisition may have won a battle, but in the end science would win out. Galileo is widely considered as one of the founders of the scientific method and modern science. And indeed the Earth not only moves around the Sun but they move together through an ever-expanding universe.

Scientific insight is increasingly called upon in an era where technological, medical, and geopolitical developments occur at a rapid pace. This implies that, more and more, science communication takes, or rather has to take, center stage. The Press Library of the Australian National University (ANU) has an interesting publication on Science Communication (Gascoigne et al., 2020). From the Introduction (Chapter 1) we learn:

> This book is a comprehensive attempt to chart the history of science communication as it developed in the modern era. It tells the story from the perspective of researchers and practitioners in the field, collecting

accounts of how modern science communication has developed internationally. The book contains 40 chapters: two introductory chapters, 36 chapters focusing on a single country, one covering the three Scandinavian countries, and one describing the communication of health issues in a region of Africa. It involves 108 authors. The results are astounding, a unique dataset to be explored and a rich cornucopia of information.

We fully agree with the last sentence, the book is a treasure trove on the subject. Its Chapter 20 on Italy contains a brief discussion of the L'Aquila trial.

Throughout the cautionary tales discussed so far, we have encountered numerous challenges faced by the public communication of scientific evidence. Galileo tried to solve the problem through a dialogue between relevant protagonists, published in book format. Fast-forward in time from the seventeenth century to the current century. We thus move from communication in paper format to a restricted audience (those able to read) to an almost world-encompassing audience that can be reached by every medium imaginable. Beyond the still-important printed format, an increasing dissemination of information, on whichever topic, can be achieved in real time via social media. It is impossible to avoid misinformation. The latter travels through the social media landscape at lightning speed, whereas science communicates at a much slower pace. New scientific findings need confirmation, refereeing and peer reviewing before they can be published, let alone communicated. Of course, science needs to adjust to this new technological landscape. It is however fair (but most unfortunate) to say that fake news travels much faster than scientifically corroborated information; and yet there is no way back. At the same time, politicians and journalists need sufficient training in order to be able to communicate their views in a clear and understandable way to a broader public; this goes to the heart of any democratic system. For an edited volume on the topic of social media and democracy, see Persily and Tucker (2020). We quote from the Foreword of that volume:

Over the last five years, widespread concern about the effects of social media on democracy has led to an explosion in research from different disciplines and corners of academia. This book is the first of its kind to take stock of this emerging multi-disciplinary field by synthesizing what we know, identifying what we do not know and obstacles to future research, and charting a course for the future inquiry. Chapters by leading scholars cover major topics – from disinformation to hate speech to political advertising – and situate recent developments in the context of key policy questions. In addition, the book canvasses existing reform proposals in order to address widely perceived threats that social media poses to democracy.

Political events worldwide, as well as the way in which the coronavirus pandemic (see Chapter 6) was addressed by different countries around the globe offer almost daily proof that this threat is real. Born out of the need for clear and evidence-based communication in the early stages of the pandemic, Blastland et al. (2020) offered five guiding rules. They are: (Rule 1) Inform, not persuade; (Rule 2) Offer balance, not false balance; (Rule 3) Disclose uncertainties; (Rule 4) State evidence quality; and (Rule 5) Inoculate against misinformation. We learn from the paper that:

We recognize that the world is in an "infodemic", with false information spreading virally on social media. Therefore, many scientists feel they are in an arms race of communication techniques. But consider the replication crisis [in medical statistics], which has been blamed in part on researchers being incentivized to sell their work and focus on a story rather than on full and neutral reporting of what they have done. We worry that the urge to persuade or to tell a simple story can damage credibility and trustworthiness.

Of the five rules listed above, the one that is closest to the main thread of our book is Rule 3. This rule is not just about communicating statistical uncertainties through the presentation of confidence intervals (CIs, see Section 8.6.3) but also at the level of "we do not know". The latter turned out to be one of the main shortcomings in the communication that took place prior to the L'Aquila earthquake. Rule 1 turned out to give key advice in trying to achieve a higher percentage of the penetration of COVID vaccinations. As part of Rule 2 we see not only statistical or scientific balance but also societal balance. Of course, every major task of evidence communication will involve a combination of all five rules. If one wants to reach a broad spectrum of a population at risk from a certain event, then it is absolutely crucial to have a sufficiently broad spectrum of stakeholders (not only scientists) involved in the formulation of measures to be taken against the risk at hand, be it environmental, technical, economic or medical.

The above discussion also stresses the importance of a balanced relationship between scientists and the media. Whereas a short course on "media communication" is not a standard offer in most technically oriented study programs, it ought to be. McConway and Spiegelhalter (2021) give some useful advice about media communication in their blog. Although in this case the presentation is based on pandemic-related experience, their findings hold quite generally. We quote the headings:

1. Get media training
2. Don't be lured out of your comfort zone
3. Beware of 'just a chat'
4. Sound human
5. Keep off the (statistical) jargon
6. Being a statistician means that you must know every number...
7. Keep the nit-picking under control
8. Don't be pulled into someone else's argument
9. You'll get things wrong. Don't agonise about that
10. Don't try and go it alone – work with others
11. Make friends with journalists...
12. ... but don't be shy about complaining

As statisticians, we do like the blog's ending:

> Statisticians have a vital role in explaining numbers and evidence. They tend not to have strong personal agendas apart from wanting things to be clear and open, but they come down hard on selection or manipulation of evidence. The media love all this, and so do the public. So we need more statisticians to get out there, put their heads above the parapet, make mistakes and learn from them, and improve the communication of lessons learned from data. We hope this list has not put you off, and instead will encourage you to engage with the media.

In a modest way, we do hope that our book does justice to this quote.

5.5 Epilogue

In this chapter we have encountered various problems around risk communication. The following quote from the American psychologist Carl Rogers (1902–1987) comes to mind: "Man's inability to communicate is a result of his failure to listen effectively." It reminds us of the 2021 movie *Don't Look Up* about two scientists who attempt to warn humanity about an approaching comet that will destroy human civilization. As we read on Wikipedia: "The impact event is an allegory for climate change, and the film is a satire of government, political, celebrity, and media indifference to the climate crisis." We return to the latter in more detail in Chapter 10. Figure 5.4 shows a cartoonist's view on this issue.

Figure 5.4 "Man's inability to communicate is a result of his failure to listen effectively" (Carl Rogers). Source: Enrico Chavez

5.6 Lessons Learned

At the level of earthquake damage to buildings, the L'Aquila earthquake highlighted the difficulties faced in improving older, often historic, buildings and bringing them to modern safety standards in earthquake-prone regions. The key lesson learned is however the lack of efficient evidence-based communication just prior to the main earthquake. Scientists were put in the dock and initially received jail sentences. The chapter navigates between popular beliefs and scientific facts about earthquake prediction; in Section 9.6 we will discuss in more detail the statistical modeling of earthquake data. In particular, we will have a closer look at the L'Aquila data. The current chapter has concentrated more broadly on the communication of scientific evidence. We revisited the famous court case of the seventeenth century against Galileo Galilei's views on heliocentrism. Galileo's *Dialogue* offers a wealth of material on guidance for science communication that still very much holds true today. We update the *Dialogue* with rules of communication that mainly come out of experiences made during the coronavirus pandemic, discussed in the next chapter. We finally touched upon a topic of increasing importance but rarely stressed: talking to journalists.

6

The Coronavirus Pandemic

Je ne sais pas ce qui m'attend ni ce qui viendra après tout ceci.
Pour le moment il y a des malades et il faut les guérir.

Albert Camus (1913–1960) in *La Peste*, © 1947, Éditions Gallimard.

I have no idea what is awaiting me, or what will happen when this all ends.
For the moment I know this: there are sick people and they need curing.

Translation by Stuart Gilbert (1883–1969)

6.1 The Early Days

A new type of severe lung disease may have been linked to a resident, aged 55, of Hubei province in China, going back to November 17, 2019. In any case, invisible to the world at large, in the mid to late November of 2019 a mysterious virus had set its path of destruction on the world. That this virus indeed corresponded to a new disease was only realized by the medical profession in China at the end of December. As reported in the following newspaper article:

On December 27, Zhang Jixian, a doctor from Hubei Provincial Hospital of Integrated Chinese and Western Medicine, told China's health authorities that the disease was caused by a new coronavirus. By that date, more than 180 people had been infected, though doctors might not have been aware of all of them at the time. By the final day of 2019, the number of confirmed cases had risen to 266. On the first day of 2020 it stood at 381.

Not being aware of these events, the world celebrated the end of 2019, welcoming the New Year 2020: little did we guess what was to happen. The Chinese New Year for 2020 fell on January 25. Under normal circumstances, this would imply extensive traveling for over 450 million Chinese people and would lead to about three billion trips. The modern Silk Roads (indeed plural) would offer very effective, fast and extensive traveling possibilities for the virus to disperse from its home in China to the rest of the world. For many people outside China, an early critical report concerned the death of the eye doctor Li Wenliang of Wuhan City, Hubei province in China, who died on February 7 after contracting the virus while treating patients in Wuhan. In December, he sent a message to medical colleagues warning of a virus he thought looked like SARS – another deadly coronavirus. However, police told him to "stop making false comments" and investigated him for "spreading rumors". No doubt doctor Li Wenliang was an early hero of humanity's fight against this microscopic enemy.

On February 26, BBC News had as a headline "Coronavirus cases surge to 400 in Italy". In these early days, Italy became the worst hit country worldwide. The onset of the virus for Italy came on January 31, 2020, when two Chinese tourists in Rome tested positive for the virus. One week later an Italian man repatriated back to Italy from Wuhan was hospitalized and confirmed as the third case in Italy. An interesting article of the *South China Morning Post* edition of April 1 highlighted the case of Prato in northern Italy. Because of its eminent importance with respect to the current as well as future pandemics, we quote below the story directly from the article. One should be aware that Prato has a large ethnic Chinese population, about 50 000. This is about 20% of the city's 2019 population. These Chinese residents were originally employed by the local textile industry, an industry for which Prato is famous. At the time, the town's infection rate stood at less than half the Italian average at 62 cases per 100 000 inhabitants versus 115 for the country: Why so? Prato's Chinese community went into lockdown from the end of January, three weeks before Italy's first recorded infection. Many were returning home from Chinese New Year holidays in China, the then epicenter. They knew what was coming and spread the word: stay home! The local population did not heed these ominous warnings. Therefore, as Italians headed for the ski slopes and crowded into cafes and bars as normal, the Chinese inhabitants of Prato had seemingly disappeared. Its streets, still festooned with Chinese New Year decorations, were semi-deserted, shops closed. Another person who went into self-isolation after returning home from China was 23-year-old university student Chiara Zheng: "I was conscious of the gravity of the situation. I felt a duty to do it for other people and those close to me." The Chinese population went from being considered as the cause of all our problems to one we should have listened to and learned from.

The case of Prato contrasts sharply with that of Bergamo, an Italian town only 320 km to the north. Indeed Bergamo, occasionally referred to as Europe's Wuhan, showed the truly ugly face of the virus. On February 9, more than 45 000 supporters followed in Milan (60 km from Bergamo) the European Champions League soccer game between Atalanta Bergamo and the Spanish team of Valencia. The outcome was 4–1 for Bergamo, resulting in all-night festivities both in Milan and in the winners' home town. At the time, Bergamo, a town of about 120 000 inhabitants, accounted for about 6000 infected and 1000 diseased. The epidemiologist Dr. Francesco Le Foche referred to the above match as "game zero": a biological time bomb had exploded with repercussions well beyond the region. The return game in Valencia took place on March 10; Bergamo won again, this time by 2–1. On returning home, the famous Italian football supporters, the tifosi, however had very little to celebrate. The fact that Spain also turned out to be badly hit by the virus was, in part, to do with the Bergamo–Valencia match, for about 2000 Spanish supporters traveled to Milan. The March 27 version of the Swiss newspaper *Neue Zürcher Zeitung* (NZZ) headlined, in translation, "Hundreds of dramas in Bergamo" and "It is like during war. [. . .] In Bergamo we only hear ambulances; apart from them, quietness reigns." At that point, the following quote from Camus (1947) became relevant:

C'est au moment du malheur qu'on s'habitue à la vérité, c'est-à-dire au silence.

Translated:

It is in the thick of the calamity that one gets hardened to the truth – in other words, to silence.

Concerning the early development of the pandemic, a multitude of dramatic stories are available and indeed worthwhile remembering. You may for instance look for how the small ski resort of Ischgl in Tirol in Austria caused an outbreak in Norway and how an early, end-of-January warning from Icelandic officials was not heeded. A further early case that obtained considerable attention is that of the Diamond Princess, a cruise ship docked in the harbor of Yokohama, Japan, with passengers not allowed to disembark. "We're in a Petri dish" became a famous quote of passenger Carol Montgomery, 67, a retired administrative assistant from San Clemente, California (a Petri dish is a shallow transparent lidded dish used by biologists to culture and study cells). For South Korea, the first case was diagnosed on January 20. A sudden jump is attributed to Patient 31 who participated in a gathering at a Shincheonji Church of Jesus, the Temple of the Tabernacle of the Testimony church in Daegu, South Korea. Worldwide, there were more cases of uncontrolled spread of the virus within tightly knit religious communities. A case that caused members of choirs worldwide to be alarmed, early on, concerned a choir from Skagit County, Washington. Super-spreading of the virus within the choir was reported around mid-March with more than 80% of its members becoming infected. Referring to the above case, the NZZ ran an article on April 1, 2020, under the title, translated, "When singing kills." From these cases, and many more later on in the pandemic, we ought to have learned what went wrong at the time and to make sure that this will not happen again. Here Winston Churchill's words "Never waste a good crisis" hold value, though one may object to "good" in this case and perhaps replace it by "serious" or "dramatic".

On March 11, the WHO officially declared COVID-19 a pandemic. The word "pandemic" derives from the Greek "pan" (all) and "demos" (local people, the crowd). The new coronavirus' grip on the world's population would become suffocating. Its image greeted us every morning through our morning papers. The word "corona" refers to the typical halo or crown seen through an electron microscope (Figure 6.1). A virus (Latin for poison, slime, venom) is

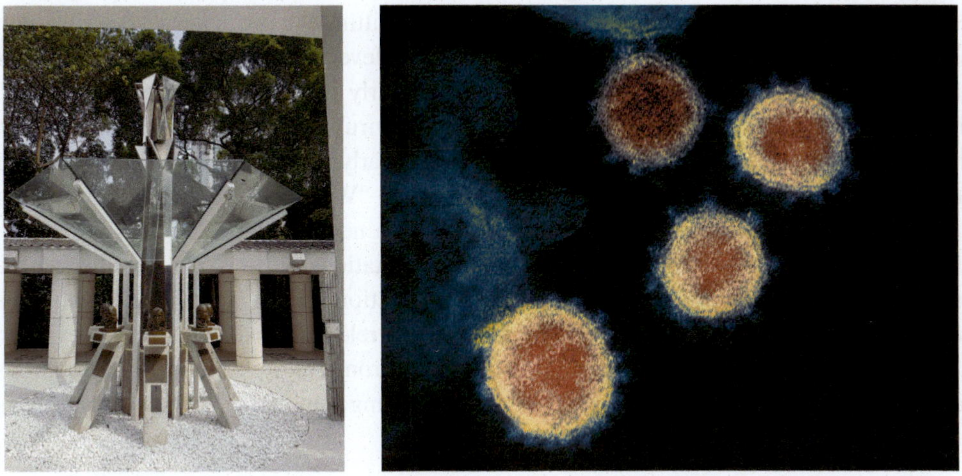

Figure 6.1 Left: The SARS Memorial in Hong Kong Park which refers to the 2002 pandemic SARS-CoV-1. This photograph was taken on November 6, 2019, just prior to the outbreak of the coronavirus pandemic SARS-CoV-2. Right: The visible "corona" of the coronavirus SARS-CoV-2 under an electron microscope. Sources: Gerda Janssens (left) and National Institute of Allergy and Infectious Diseases (right)

a sub-microscopic infectious agent that replicates only inside the living cells of an organism. Whether viruses are a form of life or whether they are organic structures that interact with living organisms we leave to virologists to debate. Viruses have been described as "organisms at the edge of life". They resemble organisms in that they possess genes, evolve by natural selection, and reproduce by creating multiple copies of themselves through self-assembly. Viruses differ importantly from bacteria; bacteria are relatively complex, single-celled living creatures, many with a rigid wall, and a thin, rubbery membrane surrounding the fluid inside the cell. Bacteria can reproduce on their own. One can think of viruses as parasite-like pieces of software affecting your defensive health or better, immune system; from this point of view stems the similarity with computer viruses. Once a virus enters a cell, by using its host cell's enzymes it is extremely efficient in setting up a reproductive assembly line producing thousands of new viruses ready to spread in its host. These copies then attack the healthy cells at neuralgic spots, like the respiratory tract. In summary, a virus typically is a set of genetic instructions surrounded by a protein coat. An important first lesson the general public had to learn was that bacterial infections, like the bubonic plague in Camus' book, can be treated through antibiotics, viral infections cannot. Examples of viral infections include the common cold, measles, chickenpox, HIV and influenza. For many viral infections there is no specific cure. The advice is certainly to drink a lot to avoid dehydration and to get plenty of rest. Of course, in some cases virus-specific medication may be available.

Rather early on, society split into factions opting for more or less rigorous restrictions on daily life, as well as being for or against vaccination. Often these fault lines ran along political lines of division. As in the case of the 2007–2008 financial crisis considered in Chapter 3, but even more so now, society became totally overwhelmed by the news media, articles, books, podcasts, influencer views, political and expert committee guidelines, you name it. Overnight, epidemiologists and virologists became media stars. As the pandemic progressed, these bearers of bad news were increasingly blamed, befitting the metaphoric phrase "Shoot the messenger." The spread of information reached a level unparalleled in the history of mankind. And yet, a good summary of the situation early in 2022 was: "We don't know." This does not mean that science had no clue what the virus was all about; on the contrary. The success of the scientific method had been proved early on through the development of vaccines and the discovery of mutations. The expression "We don't know" rather referred to how and when will it all end so that we can return to a normal life. The initial hope for a fairly quick return to normality was dashed by new mutations of the virus, together with a less than full protection through vaccination. In the sections below, we will take you for a walk through aspects of the pandemic which we deem relevant for our book. The existing literature contains numerous accounts of the pandemic from every possible angle. As things stand, the 2019–20xy coronavirus story is still developing. One thing we know for sure is that $xy \geq 22$.

6.2 Other Pandemics

The proper scientific name of the 2019 coronavirus is *severe acute respiratory syndrome coronavirus 2* (SARS-CoV-2) and the name of the disease it causes is the coronavirus disease (COVID-19). The "2" refers to the virus coming after SARS-CoV, the "19" refers to the year 2019; early on, it was referred to as the 2019 novel coronavirus. Having different names for a virus and its disease is common; think about HIV (the virus) and AIDS (the disease), for example. People often know the name of a disease, such as measles, but not the name of the virus that causes it (rubeola in English). When one communicates news on the pandemic, being precise on the terminology used early on is helpful. Of course, as a disease progresses, more colloquial as well as politically incorrect names emerge. The Spanish flu of 1918–1920 got its common name from the fact that, at the time, neutral Spain had a much less censored press and hence could more freely report on that pandemic. For instance, on May 28, 1918, a headline from the Spanish newspaper El Sol referred to a mysterious disease, "Las fiebre de los tres dias" (translated: "The three-day fever"), that already accounted for 80 000 infected in Madrid. Around 500 million people were infected and well in excess of 20 million may have died, making it one of the deadliest pandemics mankind has faced. *Spanish flu* was the first of two pandemics caused by the H1N1 influenza virus. The second example is the 2009 pandemic *H1N1 A influenza virus*, more commonly referred to as *swine flu*. From Wikipedia (2020a):

> Some studies estimated that the actual number of cases including asymptomatic and mild cases could be 700 million to 1.4 billion people – or 11 to 21 percent of the global population of 6.8 billion at the time. The lower value of 700 million is more than the 500 million people estimated to have been infected by the Spanish flu pandemic.
>
> The number of lab-confirmed deaths reported to the WHO is 18,449, though this 2009 H1N1 flu pandemic is estimated to have actually caused about 284,000 deaths (range from 150,000 to 575,000). A follow-up study done in September 2010 showed that the risk of serious illness resulting from the 2009 H1N1 flu was no higher than that of the yearly seasonal flu. For comparison, the WHO estimates that 250,000 to 500,000 people die of seasonal flu annually. Unlike most strains of influenza, the pandemic H1N1/09 virus does not disproportionately infect adults older than 60 years; this was an unusual and characteristic feature of the H1N1 pandemic.

We think it is fair to say that the 2009 swine flu pandemic, which was indeed fairly recent, has all but disappeared from our collective memory. For an excellent graphical display of past epidemics and pandemics, see LePan (2020). Known examples of related viruses are SARS and *Middle East Respiratory Syndrome Coronavirus* (MERS, MERS-CoV). For the purpose of our discussion, we will often refer to the COVID-19 disease as well as the virus as *coronavirus*, or simply the *virus*.

6.3 Exponential Growth

Exponential growth is a key characteristic of a pandemic, typically in its early stages. It enters the pandemic stage through the notion of reproduction number. The *basic reproduction number* R_0 is the expected number of infected directly generated by one case in a fully susceptible population, that is, all individuals in the population are at risk of the disease. Another such number is the *effective reproduction number* R_e, which is the expected number of infected individuals generated by one case in an ongoing outbreak. If $R_0 > 1$ then the pandemic spreads exponentially. Early analyses put a value of 2 to 2.5 on R_0; later values reported were more in the range of 1.5 to 3.5, so certainly greater than 1 and the aim of later interventions was definitely to reduce R_e to below 1. How does this number compare with other infectious diseases? Here are some approximate values: seasonal flu ($R_0 < 1$), hepatitis and ebola ($R_0 \approx 2$), HIV and SARS ($R_0 \approx 4$), mumps ($R_0 \approx 10$) and measles ($R_0 \approx 18$); the latter are two highly contagious diseases. Let us consider a disease still very much in our mind, SARS (severe acute respiratory syndrome), officially referred to as SARS-CoV, another coronavirus. It started in the Chinese province of Guangdong and became a pandemic over the period 2002–2003. SARS was a rather deadly disease with $R_0 \approx 4$. On the positive side, however, SARS-infected people first showed symptoms (during the incubation period) before they themselves became contagious (the latent period). With SARS-CoV-2 this is typically the other way around. Overall SARS, termed the "first pandemic of the twenty-first century", killed 774 people from a total of 8098 infected across 29 countries worldwide. In Hong Kong Park, a monument pays tribute to the healthcare workers, doctors and nurses who died while caring for SARS patients (Figure 6.1).

The meaning of exponential growth can easily be explained by the historical example of the Persian chessboard. There are various versions related to the introduction of the game of chess to a thirteenth-century Persian court. A passing visitor presented the local ruler a new game, chess, played on an 8×8 checkered board. The ruler was so delighted by this present that he wanted to give to the visitor, in return, a present of his choice. The visitor told the ruler that as a man of little affinity to monetary wealth, he would be content being paid in grains of wheat. For the amount of grain he referred to the chessboard and asked for the total number of grains one would obtain after successive doubling and adding all the 64 fields of the board: $1 + 2 + 4 + 8 + \cdots (+2^{63})$. On purpose, we put the correct 2^{63} in parentheses because the ruler just thought of "$1+2+4+8+\cdots$". Bemused by so much humbleness, the ruler most happily agreed. Great was his surprise when soon after his finance minister came by informing him that the amount of wheat needed would deplete more than the country's full yearly production as well as all reserves.

So how much wheat is needed? This we can calculate with a mathematical result that perhaps goes back to your high school days. We state it as a *lemma*, which in mathematics means a minor result, a kind of stepping stone. You may skip this lemma without losing the message of the story. It concerns so-called geometric series. For a real number r (written by mathematicians as $r \in \mathbb{R}$), successive powers $r^0 = 1, r^1 = r, \ldots, r^n = r \times r \times \cdots \times r$ (with n factors r) are referred to as a *geometric sequence*. The following lemma yields the corresponding sum of such numbers, a *geometric sum*. The first equality is just the mathematical notation for the sum; the second yields the desired result.

Lemma 6.1 (Geometric sum)**.** *Suppose $r \in \mathbb{R}$ and $r \neq 1$, then*

$$1 + r + r^2 + \cdots + r^n = \sum_{k=0}^{n} r^k = \frac{1 - r^{n+1}}{1 - r}.$$

Proof Multiplying $\sum_{k=0}^{n} r^k$ by $1 - r$ we obtain

$$\begin{aligned}(1-r)\sum_{k=0}^{n} r^k &= \left(\sum_{k=0}^{n} r^k\right) - r\left(\sum_{k=0}^{n} r^k\right) = \left(\sum_{k=0}^{n} r^k\right) - \left(\sum_{k=0}^{n} r^{k+1}\right)\\ &= (1 + r + r^2 + \cdots + r^n) - (r + r^2 + r^3 + \cdots + r^{n+1}) = 1 - r^{n+1}\end{aligned}$$

from which the result follows by dividing this identity by $1 - r \neq 0$ on both sides. □

With Lemma 6.1 at hand, we can now determine how much wheat there would need to be on the entire chessboard in the Persian chessboard story. There would be $2^{64} - 1 =$ 18 446 744 073 709 551 615 grains of wheat, weighing about 1.2 trillion tons (on average, 15 432 grains of wheat weigh 1 kg). The yearly production of wheat in 2019 stood at about 764.46 million tons so that the visitor's request amounted to more than 1500 times the 2019 worldwide production! We leave the rude awakening of the ruler and the ending of the tale to your imagination. A very nice version of the story is told by Tahan (1993, Chapter 16).

As already stated, as soon as $R_0 > 1$, an epidemic follows, for a certain time, a devastating exponential growth. If we denote by y the number of infected individuals and by x the number of discrete time steps elapsed (days, or weeks, say), then for $R_0 = 2$ (approximately the case for the coronavirus), we have $y = 2^x$, whereas for $R_0 = 4$ (approximately the case for SARS) this becomes $y = 4^x$. Both functions grow very fast and plotting them together is often useless; therefore it helps to apply a logarithmic transform denoted log (the natural logarithm we have already seen as "ln" in the van Dantzig formula (1.1)). The resulting functions then become $\log(y) = \log(2)\,x \approx 0.7x$ and $\log(y) = \log(4)\,x \approx 1.4x$: two linear functions with different slopes, 0.7 and 1.4. Visual comparison on this scale is much easier. We demonstrate this in Figure 6.2. The top-left plot shows the fast growth of $y = 2^x$ as a function of x. The top-right plot shows $y = 4^x$ and $y = 2^x$ as functions of x on the same graph and demonstrates how much faster the former grows in comparison to the latter; in fact, one might now be tempted to think that the dashed line (representing the same 2^x as in the top-left plot!) is constant. The bottom-left plot shows $\log(2^x) = \log(2)\,x$ and $\log(4^x) = \log(4)\,x$ as functions of x and allows for a much easier comparison between the two functions. However, to read off y-values of, say 4^x, we would need to compute the exponential since $\exp(\log(4)\,x) = \exp(\log(4^x)) = 4^x$ (the value in which we are interested). The bottom-right plot of Figure 6.2 uses a logarithmic y-axis. We see the same shape of the curves (here: straight lines) as in the bottom-left plot, but the y-axis labels are the actual values of 2^x and 4^x. This allows one to both compare the two functions and also read off individual values. Note that the numbers are so large here that the labels are given in scientific notation, for example 8e+5 stands for $8 \times 10^5 = 800\,000$; more on that in Section 6.6. Logarithmic axes are widely used in practice to identify power-like behavior, since if $y = x^\alpha$ for some $\alpha \in \mathbb{R}$ then $\log(y) = \alpha \log(x)$, so that when y is plotted as a function of x with both the x-axis and the y-axis in logarithmic scale (that is, plotting $\log(y)$ against $\log(x)$) we see a straight line with slope α. As the human eye can easily identify linearity from a plot, power-like behavior can easily be identified. It should now also be clear why the coronavirus chieftain in Figure 6.3 praises his (or her/its) colleagues for the linear growth in the graph with logarithmic y-axis shown, which really means exponential growth of the pandemic.

An important side to the Persian chessboard story is highlighted in Sagan (1997) (where Chapter 2 is titled "The Persian chessboard" and refers to the growth of bacterial populations). Sagan states the important fact that, as in the chessboard story, "Exponentials can't go on forever, because they will gobble up everything." Of course, the same is true for the exponential growth of pandemics caused by viruses. Here we are back at the crucial point of the coronavirus pandemic, how we could have made sure to break the exponential growth and buy ourselves time and resources. The same conclusion holds for exponential growth on stock markets, a lesson we should have learned for instance from the 2007–2008 financial crisis; see Chapter 3. Perhaps the best-known collection of exponential growth predictions and the consequential warnings, given finite resource-constraints, is a famous Club of Rome publication (Meadows et al., 1972); see also Section 10.1. From looking at the consequences of compound interest rates on mortgage or debt payments, we all should be well acquainted with the phenomenon of exponential growth. Albert Einstein famously said that compound

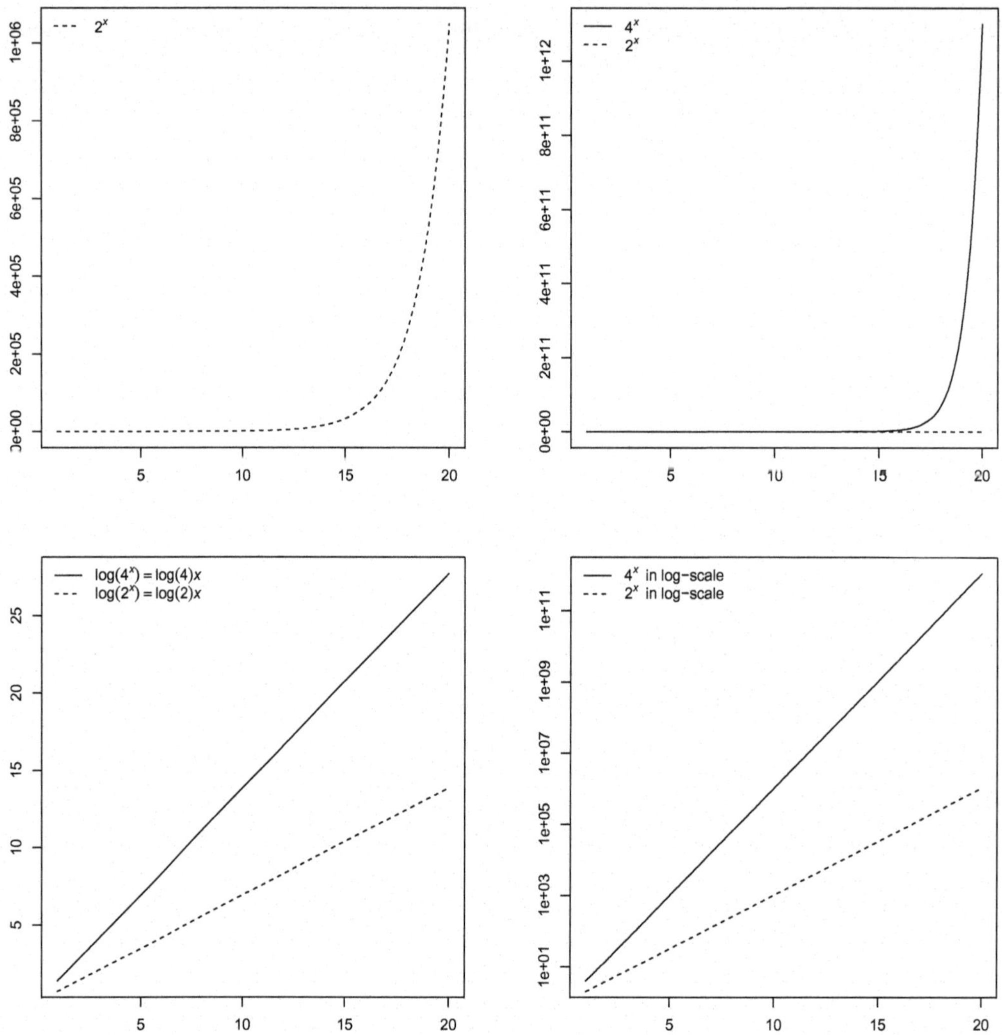

Figure 6.2 The functions $y = 2^x$ and $y = 4^x$ (top row), their logarithms (bottom left) and with logarithmic y-axis (bottom right). The last figure in particular not only allows us to compare the two functions but also to read off the original values (see the tick labels on the y-axis). Source: Authors

interest is the most powerful force in the universe: "Compound interest is the eighth wonder of the world. He who understands it, earns it; he who doesn't, pays it."

6.4 The Precautionary Principle

The *precautionary principle* states that, when facing a new and still widely unknown risk, it is sometimes necessary to act without definitive evidence. For the pandemic, this implied taking precautionary measures such as washing hands thoroughly, social and physical distancing, avoiding of large crowds, using sufficient ventilation indoors, wearing medically accepted

Figure 6.3 A cartoon explaining that linear behavior in logarithmic scale means exponential growth. Source: Enrico Chavez

face masks where and when required and self-quarantining. In a way, these measures make common sense and were also widely recommended, even partly enforced, during the Spanish flu. Unfortunately, large sections of society kept resisting these rules of conduct.

From the glossary of *Johns Hopkins Medicine* on the terminology used during the pandemic, we learn that

> The practice of *social distancing* means staying home and away from others as much as possible to help prevent the spread of COVID-19. The practice of social distancing encourages the use of things such as online video and phone communication instead of in-person contact. *Physical distancing* is the practice of staying at least 6 feet away from others to avoid catching a disease such as COVID-19. People who have been exposed to the new coronavirus and who are at risk for coming down with COVID-19 might practice *self-quarantine*. Health experts recommend that self-quarantine lasts 14 days. Two weeks provides enough time for them to know whether or not they will become ill and be contagious to other people.

The above mitigation measures typically apply at the individual level; we as individual persons are asked to take various precautionary measures in order to restrict the spreading of the virus. Early on in the pandemic, this would prevent a collapse of the medical system, especially at the level of intensive care units for COVID patients. The concept of *flattening the curve* from Figure 6.4 summarizes the envisaged aim.

Here, an historical example may be useful. The Hungarian physician Ignaz Semmelweis (1818–1865), see Figure 6.5, is known as an early pioneer of antiseptic procedures. While working in Vienna's General Hospital, he was able to demonstrate that the incidence of childbed fever could be drastically cut with sufficient hand sanitation. Despite extensive experimental proof, what today we would refer to as evidence-based medicine, the medical

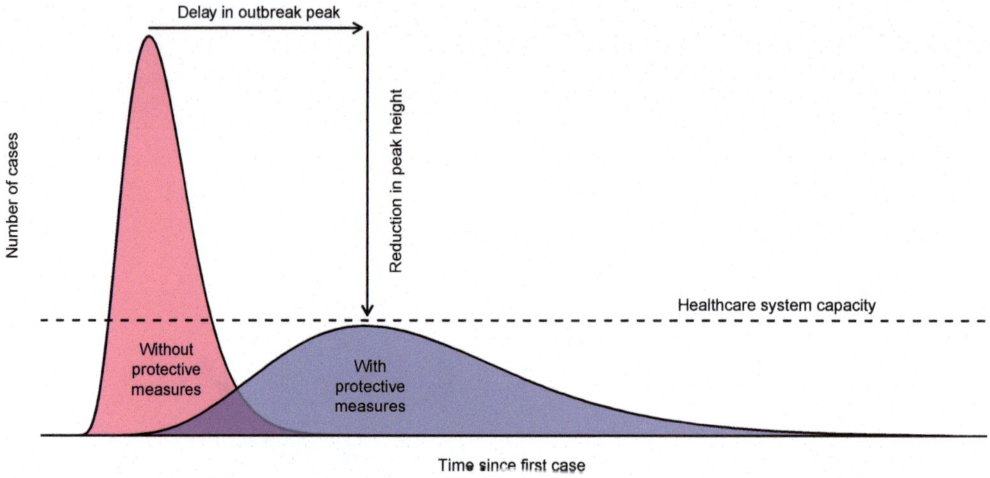

Figure 6.4 Flattening the curve. Source: Authors

Figure 6.5 The virus will do whatever it can to get you infected (left) whereas the precautionary principle, which was very much influenced by the work of Ignaz Semmelweis (1818–1865) (right), helps you to avoid infection. Sources: Enrico Chavez (left) and Wikimedia Commons (right)

profession at the time rejected his findings. As any hospital visit makes clear, today his findings are not only fully accepted, but it is difficult to imagine that it could ever have been thought otherwise.

6.5 Early Warnings

As in the case of the 2007–2008 financial crisis, considered in Chapter 3, we were all warned on multiple occasions, well before the pandemic struck. Politicians almost all over the world were caught sleeping at the wheel, though some newspapers claimed that "The Black Swan [a totally unforeseen event with severe consequences] has actually arrived." This was definitely not the case; see, for example, the following statement from Taleb (2007, p. 317):

> As we travel more on this planet, epidemics will be more acute – we will have a germ population dominated by a few numbers, and the successful killer will spread vastly more effective. [...] I see the risks of a very strange acute virus spreading throughout the planet.

An interesting document, covering the financial sector in the UK, is FBIIC (2008). This document comments on the frequency of pandemics, and it contains a very concrete warning of pandemic risk:

> An influenza pandemic is recognised as one [of] the biggest risks that could face the UK. The persistence and growing spread of H5N1 [bird flu] outbreaks in poultry sustains the risk that this or another highly pathogenic related flu virus will develop the ability to spread efficiently from person to person, leading to a global influenza pandemic. Previous flu pandemics have occurred at 10–40 year intervals. It is now 39 years since the last pandemic [the 1968–69 Hong Kong flu pandemic caused an estimated 2 million deaths worldwide]. The international consensus is that a flu pandemic could occur at any time.

Finally, on March 4, 2015, Bill Gates, the co-founder of Microsoft Corporation, gave a Technology, Entertainment and Design (TED) talk to world leaders on the threat of a pandemic on the basis of his experience from the Ebola crisis in West Africa. In this prophetic talk he clearly stated that "we are not ready for the next epidemic". He ends the talk with the words "if we start now, we can get ready for the next epidemic". Well, we (most countries) did not start preparing in 2015, and not in any year leading up to 2020. Besides these visible warnings, in the scientific literature there were numerous publications prior to 2019 that warned of virus transmission through bats and wet markets in Asia. The Wuhan seafood market, a so-called *wet market*, became a debated potential origin of the pandemic's patient zero. Several earlier publications, mainly in China, had warned that these wet markets could act as source and amplification centers for emerging infections. The "wet" in wet markets comes from the fact that meat and seafood in several Asian markets are sold as fresh as possible and, in many cases, for example birds, fish and reptiles, still living. Consequently, floors and stalls are thoroughly cleansed with water, hence "wet". If such food markets sell exotic live animals then there could easily lurk in them a viral threat. When a virus is able to transmit an infection between animals and humans, it's referred to as a *zoonotic virus*: SARS-CoV-2 is such a virus. It is no coincidence that the spread of the (bacterial) plague, which peaked in Europe around 1347 to 1351, was facilitated by the historical Silk Road. Whereas the fourteenth-century Silk Road passed over land and sea, the current Silk Roads not only use physical means of transportation (land, sea, air); they also very much exist in the virtual universe of social media. Although a biological virus cannot travel through the internet, communication about it can very much do so. We often have the impression that the virus is able not only to successfully program a viral assembly line in our infected cells, but also to reprogram the minds of some people on and via social media...

Through the pandemic, the topic of viruses has come to the forefront of ongoing societal discussions. As we already stated, viruses exist at the thin dividing line between life and non-life. There is still much for science to discover on this subject. For sure, life as we know it would not be possible without viruses. That they typically achieve notoriety as killer diseases distorts our view on their potential usefulness. It would therefore be good if in the public domain the "bad" reputation of viruses could be somewhat downplayed so we can enter into a dialogue on the many "good" things that they can do. Of single virus particles, also referred to as virions, there are many, very many. For instance, each SARS-CoV-2 infected person carries about 10^9 to 10^{11} virions. A rough estimate of the total number of all virions on our planet is about 10 (USA-)nonillion. In the USA, a nonillion is a 1 followed by 30 zeros. Elsewhere, however, it can mean a 1 followed by 54 zeros. This potential confusion leads us very naturally to the next section.

6.6 Large Numbers and Units of Measurement

When talking about exponential growth, we necessarily have to talk about *large numbers*; see the y-axis scales in all plots in Figure 6.2 except the bottom left one. We all know that a 1 followed by six zeros means a million. If we go beyond millions to nine to 12, 15, 18 and more zeros then, when communicating these numbers, one has to be more careful about where one lives and which language one uses. Some countries use the so-called *short scale* (such as the United States, Australia and English-speaking Canada) whereas others use the *long scale* (such as France and Germany). A "short scale billion" (10^9) equals a "long scale milliard", a "short scale trillion" (10^{12}) equals a "long scale billion", and, finally, a "long scale trillion" is a "short scale quintillion" (10^{18}). On March 27, 2020, US President Donald Trump announced that "The US supports its economy with a 2.2 trillion dollar package"; that means 2.2 multiplied by a 1 with 12 zeros (a short scale trillion). On April 10, 2020, media reported that the EU finance ministers had in turn agreed on a 540 billion euro rescue package to combat the economic fallout of the coronavirus pandemic. Short scale or long scale? When you find all this confusing, rest assured, so do we! It turns out that short scale applies here, which makes the amount 540 multiplied with a 1 with nine zeros. Converting euros to dollars, we obtain about 580 billion dollars in short scale, so the USA's support is about four times as large as that of the EU. As so often, clear mathematical notation avoids linguistic confusion. Early on in the crisis, the USA supported its economy with 2.2×10^{12} USD, whereas for the EU this was 540×10^9 EUR. A number represented in the form $a \times 10^b$ is said to be in *scientific notation*. Scientific notation is often used in or by computer software. For example, 540×10^9 in scientific notation would be written as 540e+9 in software, or as 5.4e+11 in *normalized scientific notation* (where $1 \leq |a| < 10$). The latter notation was used along the y-axes in all plots of Figure 6.2 except the bottom left one.

On July 21, 2020, the EU reached agreement on a 1.8 trillion euros recovery package. Even the serious Swiss newspaper *NZZ am Sonntag* stumbled over the long versus short confusion. In its July 26 edition, it referred to the deal as (in German) "Es geht hier um enorme Summen. Wir reden über 1,8 trillionen Euro." Indeed an enormous sum of money, a 1 followed by 18 zeros [sic]! In its August 2 edition it had to publish a correction: the "1,8 trillionen Euro" should have been "1,8 billionen Euro". As we know by now, a short scale trillion (in US English) equals a long scale billion (in German), a 1 followed by 12 zeros. Admittedly, still a large sum! These problems also remind us of the different ways in which numbers are expressed across different languages. Like 64, which is "sixty-four" (first six, then four) in English, as well as in French "soixante quatre", but "vierundsechzig" in German, hence first the "four", then the "six". Interestingly, if you pick up the phone in the UK, a number is transmitted as a sequence of cyphers, in Germany and Switzerland in groups of numbers. This for instance has led banks to impose the rule that, when you communicate longer numbers by phone, you must always do this by stating each cypher separately. This surely is a useful rule towards risk minimization. You may also be interested to learn that the name "Google" stems from the word "googol". The latter is 10^{100} (a 1 with 100 zeros), a truly large number; there are about 10^{80} atoms in the observable universe. It is appropriate given the fact that Google handles enormous amounts of data, estimated at about 2 trillion (2×10^{12} in short scale) searches each year. From Wikipedia we learn that the term "googol" was coined in 1920 by the nine-year old Milton Sirotta (1911–1981). He may have been inspired by the contemporary comic strip character Barney Google. Other names for this quantity include ten duotrigintillion on the short scale and ten thousand sexdecillion on the long scale. If your appetite for large numbers is not yet stilled, you may search the web for the (Ron) Graham Number which is much, much larger than even a googolplex ($= 10^{\text{googol}} = 10^{10^{100}}$). The Graham number appeared in work by Ronald Graham (1935–2020) in the field of combinatorics. Of course writing down ever larger numbers is nothing in itself, meeting ridiculously large numbers in a specific scientific context is however exciting. In an obituary for Ron Graham, *The New York Times* wrote about the Graham Number that it is "[. . .] a number so huge that there is not enough space in the entire universe in which to write all of the digits".

When it comes to risk communication, another area of possible confusion concerns *units of measurement*. An important example is found in the reporting of radiation risk. During the Chernobyl disaster of 1986, radioactive activity was mainly reported in becquerel (Bq), units named in honor of Henri Becquerel (1852–1908); in 1975 the Bq became the official standard under the International SI unit system, replacing the curie (named for Marie Skłodowska-Curie, 1867–1934). Europe, for instance, follows this convention, whereas the United States still uses the curie (Ci). The conversion between the two units is 1 Ci = 3.7 Bq. Further measures used for (ionizing) radiation are the gray (Gy) and the sievert (Sv), named, respectively, in honor of Louis Harold Gray (1905–1965) and Rolf Maximilian Sievert (1896–1966). The gray measures the dose of radiation that a body absorbs divided by its weight; the sievert on the other hand also takes into account the type of radiation, the tissues exposed and how much energy they could have absorbed. As such, the sievert is much closer to a measure of true health risk. Finally, there is the rem – here the initial "r" stands for Wilhelm Conrad Röntgen (1845–1923), who first detected X-rays (or Röntgen rays) and their useful properties for medicine; 100 rem = 1 Sv, so one millisievert corresponds to 0.1 rem. The millisievert is a standard unit in Europe and Asia; the rem in the United States. But how do we communicate which dose of radiation is (un)safe? As a first, but absolutely not final, step, a calibration with respect to natural background radiation is useful. For instance, a person's exposure to all natural sources of radiation typically ranges from 1.5 to 3.5 millisievert (0.15 to 0.35 rem) a year. One sievert will cause illness, eight sieverts causes death even if treated. We are very well aware that the above may sound bewildering, but most unfortunately it is only the beginning of a very complicated and hot debate on radiation and illness, mainly cancer. Similar discussions are taking place when we look at the radiation risk from computers, cell phones, Wi-Fi and transmission antennas such as those used for 5G in wireless communication. In this case we are not looking at the high-energy ionizing radiation (such as X- and gamma-rays), which is powerful enough to remove an electron from an atom or molecule (hence the latter become ionized), but at so-called radio-frequency radiation and the health effects of exposure to it. Thus, the possible suspects are non-ionizing low-frequency electromagnetic fields. An important task for public health authorities concerns the setting of limits in order to protect the overall population from too much exposure to radiofrequency electromagnetic fields. It is of paramount importance that these more technical risks are clearly communicated. This goal can only be achieved when different experts work closely together.

We would like to end this section with a personal anecdote. In the immediate aftermath of the Chernobyl disaster of April 26, 1986, eating spinach in Zutendaal, a village in Northeast Belgium located just 10 km from the Dutch border, was still allowed. However, over on the Dutch side, this was not the case. We wonder how the fictional character Popeye the Sailorman would have reacted. An interesting article that puts the above discussion on units of radiation measurement into perspective for both Chernobyl and Fukushima is Spiegelhalter (2011), which also adds another twist to the spinach story.

6.7 Lessons Learned

Without any doubt, the main lesson learned from the coronavirus pandemic is how totally unprepared modern society all over the world was for the potentially devastating consequences of a pandemic. We learned that warnings were widely and amply available for all those who cared to listen. The political negligence concerning pandemic risk was truly mind-boggling. As a consequence, we all had to learn as we went along. Clear communication, especially in the early stages of the pandemic concerning precautionary measures to be taken in the face of COVID-19, was not available. This caused great confusion. Society was bombarded with epidemiological jargon. Reproduction numbers larger than 1 were mapped into exponential growth of the pandemic and the consequential capacity problems in hospitals. Precautionary measures were aimed at flattening the curve as a remedy. The resulting time saved also helped in the development of vaccines. With a return period in the order of 30 years, the next pandemic may not be very far around the corner. It is to be hoped that the political establishment will not be caught sleeping at the wheel a second time.

When risk is talked about, one quickly runs into the communication of numbers. On the one hand, large numbers may play a role; this we experienced in Chapter 3 on the 2007–2008 financial crisis and in Section 6.3, where we discussed the power of exponential growth. It is rather disturbing that, whereas across some of the main international languages, such as English, French and German, similar-sounding names are used, they often correspond to different numerical values. A further problem arises when one moves from scientific notation to a linguistic version. We gave several examples to highlight these issues. A further source of possible confusion is the changes in units of measurements for physical quantities, as well as the multiple versions used across various countries. Here also, we gave several examples. Perhaps the main lessons to be learned are "Always be very careful when you communicate large numbers and use specific units of measurement. Be aware of the lack of uniformity across the world and always double check!" We learned that minimal risk prevention does not always need rocket science. A major problem throughout the pandemic concerned the (non-)availability of high-quality data. Even when, gradually, good data became more and more available, their interpretation as a source for policy communication remained difficult. Examples include the estimation and reporting of key epidemiological parameters and the analysis of excess mortality. By now, libraries are filled with publications on the pandemic; surely many more will keep on coming. An excellent book, indeed a must-read on the statistical communication and understanding of coronavirus data, is Spiegelhalter and Masters (2021). In Example 8.7 we will discuss the important issue of testing for the virus. Remark 8.8 contains an example on the (mis)interpretation of basic COVID-19 statistics.

7

Mathematical Wonderland

Mathematics reveals its secrets only to those who approach it with pure love, for its own beauty.

Archimedes of Syracuse (c. 287 to c. 212 BCE)

Up to now we have taken walks through several tales involving risk, while mainly keeping mathematics at bay. This will soon change. We will switch from a "walk" to a "hike", just like going from walking in a flat country such as The Netherlands to hiking in the mountains, as one would typically do in Switzerland. However, before setting off on our hike, let us add some extra provision in our backpack. If, while reading the following pages, you wonder occasionally whence the hike in this chapter is leading, rest assured: in Section 7.4 we return to David van Dantzig, whom we met in Chapter 1. We do however take some random turns left and right before arriving there. We hope that you enjoy the stories on the way, they are part of mathematical folklore. So please consider this chapter as evening campfire stories told before we start the next morning on our more serious hikes in Chapters 8 and 9.

7.1 From Pythagoras and Fermat to Wiles

Every four years, mathematicians from all over the world gather at the International Congress of Mathematicians (ICM). The International Mathematical Union (IMU) supports and assists the official mathematical society of the country where the ICM takes place. The first ICM was held from August 9 to August 11, 1897, in Zurich, Switzerland. From Frei and Stammbach (2007), we learn that "[it] was [...] generally considered to be appropriate for the first attempt to be made by a country that, in light of its geographic location, its underlying conditions and its traditions, was ideally suited to fostering international relations. All eyes quickly turned to Switzerland, and in particular to Zurich." The left-hand side of Figure 7.1 shows the original card for the congress within it there are portraits of Daniel Bernoulli (top left), Jacob Bernoulli (top center), Johann Bernoulli (top right), Leonhard Euler (middle left), and the Swiss mathematician Jakob Steiner (1796–1863, middle right). The vignette at the bottom shows the central part of the western façade of the Semper building of the Eidgenössisches Polytechnikum (ETH Zurich). Since 1900 the ICM has followed a four-year pattern. After World War I, it took a while before large international conferences could take place in an atmosphere of full acceptance of different national delegations. This led once more to Zurich for the ICM of 1932. We again quote from Frei and Stammbach (2007): "[...] a great deal of emphasis was placed on selecting a location for the 1932 Congress that could offer a guarantee of a completely neutral treatment of the various countries, to ensure a complete international attendance." During an ICM, several important prizes are awarded, including the highly prestigious Fields Medals, named after Canadian mathematician John Charles Fields (1863–1932): see the right-hand side of Figure 7.1. At every ICM, there are at least

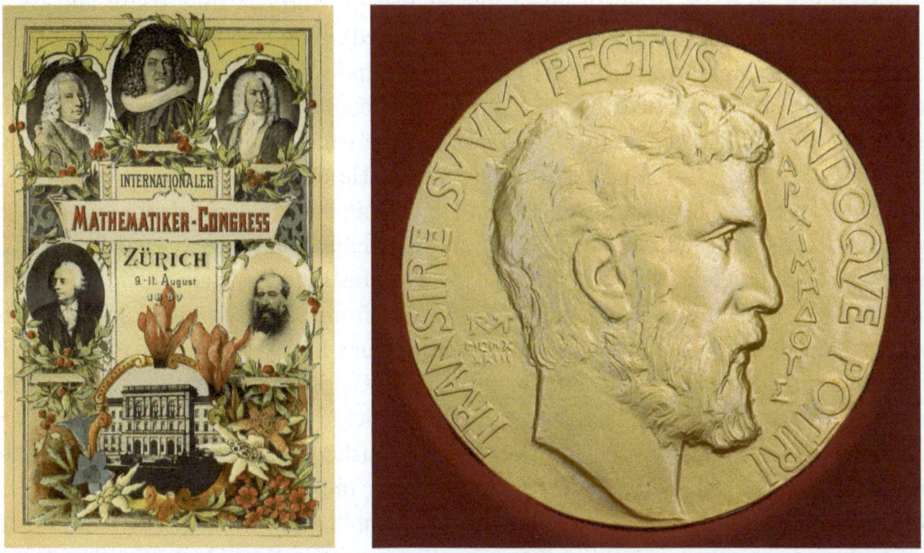

Figure 7.1 Card for the first International Congress of Mathematicians (left) and Fields Medal (right). Source: Wikimedia Commmons

two Fields Medals given, preferably four, to mathematicians whose 40th birthday did not take place before January 1 of the year of the Congress at which the Fields Medals are to be awarded. In distinction to the Nobel Prize, Fields Medals aim at honoring young talents. The first Fields medals were given on the occasion of the 1936 ICM in Oslo, to the Finnish mathematician Lars Ahlfors (1907–1996) and Jesse Douglas (1897–1965) from the USA.

Another prestigious mathematical award is the Abel Prize. Typically, but not exclusively, it covers lifetime achievements within mathematics. This yearly prize was initiated in 2003 by the Government of Norway as an equivalent to the Nobel Prize, which does not exist for mathematics. The Abel Prize is usually given by the King of Norway; it was established in memory of Niels Henrik Abel (1802–1829), the brilliant Norwegian mathematician who most unfortunately died very young, aged 26, of tuberculosis. His work had a fundamental and lasting influence on mathematics in general.

Our first campfire story concerns a theorem that you know very well from your school days, *Pythagoras' theorem*, stating that in a right-angled triangle with sides a and b and hypotenuse c, $a^2 + b^2 = c^2$. Numerous versions and proofs of Pythagoras' theorem exist, one of them even by a former, the 20th, president of the USA, James A. Garfield (1831–1881). It is also worth mentioning that the existence of so-called Pythagorean numbers, such as $(3, 4, 5)$, were discovered well before Pythagoras. Indeed, in the so-called Text BM 85196 from the Babylonian time of Hammurabi (c. 1700 BCE) one finds the start of a theory as well as some decidedly non-trivial examples of Pythagorean numbers, such as $(3367, 3456, 4825)$

and (12 709, 13 500, 18 541), as you can check. This historic discovery led to the so-called Babylonian formulas; see Hoehn and Huber (2005).

A very natural mathematical question coming out of Pythagoras' theorem is whether one can find (strictly) positive integer numbers a, b, c such that for instance $a^3 + b^3 = c^3$, or more generally $a^n + b^n = c^n$ for a given integer $n > 3$. You can now pause and try... and try... and try.... You will surely (it is to be hoped) give up after a while, not being able to find any example a, b, c for any $n > 2$. The famous French mathematician Pierre de Fermat (1607–1665) claimed in one of his papers, published posthumously (hence the name "Fermat's Last Theorem" (FLT)) to have found a truly beautiful proof of the above fact that indeed no such solutions exist for any $n > 2$. He did not publish a proof, merely stating that the margin of his paper was too small to contain the details. Fermat's "proof" was never found. Ever since (hence for over 350 years) mathematicians have tried to prove FLT. Our story picks up at a conference at the Isaac Newton Institute for Mathematical Sciences of the University of Cambridge. On June 23, 1993, in front of a spellbound audience, the British mathematician Andrew Wiles gave a final lecture in a series of three blackboard talks with the title "Modular forms, elliptic curves, and Galois representations". Notice that there is no mention of Fermat here, but specialists quickly realized that "something was afoot". About 20 minutes into this third lecture he finished by simply scribbling on the board "⇒ FLT" ("⇒" in mathematics stands for "this implies" or "it follows that") and almost apologetically added the by now famous statement "I think I'll stop here"; see Figure 7.2 for a photograph taken exactly at that historic moment in time. Instantly, thunderous

Figure 7.2 Andrew J. Wiles announcing the solution to Fermat's Last Theorem at the Isaac Newton Institute for Mathematical Sciences in Cambridge (UK) on June 23, 1993. Source: Isaac Newton Institute

applause greeted the announcement. From here, however, the story takes a rather dramatic turn. As is common in mathematics, every new discovery is checked and peer reviewed in all detail by experts. Soon after the announcement of the lengthy and very difficult proof, in August 1993, Nick Katz from Princeton University found a serious gap in one of the arguments. Needless to say, this was a massive blow to Wiles. With the help of his former doctoral student Richard Taylor, he immediately started to try to fill the gap. The world of mathematics was little aware of progress made, if any, towards a possible success of this daunting task.

For the next stage of the FLT drama, we find ourselves at the 1994 ICM, which took place in Zurich during August 3–11. At the Congress, Andrew Wiles was scheduled as the Closing Plenary Speaker with as title for his lecture "Modular forms, elliptic curves and Fermat's Last Theorem". During the conference, speculations ran high: "did he fill the gap?" By then the discovered gap was almost a year old. In his ICM talk, Wiles was not able to offer a solution to FLT. However, shortly afterwards, on September 19, 1994, he announced that he had managed to complete the proof. The proof was carefully checked by reviewers and concluded to be correct. The immense tension and final relief at the time are best summarized in Wiles' own words. We quote from Brown (2015):

> By September 1994 [hence shortly after the ICM in Zurich], they still hadn't closed the gap. On the verge of admitting defeat, Wiles took "one last look" at his original approach to try to see precisely where he had gone wrong. In the BBC documentary "The Proof", he describes what happened next. "Suddenly, totally unexpectedly, I had this incredible revelation." There, "out of the ashes" of his failed technique, appeared the very tools he needed to prove another conjecture. He called it "my Iwasawa theory", an approach he had abandoned three years earlier. Now he could use that theory to prove Fermat's Last Theorem once and for all. "It was so indescribably beautiful; it was so simple and so elegant, and I just stared in disbelief."

The proof of FLT appeared in 1995 in two scientific papers, one written by Wiles and one written together with Taylor. The printed manuscript added up to about 130 pages of highly intricate mathematics. It certainly was a proof that would not have fitted in the margin of a page. The famous mathematician John Conway (1937–2020) called it "the proof of the 20th century". As so often in science, in his ultimate achievement, Andrew Wiles also "stood on the shoulders of giants": mathematicians like Gerhard Frey, Jean-Pierre Serre, Ken Ribet and many more. Afterwards, numerous honors were bestowed on Andrew Wiles. Among others, he was knighted by the Queen in 2000 and was given the Abel Prize for 2016. If you are interested in more details of the above story, we can highly recommend the 1996 BBC Horizon documentary "Fermat's Last Theorem" as well as the book by Singh (1997). Two questions that are often asked by non-mathematicians in connection with FLT are: (Q1) whether FLT is now finally proven; and (Q2) whether Fermat himself could have had a simple though ingenious proof? Concerning (Q1) the mathematical expert community is absolutely satisfied with the proof; it has been checked in all possible detail. It is correct as it stands. Concerning (Q2), of course, one will never really know but the chances are almost nil that Pierre de Fermat, with the mathematics available at the time, could have come up with a proof; it is very much a twentieth-century proof.

An excellent talk by and interview with Andrew Wiles can be found on the blog of the Science Museum in London, November 17, 2017. That blog `blog.sciencemuseum.org.uk` contains several educational presentations on science and society.

7.2 Probability Becomes a Mathematical Discipline

We very much hope that you have enjoyed this short excursion into the realm of mathematics. In Chapter 8 we will introduce the basic mathematical language underlying an important branch of mathematics, namely probability theory. In line with the theme of the current chapter, we will make a short detour and tell you some campfire stories about probability theory. As promised, we will meet David van Dantzig and the Dutch dikes again later in that chapter.

From antiquity, probabilistic thinking has been prevalent, as in games of chance, through throwing dice, say (the oldest known dice are estimated to be from 2800 to 2500 BCE). A much-needed mathematical development of the field eluded researchers for a long time. A problem is that with questions about chance, our intuition does play a role. As we will discuss in Section 8.1, our intuition unfortunately rather often gets it wrong. There clearly is a need for scientific guidance when it comes to questions of chance in particular and of risk more broadly. An important early discourse on chance took place between Pierre de Fermat, indeed the same Fermat as we met before, and Blaise Pascal (1623–1662). The topic concerned gambling questions such as how to divide stakes in a fair game of chance when that game is stopped prematurely. Several authors contributed early on to this emerging field of science. The Swiss mathematician Jacob Bernoulli (1654–1705) wrote a truly fundamental treatise on the topic, called the *Ars Conjectandi* (the Art of Conjecture), see Bernoulli (1713) and Figure 7.3, published posthumously in 1713 by his nephew Nicolaus I. Bernoulli (1687–1759). In

Figure 7.3 Left: *Ars Conjectandi* (1713). Center: Commemorative stamp for the 1994 ICM in Zurich depicting Jacob Bernoulli and the law of large numbers. Right: The epitaph on the tombstone of Jacob Bernoulli at the Basel Minster in Basel, Switzerland. Sources: Wikimedia Commons (left), authors (center) and Wikimedia Commons, Wladyslaw Sojka (`www.sojka.photo`) (right)

this work, Jacob Bernoulli proved his famous law of large numbers (LLN). The LLN gives a mathematical version of the much-observed fact that, by averaging n independent repeats of the same experiment, the average converges for n large to some number describing the mean value of an individual experiment. As we lack the mathematical notation for probabilities at this stage of the book, we have to be vague in the formulation of the LLN. However, all of us have experienced the LLN in its easiest form when one tosses a fair coin repeatedly and observes that the relative frequency of heads, say, tends to $1/2$. Because of the importance of Jacob Bernoulli and the LLN, the Swiss Post Office commemorated the 1994 ICM in Zurich with a stamp. It may be somewhat amusing to note that the artistic impression of the LLN on the stamp is rather different from a real or simulated series of coin tosses.

Jacob Bernoulli contributed to physics and mathematics well beyond the *Ars Conjectandi*. An interesting story concerns the epitaph on his tombstone in the Basel Minster containing his famous motto "Eadem mutata resurgo" ("Although changed, I rise again the same"). This motto reflects Bernoulli's appreciation of the beauty and importance of the logarithmic spiral, the kind of spiral one often encounters in nature, for example, in shells and galaxies. He referred to this curve as *spira mirabilis*, the marvelous spiral. Unfortunately, the artist

Figure 7.4 Two versions of Bernoulli's logarithmic spiral. Left: In the nautilus shell. Right: In the logo of the Bernoulli Society. Sources: Wikipedia (left) and the Bernoulli Society (right)

chiseled the much less exciting Archimedean spiral on his tombstone. Moving from the tombstone to the postage stamp, history appears to repeat itself, exactly as in Bernoulli's words! The logarithmic spiral that Bernoulli wanted for his tombstone can be seen twice in Figure 7.4: once in the growth of shells and once in the logo of the Bernoulli Society for Mathematical Statistics and Probability, which was established in 1975 as a section of the International Statistical Institute.

The famous Swiss mathematician Leonhard Euler (1707–1783) was a student of Johann Bernoulli, a brother of Jacob. Towards the end of this chapter we will explain how Euler's name is forever linked to the most beautiful equation in mathematics. In Chapter 4, we have already encountered Daniel, a son of Johann. As we explained in Section 4.2, Daniel's contributions to the study of incompressible fluids are crucial to the understanding of tsunami risk. The Bernoulli family links go even further, including the physicists Pierre and Marie Curie, as well as Hermann Hesse, a famous German-born Swiss poet, novelist, and painter; see Figure 7.5. The Bernoullis were well-known merchants in the city of Antwerp, which in the early sixteenth century had become one of, if not *the*, cultural and trading center of the western world. The upcoming Protestantism, through followers of Martin Luther and especially John Calvin, clashed with enshrined Catholicism under the Spanish rule of the Low Countries – that is, Flanders and The Netherlands. This tension exploded on August 20, 1566 with the start of the first "Beeldenstorm" (storm of the statues). The following days, churches were pillaged during an iconoclastic attack on the outer manifestations of Catholicism. These destructions of church interiors were particularly severe in Antwerp.

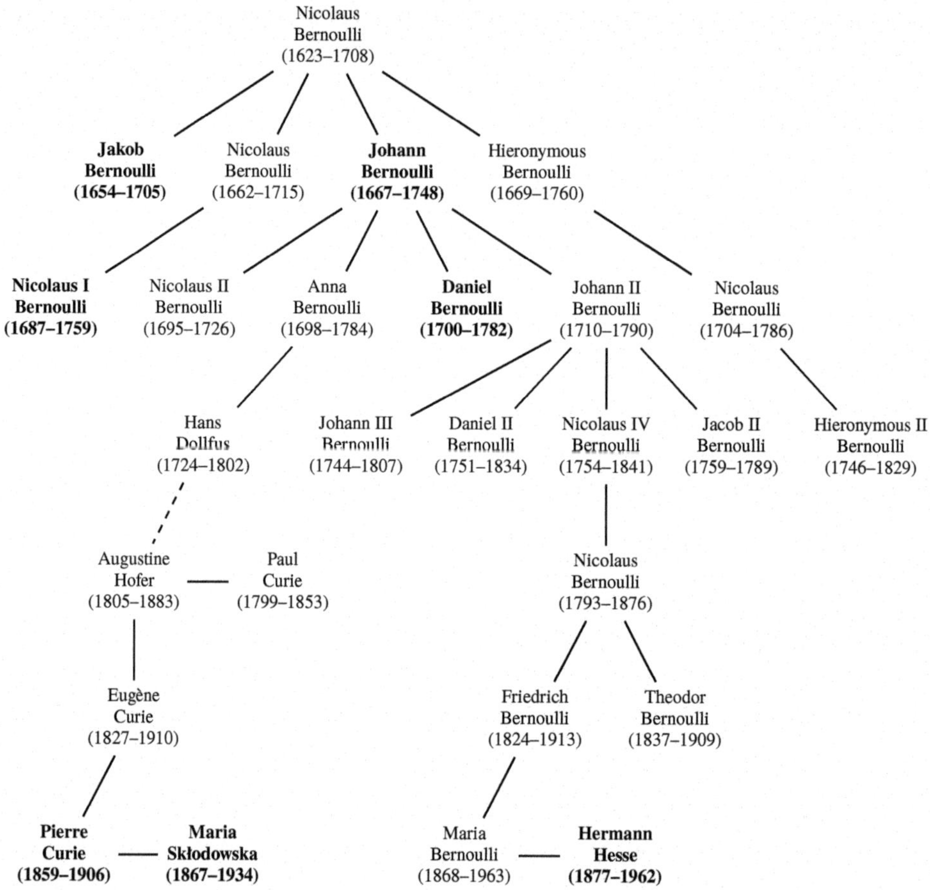

Figure 7.5 The Bernoulli genealogy. We have highlighted Jakob, Johann (doctoral advisor of Leonhard Euler), Nicolaus I and Daniel Bernoulli. We also highlighted three names that you may not have immediately expected in the Bernoulli family tree: the physicists Pierre and Marie Curie (née Maria Salomea Skłodowska), as well as the German-Swiss writer Hermann Hesse. Source: Authors

A violent repercussion from King Philip II of Spain, through his regent, the Count of Alba, became known as The Spanish Fury (1572–1579). The Bernoullis left Antwerp not long after that tumultuous period. From 1620, and with a detour to Frankfurt am Main, in Germany, the relevant, for us, branch of the Bernoulli family established itself in Basel. With eight notable academics, the Bernoullis from Basel constitute by far the most successful family dynasty in the realm of mathematics, physics and engineering. Their influence on these fields has been considerable.

A proper mathematical theory of probability came about only well into the twentieth century. For this to happen, first another ICM congress must enter the stage. At the turn of the century, in 1900, an ICM took place in Paris. There, the famous German mathematician David Hilbert gave a visionary plenary lecture summarizing mathematics' most important open problems (which means that they are not yet decided) at the time. This eventually amounted to the famous list of 23 Hilbert problems. Several of these problems are still open today. Hilbert's sixth problem concerned the need for a mathematical theory of probability:

> As to the axioms of the theory of probabilities, it seems to me desirable that their logical investigation should be accompanied by a rigorous and satisfactory development of the method of mean values in mathematical physics, and in particular in the kinetic theory of gases.

In the late nineteenth and early twentieth century, physics, in particular, badly needed a mathematical theory of probability. A full axiomatic solution of this problem, however, had to wait until the work of the brilliant Soviet mathematician Andrei Nikolajewitsch Kolmogorov (1903–1987); see Figure 7.6. It is interesting to note that the original version of this very

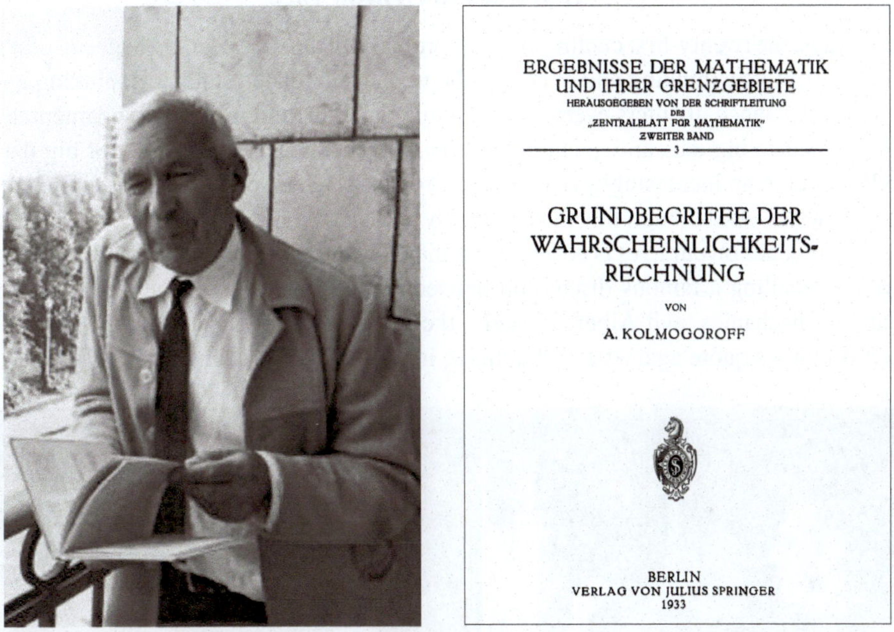

ERGEBNISSE DER MATHEMATIK
UND IHRER GRENZGEBIETE
HERAUSGEGEBEN VON DER SCHRIFTLEITUNG
DES
„ZENTRALBLATT FÜR MATHEMATIK"
ZWEITER BAND
3

GRUNDBEGRIFFE DER
WAHRSCHEINLICHKEITS-
RECHNUNG

VON

A. KOLMOGOROFF

BERLIN
VERLAG VON JULIUS SPRINGER
1933

Figure 7.6 Left: Andrei Nikolajewitsch Kolmogorov (1903–1987). Right: His 1933 book *Grundbegriffe der Wahrscheinlichkeitsrechung*, Kolmogorov (1933). Sources: Wikimedia Commons (left) and Library, Department of Mathematics, ETH Zurich (right)

important book was in German, which at the time was not uncommon in mathematics. Other-language editions appeared in 1936 in Russian and in 1950 in English.

A somewhat amusing story, told to us by Albert N. Shiryaev, a former doctoral student of Kolmogorov and now a famous mathematician in his own right, is worth retelling. If one reads the introduction in the original German 1933 version of Kolmogorov's *Grundbegriffe* and compares it with the wording in the 1936 Russian translation, then one detail jumps out. The German version gives as official date "Ostern [Easter], 1933", whereas the Russian translation has "May 1, 1933". In 1918, in the Soviet Union, May 1 became an important public holiday, known as the Day of the International Solidarity of Workers. For the Gregorian calendar, the latest possible Catholic Easter is April 25 whereas for the Russian Orthodox Church it is May 8. In 1933, however, for both, Easter fell on April 16. In 1932, Russian Orthodox

Easter indeed fell on May 1, but interestingly not in 1933. We leave it to you to draw your own conclusions. April 25, 2003, marked the 100th anniversary of Kolmogorov's birth. For this occasion Glenn Shafer and Vladimir Vovk wrote an excellent historical appreciation of Kolmogorov's book; see Shafer and Vovk (2018). In Chapter 8 we give a gentle introduction to some of the basics coming out of Kolmogorov's theory.

So far we have only very briefly touched upon the history concerning the birth of probability as a proper mathematical discipline. If you would like to delve deeper into the fascinating pre-1900 history of probability and statistics, we recommend Stigler (1990).

7.3 God Does Not Throw Dice

In many ways, the twenty-first century is when probability theory became of age, insofar that, by now, probabilistic thinking and of course the very much related statistical thinking appear in all fields of science. It enters every aspect of our daily life, and in a truly fundamental way. Examples include big data and quantum physics. We encounter the influence of big data all around us, and in an increasingly visible way. Probabilistic reasoning belongs to the core of quantum physics. And though this theory mainly applies to the subatomic level of matter, its pervasive influence on our lives is no less important. One example here is quantum computing. It is worth recalling a famous discussion between Niels Bohr, one of the founding fathers of quantum mechanics, and Albert Einstein, the father of relativity theory; see Figure 7.7. Einstein's famous quote against a probabilistic interpretation of quantum physics was "I, at

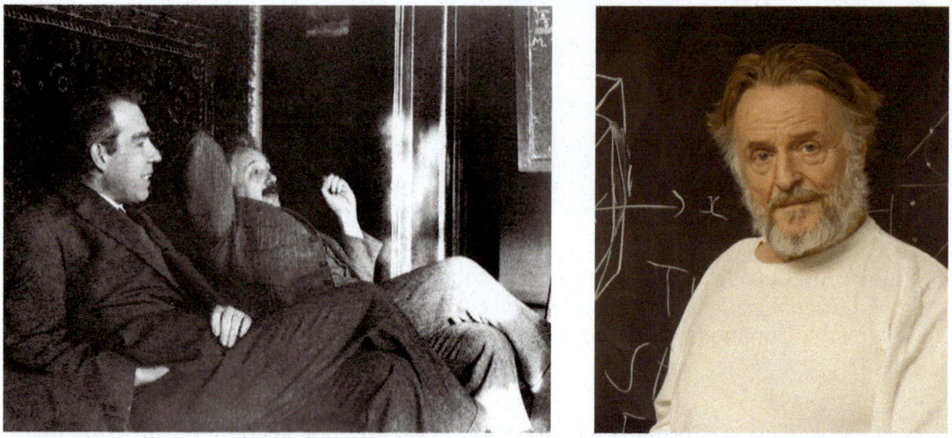

Figure 7.7 Left: The famous physicists Niels Bohr (1885–1962) and Albert Einstein (1879–1955) at Paul Ehrenfest's home in Leiden, 1925. Right: John Horton Conway (1937–2020). Sources: Wikimedia Commons (left) and Princeton University, Denise Applewhite (right)

any rate, am convinced that He [God] does not throw dice" to which Bohr replies "Einstein, stop telling God what to do." Today we know that Niels Bohr has won the argument and that indeed "God throws dice!" (or at least nature does). On the other hand, relativity theory and quantum theory live side by side and are fundamental to our understanding of our physical world and the universe we live in, often in complementary ways.

During the early stages of the coronavirus pandemic, we became more and more aware of people close to us who contracted the virus and passed away as a consequence. On April 11, 2020, the world of mathematics learned the sad news that John Horton Conway had succumbed to the virus at age 82. Our field lost one of its guiding stars. John Conway made profound contributions to wide-ranging topics in mathematics, including the so-called free will theorem (with Simon Kochen) and the surreal number system (a number system consisting of all real numbers including infinite and infinitesimal quantities). The free will theorem has considerable consequences for our understanding of quantum physics. It bears importantly on the above discussion between Albert Einstein and Niels Bohr. In the words of Conway and Kochen (2006):

> If we have free will then the indeterminacy of the particles can't be explained by randomness. Einstein's statement that God plays dice with the universe, well that doesn't matter. Einstein said that because he thought that the opposite of determinism was randomness in some way and it's not. Randomness does not help.

They also formulated it in a somewhat different way:

> Einstein could not bring himself to believe that God plays dice with the world, but perhaps we could reconcile him to the idea that God lets the world run free.

For the general public, however, John Conway will always be remembered for the creation of the *Game of Life* as immortalized by Martin Gardner in his column on recreational mathematics in Gardner (1970a). The Game of Life played its last game for John Horton Conway who once said, "Mathematics is the simple bit. It's the stuff we can understand. It's cats that are complicated."

7.4 The Language of Nature

Now for our last campfire story. We already mentioned before that, for a mathematician, being asked to give a main (one-hour) invited lecture at an ICM meeting is a very high honor. In 1954, the ICM was held in Amsterdam. On Monday, September 6, 10:20–11:20, David van Dantzig gave such a lecture, on "Mathematical problems raised by the flood disaster 1953."

This brings us full circle. We started our discourse with the 1897, 1932 and 1994 ICMs in Zurich, visited the 1900 ICM in Paris, before finally arriving at the 1954 ICM in Amsterdam. We hope that the above stories show how mathematics offers fascinating proof of exceptional human endeavor but also highlights its societal relevance. Over the recent years, several movies have highlighted the lives of mathematicians and their achievements. Examples include John Nash ("A Beautiful Mind"), Srinivasa Ramanujan ("The Man Who Knew Infinity") and Alan Turing ("The Imitation Game"). In subsequent chapters, we will show you how mathematics can fundamentally contribute to a better understanding of risk. In doing so, we may on occasion cause stress to the non-mathematical reader. We will, however, always try hard to return from a more mathematical, methodological, level to real-life examples where we can show how certain techniques can play an important role and lead to a better understanding of the problem at hand. There are two statements on mathematics we would like to take along with us on this hike. First, almost four centuries ago, Galileo Galilei (1564–1642) wrote, "The laws of nature are written by the hand of God in the language of mathematics" or in a somewhat different wording, "Nature's great book is written in mathematical language." The "language of mathematics" in Galileo's quote is highly effective, which we can also learn from the Richard Courant Lecture in Mathematical Sciences delivered at New York University on May 11, 1959 by the physicist Eugene Wigner (1902–1995). Its telling title was "The unreasonable effectiveness of mathematics in the natural sciences." For the published version of this lecture, see Wigner (1960). It is our firm belief that, for handling risk, mathematics offers a highly effective language indeed!

7.5 Epilogue: The Most Beautiful Equation of Mathematics

If one were asked to write down the most important formula of science, then the chances are that out would come Einstein's famous equation $E = mc^2$. It would be interesting to find out how many people worldwide know this formula or indeed have seen it. In its delightfully simple form, it is also extremely beautiful. A mathematical formula considered similarly important and also very beautiful is intimately linked to Leonhard Euler (1707–1783) and known as *Euler's identity*: $e^{i\pi} + 1 = 0$ (Figure 7.8). It combines Euler's number $e = 2.7182\ldots$ (used, for example, in compound interest calculations), the complex number $i = \sqrt{-1}$ (correctly defined), the fundamental circle constant $\pi = 3.1415\ldots$ and the two important integers 0 and 1 in a single formula.

Figure 7.8 Leonhard Euler juggling his famous identity $e^{i\pi} + 1 = 0$. Source: Enrico Chavez

It would be a pity to leave Euler's identity just standing in the above paragraph without paying to it the tribute it deserves as "the most beautiful equation in mathematics." Keith Devlin (2002) writes:

Like a Shakespearean sonnet that captures the very essence of love, or a painting that brings out the beauty of the human form that is far more than just skin deep, Euler's equation reaches down into the very depths of existence. It brings together mental abstractions having their origins in very different aspects of our lives, reminding us once again that things that connect and bind together are ultimately more important, more valuable, and more beautiful than things that separate.

Euler's equation also enjoys a guest appearance in a novel (Ogawa, 2009) about a retired mathematics professor who, after an accident, only has a memory time span of eight minutes.

Another Eulerian story involves Ernst Paul Specker (1920–2011), a professor of mathematics at ETH Zurich. His humorous and sharp personality contributed to the following story. When, on a Sunday, Professor Specker once noticed a young man in the almost empty Main Hall of ETH Zurich, he asked what he was doing there. The student replied that he was a mathematician wanting to look up something in the library. To which Specker asked: "What is i^i?". Receiving the correct answer $e^{-\pi/2}$ he let the student continue.

8

Stochastic Modeling

The great body of physical science, a great deal of the essential fact of financial science, and endless social and political problems are only accessible and only thinkable to those who have had a sound training in mathematical analysis, and the time may not be very remote when it will be understood that for complete initiation as an efficient citizen of one of the new great complex worldwide States that are now developing, it is as necessary to be able to compute, to think in averages and maxima and minima, as it is now to be able to read and write.

Herbert George Wells (1866–1946)

A word that somehow appears implicitly in the above epigraph, from Wells (1903, p. 204) is *stochastics*. It has its origins in the Greek word *stochos*, which can be translated as target, aim, guess. In more modern interpretations, "stochastics" typically corresponds to both probability theory and statistics. In this chapter we concentrate on the very basic principles of probability theory.

We have already alluded in Section 7.4 to a quote from Galileo: see Galileo (1623):

> Philosophy [interpreted as "natural philosophy" here, so the study of nature and the physical universe] is written in this grand book, which stands continually open before our eyes (I say the "Universe"), but cannot be understood without first learning to comprehend the language and know the characters as it is written. It is written in mathematical language, and its characters are triangles, circles and other geometric figures, without which it is impossible to humanly understand a word; without these one is wandering in a dark labyrinth.

To be able to find your way through the labyrinth, we recommend you to slow down your reading speed considerably in order to be able to complete our hike. Stop where you have to and try to understand the concepts presented – after all, we are trying to learn a new language (the "mathematical language" as Galileo wrote) as opposed to reading a novel; if necessary, revisit parts already read (the index can provide you with the relevant pointers). Do not overlook the fact that we just said "try to understand" as opposed to "understand"; the former is indeed more important than the latter. Unless you spend time trying to understand a mathematical concept, the chances of not grasping it are high. This is well known to mathematics students, who need to reinforce, at their own pace, the methodology presented in a mathematical lecture, work through assignment problems and sit exams on the covered material. As solutions to assignment problems are typically shared after the due date, students might be tempted to just read the solutions and expect (faster!) progress this way. This strategy has repeatedly failed for generations of students. If you try to solve an assignment by yourself, the likely scenario is that you will get stuck at some point and maybe try for another day or two to solve the problem (replace that by weeks, months and years when it comes to actual mathematical research problems, sometimes even a lifetime as in the case of Andrew Wiles;

see Section 7.1). Eventually seeing the solution helps you to progress in your study as indeed you painfully remember those hours or days, say, when you tried but could not solve the problem. Even when failing to solve the problem before the due date, the mathematical trick behind its solution, once seen, will stay with you and the proverbial "click" will kick in when seeing a problem of a similar nature later in an exam.

We include the occasional historical example in order to highlight the struggle humankind has encountered, and no doubt continues to encounter, when trying to tame abstract concepts such as luck, risk, randomness, uncertainty, hazard, odds, chance and the like. These concepts often need special attention, depending on the ambient field of application, be it medical, engineering or financial, say.

We add a further word of advice before you start on the more serious hike through this rather long Chapter 8. As a broad guideline, first look at this chapter through its various constituents, summarized below. As with the supporting material for a physical hike, we use the "From here to there" format, breaking up the hike in different parts of varying degree of difficulty. On each leg, do take the occasional pause, look around and enjoy the view of this particular landscape of risk.

From Randomness to Probability Spaces

From Probability Spaces to Random Variables

From Random Variables to Distribution Functions

From Distribution Functions to Summary Statistics

From Quantile Functions to Sampling

From Sampling to Limit Theorems and Beyond

Lessons Learned

With this map in mind, we are now ready to go to the starting point of our hike.

8.1 From Randomness to Probability Spaces

What is the chance of obtaining a 1 or a 6 when rolling a fair die once? What is the chance of getting a mixed outcome when tossing two fair coins? Which of the following two games would you prefer to bet on: obtaining at least one 6 in four rolls of a fair die, or having at least a double 6 in 24 throws of a pair of fair dice? What is the probability of rain tomorrow? What is the chance a 40 year old person will eventually reach the age of 85? Given its historical performance, what is the chance that the price of a particular stock will drop by more than 3% by the end of next week? What is the chance of getting severe COVID-19 symptoms given that you caught the coronavirus? What is the probability that next year's maximal recorded sea level at Neeltje Jans exceeds 3 m above NAP? What is the dike height that we expect the sea level to exceed once every 100 years? What is the probability that the increase in global average temperature can be limited to 1.5 °C above pre-industrial levels by 2100 (see Swain, 2021)?

To be able to answer such questions in a mathematically rigorous way, Kolmogorov (1933) introduced the concept of a probability space. Here and in what follows, when we introduce a mathematical notion we speak of a "concept" rather than a "definition" (the latter would require more mathematical rigor in its treatment). Despite this statement, we really are starting more seriously on our hike. Don't let the initial formalism deter you. As we already stressed above, through examples we will always make sure that we land back on solid ground.

Concept 8.1 (Probability space). A *probability space* is a triple $(\Omega, \mathcal{F}, \mathbb{P})$ (read: "Omega, F, P") which consists of the following three components:

(1) A *sample space* Ω, that is, a set containing all possible outcomes ω of the random phenomenon modeled.
(2) A set $\mathcal{F}$ containing all subsets A of Ω to which a probability can be assigned. Any such A is called an *event*.
(3) A *probability measure* $\mathbb{P}$ which assigns probabilites to events A, so $\mathbb{P}(A)$ denotes the probability of the event A.

As examples of the concepts thus introduced, we often roll fair dice or toss fair coins since these examples are straightforward to imagine, and, indeed, could physically (or by means of simulation on a computer) be replicated to confirm certain statements empirically. From a mathematical perspective, modeling a risk in the form of a loss or gain (each with 50% chance) at the end of a fixed time horizon is indistinguishable from tossing a fair coin. Such connections through abstraction are the real beauty of mathematics and one of the reasons for the widespread applicability and far-reaching consequences of mathematics. More realistic applications to risk management are thus indeed not far around the corner. Note the information "each with 50% chance" is given for the loss and gain of the said risk. In real applications such probabilities are unknown and have to be determined on the basis of assumptions, scenarios, historical data, etc. This, then, is where statistics plays a major role.

Remark (Not dice and fair coins again!). As we will see, probability theory very much grew out of gambling questions. Even today, the latter constitute one of the most relevant contact platforms for you to meet randomness (think, for example, of playing games with your family or of playing the national lottery). We shall discuss these early attempts of trying to tame the elusive concept of chance. Pedagogically it can be questioned whether a better entrance to the world of randomness than gambling questions is to be preferred; certainly there is not one best way, and we are well aware of this. For an alternative entrance to the Secret Garden of Chance, we suggest Gage and Spiegelhalter (2018). The authors make considerable use of spinners (as in the game show "Spin the Wheel") in order to motivate and illustrate new concepts.

Example 8.2 (Rolling a fair die once). Consider the first of the above questions, namely rolling a fair die once and determining the probability of obtaining a 1 or a 6. In this case, the sample space is $\Omega = \{1, 2, \ldots, 6\}$, the set of integers from 1 to 6 and thus indeed the set of all possible outcomes ω when rolling a fair die once. The event of interest here is written as $A = \{1, 6\}$ and $\mathbb{P}(A)$ is the corresponding probability of rolling a 1 or a 6 with the fair die. Since the die is fair, we know that each number ω in Ω (also written as $\omega \in \Omega$) appears with equal probability $1/6$. In two of the six possible outcomes (namely if $\omega = 1$ or $\omega = 6$) the event A happened, so we have $\mathbb{P}(A) = 2/6 = 1/3$. Of course, you could have obtained this result without any formalism, but just wait: more interesting and less trivial examples are just around the corner.

Note that we could have also used a defining property of $\mathbb{P}$ to calculate $\mathbb{P}(A)$, namely that if we can *partition* A, that is, split up A into sets whose elements do not overlap (think of the pieces of a finished jigsaw puzzle), then we can calculate $\mathbb{P}(A)$ as the sum of the individual probabilities. Specifically, since $A = \{1,6\}$ can be split up into the two sets $A_1 = \{1\}$ and $A_2 = \{6\}$, we could have also calculated $\mathbb{P}(A)$ as $\mathbb{P}(A) = \mathbb{P}(A_1) + \mathbb{P}(A_2) = 1/6 + 1/6 = 1/3$. This is a special case of a more general property of $\mathbb{P}$ known as *σ-additivity* (read: "sigma-additivity") and it often allows us to calculate more complicated probabilities from several much simpler ones.

There are two trivial cases of events for which we can also calculate the probabilities. The first is if the event is $A = \{1,2,\ldots,6\}$, so Ω itself. This happens with probability $\mathbb{P}(\Omega) = 1$ since we always obtain a number from 1 to 6 when rolling a fair die. The second event is the empty set $A = \{\} = \emptyset$. It has probability $\mathbb{P}(\emptyset) = 0$ since obtaining no number when rolling a fair die is considered impossible. In the language of probability, events A with $\mathbb{P}(A) = 0$ are called *null sets* and events A with $\mathbb{P}(A) = 1$ are said to happen *almost surely*. So we can also say that a die landing on its edge is a null set since rolling a die almost surely gives us an integer from 1 to 6.

So far you may look at the above discussion as rather mathematically pedantic and somewhat over the top. We agree, for the example of rolling a die. But as we briefly mentioned, the same formalism will allow us to treat the most complicated problems involving randomness.

In Example 8.2, we set up the sample space, defined the event of interest A and computed its probability. We skipped defining the set $\mathcal{F}$ from Concept 8.1(2). Here, it can simply be taken as the set of all possible subsets of Ω. A set with n elements allows for 2^n subsets and so $\mathcal{F}$ has $2^6 = 64$ elements in this case (the empty set $\emptyset$, the six sets $\{1\},\{2\},\ldots,\{6\}$, the fifteen sets $\{1,2\},\{1,3\},\ldots,\{5,6\}$, and so on, until $\{1,2,\ldots,6\} = \Omega$). As is intuitive, we can assign probabilities to any possible event of interest in this example, not just $A = \{1,2\}$, $A = \Omega$ or $A = \emptyset$. This can always be done if the sample space Ω has at most a *countable* number of elements (if you can assign distinct positive integers to all elements of Ω). However, this is in general no longer possible if Ω contains uncountably many elements, for example if Ω is an interval such as $[0,1]$. The problem is that there are too many possible subsets of such sets Ω and one cannot assign probabilities to all of them without creating contradictions to the defining properties of $\mathbb{P}$. The mathematical proof of this result, known as *Vitali's theorem*, named after Giuseppe Vitali (1875–1932), is involved but a more visual example is given by the so-called *Banach–Tarski paradox*, named after Stefan Banach (1892–1945) and Alfred Tarski (1901–1983). A *paradox* refers to a mathematical truth which is difficult to square with intuition; as we will see later, probability theory has several such examples. An informal version of the Banach–Tarski paradox states that given two solid objects such as a small ball and a large ball, one can take either object, cut it into pieces and reassemble it into the other (so that the two have the same volume). At this informal level, the Banach–Tarski paradox is also called *the pea and the Sun paradox*, according to which one can disassemble a pea and reassemble its pieces into the Sun (obviously difficult to square with intuition). Now before you rush to learn the mathematics to understand how this can be done (in order to apply it to a small piece of gold to get a larger one!), do not be disappointed to learn that the proof is not constructive. This means that one can prove the existence of such a decomposition of a pea to be reassembled into the Sun, but the proof does not give you a set of instructions how this can actually be done step by step; if such steps were known it is safe to assume that certain protagonists met in Chapter 3 would be highly interested in them (see Remark 8.3 below). To cut a long story short, to exclude such paradoxes from probability spaces we cannot permit the computation of probabilities $\mathbb{P}$ for all possible subsets of Ω and this is why the set $\mathcal{F}$, containing the subsets to which a probability can be assigned, is an integral part of a probability space. Since the types of $\mathcal{F}$ typically used still contain an unimaginably large number of sets, in particular all sets to which we could ever hope to assign probabilities, we do not consider $\mathcal{F}$ from this point on and mostly focus on working with Ω and $\mathbb{P}$.

Remark 8.3 (Banach–Tarski on Wall Street). Das et al. (2013) contains the following story:

In an interview in 1999, the second author [Paul Embrechts] made the following statement: *Die Finanzwelt ist die einzige Welt, wo die Leute immer noch glauben, dass sich Eisen in Gold verwandeln lässt*; see Embrechts (1999). Translated into English: "The world of finance is the only one in which people still believe in the possibility of turning iron into gold." The above statement was made in the wake of the 1998 LTCM [Long-Term Capital Management] hedge fund crisis; little did we know at the time how true this statement would become about ten years later! [. . .] [Before the 2007–2008 financial crisis], asset-backed securities, like CDOs, were in the popular press often likened to magical financial engineering tools allowing [one] to cut a pizza into several pieces, reassemble them and end up with two pizzas each in size equal to the one we started from. In the language of modern finance, the second pizza would be referred to as "a free lunch". This is the point where Banach–Tarski [as mentioned above] enters; the lesson to be learned is to always understand in detail the conclusions of a mathematical theorem [a *theorem* is a major mathematical result].

8.1.1 Chevalier de Méré Paradox

Antoine Gombaud (1607–1684), a writer and professional gambler, became known for the so-called *Chevalier de Méré paradox*. He considered the following two games.

Game 1: Roll a fair die four times and note whether *at least one* 6 occurs.
Game 2: Roll two fair dice 24 times and note whether *at least one* double 6 occurs.

Gombaud's intuition told him that both probabilities should be equal. His (fallacious) reasoning is as follows. For a die as in Game 1, the probability of a 6 in one roll is, of course, $1/6$ (correct). For a pair of dice as in Game 2, the probability of a double 6 in one roll is $(1/6)\times(1/6) = 1/36$, which is $1/6$ of the probability of throwing a 6 with a single die (correct).

There is an important concept underlying our calculation here. Because the outcomes of the two throws do not depend on each other, we can multiply the corresponding probabilities (the two probabilities $1/6$). More formally, two events A, B are *independent* if $\mathbb{P}(A \cap B) = \mathbb{P}(A)\mathbb{P}(B)$ (read: "the probability of A and B both happening is the product of the probability of A happening and the probability of B happening") which means that the probability of both A and B happening is the product of the two individual probabilities. So far, so good (and, all correct).

Now where does the number 24 in Game 2 come from? Gombaud's intuitive reasoning is as follows. The probability of success of a single roll of two dice in Game 2 ("one double 6 occurs") is one sixth of the probability of success for a single roll of one die in Game 1 ("one 6 occurs"); thus the number of rolls needs to be six times as large for Game 2 than for Game 1, hence $4 \times 6 = 24$ times. However, after having played Game 2 sufficiently often and betting on that 'knowledge', he was consistently losing. After convincing himself that the odds for Games 1 and 2 are different, he had difficulty squaring this knowledge with his professional gambler's intuition. For the solution to his predicament, he turned to the famous

French mathematician, writer, physicist and philosopher Blaise Pascal (1623–1662). Pascal in turn started a conversation with Pierre de Fermat (1607–1665) and the exchange of their letters is commonly viewed as the birth of probability theory.

Let us now calculate the two probabilities properly. We follow the mathematician's (and indeed Pascal's and Fermat's) mind and abstract the two games to simplify the calculation of their probabilities.

Consider Game 1. The basic underlying event in this game is the event A_1 of rolling a 6 in a single throw. Denote its probability by $\mathbb{P}(A_1)$. The event *complementary* to A_1 is A_1^c (read: "*A* one complement" or "not *A* one") and it stands for rolling "not a 6". Since we either roll a 6 or not a 6 in a single throw, we must have $\mathbb{P}(A_1) + \mathbb{P}(A_1^c) = 1$ and so $\mathbb{P}(A_1^c) = 1 - \mathbb{P}(A_1)$. We thus have a rule for computing the probability of complementary events which can make a probabilist's life significantly easier. As a direct application, let us now turn to the four throws in Game 1. The probability of obtaining at least one 6 is the probability of obtaining exactly one 6 after four throws, plus the probability of obtaining exactly two 6s after four throws, plus the probability of obtaining exactly three 6s after four throws, plus the probability of obtaining four 6s after four throws. This seems rather complicated! In fact the correct probability can be calculated most easily by using the idea of complementary events. So, we consider the complementary event "no 6 occurs in four throws" instead. Going for the complement should become a Pavlov (that is, an immediate, automatic) reaction in such situations. Now we use the fact that the four throws happen independently of each other, which means we can get the probability of "no 6 occurs in four throws" by multiplying the individual probabilities from each throw, so $\mathbb{P}(\text{"no 6 occurs in four throws"}) = \mathbb{P}(A_1^c) \times \mathbb{P}(A_1^c) \times \mathbb{P}(A_1^c) \times \mathbb{P}(A_1^c) = \mathbb{P}(A_1^c)^4 = (1 - \mathbb{P}(A_1))^4$. The probability of our event of interest ("at least one 6 in four throws") is thus $1 - \mathbb{P}(\text{"no 6 in four throws"})$ and so we obtain

$$\begin{aligned}\mathbb{P}(\text{"at least one 6 in four throws"}) &= 1 - \mathbb{P}(\text{"no 6 in four throws"})\\ &= 1 - (1 - \mathbb{P}(A_1))^4. \end{aligned} \tag{8.1}$$

Note that we have derived a formula for the probability of the event of interest in Game 1, but we have not yet used actual numbers. The use of numbers was indeed not necessary in order to derive (8.1), and such types of formulas are what mathematicians are interested in much more than the actual number since a formula reveals the structure of a solution instead of just its numerical value. We can now indeed see from (8.1) how repeated throws (here: four) enter the probability, namely in the form of a power (here: the fourth power) rather than a multiplication as Gombaud thought. Of course, the numerical solution is easily obtained from (8.1), too. By using $\mathbb{P}(A_1) = 1/6$, we obtain $1 - (1 - 1/6)^4 \approx 0.5177$ for the probability of the event described in Game 1, whereas Gombaud thought the answer was $(1 - (1 - 1/6)) \times 4 = 2/3 \approx 0.6667$.

How about Game 2? The fact that we derived the probability of the event in Game 1 in abstract terms now again pays out since we can follow the same strategy. The basic underlying event in Game 2 is the event A_2 of rolling a double 6 in a single throw, so $\mathbb{P}(A_2) = 1/36$. The probability of "no double 6" is thus $1 - \mathbb{P}(A_2)$ and the probability of "no double 6 in 24 throws" is then $(1 - \mathbb{P}(A_2))^{24}$, so the probability of the event of interest is

$$\begin{aligned}\mathbb{P}(\text{"at least one double 6 in 24 throws"}) &= 1 - \mathbb{P}(\text{"no double 6 in 24 throws"})\\ &= 1 - (1 - \mathbb{P}(A_2))^{24}.\end{aligned}$$

We therefore obtain that the probability of the event in Game 2 is $1 - (1 - 1/36)^{24} \approx 0.4914$, whereas Gombaud thought this probability was $(1 - (1 - 1/36)) \times 24 = 2/3 \approx 0.6667$. The correct probabilities for Games 1 and 2 are quite close together, offering an acknowledgement of Gombaud's intuition as a gambler, but they are indeed mathematically different. Such discrepancies make all the difference in the long run when gambling for real!

The formulas for the probabilities of the two games are conceptually identical (in the way they are logically derived; we will meet the same formula again in Section 9.5.2), and so the trained eye would have spotted the solution to Game 2 from the solution to Game 1 (thus we see the advantage of a formula revealing the structure of a solution, as mentioned earlier). The word "trained" is important here; this is why mathematics students spend hours, days, months and years with such (and, of course, much harder) problems. Only then can the necessary creativity to spot similarities and connections between problems flourish, an aspect perhaps more common to arts than to science. Similar to many other subjects of intense study, this can have "side effects"; for example, it can lead to mathematical formulas appearing in dreams when the brain subconsciously continues to work on a problem or learns from the exposure to it. Although there are stories known about scientific discoveries during dreams (for example, the discovery of the ring shape of the benzene molecule by August Kekulé (1829–1896)), mathematical dreams are normally rather blurred and do not lead to clear solutions when waking up. Nevertheless, it is hard to imagine that the brain's work would not at least increase the understanding of a problem and thus lead to higher chances of finding a solution once one is fully awake. A less deep but similar phenomenon is also often experienced by mathematics students after having spent hours on a specific assignment problem in the hope of solving it after investing just "a bit more" time. If one is then forced to put the problem aside (for example, because of an appointment, a sports hobby, etc.), one often rather immediately spots a solution when being reunited with the problem a couple of hours later. Such forced distractions can occasionally actually turn out to be fruitful in the sense of giving creativity enough room to flourish in the meanwhile. An extreme example of this in the case of a professional mathematician was encountered in the story of Andrew Wiles' proof of Fermat's Last Theorem; see Section 7.1. Wiles strongly believes in the importance of the three Bs: bus, bath and bed. Time when your mind is free to wander away from the immediate problem. Surely this holds true well beyond mathematics.

8.1.2 Mathematical Correctness versus Intuition

The story of Jean-Baptiste le Rond d'Alembert, see Figure 8.1, goes in a similar direction, though it took place a century later: mathematical correctness versus intuition in situations where randomness plays a role. In this case the problem at hand involves coin tosses. Tossing a fair coin twice, what is the probability that heads (H) turns up at least once? Denoting tails by T, fairness of the coin means that $\mathbb{P}(H) = \mathbb{P}(T) = 1/2$. D'Alembert reasoned as follows: either H turns up in the first toss; in that case we can stop. But when T turns up in the first toss, we continue and either H or T occurs in the second toss. From these three events (the three ways of obtaining H at least once), two are *favorable* in the sense that H occurs at least once (our event of interest here) hence the required probability is two out of three, so $2/3$. However, this deduction is wrong. The problem is that the three events do not have the same probability of occurring, so determining the probability as the number of favorable events out of the number of all events does not give the correct answer.

There are two ways out. Either we correctly determine the probabilities of each of the three events (this is often the more complicated approach) or we choose our probability space in

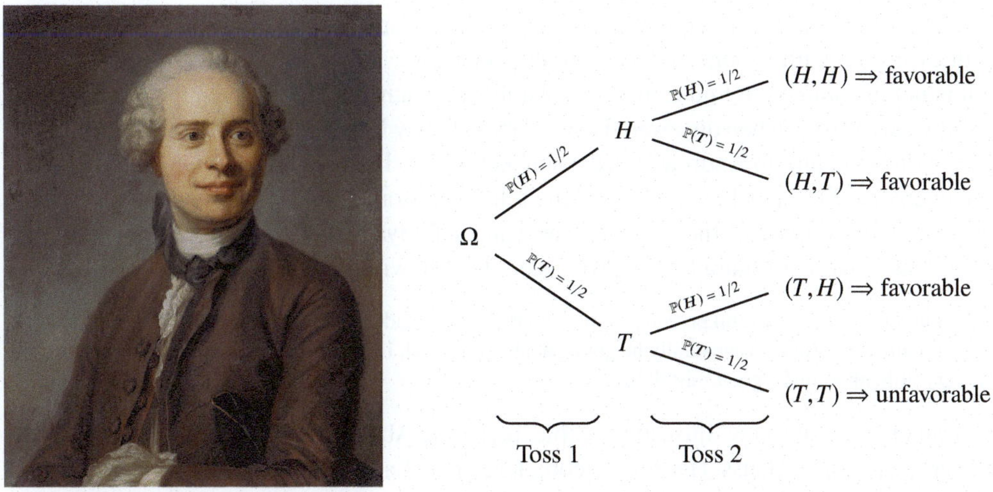

Figure 8.1 Left: Jean-Baptiste le Rond d'Alembert (1717–1783). Right: a tree diagram leading to the correct solution $3/4$ of the "croix ou pile" (translated: "heads or tails") problem to which he gave the wrong answer of $2/3$. Sources: Wikimedia Commons (left) and authors (right).

such a way that the occurring events have equal probabilities. We will do the latter. To this end, we define the sample space as $\Omega = \{(H,H),(H,T),(T,H),(T,T)\}$ containing all possible pairs of outcomes, where (H,T) stands for the event "the first toss is heads, the second one is tails", for example. All four possible outcomes ω in Ω have equal probability $(1/2)(1/2) = 1/4$ (as before we can multiply the probabilities of $1/2$ for two independent tosses to see that), and this is why we can now count the (three) favorable events (H,H), (H,T) and (T,H) as a fraction of all four events to obtain the correct probability $1/4 + 1/4 + 1/4 = 3/4$ and not $2/3$ as given by d'Alembert. d'Alembert's (wrong) approach appeared in the famous encyclopedia article "croix ou pile" ("heads or tails"); see d'Alembert (1754).

Note that solutions to problems of these types are often easiest to spot by utilizing tree diagrams. Figure 8.1 shows an example. The tree starts from Ω as root and then branches out to the right according to the outcome heads (with probability $\mathbb{P}(H) = 1/2$ as indicated on the edge from Ω to H) or tails (with probability $\mathbb{P}(T) = 1/2$ as indicated on the edge from Ω to T) of the first coin toss. Given this outcome, the tree branches out further according to the outcome heads or tails (each with probability $1/2$) of the second coin toss. This leaves us with the four outcomes $\{(H,H),(H,T),(T,H),(T,T)\}$ which make up Ω. Each of these four outcomes has equal probability (of $1/4$) of occurring and three of the four outcomes are favorable, which leads to the probability $3/4$.

Gottfried Wilhelm Leibniz (1646–1716), who, with Isaac Newton (1643–1727), was one of the fathers of integral and differential calculus, is said to have believed that "Par exemple, avec deux dés, il est tout aussi probable de lancer 12 points que de lancer 11; car l'un & l'autre ne peuvent se réaliser que d'une seule manière"; see Leibniz (1768, p. 217), adapted to modern French. Translated: "[...] for example, with two dice, it is equally as likely to throw 12 points as to throw 11; because one and the other can be done in only one manner". In fact, it is twice as likely to throw 11 as to throw 12, as 12 can only be obtained by throwing a double 6 but 11 can be obtained by throwing a 5 with one die and a 6 with the other die or by throwing a 6 with the one and a 5 with the other die.

It is not uncommon to find errors in probabilistic, mathematical or scientific work. Occasionally such errors, when discovered and corrected, may lead to exciting new developments. Once more we return to Andrew Wiles from Section 7.1. In the excellent 1996 BBC Horizon Documentary "Fermat's Last Theorem" the following story is told. One important contributor to the final solution of the problem was Yutaka Taniyama (1927–1958). In his obituary, his colleague Goro Shimura (1930–2019) said the following:

> Taniyama was not a very careful person as a mathematician. He made a lot of mistakes, but he made mistakes in a good direction. And, so eventually he got right answers. And, I tried to imitate him, but I found out that it is very difficult to make good mistakes.

Lecat (1935) lists many illustrious people, including Abel, Cauchy, Cayley, Descartes, Euler, Fermat, Galileo, Gauss, Hermite, Jacobi, Lagrange, Laplace, Legendre, Leibniz, Newton, Poincaré and Sylvester, who have made mistakes. So, when students make mistakes, they may find themselves in the company of giants of the field. The key point is that errors should not remain undetected and should be corrected. Occasionally we do encounter amusing errors, like the one made by the English amateur mathematician William Shanks (1802–1882), who in 1873 claimed to have computed π to 707 decimal places, at the time a highly non-trivial feat. Recall that π as a non-rational (that is, *irrational*) number has an infinite non-repeating decimal expansion. Unfortunately, as much later turned out, an error occurred at the 528th decimal place. You may remember this story when you celebrate pi(e)-day on the next March 14! When written as the date "3/14", we see the beginning "3.14…" of π, and hence eat a real pie on that day. Sometimes the celebrations are more involved, for example, at University of Waterloo, Canada, the event starts at 1:59 pm (resembling "3.14159…", so more digits of π). As Waterloo has three terms per year and not all mathematics students are on campus in the first term in the year, there are also celebrations on July 22 (as $22/7 \approx 3.14$; July 22 falls in the second term of the year) and on the 314th day of the year (November 10, or November 9 on a leap year; this date falls in the third term of the year). July 22 is known as "pi Approximation Day" on which students eat cake instead of pie.

For a discussion on early historical probabilistic faux pas we refer to Gorroochurn (2011). Here we copy the conclusion of this article as it fully captures our aims in presenting the very basics of probability theory:

> We have outlined some of the more well-known errors that were made during the early development of the theory of probability. The solution to the problems we considered would seem quite elementary nowadays. It must be borne in mind, however, that in the times of those considered here and even afterwards, notions about probability, sample spaces, etc. were quite abstruse. It took a while before the proper notion of a mathematical model was developed, and a proper axiomatic model of probability was not developed until Kolmogorov (1933). Perhaps, then, the personalities and their errors discussed in this article should not be judged too harshly.

Further support for the above statement is found in the following quote from a historian of science, Stephen Jay Gould (1992): "Tversky and Kahneman argue, correctly I think, that our minds are not built (for whatever reason) to work by the rules of probability." Amos Tversky (1937–1996) and Daniel Kahneman were recipients of the 2002 Nobel Memorial Prize in Economic Sciences "for having integrated insights from psychological research into economic science, especially concerning human judgment and decision-making under uncertainty". Also, as the English poet Alexander Pope (1688–1744) wrote in 1711: "To err is human [. . .]". What is important is to fix the errors and learn from them. There are many quotes in this regard, for example, "Irrend lernt man" (translated: "By erring do we learn") from Johann Wolfgang von Goethe (1749–1832). A quote by former First Lady of the United States Anne Eleanor Roosevelt (1884–1962) applies very much to risk management, and we hope what our book contributes to this: "Learn from the mistakes of others. You can't live long enough to make them all yourself."

8.1.3 The Problem of Points

There is no doubt that the most important problem coming out of Gombaud's pen entered history as *the problem of points*. Its solution again involved Blaise Pascal and Pierre de Fermat. The original version goes as follows:

> Two players A and B play a fair game such that the player who is the first to win a total of six [independent] rounds wins a prize. Suppose the game unexpectedly stops when A has won a total of five rounds and B has won a total of three rounds. How should the prize be divided between A and B?

You may consult Gorroochurn (2011) for the full historical details on the various erroneous answers in existence at the time, such as 5:3 (by Fra Luca Pacioli (1447–1517), the father of double-entry bookkeeping) and 6:1 (by Gerolamo Cardano (1501–1576) whom you may remember from your school days for his general solution of cubic equations) for a prize division between A and B. Both Pacioli and Cardano were eminent Renaissance mathematicians; the Tversky–Kahneman quote from Stephen Jay Gould, mentioned above, resonates once more. The correct solution, 7:1, forms part of the celebrated exchange of letters between Pascal and Fermat in the mid seventeenth century.

Let us write A_i, respectively B_i, for the event that player A, respectively player B, wins in round i. The information available at the time of stopping is that A won five and B won three rounds. The order in which this happened is irrelevant for the problem, which is why we can assume for the sake of argument that $A_1, A_2, A_3, A_4, A_5, B_6, B_7, B_8$ happened, so that A won the first five and B the last three rounds. To determine the division of the prize between A and B we compare the probabilities of A and of B winning if the game were continued to the end. Player B can only win when (s)he wins three times in a row (as another win by A will terminate the game, so now the order is relevant), hence we must have B_9, B_{10}, B_{11}. As the rounds are played independently of each other, the probability for this to happen, as we learned, is obtained by taking the third power, so it is given by $(1/2)^3 = 1/8$. We thus have $\mathbb{P}(\text{"B wins"}) = 1/8$ and hence $\mathbb{P}(\text{"A wins"}) = 1 - \mathbb{P}(\text{"B wins"}) = 7/8$. Therefore, the prize should be divided according to the odds 7:1 (not 5:3 or 6:1).

Historically, the problem in the above analysis was that the sample space used was $\Omega = \{A_9, B_9 \cap A_{10}, B_9 \cap B_{10} \cap A_{11}, B_9 \cap B_{10} \cap B_{11}\}$; recall that the intersection symbol $\cap$ stands for "and". As already mentioned for d'Alembert's coin toss problem, this is not a problem *per se* as long as one assigns the right probabilities to each of the four continuations of the game. These cannot be equal though, since, for example, the chance of A_9 occurring is larger than that of $B_9 \cap A_{10}$ (both B_9 and A_{10} taking place). And, as said before, it is often easier to define the sample space in such a way that all its elements (game outcomes, that is, combined round outcomes) have the same probability, as we can then obtain the probability of a game outcome by simply counting the favorable outcomes as a fraction of all possible outcomes. To this end, we imagine continuing the game until 11 rounds have been played (the least number of rounds needed for either A or B to be able to win) and only then checking who has won. This leads to the sample space

$$\Omega = \{A_9 \cap A_{10} \cap A_{11}, A_9 \cap A_{10} \cap B_{11}, A_9 \cap B_{10} \cap A_{11}, A_9 \cap B_{10} \cap B_{11},$$
$$B_9 \cap A_{10} \cap A_{11}, B_9 \cap A_{10} \cap B_{11}, B_9 \cap B_{10} \cap A_{11}, B_9 \cap B_{10} \cap B_{11}\},$$

where each element now has the same probability of happening. If any of the first seven elements of Ω happens, player A turns out to be the winner; B only wins in the last event, so we again obtain the odds 7:1. It is not difficult to imagine that such kind of calculations become increasingly involved when the number of players increases. This is exactly where Pascal's genius entered. You may recall from your schooldays Pascal's triangle, which is not far around the corner here. As with d'Alembert's coin tosses, we can utilize a tree diagram to solve the problem of points. In Figure 8.2 we see on the right-hand side who wins the game if it were continued until 11 rounds have been played. We (again) see that only in one out of eight possible outcomes B wins, so we (again) obtain the odds 7:1.

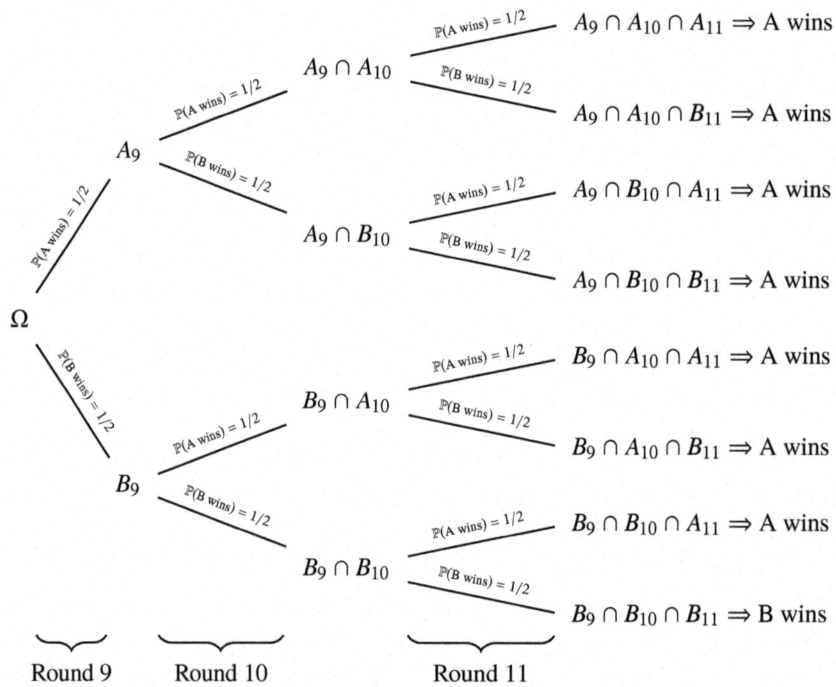

Figure 8.2 Tree diagram showing the idea of the (correct) solution to The Problem of Points. Along all eight paths of possible outcomes in rounds 9, 10 and 11 (all with equal probability), player B only wins the game overall (as indicated on the very right) if (s)he wins all three rounds. Source: Authors

Remark (Jacob Bernoulli, choosing a new Pope, Swiss federal judges). It is interesting that Jacob Bernoulli, trained as a theologian, was ordained as a minister of the Reformed church at Basel in 1676. This may perhaps explain why, in the first lines of Bernoulli (1713), he philosophizes on the objective versus subjective nature of the existence of past/present/future events. Whereas he states that objectivity is clear for humans concerning past and present events, for future events this remains only so for the Creator, God. As such, working out a theory of probability is a step into the direction of learning about the thoughts of God. Jacob Bernoulli apologizes for his progress on this theory and distances himself from its interpretation. He leaves those kinds of religious interpretations to others to be argued about; nevertheless, he does mention the issue. If you are interested in these early views on the contrast between a theory of chance and Divine Providence you can consult Schneider (1984); in particular, see p. 86 of the latter publication. The paper also contains an interesting historical discussion on the history of probabilistic thinking, with a special view on the discussions between Leibniz and Bernoulli. We will briefly revisit this historical discourse when we introduce you to Johan de Witt in Chapter 13.

This brief discussion highlights the fact that, throughout the ages, chance events were often considered as predetermined by God, their outcomes indeed being decided by that Creator. An interesting modern-day example where divinity plays a crucial role in chance outcomes is to be found in the election of the pope of the Coptic Orthodox Church of Alexandria in Egypt. First, an elaborate process leads to a proposal of three candidates, whose names are written on separate paper scrolls; each one is placed in a ball, these are put together in a bowl. A blindfolded young boy, chosen from the congregation, then draws from the bowl a ball with the name of the new pope. As such, God has made the determining choice of the new Patriarch, not chance, and surely not humans (beyond suggesting the three candidate names). The last time this election took place was with the choice of Tawadros II in 2012.

Switzerland's direct democracy includes the right of the electorate to have popular votes through referendums and initiatives. An initiative allows citizens to propose changes to the Swiss Federal Constitution. On November 28, 2021, the so-called Judge Initiative was voted on. The initiative aimed to introduce a new election process for federal judges. Rather than being chosen by Parliament, in a first step a group of eligible candidates would be proposed by a committee of experts. In a second step, the successful candidate would be selected from this group of candidates by lottery, very much in line with the above example of the Coptic Orthodox Church. The vote on this initiative turned out to be negative, with a 68.1% no vote. So rolling dice for choosing Swiss federal judges is not yet allowed.

8.1.4 Conditional Probabilities

The problem of points is inherently a conditional one. For example, whether player B can still win in the tenth round depends on the outcome of the ninth round. If A has won the ninth round (so A_9 has occurred), then B can no longer win overall, whereas if B has won the ninth round (so B_9 has occurred), then B can still win overall ((s)he just needs to win the two remaining rounds, so that $B_{10} \cap B_{11}$). Such situations can be modeled with *conditional probabilities*.

Concept 8.4 (Conditional probability). If $\mathbb{P}(B) > 0$ then

$$\mathbb{P}(A \mid B) = \frac{\mathbb{P}(A \cap B)}{\mathbb{P}(B)} \tag{8.2}$$

(read: "the probability of A given B is the probability of A and B divided by the probability of B") is the *conditional probability of A given B*.

In terms of conditional probabilities, we can express the probability of B winning the game via

$$\begin{aligned}\mathbb{P}(\text{“B wins”}) &= \mathbb{P}(B_9 \cap B_{10} \cap B_{11}) = \mathbb{P}(B_9)\,\frac{\mathbb{P}(B_9 \cap B_{10})}{\mathbb{P}(B_9)}\,\frac{\mathbb{P}(B_9 \cap B_{10} \cap B_{11})}{\mathbb{P}(B_9 \cap B_{10})} \\ &= \mathbb{P}(B_9) \times \mathbb{P}(B_{10} \mid B_9) \times \mathbb{P}(B_{11} \mid B_9 \cap B_{10}) \\ &= \mathbb{P}(\text{“B wins round 9”}) \\ &\quad \times \mathbb{P}(\text{“B wins round 10 given that (s)he won 9”}) \\ &\quad \times \mathbb{P}(\text{“B wins round 11 given that (s)he won 9 and 10”}). \end{aligned} \tag{8.3}$$

As every new round between the players A and B is fair, B wins every new round with probability $1/2$, so the latter three probabilities are all $1/2$ and we (again) obtain $\mathbb{P}(\text{“B wins”}) = (1/2)^3 = 1/8$. The three probabilities in (8.3) (two of which are conditional) can also be found in the tree diagram in Figure 8.2: we could have labeled the edge from Ω to B_9 by $\mathbb{P}(B_9) = 1/2$, the edge from B_9 to $B_9 \cap B_{10}$ by $\mathbb{P}(B_{10} \mid B_9) = 1/2$ and the edge from $B_9 \cap B_{10}$ to $B_9 \cap B_{10} \cap B_{11}$ by $\mathbb{P}(B_{11} \mid B_9 \cap B_{10}) = 1/2$ (and similarly for the remaining edges). We can thus find the calculation (8.3) in the tree diagram simply by multiplying these conditional probabilities along their edges. And this also holds for the other seven possible outcomes of rounds 9 to 11.

Note that every event in the tree in Figure 8.2 is partitioned into all events immediately to the right of it. For example, the event B_9 is partitioned into the two sets $B_9 \cap A_{10}$, $B_9 \cap B_{10}$, and also into the four sets $B_9 \cap A_{10} \cap A_{11}$, $B_9 \cap A_{10} \cap B_{11}$, $B_9 \cap B_{10} \cap A_{11}$, $B_9 \cap B_{10} \cap B_{11}$. As a consequence, we must have

$$\begin{aligned}\mathbb{P}(B_9) &= \mathbb{P}(B_9 \cap A_{10}) + \mathbb{P}(B_9 \cap B_{10}) \\ &= \mathbb{P}(B_9 \cap A_{10} \cap A_{11}) + \mathbb{P}(B_9 \cap A_{10} \cap B_{11}) + \mathbb{P}(B_9 \cap B_{10} \cap A_{11}) + \mathbb{P}(B_9 \cap B_{10} \cap B_{11}).\end{aligned}$$

Such representations of $\mathbb{P}(B_9)$ through a partition of events follow more formally from the so-called law of total probability, which we now state as a *proposition*, a medium type of result, typically more involved than a lemma.

Proposition 8.5 (Law of total probability). *Let $B_1, B_2, \ldots$ be events that partition Ω. Then, for any event A, one has*

$$\mathbb{P}(A) = \sum_{i=1}^{\infty} \mathbb{P}(A \cap B_i); \tag{8.4}$$

this is known as the law of total probability. *Using* (8.2) *for the summands in* (8.4), *we have* $\mathbb{P}(A \cap B_i) = \mathbb{P}(A \mid B_i)\mathbb{P}(B_i)$ *(interpreted as* 0 *if* $\mathbb{P}(B_i) = 0$*) and thus obtain the alternative formulation, in terms of conditional probabilities,*

$$\mathbb{P}(A) = \sum_{i=1}^{\infty} \mathbb{P}(A \mid B_i)\mathbb{P}(B_i) \tag{8.5}$$

of the law of total probability.

Similarly to what was mentioned in Example 8.2, conditional probabilities and the law of total probability allow us to express probabilities of more complicated events in terms of simpler ones and so facilitate the calculation of probabilities, even though the formulas might suggest that we are making a probability such as $\mathbb{P}(A)$ more complicated.

Example 8.6 (Monty Hall problem). Monty Hall (1921–2017) was known as the long-running host and producer of the American television game show "Let's Make a Deal" in which a selected audience member, the *trader*, is given something of medium value and the host offers the trader the opportunity to trade for an unknown prize (which could be worth much more, or much less). In the final part of the show (the *Big Deal*), the highest winner among the traders is asked whether (s)he wants to trade in everything won so far in exchange for the opportunity of winning a larger prize. If the trader chooses this option, (s)he needs to pick one of three numbered doors on the stage, each of which conceals either a prize or something of little value known as the *zonk*. As the Big Deal is behind only one of the doors, one of the doors not picked by the trader can then be opened by the host to reveal a zonk. The host then offers the trader to switch between the door (s)he picked (but which is still unopened) and the remaining unopened door. The question is: should the trader switch?

At first glance, one might think that switching the doors has no impact on the probability of the trader winning the Big Deal. Indeed the prize is behind a fixed door before the game started and the choice of the trader winning the Big Deal should be $1/3$ as (s)he randomly picks among the three equal doors. We leave any psychological aspects of the choices made aside. Obviously the way in which the host interacted with the trader and the effect of being on television might have influenced the trader's decision-making process but that is not something we can easily model.

To fix our ideas, Figure 8.3 shows the three doors. As the problem is symmetric in the doors, we can assume "without loss of generality" (as mathematicians say) that the trader picks the first door. The figure shows the state where the host opens the second door with a zonk behind. The question is then whether the trader should switch from the first to the third door.

As the trader picks the first door, the only two pieces of information to determine the outcome of the game are behind which door is the Big Deal and which door does the host open. If B_i stands for the Big Deal being behind door $i = 1, 2, 3$ and H_j stands for the door $j = 1, 2, 3$ that the host opens with the zonk behind, we have

$$\Omega = \{B_1 \cap H_2, B_1 \cap H_3, B_2 \cap H_3, B_3 \cap H_2\},$$

where, for example, $B_1 \cap H_3$ means that the Big Deal is behind the first door (the one the trader chose) and a zonk is behind the third door (the door opened by the host). Figure 8.4 shows the relevant tree diagram, with the elements of Ω as leaves and the probabilities along the edges as before. We also include the probabilities of all elements of Ω on the right, which, as before, can be obtained by multiplying the probabilities along the edges leading to the

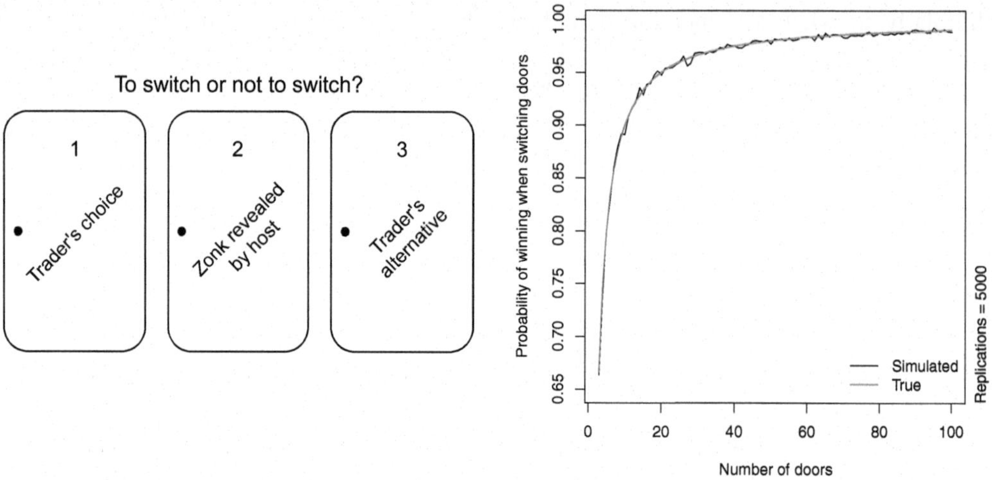

Figure 8.3 Left: Three doors to choose from for the Big Deal: the trader picks the first, the host opens the second with a zonk behind and the question is whether the trader should switch from the first to the third door. Right: Simulated probability of winning the Big Deal when switching doors as a function of the number of doors. Source: Authors

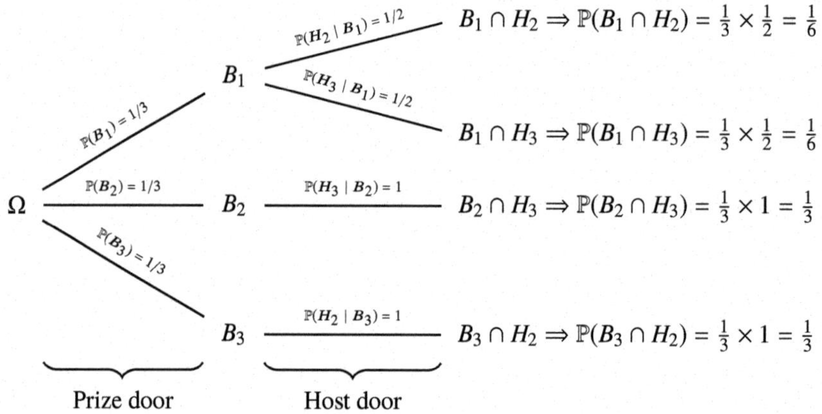

Figure 8.4 Tree diagram showing the idea of the solution to the Monty Hall problem. Source: Authors

corresponding leaf. The trader wins the Big Deal by sticking to her/his initially chosen door precisely if B_1 happens, that is if $B_1 \cap H_2$ or $B_1 \cap H_3$ happens. We thus obtain that

$$\begin{aligned}\mathbb{P}(\text{“trader wins when not switching”}) &= \mathbb{P}(B_1) = \mathbb{P}(B_1 \cap H_2) + \mathbb{P}(B_1 \cap H_3)\\ &= \frac{1}{6} + \frac{1}{6} = \frac{1}{3}.\end{aligned}$$

The probability of the trader winning when switching doors is thus

$$\begin{aligned}\mathbb{P}(\text{“trader wins when switching”}) &= 1 - \mathbb{P}(\text{“trader wins when not switching”})\\ &= 1 - \frac{1}{3} = \frac{2}{3};\end{aligned}$$

note that this is also equal to $\mathbb{P}(B_2 \cap H_3) + \mathbb{P}(B_3 \cap H_2)$. Therefore the trader is indeed twice as likely to win the Big Deal if (s)he switches!

Besides the precise calculation, why is this result counter-intuitive? If the trader sticks to her/his choice of door, then (s)he does not use any additional information than (s)he had when initially choosing that door. In this case one would indeed expect a probability of winning of $1/3$ as the Big Deal can be behind any of the three doors. By switching, the trader utilizes additional information from the host, as Monty Hall made his decision about which door to open according to behind which door the Big Deal is located. For example, in two of the three possible locations of the Big Deal (namely if it is behind the second or third door), he does not even have a choice but has to pick the remaining door with the zonk (the second door if the Big Deal is behind the third and the third door if the Big Deal is behind the second). The fact that by switching doors the trader exploits the fact that Monty Hall made the decision about which door to open, knowing where the Big Deal is located, thus helps increase the trader's chances of winning.

If this still sounds strange to you although we have mathematically computed the result in front of your eyes, we do not blame you. These things are not easy to grasp, which is why it is so important to have conditional probabilities available to formally compute answers to such questions (which is by no means a hard or complex mathematical problem). What often helps in such situations is to consider (more) extreme cases. Suppose there are $d \geq 3$ doors, the trader picks the first, and Monty Hall opens all but one of the remaining doors. The chance that the Big Deal is behind the door the trader picked is $1/d$, so the chance that it is behind doors 2 to d must be $1 - 1/d$. But Monty Hall opened doors 2 to d except one, so this remaining door, to which the trader can switch, must contain the Big Deal with probability $1 - 1/d$. Clearly, for large d, switching the door is favored. For $d = 3$, we recover the same probabilities as derived before, but for large d the argument seems much more immediate.

Another way of reaching the right answer is simulation, which can also be a good way to check one's intuition about a solution; see also Section 8.5 below, where we discuss this in more detail. Figure 8.3 shows on the right the simulated probability (computed as the average over $n = 5000$ replications of the game) of winning when switching doors and the true probability $1 - 1/d$ as a function of the number of doors $d = 3, 4, \ldots, 100$. The simulated probability very much agrees with the true one across all d.

From a conditional probability of the form $\mathbb{P}(A \mid B)$, one can also obtain the conditional probability $\mathbb{P}(B \mid A)$ for $\mathbb{P}(A) > 0$. This result is known as *Bayes' theorem*, named after Thomas Bayes (c. 1701 to 1761), and can easily be derived from (8.2) via

$$\mathbb{P}(B \mid A) = \frac{\mathbb{P}(B \cap A)}{\mathbb{P}(A)} = \frac{\mathbb{P}(A \cap B)}{\mathbb{P}(A)} = \frac{\mathbb{P}(A \mid B)\mathbb{P}(B)}{\mathbb{P}(A)}. \tag{8.6}$$

This is one of those results that students are encouraged *not* to learn by heart but rather derive on the fly when needed. Whereas this result is indeed trivial to derive and compared to (8.2) is a tautology, its consequences for statistics are considerable and through various generalizations lead to the highly relevant field of *Bayesian statistics*. The crux of (8.6) with respect to applications lies in the switching from the determination of a probability of an event A given B to one of B given A. This may sound innocuous but is often crucial in practice; the next example is highly relevant in this regard.

Example 8.7 (COVID-19). During the coronavirus pandemic, reverse transcription polymerase chain reaction (RT-PCR) or simply PCR tests were frequently used to ascertain the presence of the coronavirus at the time of the test. These tests were typically conducted by a nasal swab and results were available within 24 hours.

The actual numbers used in this example are estimates which largely depend on the underlying population and other factors such as the time between infection and when the test took place, etc. They therefore only serve as examples; manufacturers of tests know such numbers more accurately and may even be required to include them with the tests.

For simplicity, let C denote the event that a randomly selected person has the coronavirus (so C^c is the event that the person does not have the coronavirus) and let P denote the event that the person's nasal swab PCR test is positive (so P^c is the event that the test is negative). Cohen (2020) reported the estimated fraction of infected people testing negative as 25%, so the *false-negative rate* is $\mathbb{P}(P^c \mid C) = 0.25$, and an estimated fraction of non-infected people testing positive as 0.5%, so the *false-positive rate* is $\mathbb{P}(P \mid C^c) = 0.005$.

Now we can answer several questions:

(1) What is the *sensitivity* or *true positive rate* $\mathbb{P}(P \mid C)$ of the test, that is, the probability of an infected person being correctly identified by the test? The rule for computing probabilities of complementary events that we derived in Section 8.1.1 also applies to conditional probabilities. From the false-negative rate we thus obtain the sensitivity

$$\mathbb{P}(P \mid C) = 1 - \mathbb{P}(P^c \mid C) = 1 - 0.25 = 0.75.$$

(2) What is the *specificity* or *true negative rate* $\mathbb{P}(P^c \mid C^c)$ of the test, that is, the probability of a non-infected person being correctly identified by the test? From the false-positive rate we obtain

$$\mathbb{P}(P^c \mid C^c) = 1 - \mathbb{P}(P \mid C^c) = 1 - 0.005 = 0.995.$$

(3) What is the probability that a test shows a wrong result? The test is wrong if it shows a negative result for a person with the coronavirus or if it shows a positive result for a person without the coronavirus. We thus have

$$\begin{aligned}\mathbb{P}(\text{“wrong test result”}) &= \mathbb{P}(P^c \cap C) + \mathbb{P}(P \cap C^c)\\ &= \mathbb{P}(P^c \mid C)\mathbb{P}(C) + \mathbb{P}(P \mid C^c)\mathbb{P}(C^c)\\ &= 0.25\mathbb{P}(C) + 0.005(1 - \mathbb{P}(C)).\end{aligned}$$

The *prevalence* $\mathbb{P}(C)$ is the probability that a randomly selected person is infected. This would need to be estimated for a given population of interest, for example, all air travelers at an airport. For educational purposes, we leave it as a variable. Figure 8.5 shows the probability of obtaining a wrong test result as a function of the prevalence.

(4) What is the *positive predictive value (PPV)* $\mathbb{P}(C \mid P)$ of the test, that is, the probability of

a positively tested person indeed being infected? Here we can use Bayes' theorem (8.6) and the law of total probability to obtain

$$\mathbb{P}(C \mid P) = \frac{\mathbb{P}(P \mid C)\mathbb{P}(C)}{\mathbb{P}(P)} = \frac{\mathbb{P}(P \mid C)\mathbb{P}(C)}{\mathbb{P}(P \mid C)\mathbb{P}(C) + \mathbb{P}(P \mid C^c)\mathbb{P}(C^c)}$$
$$= \frac{0.75\mathbb{P}(C)}{0.75\mathbb{P}(C) + 0.005(1 - \mathbb{P}(C))}.$$

The left-hand side of Figure 8.5 shows the PPV as a function of the prevalence; the right-hand side shows the same plot with a logarithmic x-axis. For a person under test,

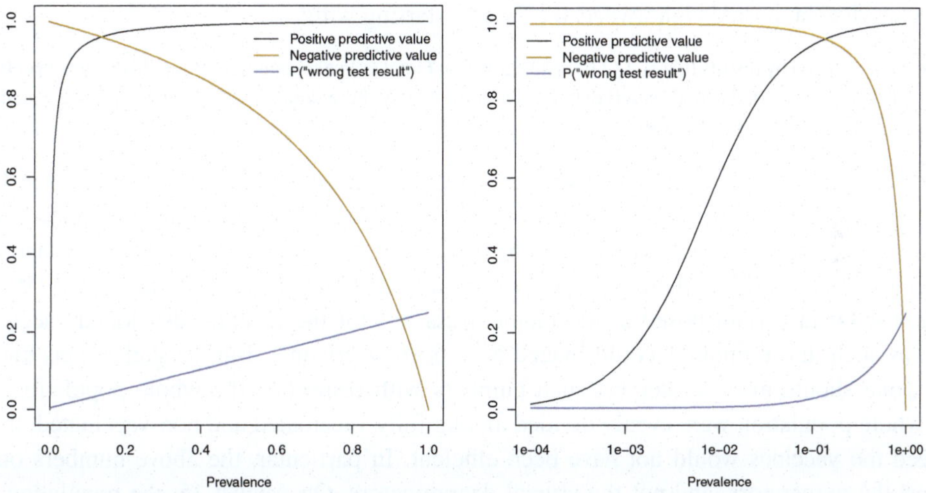

Figure 8.5 Left: The positive predictive value (PPV) $\mathbb{P}(C \mid P)$, the negative predictive value (NPV) $\mathbb{P}(C^c \mid P^c)$ and the probability of a wrong test result as functions of the prevalence $\mathbb{P}(C)$. Right: The the same with a logarithmic x-axis. Source: Authors

the positive predictive value PPV is all-important. Beyond COVID-19 you can think for example of a pregnancy test, a cervical smear test which is widely used to prevent the incidence of cervical cancer in women, a PSA test to screen for prostate cancer in men or a test for HIV. The striking fact is that, even for excellent tests, that is, tests with high sensitivity as well as specificity, when used for a population-wide screening for a rare disease (low prevalence) the PPV may in fact be rather small. You can already notice this feature on the right-hand side of Figure 8.5. For example, for $\mathbb{P}(C) = 0.01$ (see the value 1e-02 on the x-axis of the right-hand side plot of Figure 8.5), the PPV is only about 0.6.

(5) What is the *negative predictive value (NPV)* $\mathbb{P}(C^c \mid P^c)$ of the test, that is, the probability that a negatively tested person is indeed not infected? In the same way as for the PPV, we obtain

$$\mathbb{P}(C^c \mid P^c) = \frac{\mathbb{P}(P^c \mid C^c)\mathbb{P}(C^c)}{\mathbb{P}(P^c \mid C)\mathbb{P}(C) + \mathbb{P}(P^c \mid C^c)\mathbb{P}(C^c)}$$
$$= \frac{\mathbb{P}(P^c \mid C^c)(1 - \mathbb{P}(C))}{(1 - \mathbb{P}(P \mid C))\mathbb{P}(C) + \mathbb{P}(P^c \mid C^c)(1 - \mathbb{P}(C))}$$
$$= \frac{0.995(1 - \mathbb{P}(C))}{0.25\mathbb{P}(C) + 0.995(1 - \mathbb{P}(C))}.$$

Figure 8.5 shows on the right the NPV as a function of the prevalence; the PPV and NPV curves meet at about 14%.

We end this example with a recommendation to watch the excellent Youtube broadcast Spiegelhalter (2020) on false positives (and related fallacies) in the case of COVID-19. If you are interested in making statistical sense out of the multitude of COVID-19 data, Spiegelhalter and Masters (2021) is a must.

Remark 8.8 (On the (mis)interpretation of basic COVID-19 statistics). On August 16, 2021, Jeffrey S. Morris, Professor of Biostatistics at the Perelman School of Medicine, University of Pennsylvania, posted eight tweets on Twitter, starting with:

Many are confused by results that > 1/2 of hospitalized in Israel are vaccinated, thinking this means vaccines don't work. I downloaded actual Israeli data [. . .] and show why these data provide strong evidence vaccines strongly protect vs. serious disease.

The data referred to in the tweets were reported on August 15, 2021, one day earlier. On that day, the numbers of severe COVID-19 cases in Israel were 301 fully vaccinated and 214 non-vaccinated. As he explained in his second tweet, some took this to mean that the vaccines "don't work" as there are more severe fully vaccinated than non-vaccinated. At the time, the Delta variant posed a first major stress test for the recently developed vaccines and, as such, these findings could indeed have been worrisome. But, as statisticians know well, one should always interpret such numbers with respect to the whole population. If the whole population were evenly divided among fully vaccinated and non-vaccinated, then indeed the vaccines would not have been efficient. In particular, the above numbers only report the numerators, and not the crucial denominators. On August 15, the population in Israel split into 5 634 634 fully vaccinated and 1 302 912 non-vaccinated. This results in the probabilities $301/5\,634\,634 \approx 0.0053\%$ for a fully vaccinated person to turn into a severe case and $214/1\,302\,912 \approx 0.0164\%$ for a non-vaccinated to turn into a severe case. It is therefore $0.0164\%/0.0053\% \approx 3.0943$ times as likely for a non-vaccinated person to become a severe case than a fully vaccinated person! There is an easier argument than the actual calculation. Just assume that the whole population were fully vaccinated. Then every single severe case would necessarily be a fully vaccinated person (and, again, this does not mean that the vaccines "don't work" as the absolute number of severe cases would still be much lower than if everyone were not vaccinated). Note again here that we utilized the principle from Example 8.6, namely, we considered extreme cases (all fully vaccinated or all non-vaccinated) as a sanity check of a statement. Unfortunately, 'forgetting' to report the denominator happened over and over again on social media during the pandemic.

The *vaccine efficacy* is the signed relative error of these two probabilities in percentages, so $(0.0164 - 0.0053)/0.0164 \approx 67.68\%$. It provides the percentage reduction of disease-cases among the fully vaccinated in comparison to the non-vaccinated; a positive value indicates the effectiveness of the vaccine. The value of about 68% is the same across all age groups. In his tweet on August 16, 2021, Jeffrey Morris shows the same calculations but now depending on age group (one also says *stratified* by age group). The results are depicted in Figure 8.6. The left-hand side shows estimated probabilities of developing severe COVID-19 symptoms in Israel on August 15, 2021 for a fully vaccinated and for a non-vaccinated person, stratified by age group; note that the seemingly rather large probabilities displayed are not in contradiction

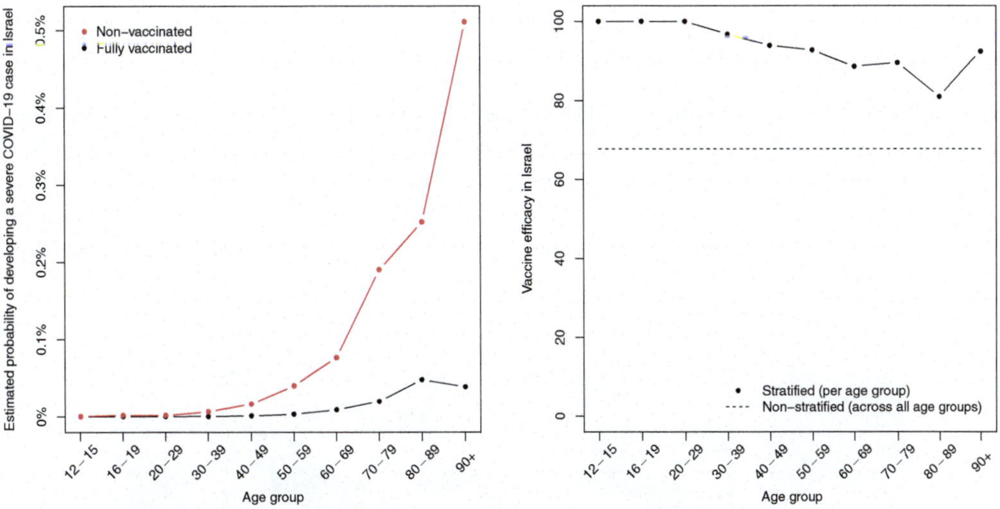

Figure 8.6 Left: Estimated probabilities of developing severe COVID-19 symptoms in Israel on August 15, 2021 for a fully vaccinated and for a non-vaccinated person. Right: The corresponding vaccine efficacy. Source: Authors

with the reported 0.0053% and 0.0164% from above, owing to the age structure of Israel's population (with a large fraction of young people, especially in the non-vaccinated group). We see that the estimated probability of developing severe symptoms is much larger for the non-vaccinated than the fully vaccinated, for all age groups. The right-hand side of Figure 8.6 shows the vaccine efficacy for each age group. The dashed line indicates the efficacy of 67.68%. The phenomenon that the overall (non-stratified) vaccine efficacy is only about 68% whereas the efficacy for every single age group is quite a bit larger is known as *Simpson's paradox*; see also Wikipedia (2021d). From the data of August 15, 2021, there is thus no strong reason to believe that the vaccine is not efficient almost to the degree originally reported for the Pfizer–BioNTech vaccine (the vaccine predominantly used in Israel) before the appearance of the Delta variant. To make more definitive statements would require one to consider more days and also confidence intervals for the estimated efficacies; for the latter, see Section 8.6.3. As reported in various sources, the efficacy of the Pfizer–BioNTech vaccine did drop a bit owing to the Delta variant.

In Section 8.1.1 we introduced the concept of independence of two events A, B via $\mathbb{P}(A \cap B) = \mathbb{P}(A)\mathbb{P}(B)$. This can also be motivated from (8.2). If A and B are independent events, then whether we condition on B should not have an influence on A. We thus have

$$\mathbb{P}(A) = \mathbb{P}(A \mid B) = \frac{\mathbb{P}(A \cap B)}{\mathbb{P}(B)}$$

and so indeed $\mathbb{P}(A \cap B) = \mathbb{P}(A)\mathbb{P}(B)$.

Probability experiments offer many examples leading to surprising, even counterintuitive results. We next present two such experiments. The first one, on "nontransitive" dice (see below), surely defies one's intuition. The second example highlights how probability theory can form the basis for magic tricks. From a mathematical point of view, they are related to the calculation of conditional probabilities.

Example 8.9 (Nontransitive dice)**.** Consider the following game between two players A and B. They are given four unloaded dice, each however with a non-standard sequence of number of eyes on the six faces: D_1 has faces $0, 0, 4, 4, 4, 4$, D_2 has faces $3, 3, 3, 3, 3, 3$, D_3 has faces $2, 2, 2, 2, 6, 6$ and D_4 has faces $1, 1, 1, 5, 5, 5$; they are to be found in that order in Figure 8.7. What is special about the dice is that no two dice can show the same number when thrown

Figure 8.7 A set of nontransitive or Efron dice as they were made by one of our students in a course on Randomness and Risk. Source: Authors

and that, for a given die, numbers can occur more than once. Player A can choose a die first; B has to take one of the remaining three. Then A and B each throw their die and the higher number wins (as mentioned before, ties are not possible). The four dice are then placed on the table again and the game restarts: first A chooses whichever die (s)he now wants, followed by B, etc. The question is: who has a winning strategy? Surely, most would say, player A; (s)he can choose the "best" die. All right, let us calculate the following probabilities, denoting $\{D_i > D_j\}$ if in one throw die i shows a higher number than die j. We can now check that $\mathbb{P}(D_1 > D_2) = 2/3$ (this is easy since four out of the six faces of D_1 show a larger number than those of D_2), $\mathbb{P}(D_2 > D_3) = 2/3$ (same argument applies) and $\mathbb{P}(D_3 > D_4) = 2/3$; in the latter case we can use the law of total probability (8.5) to see that

$$\begin{aligned}\mathbb{P}(D_3 > D_4) &= \mathbb{P}(D_3 > D_4 \mid D_3 = 2)\mathbb{P}(D_3 = 2) + \mathbb{P}(D_3 > D_4 \mid D_3 = 6)\mathbb{P}(D_3 = 6) \\ &= \frac{3}{6} \times \frac{4}{6} + \frac{6}{6} \times \frac{2}{6} = \frac{2}{3}.\end{aligned}$$

Bingo, A always chooses D_1.

Well, not so fast, indeed $P(D_4 > D_1) = 2/3$ (check!). So after A commits to a die, B can always choose the "better" die and consequently has a winning strategy, very much against our intuition. For real numbers x, y, z and the relation ">", we know that if $x > y$ and $y > z$ then $x > z$, a property of ">" known as *transitivity*. But the relation "has a higher probability of winning than" is not transitive, hence the name *nontransitive dice*. The dice D_1, D_2, D_3, D_4 are also known as *Efron dice*, named after statistician Bradley Efron who came up with this example; we will meet him again when we discuss the bootstrap in Section 8.6.3. From a mathematical point of view, the example of nontransitivity is known as the Steinhaus–Trybula paradox and goes back to Steinhaus and Trybula (1959). The game was made popular by Martin Gardner (1970b). The internet contains numerous scientific publications on this topic. In particular, look for links with Condorcet's paradox and the voting paradox or Arrow's impossibility theorem from social choice theory. As we mentioned before, there is more to randomness than meets the eye!

Example 8.10 (The Kruskal count). The mathematician and physicist Martin David Kruskal (1925–2006) came up with the following card trick. Take a deck of 52 French-suited playing cards, with the classical four suits (clubs, diamonds, hearts and spades), each numbered 1 to 10 with three face cards jack, queen and king; see Figure 8.8. We associate each card with its

Figure 8.8 Deck of 52 shuffled playing cards. Starting in position 6 leads to the final position as indicated by a silver commemorative coin (from the Singapore Botanic Gardens) in the bottom-right corner. Source: Authors

numerical value, and give value 5 to each face card. The magician shuffles the pack and asks you to choose a number at random between 1 and 10, say you take 6, but to not tell her/him. (S)he then turns the cards slowly one after the other face up and tells you to look at the card corresponding to your chosen number, that is the 6th card. You now change your 6 for the value on that card, say it is a jack of hearts, hence you change from 6 to 5 and continue like that until you reach a card for which your last chosen value exceeds that of the 52 cards of the deck. You remember that last card on which you landed. For example, if the magician turns the cards as shown in Figure 8.8 (left to right, top to bottom), then the 6th card is a 5, the 5th card thereafter is a 4, the 4th card thereafter is a king (so a 5), and so on until we land on the third-last card, the jack of diamonds as indicated in Figure 8.8.

The magician claims (s)he can tell you the exact card upon which you finally landed. Well, (s)he cannot do that with 100% certainty but with a probability around 84%, which is not bad. But how? Well the magician plays in her/his head the same game, choosing a number and going through the cards as you do. In doing so, once you both land on the same card, you stay together. The latter event is known as *coupling*. We thus need to calculate the probability of a coupling to happen, which turns out to be close to 84%. This needs some work to prove; see for instance Lagarias et al. (2009). Figure 8.9 shows the outcomes of 100 simulated such games. If a game leads to no coupling (here in 15 out of the 100 cases), we

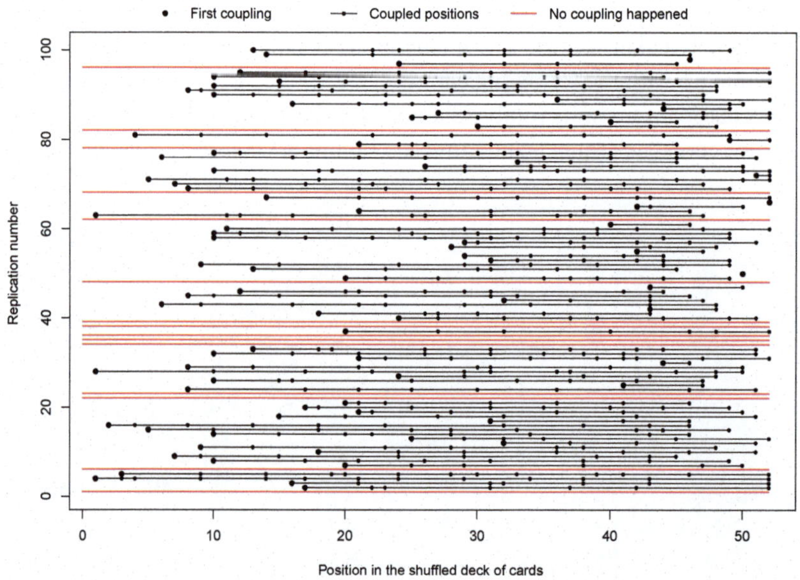

Figure 8.9 Results of 100 simulated games. For each game (shown along the y-axis), we obtain either no coupling (indicated by a horizontal red line) or a coupling (in which case we plot the positions of the couplings in the shuffled deck of cards with a dot; the position of the initial coupling is indicated by a larger dot). Source: Authors

plot a horizontal red line. And if a game leads to a coupling, we draw a black line linking a slightly larger dot (indicating the position of the initial coupling) with the final dot where both the player and the magician land together before leaving the deck. The dots in between indicate the positions they visit together between the initial coupling event and the end of the game. For the 100 replications shown in Figure 8.9, we obtain a relative frequency of 85% as the estimated probability of a coupling. For 1000 replications we obtain a relative frequency of 84.4%; interestingly, among those runs with a coupling, the average length of the coupling was about 6.3.

You do not have to go through the mathematics or the details of the simulation, just performing the trick as magician is great fun. About five out of six times you will be right (and if not you could tell the audience member you picked that (s)he is special and move on to the next person; the probability of failing twice is only about 3%). It may be interesting to learn that Kruskal's count can be used in the context of the factorization of numbers and hence code breaking; for this link, see the "Added in proof" comment in Pollard (1978).

8.1.5 Statistical Literacy

We recall the epigraph from H. G. Wells at the start of this chapter: "... it is as necessary to be able to compute, to think in averages and maxima and minima, as it is now to be able to read and write." In particular, in our increasingly data-driven society, basic statistical literacy is becoming an almost as necessary skill, as it was for our far distant ancestors to be able to correctly interpret the sounds of the surrounding savanna. These statistical skills we surely should expect from our politicians. In its category *Statistics News* of February 11, 2022, the Royal Statistical Society (RSS) communicated the findings of a 2021 poll on the statistical literacy of 101 British Members of Parliament (MPs); see Masters (2022). The poll was based on three elementary questions. The summary below is based on comments made by Christl Donnelly, the Society's Vice President for External Affairs, and Anthony Masters, its Statistical Ambassador:

Question 1: If you toss a coin twice, what is the probability of getting two heads?
Parliamentarians could choose 15%, 25%, 40%, 50%, 75%, or offer their own answer. The correct answer is 25%.

Question 2: Suppose you roll a six-sided die. The rolls are 1, 3, 4, 1, and 6. What are the mean and mode values?
There were open text boxes for respondents to fill out. There was also an option to say they did not know. The correct answers are 3 and 1, respectively.

Question 3: Suppose there was a diagnostic test for a virus. The false-positive rate (the proportion of people without the virus who get a positive result) is 1 in 1000. You have taken the test and tested positive. What is the probability that you have the virus?
The response list was 99.9%, 99%, 1%, 0.01%, "Not enough information to know" and "Don't know". The correct answer is "Not enough information to know".

Question 1 brings us back to the story of d'Alembert, told in Section 8.1.2. Question 3, concerning testing, is highly pertinent in the context of the coronavirus pandemic; see Example 8.7. For Question 1, the answers given by the MPs were 51% correct (up from 40% from a poll in 2011). For Question 3, only 16% gave the correct answer "Not enough information to know". We leave the conclusions from these success–failure rates in this (admittedly, limited) study to you. But before you formulate your verdict, do reread the final paragraphs of Section 8.1.2. The RSS plans to extend polling on statistical literacy to the general population. Of course, as noted by Anthony Masters, "Constructing a short battery of questions to measure statistical literacy is challenging. There are different conceptions of what it means for a person or society to be 'statistically literate'. It is hard to assess statistical thinking and contextual understanding, especially without full examinations." The problems encountered around risk communication during the coronavirus pandemic have clearly shown how important basic knowledge on statistical reporting and its inherent uncertainties are. One key recommendation no doubt is that society needs more education on statistics from an early age onwards.

8.2 From Probability Spaces to Random Variables

We now come to the most central concept underpinning stochastic and risk modeling, the concept of random variables. Random variables model outcomes of experiments and they often help us express events of interest and their probabilities in a more natural way. For example, we could have used a random variable, say, X, to describe the rolled number when rolling a fair die once. The probability of rolling a 1 or a 6 as in Example 8.2 can then be written as $\mathbb{P}(X \in \{1,6\})$. If we are interested in the probability of rain tomorrow, then X could be the precipitation in millimeters. If we are interested in the probability that a 40-year-old person will eventually reach the age of 85, then X could be the duration of a human life. Or X could denote a financial firm's loss over the next year in some monetary unit if that is of interest to us. Or X could be the number of people out of 100 000 to get severe COVID-19 symptoms given that they caught the coronavirus. The number of applications is endless.

A classical example of a function from high school is the quadratic function $h(x) = x^2$, $x \in \mathbb{R}$, a function returning the square of a given real number. If we write X instead of h and ω instead of x, we obtain $X(\omega) = \omega^2$. This is already an example of what a random variable is mathematically, namely a function which maps outcomes ω from Ω (here $\Omega = \mathbb{R}$) to real numbers. A random variable is only "random" in the sense that we do not know the *evaluation point* ω of the function X before we conduct an experiment. For example, before throwing the fair die, we know only that X takes on each of the values in $\{1,2,\dots,6\}$ with probability $1/6$, but we do not know which of the six numbers will appear on a throw. Only after we threw the die do we know the ω that happened and thus do we know the realized number $X(\omega)$ in this experiment.

Concept 8.11 (Random variable, realization)**.** A *random variable* X is a function which maps an outcome ω from Ω to a real number $X(\omega)$ called a *realization* of X.

Not confusing the concept of a random variable X (a function) with that of its realization $X(\omega)$ (a real value) is crucial for understanding probability and statistics. A random variable X models the outcome of an experiment and $X(\omega)$ is the realized value (the actual outcome) once ω is known (after the experiment). One also frequently encounters several, say n, random variables $X_1, X_2, \dots, X_n$. Their realizations $X_1(\omega), X_2(\omega), \dots, X_n(\omega)$ are typically written as $x_1, x_2, \dots, x_n$, the *data.* If the focus is on the random variables themselves (for example, for understanding how they behave and how they can be used to adequately model risks), one writes $X_1, X_2, \dots, X_n$. We mostly do the latter but occasionally the former in order to stress what our focus is – which concept is meant should be clear from the context.

As an example, let us redo the calculation of the probability of obtaining at least one 6 in four throws in Game 1 of the Chevalier de Méré paradox; see Section 8.1.1. It is natural to model the four throws by four random variables X_1, X_2, X_3, X_4 and so we have

$$\begin{aligned}
&\mathbb{P}(\text{“at least one 6 in four throws”}) \\
&= 1 - \mathbb{P}(\text{“no 6 in four throws”}) \\
&= 1 - \mathbb{P}(X_i \in \{1,2,\dots,5\}, i = 1,2,3,4) && (8.7)\\
&= 1 - \mathbb{P}(X_1 \in \{1,2,\dots,5\}) \cdots \mathbb{P}(X_4 \in \{1,2,\dots,5\}) && (8.8)\\
&= 1 - \mathbb{P}(X_1 \in \{1,2,\dots,5\})^4 && (8.9)\\
&= 1 - (5/6)^4 \approx 0.5177.
\end{aligned}$$

Equation (8.7) expresses the probability of obtaining at least one 6 in four throws as one minus the probability of the complementary event, so of obtaining no 6 in any of the four throws (remember Pavlov from Section 8.1.1). In (8.8), the probability of the latter event splits into the product of the probabilities corresponding to each throw since the throws happen independently of each other. Now comes an important observation. The probability of not throwing a 6 is the same in each throw. This is what is used in (8.9) to write the product of the four probabilities as the fourth power of one of them; we used $\mathbb{P}(X_1 \in \{1,2,\dots,5\})$ here without loss of generality. Moreover, not only are these four specific probabilities the same, the behavior of the random variables is identical as they all model the same phenomenon, the outcome of one throw of a fair die. If the behavior of random variables is identical, one says that they are *identically distributed* or that they have the same *distribution*. What this means mathematically will be made more precise in Section 8.3 when we learn how to describe the behavior of a random variable through its “distribution”, that is, the probability it assigns (or “distributes”) to certain events.

The importance of random variables becomes clear from the fact that we can combine simple, well-understood ones to model the outcomes of much more complicated experiments, resembling the construction of a dream house out of simple Lego bricks. Such constructions of new, typically more involved, random variables through simple ones is known as a stochastic representation, informally defined as follows.

Concept 8.12 (Stochastic representation). A *stochastic representation* is a representation of a random variable by other (typically simpler) random variables.

Stochastic representations are at the core of any stochastic model. In principle, they can be used to model experiments of interest, for example those underlying the motivating questions from the beginning of Section 8.1. Here are some examples of stochastic representations that are often of interest in applications.

Example 8.13 (Stochastic representations).

(1) The sum of n random variables $X_1, X_2, \dots, X_n$ is written as

$$S_n = X_1 + X_2 + \cdots + X_n = \sum_{i=1}^{n} X_i$$

(read: “the sum of X i for i from 1 to n”). The new random variable S_n can be used in a wide variety of applications, for example, if X_i models the loss of business line i of a financial firm in the next year then S_n models the total loss of the financial firm in its business lines 1 to n in the next year. Related to the sum is the average $(X_1 + X_2 + \cdots + X_n)/n$, also written as

$$\bar{X}_n = \frac{1}{n}\sum_{i=1}^{n} X_i,$$

which could model the average number of new daily coronavirus infections among 100 000 inhabitants of a specific area over the next n days; here X_i represents the new number of infections i days ahead. Based on actual realizations in the past $n = 7$ days, this would model the *seven-day incidence rate* reported in the media during the coronavirus pandemic. Also here, note the difference: what is reported in the media is the statistical quantity based on actual realizations (an *estimate*), whereas $\bar{X}_n$ represents the probabilistic quantity we want to study in order to infer properties of the seven day incidence rate in the hope that we can make predictions how this average will behave in the near future. Such predictions then affect political decisions, for example.

Even if we knew the aforementioned behavior (or "distribution" as we will call it later) of all of $X_1, X_2, \ldots, X_n$ together, note that it is typically challenging to determine the behavior (or distribution) of S_n, even though the sum is a rather simple stochastic representation. This is particularly a problem if the random variables $X_1, X_2, \ldots, X_n$ depend on each other, so that the outcome of one influences another, for example. Even if one assumes that the random variables $X_1, X_2, \ldots, X_n$ are independent of each other (as in independent throws of a die, for example), it is non-trivial to determine the behavior of S_n. And the assumption of independence is often not justifiable in real applications: think of the n business lines that all belong to the same financial firm and are thus affected by the same firm-wide decisions or macro-economic factors. Or think about how a high number of incidences in one day still influences the seven day incidence rate over the next six days. General solutions to this problem typically involve approximations of some sort (for example, by simulation, see Section 8.5). Depending on the application, such approximations need fine-tuning and, certainly, time to talk to the statisticians.

(2) If N is a random variable which takes on non-negative integers $\{0, 1, 2, \ldots\}$, and $X_1, X_2, \ldots$ are random variables as before, then the random sum

$$S_N = \sum_{i=1}^{N} X_i = X_1 + X_2 + \cdots + X_N$$

is a sum with a random number of random summands; for $N = 0$, S_0 is defined as 0. Such sums appear in insurance risk modeling. For example, suppose you are a car insurer and you have 100 000 clients. In a particular year, a random number N of your clients is affected by, say, car accidents and the corresponding claims that you face as an insurer are $X_1, X_2, \ldots, X_N$. So the random sum $S_N = X_1 + X_2 + \cdots + X_N$ models the total loss you face due to claims from car accidents of your clients in that particular year.

(3) If X_i models the maximal sea level at Neeltje Jans in The Netherlands on a particular day i out of $n = 365$ days, then the maximum

$$M_n = \max\{X_1, X_2, \ldots, X_n\}$$

can be used to model the maximal sea level in a year. One can then answer questions such as, what is the probability that next year's maximal recorded sea level at Neeltje Jans exceeds a given height of 3 m? Or what is the dike height we expect the sea level at Neeltje Jans to exceed once every 100 years? Such types of questions are addressed in Chapter 9.

8.3 From Random Variables to Distribution Functions

8.3.1 The Concept and Its Properties

We now address what we meant by the "behavior" or "distribution" of a random variable. It is a convenient way to characterize the probabilistic properties of a random variable.

Utilizing our example $h(x) = x^2$, $x \in \mathbb{R}$, again we see that this function is specified by the instruction that the input x is mapped to the output x^2. We could have equally well described this quadratic function by specifying to what any interval of the form, say, $(a, b]$ (all the points x greater than a and less than or equal to b) is mapped by h. For example, h maps $(0, 1]$ to $(0, 1]$, $(2, 3]$ to $(4, 9]$ and $(-1, 1]$ to $[0, 1]$. Doing this for all intervals $(a, b]$ provides a way to describe h which allows us to shift the focus from particular inputs x to how the output of h behaves. This is particularly useful for random variables X as we are not so much interested in the function X (that is, h) at a specific evaluation point ω (the input x), but rather in how X behaves on intervals. For example, if X models the maximal sea level at Neeltje Jans in meters in the next year, then we are interested in the probability that X lies in the interval $(3, \infty)$, and thus in the event that $X > 3$, the event of a high water-mark exceedance. In other words, we are interested in the event of all realizations ω such that $X(\omega)$ exceeds 3. The probability of this event can be written as

$$\mathbb{P}(\{\omega \in \Omega : X(\omega) > 3\})$$

(read: "the probability of the set of all realizations omega in Omega such that X of omega is greater than 3"). This probability is more compactly written as

$$\mathbb{P}(X > 3);$$

in justification of this and many other abbreviations in the mathematical language, mathematicians typically mention the jokey phrase "A good mathematician is a lazy mathematician"; see Morris (2016) and Figure 8.10.

The notation $\mathbb{P}(X > 3)$ is intuitive as it hides the rather abstract first two components $(\Omega, \mathcal{F})$ of the underlying probability space. For developing a gut feeling for how random variables behave, this notation is excellent and natural. Of course the 3 is linked to our sea-level example, so we now replace it by a general real number x. We can then consider $\mathbb{P}(X > x)$ as a function of x (termed the "survival function" below), and this turns out to already be one way to characterize the behavior or distribution of X. Similarly to the event $\{X > x\}$, its complementary event $\{X \leq x\}$ is frequently used. When viewing the probability $\mathbb{P}(X \leq x)$ as a function of x, one obtains the "distribution function" of the random variable X, which characterizes X. Both distribution functions and survival functions allow us to characterize random variables without explicitly referring to the underlying probability space.

Figure 8.10 A good mathematician. Source: Enrico Chavez

Concept 8.14 (Distribution function, survival function)**.** Let X be a random variable. Then the *distribution function* of X is the function

$$F(x) = \mathbb{P}(X \leq x), \quad x \in \mathbb{R}, \tag{8.10}$$

(read: "F of x is the probability that X is less than or equal to x"). The distribution function F uniquely characterizes (the behavior or distribution of) a random variable X. One thus often simply writes $X \sim F$ (read: "X follows F" or "X is distributed as F") if X is a random variable with distribution function F.

Similarly, the *survival function* of X is the function

$$\bar{F}(x) = \mathbb{P}(X > x), \quad x \in \mathbb{R},$$

(read: "F bar of x is the probability that X is greater than x"). The survival function $\bar{F}$ also uniquely characterizes the random variable X.

From now on we mostly focus on distribution functions to describe random variables. The distribution function of a random variable $X \sim F$ allows us to describe a random variable modeling a random phenomenon or experiment purely in terms of the deterministic function F, a type of function we have known since high school. As already mentioned, the distribution function F describes how probabilities are "distributed" over the realizations of X.

The Central Formula

We have already stated that F characterizes X. So can we, for example, express the probability that X falls in an interval in terms of F? That is indeed the case. Using properties of $\mathbb{P}$, one can show that

$$\mathbb{P}(a < X \leq b) = F(b) - F(a), \quad a \leq b \tag{8.11}$$

(read: "the probability that X lies in the interval from a to b is F of b minus F of a"). Equation (8.11) expresses the probability that the random variable X falls in an interval $(a, b]$ in terms of the distribution function F of X. The right-hand side is a simple difference of values of the function F, there is no notion of randomness visible, no complicated underlying sample space, no $\mathcal{F}$, etc. For those who understand the theory, shorthand expressions like (8.11) come to life and replace the cumbersome and often imprecise use of natural language. In doing so, linguistic confusion is avoided. Unfortunately, for those not initiated in the art of mathematics, such formalism often works as a deterrent. We will soon see characterizing properties and many examples of F which stress the importance of distribution functions for the stochastic modeling of risks. In principle, if one knew the distribution function of the random variable underlying each of the questions raised at the beginning of Section 8.1, one would be able to provide precise answers to these questions. Note that they are ordered in increasing difficulty. For the first couple of questions we can give precise answers (based on the probabilistic tools we present); for the remaining questions one can still give answers but those answers increasingly rely on assumptions made and are also increasingly affected by uncertainty. This is where statistics enters: what answers do different models provide, how can we compare them in terms of uncertainty, which model shall we ultimately use, etc. The last question concerns the modeling of a complex system over 100 years, which requires so many assumptions and is affected by such high uncertainty that it is fair to doubt whether it is even possible to give a meaningful answer. The problem with any answer is that it might ignite confidence that one is able to precisely model such events where there should not be much, if any, such confidence.

Limits

For discussing the properties of distribution functions and other concepts, we need to introduce the notion of a *limit*. Informally, we say that a function value $h(x)$ *converges to* or *tends to* $h(y)$ for x converging to y if $h(x)$ gets arbitrarily close to $h(y)$ for every x sufficiently close to y. For example, $h(x) = x^2$ converges to 0 for x converging to 0, or $h(x) = 1/x$ converges to 0 for x tending to ∞ (infinity). There is an abbreviation for such limits, namely $h(x) \to h(y)$ for $x \to y$ or also $\lim_{x\to y} h(x) = h(y)$ (read: "the limit of h of x for x converging to y is h of y"). For example, if $h(x) = x^2$ then $\lim_{x\to 0} h(x) = 0$, and if $h(x) = 1/x$ then $\lim_{x\to\infty} h(x) = 0$. Sometimes it matters from which side x approaches y. If x approaches y from the left that is, if $x < y$, we often replace "$\to$" by "$\uparrow$" and write $\lim_{x\uparrow y} h(x)$. Similarly, if x approaches y from the right, that is, if $x > y$, we often replace "$\to$" by "$\downarrow$" and write $\lim_{x\downarrow y} h(x)$.

Properties of Distribution Functions

If $X \sim F$, one can use properties of the probability $\mathbb{P}$ to show that F always fulfills the following three properties; we will later learn what these properties mean for X.

(1) $\lim_{x\to-\infty} F(x) = 0$; in words, if x gets smaller and smaller, then $F(x)$ converges to 0. And $\lim_{x\to\infty} F(x) = 1$; in words, if x gets larger and larger, then $F(x)$ converges to 1.
(2) F is *increasing*, that is, $F(x) \leq F(y)$ for all $x < y$.
(3) F is *right-continuous*, that is, $F(x) = \lim_{y\downarrow x} F(y)$ for all $x \in \mathbb{R}$; in words, at any x, $F(y)$ converges to $F(x)$ for y converging to x from the right.

Properties (1)–(3) follow from the axiomatic properties of $\mathbb{P}$ and need a bit more mathematics in order to explain them in detail. It suffices to grasp these properties intuitively, which is possible for the first two properties. For example, if $x < y$, then the inequality $X \leq x$ is at least as restrictive as $X \leq y$ (in the sense that all ω for which $X(\omega) \leq x$ also fulfill $X(\omega) \leq y$), hence the probability of $X \leq x$ must be less than or equal to that of $X \leq y$, so $F(x) = \mathbb{P}(X \leq x) \leq \mathbb{P}(X \leq y) = F(y)$. Properties (1)–(3) are best remembered in terms of a picture. Figure 8.11 shows three examples of distribution functions, let us call them F_1 (left), F_2 (center) and F_3 (right). We see that F_1, F_2, F_3 converge to 0 for smaller and

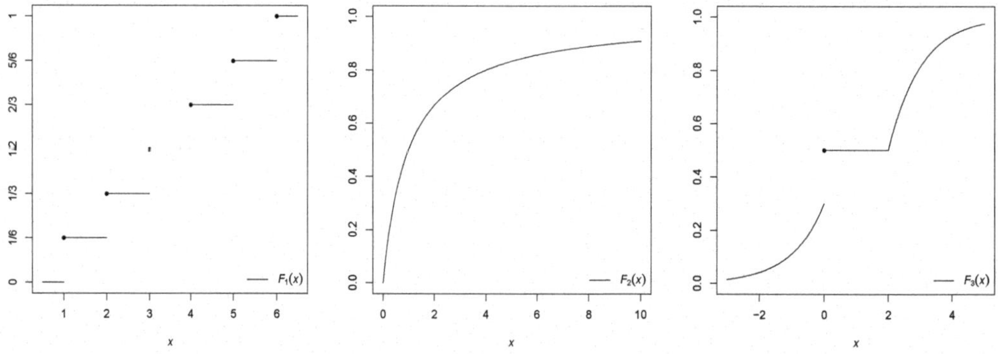

Figure 8.11 Left: A discrete distribution function F_1. Center: A continuous distribution function F_2. Right: A mixed-type distribution function F_3. Source: Authors

smaller x (F_1 and F_2 even reach 0) and F_1, F_2, F_3 converge to 1 for larger and larger x (F_1 even reaches 1). Also, we see that the graphs of F_1, F_2, F_3 are all increasing. Note that distribution functions are allowed to have flat parts; we will come back to what that means for the corresponding random variables soon. Furthermore, we can verify (3) from the graphs of F_1, F_2, F_3 by considering any two points $x < y$ and decreasing y towards x. The corresponding function value $F(y)$ then indeed decreases to $F(x)$ as it should by (3). For example, for F_3 and $x = 0$, we see from the plot on the right-hand side of Figure 8.11 that $F_3(y) = 0.5$ for all $0 < y \leq 2$ and so $F_3(y)$ indeed decreases to $F_3(x)$ (we even have $F_3(y) = F_3(x)$ for such y here). To make clear what the exact value of a distribution function F is at a point where the graph of F jumps, one often indicates the value with a small dot in the graph of F; see, for example, the graph of F_3 at $x = 0$.

Expressing Probabilities with Distribution Functions

We have already mentioned that (8.11) can be used to express any probability of interest in terms of the distribution function F of the random variable X under consideration. Although (8.11) considers the probability that X falls in an interval $(a, b]$, other probabilities (and virtually all probabilities of interest) can also be expressed in terms of F, as the following example shows.

Example 8.15 (Expressing probabilities with distribution functions).

(1) A survival function $\bar{F}(x) = \mathbb{P}(X > x)$ can also be expressed in terms of F. We have already seen the trick of expressing a probability in terms of its complementary event. Since the event that $X \leq x$ is the complementary event of the event $X > x$, we obtain

$$\bar{F}(x) = \mathbb{P}(X > x) = 1 - \mathbb{P}((X > x)^c) = 1 - \mathbb{P}(X \leq x) = 1 - F(x)$$

so the survival function $\bar{F}$ of a random variable $X \sim F$ is simply $\bar{F}(x) = 1 - F(x)$. This now also makes clear why both a distribution function and a survival function characterize a random variable: if you know one you know the other. If it is irrelevant for the context which characterizing concept is used to describe a random variable X, we simply speak of the *distribution* of X. Note that "distribution" is also often used as a synonym for "distribution function".

(2) You might have already wondered why (8.11) considers $\mathbb{P}(a < X \leq b)$ and not, say, $\mathbb{P}(a < X < b)$, $\mathbb{P}(a \leq X < b)$ or $\mathbb{P}(a \leq X \leq b)$. All these probabilities can also be expressed in terms of F if one uses limits. One has:
   - $\mathbb{P}(a < X < b) = \lim_{y \uparrow b} F(y) - F(a)$, which is written in a more compact way as $\mathbb{P}(a < X < b) = F(b-) - F(a)$. The minus sign after b conveys the idea that we approach b from the "left", so we consider the limit for $y \uparrow b$.
   - In the same vein, $\mathbb{P}(a \leq X < b) = \lim_{y \uparrow b} F(y) - \lim_{x \uparrow a} F(x)$, which is written in a more compact way as $\mathbb{P}(a \leq X < b) = F(b-) - F(a-)$.
   - And again in a similar way to before, $\mathbb{P}(a \leq X \leq b) = F(b) - F(a-)$.

   In short, by using limits of F we can express the probability that $X \sim F$ falls in any interval in terms of F.

(3) What about the probability $\mathbb{P}(X = x)$? As you might have guessed, it can also be expressed in terms of F. We have just seen that $\mathbb{P}(a \leq X \leq b) = F(b) - F(a-)$ and so by letting $a = x$ and $b = x$, we obtain

$$\mathbb{P}(X = x) = \mathbb{P}(x \leq X \leq x) = F(x) - F(x-), \tag{8.12}$$

   which corresponds to the jump height of F at x; see Figure 8.11.

A Classification of Distribution Functions

Essentially all distribution functions relevant in risk modeling and other stochastic applications can be classified into three categories: discrete, continuous or mixed type. We now provide more insight into these three cases, examples of which we have already seen in Figure 8.11.

(1) If the graph of a distribution function consists only of flat parts or jumps, we call such a distribution function *discrete*. For an example, see the graph of F_1 in Figure 8.11. Note that F_1 is the distribution function of a random variable X_1 which models the outcome when rolling a fair die once, so X_1 attains each of the numbers from 1 to 6 with probability $1/6$ (hence the jumps of the graph of F_1 by $1/6$ at 1 to 6). Informally, F_1 "distributes" the probability $1/6$ over each of the numbers 1 to 6. One also says that the *probability mass* $1/6$ is assigned to each of the numbers 1 to 6 when describing which probabilities are allocated to the values that X_1 can take on.

   Although we frequently use toy examples such as the rolling of a fair die, note that a real risk application is not far around the corner. Assume you have n observations of financial losses $x_1, x_2, \ldots, x_n$ reported over the last year and you want to use them to model next year's losses. One (out of many) way(s) to do that is to construct a distribution function solely based on the data $x_1, x_2, \ldots, x_n$ (in contrast to an assumption on the distribution function such as a certain parametric form). This is known as an *empirical distribution function* and typically denoted by $\hat{F}_n$. This distribution function $\hat{F}_n$ puts probability mass $1/n$ at each given data point $x_1, x_2, \ldots, x_n$. This is just like rolling a fair die with n sides; see Figure 8.12 for an example with $n = 10$. And, just like

Figure 8.12 A cylindrical object with ten flat sides, representing a 10-sided 'die'. Source: Authors

mimicking rolling a fair die by drawing 1 to 6 with probability 1/6 each, mimicking generating realizations from the empirical distribution function means drawing each of the losses $x_1, x_2, \ldots, x_n$ with probability $1/n$. Much as rolling a fair die serves as a running example throughout this chapter, we will revisit empirical distribution functions multiple times.

(2) Consider F_2 from Figure 8.11 at any point $y \in \mathbb{R}$. Then $F_2(x)$ also approaches $F_2(y)$ if x increases towards y. In the notation for limits introduced before, we have $F_2(y-) = F_2(y)$. If this is the case for all y then we speak of a *continuous* distribution function. Informally, a distribution function F is continuous if its graph has no jumps. It follows from (8.12) that in this case

$$\mathbb{P}(X = x) = F(x) - F(x-) = F(x) - F(x) = 0.$$

So if $X \sim F$ for a continuous distribution function F then the probability that X attains a single outcome x is 0 – and this holds for any real outcome x!

Let us think more deeply about what $\mathbb{P}(X = x) = 0$, for every x, means for the random variable X of a continuous distribution function F. If we use such an X for modeling an outcome of an experiment, then *before* conducting the experiment, we know that each outcome of X has probability zero of being attained. Yet *after* we conducted the experiment and we have obtained a realization $X(\omega)$ of X, then this particular $x = X(\omega)$ has, against all odds, appeared! So, for continuous distribution functions, just because a value has probability 0 of being attained, this does not mean it cannot appear. To phrase this in terms of the complement, just because a value has probability 1 of not appearing does not mean it cannot appear. Hence the "almost" in "almost surely" as introduced earlier. In the discrete case this may remind us of the "Een dubbeltje op zijn kant", the probability that a tossed coin will land on its edge, already mentioned in Example 8.2. Although being treated as 0 for all modeling purposes, it does not seem out of this universe to imagine a tossed coin landing on its edge. Indeed Agnes Herzberg (see Introduction) once told the story of a speaker at Imperial College London who started a lecture on probability by tossing a coin, and it landed on its side as it got stuck in a small crack of the desk. Just imagine how to continue from there!

As you can see, the wording of the mathematical concept has been chosen very carefully (here, regarding the "almost sure" equality of Example 8.2). The more difficult the concept, the more important it is for a mathematician to think about how to name it when introducing it so that other generations of colleagues, academics, practitioners, students, etc. can more easily understand it at a technical level but also develop a gut feeling for working with it at a more abstract level.

Now that we have already started to open the door to more challenging aspects of the world of mathematics, let us open it a bit further. A continuous distribution function F, like every distribution function, needs to distribute the probability mass 1 over all outcomes of $X \sim F$; think of F_1 distributing probability mass $1/6$ over each of the numbers 1 to 6 as we said above. But if, for a continuous F, we do indeed distribute probability mass 0 at each x as just explained, do we not then end up with no probability mass being distributed over all x? No. As we already said, it is correct that for $X \sim F$, with F continuous, $\mathbb{P}(X = x) = 0$ for any real x. The point is that there are uncountably many such real numbers x, and summing the probability $\mathbb{P}(X = x) = 0$ uncountably-often does not necessarily give 0. In fact, for continuous distribution functions, such a sum is always 1, as it should be. Such trap doors of uncountable sums when working with continuous distribution functions can be avoided by not thinking of events where $X = x$ but rather events where X falls in an interval, as we have done so before and as is also reflected by formula (8.11). We will come back to a similar point in Remark 8.32(6) later.

(3) The distribution function F_3 is a mixed case in the sense that its graph contains both jumps (namely at $x = 0$) and continuous parts (namely for all x not equal to 0). Such distribution functions are said to be of *mixed type*. Similarly to F_3 but with many more jumps, one can imagine a distribution function F modeling financial losses to be the empirical distribution function of historical losses in the *body* of F (that is, for those x for which $F(x)$ attains moderate values) and to be a continuous distribution function in the left and right *tails* of F (so for those x for which $F(x)$ attains values close to 0 or close to 1, respectively). We will meet such a case in Example 8.18 below, where we use the empirical distribution function in the body, combined with a particularly important model for the right tail of the distribution function.

Existence of Random Variables

We have already stated that every distribution function F satisfies properties (1)–(3). A deeper mathematical result also confirms the converse statement: every function F which satisfies properties (1)–(3) is the distribution function of some random variable X. This is a very powerful result as it allows us to describe a random variable purely in terms of a deterministic F, as mentioned after Concept 8.14. For example, one can easily verify that the function $F(x) = 1 - 1/(1 + x)$ satisfies properties (1)–(3) as long as $x \geq 0$. Therefore,

$F(x) = 1 - 1/(1+x)$, $x \in [0, \infty)$, is the distribution function of a random variable X and we can use formula (8.11) to determine any probability of interest for such an X; the graph of its distribution function F is indeed what we chose as F_2 in the center of Figure 8.11. In Section 8.5 we will see how, for a given F, we can generate realizations of $X \sim F$ with a computer. Such realizations then enter stochastic models which are evaluated on the basis of such realizations.

What the Properties of a Distribution Function Imply About Its Random Variable

For a given random variable X, (8.10) provides the distribution function F of X. We will now address the converse, namely what a certain form of a distribution function F implies for the behavior of its random variable $X \sim F$. For this, the graph of F_3 on the right-hand side of Figure 8.11 is important to keep in mind as it contains three cases that convey what properties of a distribution function mean for the underlying random variable.

(1) Consider F_3 on the interval $(a, b] = (2, 4]$. The probability that X falls in this interval, so that $2 < X \le 4$ (or $X \in (2, 4]$), is given by formula (8.11) as $\mathbb{P}(X \in (2, 4]) = F_3(4) - F_3(2)$. Since F_3 is *strictly increasing* on this interval $(a, b]$ (so that it satisfies $F_3(x) < F_3(y)$ for all $x < y$), we know that $F_3(4) > F_3(2)$ and so $\mathbb{P}(X \in (2, 4]) = F_3(4) - F_3(2) > 0$. The same logic applies to any distribution function F over an interval $(a, b]$ for which $F(b) > F(a)$. So, if F is a strictly increasing distribution function over an interval $(a, b]$, we know that $X \sim F$ has positive (that is, non-zero) probability of lying in $(a, b]$ (and the exact probability is given by formula (8.11)).

(2) Now consider F_3 on the interval $(a, b] = (0, 2]$. As before, the probability of $X \in (a, b]$ is $\mathbb{P}(X \in (a, b]) = F_3(b) - F_3(a)$. Since $F_3(b) = F_3(a) = 0.5$ we obtain that $\mathbb{P}(X \in (a, b]) = F_3(b) - F_3(a) = 0$. The same logic applies to any distribution function F that is flat over an interval $(a, b]$. So if F is flat over $(a, b]$ then its random variable $X \sim F$ has probability 0 of taking on values in $(a, b]$. In other words, X almost surely does not take on values in $(a, b]$.

    With this knowledge, we can now interpret the distribution function F_1 on the left-hand side of Figure 8.11. The corresponding random variable almost surely does not take on any values apart from the numbers 1 to 6, which is indeed what we expect when rolling a fair die.

(3) Finally, consider F_3 at $x = 0$. As we can see from the right-hand side of Figure 8.11, F_3 jumps at $x = 0$ from the value 0.3 to the value 0.5. As already discussed, the probability that $X \sim F_3$ takes on the single value x is $\mathbb{P}(X = x) = F(x) - F(x-)$. The left-sided limit of F at $x = 0$ is 0.3 and $F(x) = 0.5$. We thus see that $\mathbb{P}(X = 0) = 0.5 - 0.3 = 0.2$. This means that with probability 0.2, X takes on the single value $x = 0$. The same logic applies to any distribution function F that has a jump in a point x. So if F has a jump at x of height p, then its random variable $X \sim F$ takes on the value x with probability p.

    We can now interpret F_1 on the left-hand side of Figure 8.11 with this knowledge. Since F_1 jumps by $1/6$ at each of the numbers 1 to 6, the random variable $X \sim F_1$ takes on each of these six values with probability $1/6$, again what we expect when rolling a fair die.

One example where discrete distribution functions appear naturally is through given data $x_1, x_2, \ldots, x_n$ in terms of the corresponding empirical distribution function. We have already mentioned empirical distribution functions before, but now we introduce the concept more formally. To this end, we introduce a simple yet important mathematical function, the *indicator (function)*. For a set A, the indicator $I_A(x)$ is 1 if $x \in A$, and 0 if $x \notin A$. As the name suggests, $I_A(x)$ thus indicates (with the value 1) whether $x \in A$. We also frequently write $I_{\{x \in A\}}$ for $I_A(x)$. There are several other ways of writing indicators in mathematics, so do not get confused if you see a different notation.

Concept 8.16 (Empirical distribution function)**.** Given n data points $x_1, x_2, \ldots, x_n$, the corresponding *empirical distribution function* is

$$\hat{F}_n(x) = \frac{\text{number of data points} \leq x}{n} = \frac{1}{n} \sum_{i=1}^{n} I_{\{x_i \leq x\}}, \tag{8.13}$$

where $I_{\{x_i \leq x\}} = 1$ if $x_i \leq x$ and $I_{\{x_i \leq x\}} = 0$ if $x_i > x$.

Remark 8.17 (Understanding empirical distribution functions)**.** If we order the data points $x_1, x_2, \ldots, x_n$ and label the smallest as $x_{(1)}$, the second smallest as $x_{(2)}$ and so on, we obtain the data $x_{(1)} \leq x_{(2)} \leq \cdots \leq x_{(n)}$ in increasing order; these are the *order statistics* of $x_1, x_2, \ldots, x_n$. Since the order of summation in (8.13) does not matter, we can also write the empirical distribution function $\hat{F}_n(x)$ as

$$\hat{F}_n(x) = \frac{1}{n} \sum_{i=1}^{n} I_{\{x_{(i)} \leq x\}}.$$

Slowly increasing the evaluation point x from below the smallest data point $x_{(1)}$ to beyond the largest data point $x_{(n)}$, we see that $\hat{F}_n(x) = 0$ for $x < x_{(1)}$, then $\hat{F}_n(x)$ jumps by $1/n$ at each data point, and finally $\hat{F}_n(x) = 1$ for $x \geq x_{(n)}$. Figure 8.13 shows plots of $\hat{F}_n$ for $n = 5$ data points. Put differently, at the kth-smallest data point $x = x_{(k)}$, $\hat{F}_n(x)$ jumps from $(k-1)/n$

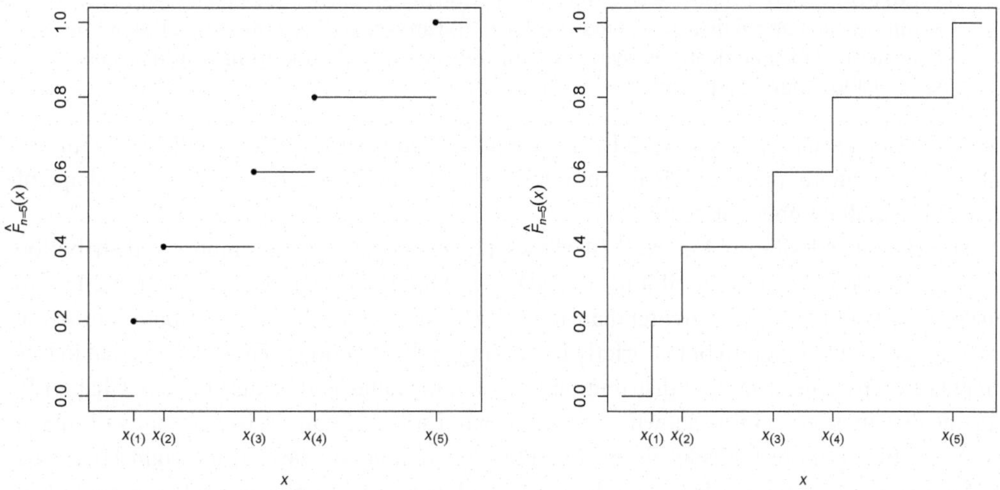

Figure 8.13 Stylized empirical distribution function based on $n = 5$ data points. The values of $\hat{F}_n$ at jumps are indicated by dots (left) and by vertical lines (right); the latter are often used when n is large, for better visibility. Source: Authors

to k/n; in particular, $\hat{F}_n(x_{(k)}) = k/n$ for all $k = 1, 2, \ldots, n$. Apart from their order, we can therefore reconstruct the data points $x_1, x_2, \ldots, x_n$ from $\hat{F}_n$, and so the empirical distribution function $\hat{F}_n$ contains all information about the n data points (except their order).

Example 8.18 (Distribution functions of disaster data)**.** To illustrate distribution functions based on real losses, we computed the yearly number of deaths worldwide due to natural disasters and the yearly total loss (all damages and economic losses, adjusted for inflation to 2020–2021 prices) in billions of USD, owing to such disasters from 1900 to 2020 from CRED (2021).

The left-hand side of Figure 8.14 shows the empirical distribution function of the 121 yearly numbers of deaths worldwide due to natural disasters, x, from 1900 to 2020. Note that

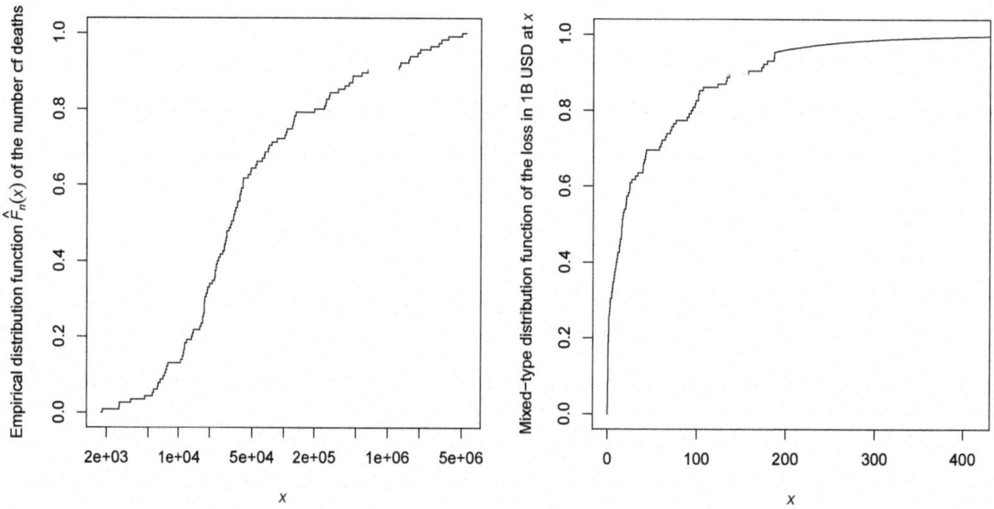

Figure 8.14 Left: Empirical distribution function of the yearly numbers of worldwide deaths due to natural disasters since 1900. Right: Empirical distribution function with continuous tail model of the yearly loss in billions of USD due to natural disasters since 1900. Source: Authors

due to the large sample size $n = 121$ here, we have used vertical lines to indicate the jumps of $\hat{F}_n$. Furthermore, the x-axis is given in logarithmic scale to see more clearly the shape of $\hat{F}_n(x)$ for small numbers of deaths x.

The right-hand side of Figure 8.14 shows the main part of the empirical distribution function of the yearly loss in billions of USD due to natural disasters from 1900 to 2020. In the right tail of this $\hat{F}_n$ (for x greater than about 190), you can see that the distribution function is no longer a step function but is strictly increasing and continuous. The reason is that in this right tail we have used another distribution function; the latter will be discussed in Chapter 9. So, the distribution function shown on the right-hand side of Figure 8.14 is actually of mixed type and thus a more realistic instance of the toy example we saw on the right-hand side of Figure 8.11. As we will learn in Chapter 9, such a strictly increasing and continuous tail of a distribution function can provide a reasonable continuation of a distribution function far beyond its largest "loss value" (that is, where the empirical distribution function would simply jump to 1 and stay there) and can thus be used to provide answers to questions concerning events well beyond the given n data points. For example, the largest yearly loss turns out to be about 414 billion USD. Asking, on the basis of the original data alone, what the probability is that next year's loss exceeds 414 billion USD is meaningless (as it would be 0). However, using the mixed-type distribution function we constructed with its continuous right tail, we can determine this probability as 0.5%. In the same vein, the methods presented in Chapter 9 will allow us to provide answers to questions one could not find answers to by just considering the given data points at hand. And this situation is very common in practice. Remember the problem of determining the dike height for the van Dantzig report; see Section 1.3. Only having a model for sea levels until the highest level observed would certainly not be sufficient to solve that problem.

We encountered similar problems related to the 2004 Boxing Day tsunami, the 2011 Tōhoku earthquake and the Fukushima nuclear disaster; see Chapter 4.

8.3.2 From Distribution Functions to Probability Mass Functions and Densities

We have so far covered random variables, their distribution functions, how the two concepts are related to each other and why they are so important for modeling risks. In this section, we cover more concrete examples of distribution functions which are widely used in stochastic modeling. In the classification introduced earlier, these examples are either discrete or of a certain continuous type. In both these cases there is another helpful concept available for characterizing the distribution function.

Examples of Discrete Distributions

One of the discrete distribution functions we have worked with repeatedly so far is that of a random variable modeling the outcome when rolling a fair die once; see the left-hand side of Figure 8.11. A distribution function F whose graph only consists of jumps and flat parts is fully determined by its jumps alone, so by specifying at which values x the distribution function F jumps and what the corresponding jump heights are. All such x values are the values that $X \sim F$ takes on almost surely and the jump height at x, as we know from (8.12), determines the probability that X equals x. In the discrete case, it thus makes sense to describe X or F by the function $f(x) = \mathbb{P}(X = x)$.

Concept 8.19 (Probability mass function)**.** For a discrete random variable X or its distribution function F, the *probability mass function* is the function $f(x) = \mathbb{P}(X = x)$, $x \in \mathbb{R}$.

For example, if X denotes the rolled number when rolling a fair die, then its probability mass function satisfies $f(x) = 1/6$ for $x = 1, 2, \ldots, 6$ (and $f(x) = 0$ for all other values x).

Remark (On notation and its recycling)**.** One also frequently finds the notation $p(x)$ for $f(x)$, especially when distinguishing the concept of a probability mass function from that of a density, typically denoted by f too, as seen later. We have deliberately and for a good purpose chosen f for probability mass functions. Apart from differences in their interpretation, which we will discuss, the two concepts are strongly related and help describe distributions in a similar fashion. It thus makes sense to *recycle* f here and reserve p for denoting probabilities or parameters of distributions; see Example 8.25 below. To leave no room for misinterpretation, mathematical publications in general try not to recycle notation too much, but in more involved areas of mathematics this is unavoidable given the myriad of mathematical concepts but only a comparably small number of meaningful letters or symbols to describe the concepts. For this reason, mathematicians combine letters with symbols and write x, $\tilde{x}$ (read: "x tilde"), $\tilde{\tilde{x}}$ (read: "x double tilde"), x^* (read: "x star"), or add subscripts or superscripts to letters, which is why mathematical publications can look a bit daunting. A rather telling example, but evolved from the necessity to describe such concepts, can be found in actuarial notation as used in life insurance mathematics; see Wikipedia (2021b). In particular, mathematicians are keen users of the Greek alphabet; we already encountered examples such as α and ω. In general, one cannot overestimate the importance of notation for learning the language of mathematics, and the same applies to the implementation of mathematical concepts, functions or algorithms in computer language. Replacing a single notation, one can typically quickly adapt to that change while not losing understanding. However, grasping a concept at the required depth when large parts of informative notation are replaced by non-informative or non-standard notation can become hopeless.

For discrete distributions, their probability mass functions characterize X or F. By definition as a probability, a probability mass function f must be non-negative (so $f(x) \geq 0$ for all x) and must sum to 1 (as we have probability mass 1 being distributed over all outcomes of X). The converse is also true, any such function f (which is non-negative and sums to 1) is the probability mass function of a discrete distribution.

Remark (Standing on the shoulders of giants). If you find this rather intuitive, then you are starting to develop a gut feeling for random variables and how they can be described. Of course, in an actual lecture, the concepts we present would be discussed in greater detail and there would be homework assignments for you to deepen your understanding of these concepts. Now imagine that you need to come up with such a concept by yourself. You would need to make sure it makes sense. So, probability mass functions need to sum to 1, as we said. How would we write this down for a particular f? Perhaps by using the summation symbol like this:

$$\sum_{x\in\mathbb{R}} f(x) = 1. \tag{8.14}$$

But how can we make sense of such a summation here if it is over uncountably many x values, so that whenever we pick a first, second, third, etc. x we have omitted many other values x in between (in fact, again uncountably many). It is in order to make sense of such situations that mathematicians are required. After they have done so, everyone else applying such concepts is then standing on their shoulders.

If possible, one avoids writing sums over uncountably many elements such as in (8.14). And, indeed, one can avoid that here, since it can be shown that there are at most countably many x for which $f(x) > 0$; see below. The set of these x where $f(x) > 0$ is known as the *support* of f (or F or $X \sim F$). Say the support of f is $\{x_1, x_2, \dots\}$; then one can simply write (8.14) as

$$\sum_{k=1}^{\infty} f(x_k) = 1.$$

The question we have not answered yet is why there are at most countably many x for which $\mathbb{P}(X = x) > 0$? Rephrased in terms of distribution functions, why can a distribution function F have at most countably many jumps? Standing on the shoulders of giants (other mathematicians), there is an easy argument for that. In every x at which F jumps, the jump leaves a gap. We know that $\mathbb{P}(X = x) = F(x) - F(x-)$, so the gap is the interval $(F(x-), F(x)]$ which is of the form $(a, b]$ for $a < b$. The shoulders of the first giant we now need are those of the Greek astronomer, mathematician and scholar Eudoxus of Cnidus (c. 408 to c. 355 BCE) who formulated in Euclid (–300, Definition 4) a result by now known as the *Archimedean property*: "[. . .] magnitudes are said to have a ratio with respect to one another which, being multiplied, are capable of exceeding one another." Translated to today's formulation: Given two numbers $y > x > 0$, there is a positive integer n such that $nx > y$. Geometrically speaking, if you draw a shorter and a longer line segment with a pencil on a piece of paper, you can repeatedly concatenate the shorter one until the length exceeds that of the longer line segment. One can use this property to show that any interval $(a, b]$ contains a rational number, that is, a number that can be represented as a fraction of integers. To see this, you can go through the following steps if you like. The proof below constitutes a short but somewhat strenuous hike in mathematical wonderland (Chapter 7).

(1) Let $x = b - a > 0$ be the length of the interval $(a, b]$ and let $y = 1$. We can assume without loss of generality that $x < y$; if the length x of the interval $(a, b]$ exceeds 1, then we certainly find an integer in it which is also a rational number. Applying the Archimedean property to this x and y we know that there exists a positive integer n such that $nx > 1$. Plugging in $x = b - a$, we obtain that $na + 1 < nb$.
(2) Since na is a real number, it is bounded by two consecutive integers which can be expressed as $m - 1 \leq na < m$ for some integer m. These two inequalities are $m \leq na + 1$ and $na < m$, so we obtain $na < m \leq na + 1$.
(3) By (1) we know that $na + 1 < nb$. Applying this inequality to $m \leq na + 1$ from (2), we obtain that $m < nb$ and thus (again by (2)) that $na < m < nb$. Dividing by n, we see that $a < m/n < b$. Since m/n is a rational number, we see that any interval $(a, b]$ contains a rational number. Mathematicians say that the rational numbers are *dense* in the real numbers.

These steps indeed constitute a mathematical proof. We use this result to infer that whenever a distribution function F jumps, we can find a rational number in the jump gap. Since F is increasing, F can only jump upwards, so the rational numbers we pick out of each jump gap are all distinct. This means that a distribution function F can jump at most as many times as there are different rational numbers, since we find a different rational number in every jump gap. Now we only have left to show that we can count the number of rational numbers (we have then established that any distribution function F can have at most countably many jumps). For this we need to stand on another giant's shoulders, those of mathematician Georg Cantor (1845–1918, the father of the theory of sets); see Figure 8.15. We start by explaining *Cantor's (first) diagonal argument*, which shows that the set of all positive rational numbers is countable. This set can be written as all fractions m/n for positive integers m and n. One might be tempted to keep one of the two numbers m, n fixed and start iterating through the other in order to count all such fractions, for example $1/1, 2/1, 3/1, \ldots$, but this sequence never stops and so we would not reach denominators other than 1. One cannot simply count to infinity in the numerator and then increment the denominator. At least mathematicians cannot do that; as the saying goes only Chuck Norris (American martial artist and actor) is known to have counted to infinity, twice. But Cantor had an idea of how to count through

Figure 8.15 Left: Georg Cantor (1845–1918). Right: Cantor's diagonal argument. Sources: Wikipedia (left) and authors (right)

all the positive rationals in one go, namely by arranging all such numbers in a matrix-like rectangular scheme, where the entry in row m and column n is m/n; see Figure 8.15. By following along the arrows diagonally through the matrix, one eventually reaches all positive rational numbers. Since we can omit those already reached (for example $3/3 = 2/2 = 1/1$), we can count all positive rationals as:

$$\frac{1}{1}, \frac{1}{2}, \frac{2}{1}, \frac{3}{1}, \frac{1}{3}, \frac{1}{4}, \frac{2}{3}, \frac{3}{2}, \frac{4}{1}, \dots$$

One can expand the trick to all rationals, by starting from 0 and alternating between positive and negative fractions:

$$0, \frac{1}{1}, -\frac{1}{1}, \frac{1}{2}, -\frac{1}{2}, \frac{2}{1}, -\frac{2}{1}, \frac{3}{1}, -\frac{3}{1}, \frac{1}{3}, -\frac{1}{3}, \frac{1}{4}, -\frac{1}{4}, \frac{2}{3}, -\frac{2}{3}, \frac{3}{2}, -\frac{3}{2}, \frac{4}{1}, -\frac{4}{1}, \dots$$

So you do not need to be as strong as Chuck Norris to count to infinity; it suffices to be as smart as Georg Cantor!

Some more remarks are in order here.

Remarks (To infinity and back)**.** The diagonal argument, which we encountered above, was published in Cantor (1891). It appears in various guises throughout mathematics and not only sits at the basis of several deep theorems but also leads to very important results in philosophy. If you want to learn more about its wide influence you should read Hofstadter (1979). The latter well-known book compares and contrasts geniuses from mathematics (Kurt Gödel (1906–1978)), art (Maurits Cornelis Escher (1898–1972)) and music (Johann Sebastian Bach (1685–1750)). Once you have worked yourself through the fascinating but demanding 777 pages, you can read about the famous Wiener Kreis (Vienna Circle) of the 1920s and 1930s of which Kurt Gödel was a member. Sigmund (2017) tells the story of how Viennese scientists and philosophers in the 'demented times' before World War II met in search for scientific guidance and truth. That book contains an extended Preface by Douglas Hofstadter. He concludes as follows:

> Though it is long gone and not so often talked about today, there is no doubt that Der Wiener Kreis was an assemblage of some of the most impressive human beings who have ever walked the planet, and Karl Sigmund's book tells its story, and their stories, in a gripping and eloquent fashion. It is a wonderful historical document, and it will perhaps inspire some readers to dream great dreams in the way that they were dreamt in the Vienna of those far-off days.

Through the formulation of set theory, Cantor not only provided mathematics with "a paradise that no one shall expel us from" (David Hilbert, whom we briefly met in Section 7.2), he also provided mathematics with an understanding of the concept of infinity.

We all have asked questions like "What is infinity?", "If I keep on counting will I finally get there?" or "What is infinity plus one?" at some point. Several (counterintuitive) properties of infinite sets are highlighted in a thought experiment known as *Hilbert's paradox of the Grand Hotel*; see Figure 8.16. Let us start simply. In mathematics, the *pigeonhole principle*

Figure 8.16 Hilbert's paradox of the Grand Hotel as a cartoon. Mathematicians will immediately recognize David Hilbert (the summer hat, the small beard and the glasses). Source: Enrico Chavez

states that if you put k items into n containers, with $k > n$, then there must be at least one container with more than one item. So if you arrive at a hotel with finitely many rooms (in fact, at any real hotel you have ever seen and will see) and all rooms are occupied, there is no room for you. Not so if you go to Hilbert's (paradox of the Grand) Hotel. Hilbert's Hotel has infinitely many rooms, and they are all still assumed to be occupied. If you arrive at Hilbert's

Hotel, is there a room for you if all of them are occupied? You might think no, but there indeed is (why would it be called a paradox otherwise?) All the receptionist needs to do is to tell the guest in each room to move one further down to the next room, then the first room is available for you! What about finitely many new guests? Well, if m new guests arrive, the receptionist would only need to tell each occupant to move m rooms further down, then the first m rooms are available. Hey, and what if a bus (Hilbert's bus) arrived with a countably infinite number of guests? You guessed it, it is also possible, the receptionist just tells the guest in room numbered i to move to room number $2i$. Then all the odd-numbered rooms are available, which are countably infinitely many and thus all newly arrived guests will have a place to stay.

After this detour of countably many steps, we now cover several concrete examples of discrete distributions. They are most directly described in terms of their probability mass functions.

Example 8.20 (Discrete uniform distribution)**.** The *discrete uniform distribution* on n points $x_1, x_2, \dots, x_n$ puts probability mass $1/n$ at each of these points. The support is therefore $\{x_1, x_2, \dots, x_n\}$. The discrete uniform distribution is abbreviated by $\mathrm{U}(\{x_1, x_2, \dots, x_n\})$ and one writes $X \sim \mathrm{U}(\{x_1, x_2, \dots, x_n\})$ for a random variable following this distribution. Its probability mass function is $f(x_k) = 1/n$, $k = 1, 2, \dots, n$ (and 0 otherwise).

Our prime example of a discrete uniform distribution is that for rolling a fair die once and noting the number, in which case $n = 6$ and the support is $x_i = i$ for $i = 1, 2, \dots, 6$; see Figure 8.12 for an example of a (non-cubical) 10-sided die representing $\mathrm{U}(\{0, 1, \dots, 9\})$. Another important example that we have already seen is the empirical distribution, which puts mass $1/n$ at each of n data points $x_1, x_2, \dots, x_n$ and is thus simply a discrete uniform distribution on these numbers. As we have already seen the empirical distribution function in (8.13), we know that this is also the distribution function of $\mathrm{U}(\{x_1, x_2, \dots, x_n\})$; see also the left-hand sides of Figures 8.11 ($n = 6$) and 8.14 ($n = 121$) for graphs of such distribution functions.

Example 8.21 (Bernoulli distribution)**.** A random variable X which can only take on the two values $0, 1$, with $\mathbb{P}(X = 1) = p$ and $\mathbb{P}(X = 0) = 1 - p$, follows a *Bernoulli distribution* with parameter $p \in [0, 1]$, denoted by $X \sim \mathrm{B}(1, p)$. As this distribution depends on a parameter, p, one also speaks of the *parametric family* of Bernoulli distributions. The probability mass function of $\mathrm{B}(1, p)$ is given by $f(0) = 1 - p$, $f(1) = p$ and $f = 0$ otherwise. If $p = 0$ then $\mathbb{P}(X = 0) = 1$, so X is almost surely 0, and if $p = 1$ then $\mathbb{P}(X = 1) = 1$, so X is almost surely 1; in such cases where a random variable X is constant almost surely (so that its distribution function F consists of a single jump of height 1) one refers to X (or F) as *degenerate*. To exclude such cases, when necessary, one speaks of *non-degenerate* random variables or distribution functions.

The distribution function F of $\mathrm{B}(1, p)$ at x is $\mathbb{P}(X \le x)$, where X can only take on the values 0 or 1. Hence, if $x < 0$ then $F(x) = \mathbb{P}(X \le x) = 0$. At $x = 0$, we know that $\mathbb{P}(X = 0) = 1 - p$, so $F(x)$ jumps to $1 - p$ and thus $F(0) = 1 - p$. Since X does not take on values in $(0, 1)$, we know that $F(x) = 1 - p$, $x \in (0, 1)$. And $F(1) = \mathbb{P}(X \le 1) = 1$, so $F(x) = 1$ for all $x \ge 1$.

Overall, we can therefore write f and F as follows:

$$f(x) = \begin{cases} 1-p, & x = 0, \\ p, & x = 1, \\ 0, & \text{otherwise}, \end{cases} \quad \text{and} \quad F(x) = \begin{cases} 0, & x < 0, \\ 1-p, & x \in [0,1), \\ 1, & x \geq 1. \end{cases} \tag{8.15}$$

To check whether we have derived F from f correctly we can determine the jump heights from F and compare them with f. The jump height of F at $x = 0$ is $1 - p$ and it is p at $x = 1$, which is indeed identical to f.

Since $X \sim \mathrm{B}(1,p)$ only takes on the values $0, 1$, we can interpret X as an indicator, namely $X = I_A$ for any event A with $\mathbb{P}(A) = \mathbb{P}(X = 1) = p$. Note that the indicator I_A of the event A is 1 if A happens (which is interpreted as *success*) and 0 otherwise (*failure*). Indicators are convenient for describing a *dichotomous* or *Bernoulli* experiment, hence an experiment with only two possible outcomes. Such experiments might seem to be of limited applicability but in fact are extremely useful for modeling purposes. The event A could be that we obtain heads when tossing a fair coin (in this case $X \sim \mathrm{B}(1, 1/2)$). For $p \neq 1/2$ one can model biased coins. Another example is obtaining a 6 when rolling a fair die, in which case we have $X \sim \mathrm{B}(1, 1/6)$. Or $X \sim \mathrm{B}(1,p)$ could indicate downward movements of a stock price ($X = 0$ with probability $1 - p$) or upward movements of a stock price ($X = 1$ with probability p). Yet another example, for small p, say $p = 1/10\,000$, is the modeling of rare events, that is, events that happen rarely (namely with probability $p = 1/10\,000$ here).

Later we will use indicators for events A that are easy to simulate on a computer (see Example 8.62 below) in order to generate realizations of $X \sim \mathrm{B}(1,p)$ and thus to be able to simulate realizations of X for all Bernoulli experiments.

As we have seen in Example 8.21, a coin toss can be modeled by a Bernoulli distribution via $X \sim \mathrm{B}(1, 1/2)$. As in the story of d'Alembert in Section 8.1.2, it is natural to ask for an experiment involving more than one coin toss. In what follows we will rely on a result that is mathematically out of our scope, but is credited to Kolmogorov: for a given random variable $X \sim F$, one can always construct a sequence $X_1, X_2, \ldots$ of (infinitely many) independent random variables with $X_i \sim F$ for each $i \in \mathbb{N} = \{1, 2, \ldots\}$. The reason we need to stand on the shoulders of Kolmogorov here lies in the "infinitely many". So, the distribution of each X_i is that of $X \sim F$ and each X_i attains its values independently of the values of the other random variables in the sequence. In the language of mathematics, one says that such a sequence of random variables $X_1, X_2, \ldots$ is *independent and identically distributed (iid)* according to F; we will use the notation $X_1, X_2, \ldots \overset{\text{iid}}{\sim} F$ for this situation. For modeling n coin tosses, for example, we can therefore simply write $X_1, X_2, \ldots, X_n \overset{\text{iid}}{\sim} \mathrm{B}(1, 1/2)$ and the whole experiment is described.

In Section 8.1.1 we computed the probability that independent events will happen as the product of the probabilities of the individual events. This property can be shown to carry over to independent random variables (that is, random variables whose outcomes are not dependent on each other) and probabilities involving them, including distribution functions. For example, if we model heads as 1 and tails as 0, and if we want to compute the probability that the first toss of a coin gives heads and the second gives tails, then we are interested in the probability $\mathbb{P}(X_1 = 1, X_2 = 0)$ (read: "the probability that X one is 1 and X two is 0"). Since the outcomes of the two tosses are independent of each other, one has $\mathbb{P}(X_1 = 1, X_2 = 0) = \mathbb{P}(X_1 = 1)\mathbb{P}(X_2 = 0)$. This is also what we used in Section 8.1.2 to infer that the probability of any combination of heads or tails in two coin tosses is $(1/2) \times (1/2) = 1/4$, but now we are just writing events in terms of random variables. More generally, if $X_1 \sim F_1, X_2 \sim F_2, \dots, X_n \sim F_n$, we are often interested in the probability

$$\mathbb{P}(X_1 \leq x_1, X_2 \leq x_2, \dots, X_n \leq x_n),$$

that is, the probability that $X_j \leq x_j$ for all $j = 1, 2, \dots, n$. This is also known as a *joint probability*, because it involves several random variables jointly. As one expects, for independent random variables $X_1, X_2, \dots, X_n$, this joint probability splits into the product of the individual probabilities, so

$$\begin{aligned} &\mathbb{P}(X_1 \leq x_1, X_2 \leq x_2, \dots, X_n \leq x_n) \\ &= \mathbb{P}(X_1 \leq x_1)\mathbb{P}(X_2 \leq x_2) \cdots \mathbb{P}(X_n \leq x_n) \end{aligned} \tag{8.16}$$

for all $x_1, x_2, \dots, x_n$. The converse is also true in the sense that if, for random variables $X_1 \sim F_1, X_2 \sim F_2, \dots, X_n \sim F_n$, equation (8.16) holds for all $x_1, x_2, \dots, x_n$ then the random variables $X_1, X_2, \dots, X_n$ are independent. As we have only intuitively (but not formally, in a mathematically precise way) stated what independence of random variables means, we take this characterization as a definition of independence and thus obtain the following concept.

Concept 8.22 (Independence of random variables)**.** Random variables $X_1, X_2, \dots, X_n$ are *independent* if and only if (8.16) holds for all $x_1, x_2, \dots, x_n \in \mathbb{R}$.

An immediate application of (8.16) is the calculation of the distribution function of the maximum of iid random variables.

Example 8.23 (Distribution of the maximum of iid random variables)**.** If $X_1 \sim F_1, X_2 \sim F_2, \dots, X_n \sim F_n$, we have already motivated in Example 8.13(3) that the random variable $X = M_n = \max\{X_1, X_2, \dots, X_n\}$ is often of interest; think of determining the maximal out of n losses or the maximal sea-level height over n days. To determine the distribution function F of X, we can utilize the observation that the maximum M_n of n numbers $x_1, x_2, \dots, x_n$ is less than or equal to x if and only if each number x_i is less than or equal to x, so if and only if $x_1 \leq x, x_2 \leq x, \dots, x_n \leq x$. With this observation we have

$$\begin{aligned} F(x) &= \mathbb{P}(X \leq x) = \mathbb{P}(\max\{X_1, X_2, \dots, X_n\} \leq x) \\ &= \mathbb{P}(X_1 \leq x, X_2 \leq x, \dots, X_n \leq x). \end{aligned}$$

If the n random variables are independent, we obtain from (8.16) that

$$\begin{aligned} F(x) &= \mathbb{P}(X_1 \leq x, X_2 \leq x, \dots, X_n \leq x) \\ &= \mathbb{P}(X_1 \leq x)\mathbb{P}(X_2 \leq x) \cdots \mathbb{P}(X_n \leq x) \\ &= F_1(x)F_2(x) \cdots F_n(x). \end{aligned}$$

And if, additionally, all random variables share the same distribution function, say F_1 (so $X_1, X_2, \dots, X_n \overset{\text{iid}}{\sim} F_1$), then we have

$$\begin{aligned} F(x) &= F_1(x)F_2(x) \cdots F_n(x) = F_1(x)F_1(x) \cdots F_1(x) \\ &= F_1^n(x), \quad x \in \mathbb{R}, \end{aligned}$$

so the distribution function of the maximum of n iid random variables is the distribution function of one of them to the nth power.

It becomes clear from this example that the iid assumption allows us to express the distribution function F of the maximum via the distribution function F_1 of one of the random variables. Mathematically, the iid assumption for the underlying random variables $X_1, X_2, \dots, X_n$ is therefore very convenient. Whether this assumption is justifiable in a real application is an entirely different story. For example, if $X_1, X_2, \dots, X_n$ are losses of a financial firm in n business lines over the next year, then one cannot expect these random variables to be independent because all business lines are part of the same firm and thus affected by the financial health of the firm overall, its management decisions, etc. Even more so, the assumption of an equal distribution of the losses in each of the n business lines is typically not fulfilled, for example if in one business line the losses are known to be larger than in another then that is already a violation of the assumption of an identical distribution. Another example is the modeling of the maximal sea-level height at Neeltje Jans over n consecutive days. Clearly, if a several-day-long storm surge is observed, the sea levels on consecutive days are not independent. It always depends on the application of interest to what extent one can justify (formally or empirically) certain assumptions. Although there are more sophisticated modeling techniques available if the iid assumption is violated, it will be lurking in the background of some models we discuss in this and the next chapter. What is important for us at the moment is simply to know and understand such a crucial assumption and then to focus on what we can learn from it about the modeling of risks. We need to learn how to walk before we can run.

We now return to the modeling of repeated coin tosses. The resulting distribution of the number of, say, heads, is the binomial distribution.

Example 8.24 (Binomial distribution)**.** Having the notions of the Bernoulli distribution, independence and stochastic representations at hand, we can introduce the binomial distribution. If $X_i \overset{\text{iid}}{\sim} \mathrm{B}(1,p)$, $i = 1, 2, \dots, n$, then

$$X = S_n = \sum_{i=1}^{n} X_i \tag{8.17}$$

follows a *binomial distribution*, denoted by $X \sim \mathrm{B}(n,p)$. Since each X_i can only take on the values 0 or 1, we immediately see from (8.17) that X can only take on values in $\{0, 1, \dots, n\}$, almost surely. In other words, the support of $\mathrm{B}(n,p)$ is $\{0, 1, \dots, n\}$. The binomial distribution models the number of successes when one independently repeats the same experiment with outcomes "success" (with probability p) and "failure" (with probability $1 - p$). Such success–failure experiments with equal success probability p are also known as *Bernoulli trials*.

To derive the probability mass function of $X \sim \mathrm{B}(n,p)$, we need to count the number of ways in which we can assign $x \in \{0, 1, \dots, n\}$ successes to n Bernoulli trials when the order of the successes is irrelevant. This corresponds conceptually to drawing x balls from an urn containing n balls numbered $1, 2, \dots, n$ when no ball can be drawn more than once (so without replacement) and when the order in which the balls are drawn is irrelevant (so without considering the order). This is one of the urn models typically discussed in combinatorics in high school, and the said number is the *binomial coefficient* $\binom{n}{x}$ (read: "n choose x"), given by

$$\binom{n}{x} = \frac{n(n-1)\cdots(n-x+1)}{x!}.$$

A classic example application of binomial coefficients is to calculate the number of possible outcomes in a lottery where $x = 6$ balls are drawn from an urn with $n = 49$ numbered balls. This lottery, known as *Lotto 6/49*, has indeed $\binom{49}{6} = 13\,983\,816$ possible outcomes. The chance that you pick the right six numbers in such a lottery is therefore 1 in 13 983 816, so roughly 7.1511×10^{-8}. Although the official numbers vary, the National Weather Service of the United States estimates the probability of being struck by lightning once in your lifetime as about 1 in 15 300 or $1/15\,300 \approx 6.5359 \times 10^{-5}$, which is about 914 times larger than the chances of winning the lottery. A joke among probabilists says that until you have been struck that many times by lightning, there's no point in starting to play the lottery!

With the binomial coefficient at hand, we can now derive the probability mass function of $X \sim \mathrm{B}(n, p)$. For $x \in \{0, 1, \ldots, n\}$, the probability that $X = x$ is the probability that we have x successes and $n - x$ failures. There are $\binom{n}{x}$ possible ways of obtaining x successes: x successes happen with probability p^x and the remaining $n - x$ Bernoulli trials give failures, which happen with probability $(1 - p)^{n-x}$. We thus obtain the probability mass function

$$f(x) = \begin{cases} \binom{n}{x} p^x (1-p)^{n-x}, & x \in \{0, 1, \ldots, n\}, \\ 0, & \text{otherwise.} \end{cases} \tag{8.18}$$

Bernoulli trials can be used to model, for example, n independent tosses of a fair coin. If $X_1, X_2, \ldots, X_n \overset{\text{iid}}{\sim} \mathrm{B}(1, 1/2)$ denote the outcomes of the n tosses (with $X_i = 1$ if the ith toss shows heads and $X_i = 0$ if it shows tails) then X in (8.17) is the total number of heads obtained in the n tosses, and the probability of obtaining a certain number of heads can be determined with (8.18).

Another example occurs when $X_1, X_2, \ldots, X_n$ indicate whether the yearly maximal sea level at Neeltje Jans in n consecutive years exceeds a 3 m dike, so whether ($X_i = 1$) or not ($X_i = 0$) there was a high-water-mark exceedance in a particular year over n consecutive years. In order for X in (8.17) to follow a binomial distribution, we would then need to assume that the maximal sea levels per year (that is, the $X_1, X_2, \ldots, X_n$) are independent across different years (a less restrictive assumption than if we considered n consecutive days, as mentioned before) and that all follow the same distribution $\mathrm{B}(1, p)$ and in particular share the same probability p of exceeding 3 m.

Note that in the last example we used $X_i = 1$ to indicate whether there was a high-water-mark exceedance, so the modeling 'success' is a natural disaster in this case. This is mathematically most natural as a disaster is our event of interest, but, of course, we could have equally well modeled an exceedance as $X_i = 0$ and no exceedance as $X_i = 1$.

The following example considers a discrete distribution with infinite support, so it is possible to assign non-zero probability mass to each of countably infinitely many points in such a way that all probabilities add up to 1.

Example 8.25 (Geometric distribution)**.** Suppose we do not have a fixed number of independent Bernoulli trials but we repeat the experiment indefinitely and count the number of trials it takes until the first success. For example, we toss a coin until we see heads for the first time. Or we count the number of days before the 3 m sea-level threshold at Neeltje Jans is exceeded; see Figure 1.6. The random variable X modeling this number of independent Bernoulli trials (each with success probability $p \in (0,1]$) that have taken place until the first success follows a *geometric distribution*, denoted by $X \sim \text{Geo}(p)$. Its probability mass function has as support the positive integers $\mathbb{N} = \{1, 2, \dots\}$ and is given by

$$f(x) = \begin{cases} p(1-p)^{x-1}, & x \in \mathbb{N}, \\ 0, & \text{otherwise.} \end{cases}$$

For positive integer x, the event $X = x$ means that we see the first success on the xth trial. In other words, the first $x - 1$ trials must have been failures and the xth trial was a success. Since the trials are independent, $\mathbb{P}(X = x)$ is the product of the probabilities over all trials. Because the first $x - 1$ trials were failures, they happened with probability $1 - p$ each, so we obtain the factor $(1 - p)^{x-1}$, and because the xth trial was a success, which happened with probability p, we obtain $\mathbb{P}(X = x) = p(1 - p)^{x-1}$. This is reflected correctly by the probability mass function of Geo(p). Note that if $p = 1$ then the success probability is 1, so the first success will always happen on the first trial. In this degenerate case we correctly have $f(1) = 1(1 - 1)^0 = 1$, so $X = 1$ almost surely.

The left-hand side of Figure 8.17 shows the graphs of the probability mass function f for different parameters p. Note that only the dots provide the correct function values of f; the lines connecting the dots merely serve as a visual aid to better distinguish the function values. The correct value of $f(x)$ for non-integer x is of course $f(x) = 0$.

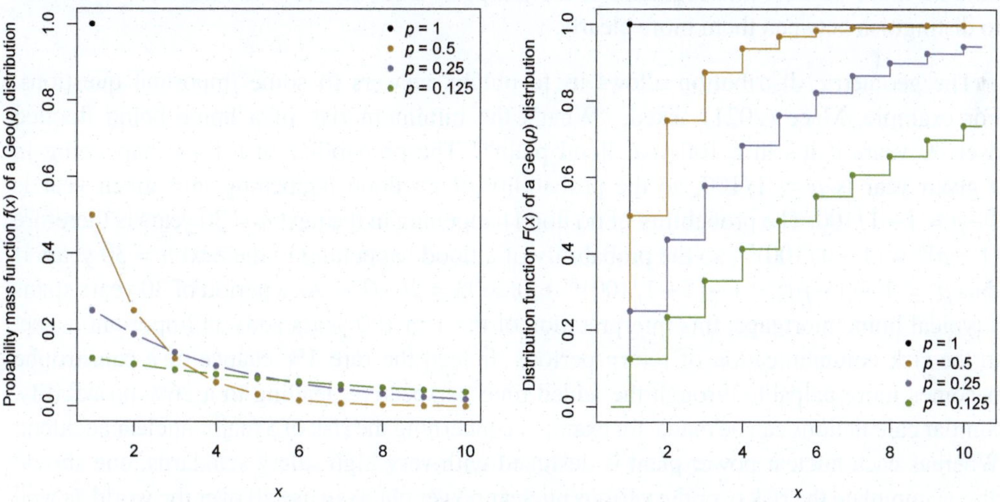

Figure 8.17 Probability mass functions (left) and distribution functions (right) of Geo(p) distributions for $p \in \{1, 1/2, 1/4, 1/8\}$. The lines connecting the dots in the left plot and the vertical lines in the right plot are only there to visually distinguish the curves. Source: Authors

The probability mass function f of the Geo(p) distribution is simple enough that we can express its distribution function F in a closed-form expression, that is, in *analytical form*.

In other words, the distribution function is *analytically tractable*. To express it, we need the *floor function* $\lfloor x \rfloor$ (read: "the floor of x") of a real number x, that is, the largest integer less than or equal to x; for example $\lfloor 2 \rfloor = 2$, $\lfloor 2.1 \rfloor = 2$ and $\lfloor 3.9 \rfloor = 3$. For deriving F, we already know that X almost surely takes on positive integer values, so $\mathbb{P}(X \le x) = \mathbb{P}(X \le \lfloor x \rfloor)$. This holds because no probability mass lies in the interval $(\lfloor x \rfloor, x]$ since X does not take on values there almost surely; for these types of discussion, Section 8.3.1 is crucial. Since $\lfloor x \rfloor$ is an integer we can write $\mathbb{P}(X \le \lfloor x \rfloor)$ as $\mathbb{P}(X = 1) + \mathbb{P}(X = 2) + \cdots + \mathbb{P}(X = \lfloor x \rfloor)$, which is $f(1) + f(2) + \cdots + f(\lfloor x \rfloor)$ and so by plugging in the form of f from above we obtain

$$F(x) = \mathbb{P}(X \le x) = \mathbb{P}(X \le \lfloor x \rfloor) = \sum_{k=1}^{\lfloor x \rfloor} f(k) = p \sum_{k=1}^{\lfloor x \rfloor} (1-p)^{k-1}$$

$$= p((1-p)^0 + (1-p)^1 + \cdots + (1-p)^{\lfloor x \rfloor - 1}) = p \sum_{k=0}^{\lfloor x \rfloor - 1} (1-p)^k. \tag{8.19}$$

How can we compute the latter sum? By realizing that it is a geometric sum. Applying Lemma 6.1 (with $r = 1 - p$ and $n = \lfloor x \rfloor - 1$), we obtain that

$$F(x) = p\frac{1-(1-p)^{\lfloor x \rfloor}}{1-(1-p)} = 1 - (1-p)^{\lfloor x \rfloor}, \quad x \ge 0.$$

We see from this formula that, for any $p \in (0, 1]$, we indeed have $F(x) \to 1$ for $x \to \infty$, for example.

The right-hand side of Figure 8.17 shows the graphs of F for different p. As before, we have included vertical line segments at the jumps of these distribution functions to be able to distinguish between them more clearly.

The geometric distribution allows us to obtain answers to some important questions. For example, Mace (2021) asked: "What's the minimum risk of a home being flooded over 30 years if it's in a 100-year flood plain?" The probability of a flood happening in a given year is $p = 1/100$, so the probability of no flood happening in a given year is $1 - p = 1 - 1/100$. The probability of no flood happening in the next $n = 30$ years is therefore $(1-p)^n = (1 - 1/100)^{30}$, so the probability of a flood happening in the next $n = 30$ years is thus $p^* = 1 - (1-p)^n = 1 - (1 - 1/100)^{30} \approx 0.2603 = 26.03\%$. As a period of 30 years spans a typical home mortgage, this interpretation of the 1 in 100 years adds an important aspect to the risk communication of return periods. Indeed the rare 1% chance of a catastrophe becomes more palpable through the added time perspective, leading to a 26% probability. Similar calculations can be made, for example concerning the risk of a major nuclear accident. Whereas each nuclear power plant is designed with very high safety standards, one should also accumulate the risk over the various nuclear power plants in use all over the world as well as over their average lifespan, referred to as reactor years. An almost negligible probability of a severe accident for one particular plant then quickly becomes non-negligible; there are 440 reactors operable as of April 2020. We highly recommend reading the interesting paper of Rose and Sweeting (2016), which also highlights the problem of obtaining reliable data. The authors come to the following chilling conclusion: "By our calculations, the overall probability of a core-melt accident in the next decade, in a world with 443 reactors, is almost 70%. (Because of statistical uncertainty, however, the probability could range from about

28% to roughly 95%.) The United States, with 104 reactors, has about a 50% probability of experiencing one core-melt accident within the next 25 years." We will discuss the reporting of a statistical uncertainty range in the context of confidence intervals, in Section 8.6.3. The following derivation addresses a problem of a similar nature as the flood problem above, but where p and p^* are given and we are interested in n. These examples highlight the importance of basic tools from probability and statistics in understanding and communicating risk.

In the above calculations we assumed that the individual Bernoulli events are iid. In practice this is rarely the case and needs to be addressed. Often, iid-based results turn out to yield lower bounds for the real probabilities; see, for instances, Section 9.5.3. As such, these calculations yield a first, though important, step on the modeling ladder.

Derivation 8.26 (How many years?). As mentioned in Example 8.24, let us assume that the yearly maximal sea levels at Neeltje Jans are independent. If we know the probability $p \in (0, 1]$ of a high-water-mark exceedance occurring in a particular year at Neeltje Jans, we can answer the following question: what is the maximal number of years for which the probability of such an exceedance is at most $p^* \in (0, 1]$?

Under the assumption stated, we have independent Bernoulli trials, one per year, with "success" if there is a high-water-mark exceedance. As such, we know that the number of years X until we see the next exceedance is Geo(p)-distributed. The question asks for the largest n such that the probability of the next exceedance happening in the next n years is at most p^*. Translating this question into mathematical language, we need to determine the largest integer n such that $\mathbb{P}(X \leq n) \leq p^*$. Since we have $\mathbb{P}(X \leq n) = F(n) = 1 - (1 - p)^{\lfloor n \rfloor} = 1 - (1 - p)^n$, we are looking for the largest integer n such that $1 - (1 - p)^n \leq p^*$. This is equivalent to $1 - p^* \leq (1 - p)^n$, which holds if and only if $\log(1 - p^*) \leq n \log(1 - p)$. Since $\log(1 - p) < 0$, the latter inequality is equivalent to $n \leq \log(1 - p^*)/\log(1 - p)$. The left-hand side of this inequality is the integer n. Therefore we can round down the right-hand side $\log(1 - p^*)/\log(1 - p)$ without changing the numbers n that solve this inequality. We are thus looking for the largest n such that $n \leq \lfloor \log(1 - p^*)/\log(1 - p) \rfloor$. Every n at most as large as $\lfloor \log(1 - p^*)/\log(1 - p) \rfloor$ fulfills this inequality. So, the answer to the question of the maximal number n of years over which an exceedance happens with probability at most $p^* \in (0, 1]$ is

$$\left\lfloor \frac{\log(1 - p^*)}{\log(1 - p)} \right\rfloor. \tag{8.20}$$

From a mathematical point of view, we have obtained a formula for general parameters p and p^*. We see from this formula that the smaller is p (so the less likely that a high-water-mark exceedance occurs) or the larger is p^* (the more likely we allow an exceedance to occur), the larger is the number of years n. As already stressed in Section 8.1.1, such model properties are important to know besides answers for concrete values of p and p^*. If a model behaves fundamentally differently from our gut feeling, then one of the two may be (or: is) wrong, which is a warning sign; recall AAA-rated AIG from Section 3.4 on the one hand and subprime/ninja loans on the other. This can be a particular problem for highly complex models (such as CDOs, CDO-squared, CDO-cubed, as we briefly met in Chapter 3, in particular, in Section 3.3), for which developing a gut feeling is challenging and a more objective assessment is not easily available. In such cases, it can sometimes be preferable to use a simpler model with known limitations than a more complex one with unknown limitations. To find this balance, one often compares different models of increasing sophistication.

From a practical point of view, we should ask where p and p^* come from. The probability p of a high-water-mark exceedance in a particular year can be approximately computed (or *estimated*) from yearly maximal sea-level data measured at Neeltje Jans. Alternatively, a scenario analysis could determine that value ("What if p is equal to . . . "), for example by asking experts to judge. The choice p^* is something on which politics and the public have to agree. With $p^* = 1/2$, the above question essentially reads: how many years will it take until the next exceedance is likely to occur? Depending on how many lives are at stake, one would hope that a much smaller p^* is chosen, though. Overall, this simple example with two parameters already shows that the proper modeling of risk requires people from various backgrounds, such as mathematicians, statisticians, engineers, economists, politicians and the public.

Say that one found $p = 1/10\,000$ and $p^* = 1/100$, for which (8.20) leads to 100 years. We already see the media reporting "Hooray, we are safe for the next 100 years!". In the communication of such results, we need to be very careful, stressing again the underlying assumptions and ideally even anticipating misinterpretations. The computed n provides the largest number of years for which the probability of the next high-water-mark exceedance is at most p^* *if* yearly exceedances can be modeled as independent Bernoulli trials with success probability p. Neither does n provide the number of years for which you are safe from a high-water-mark exceedance (an exceedance can still occur with probability p in every year according to our Bernoulli trials), nor is n accurate if one of our modeling assumptions is violated. For example *what if* yearly exceedances are not independent or do not happen with the same probability p every year? Or *what if* p is estimated using assumptions that are not justifiable?

The following lemma addresses a standard result in compound interest calculations which dates back to Euler, hence the e. We will use it on several occasions.

Lemma 8.27 (Convergence to the exponential). *If $a_1, a_2, \ldots$ is a sequence of numbers converging to a, then*

$$\lim_{n\to\infty}\left(1 + \frac{a_n}{n}\right)^n = e^a.$$

Proof The argument we give here is correct and rather short, but it takes a bit more work to make it mathematically precise. We first show that $\log((1 + a_n/n)^n) \to a$ for $n \to \infty$. By the power formula for the logarithm, this is equivalent to showing that $n\log(1 + a_n/n) \to a$. Since $a_n \to a$, we know that $a_n/n \to 0$ for $n \to \infty$. By using that $\log(1 + x) \approx x$ for x near 0, we obtain for $n \to \infty$ that

$$n\log\left(1 + \frac{a_n}{n}\right) \approx n\frac{a_n}{n} = a_n \to a.$$

The result as stated now follows on applying the exponential function $\exp(x) = e^x$ and using the fact that if $b_n \to b$ then $g(b_n) \to g(b)$ for any continuous function g (we apply this result for $g(x) = \exp(x)$). □

Derivation 8.28 (Poisson limit theorem). On the basis of the binomial distribution we will now derive another discrete distribution with infinite support, the Poisson distribution.

The Poisson distribution depends on the parameter $\lambda > 0$ (another Greek letter) and this distribution is used to model the number of events over time, if such events occur at a constant rate per fixed unit of time and independently of the last time an event happened. For example, if we receive about five emails per hour, then the unit of time is one hour and $\lambda = 5$; see Figure 8.18 for a visualization. Or if we check the exceedance of a given high sea water level

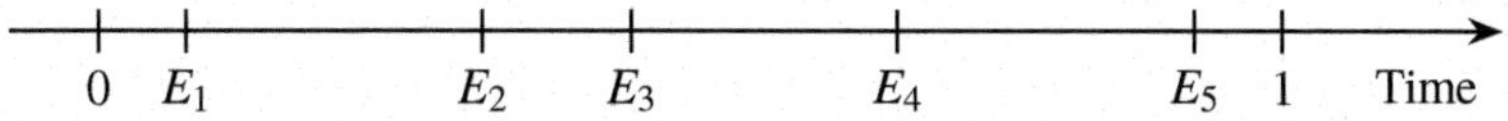

Figure 8.18 The times $E_1, E_2, \ldots, E_5$ at which five emails arrive within one hour. Source: Authors

at Neeltje Jans on average once every three hours, independently of the last time we checked, then it would be natural to take as a unit of time one day and $\lambda = 24/3 = 8$ (as we measure eight water heights per day). Alternatively, we could fix the unit of time as one hour and take $\lambda = 1/3$. Either way, our set-up is that we fix some unit of time and that we have a parameter λ which describes how many events on average we see per this unit of time.

It might seem a bit confusing (and it indeed is for some students) that we suddenly have a time component, but this is only because we have covered rather basic mathematical models so far. The notion of time adds an often unnecessary additional layer of complexity at this stage and is thus omitted. However, in many applications a time component is unavoidable and indeed crucial for the phenomena being modeled, and the Poisson distribution is an example of a distribution that becomes important when random phenomena are modeled over time. As we will see, the Poisson distribution most naturally appears from the binomial in this set-up and this may well be the first time one typically encounters time in a probabilistic context.

If we divide the unit time interval into n equal parts such that in each of the n time sub-intervals one sees at most one event taking place (think of a sufficiently large n), then we have n independent Bernoulli trials and are thus in the set-up of the binomial distribution; see Figure 8.19 for a visualization in terms of the email example. The probability that we

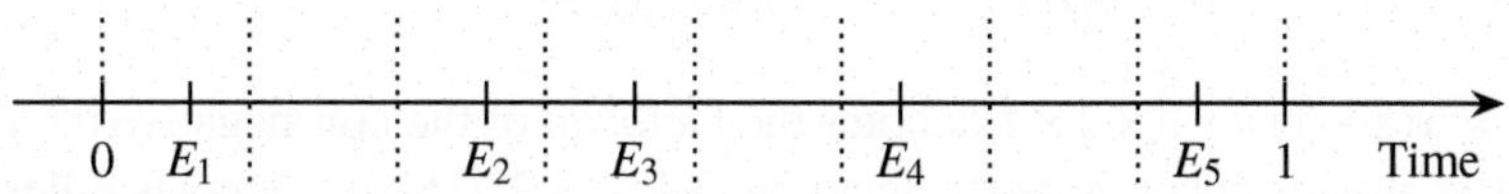

Figure 8.19 The times at which the five emails $E_1, E_2, \ldots, E_5$ arrive, now with divisions of the time line into $n = 8$ equidistant parts. In each part, either zero emails arrive or one email arrives (independent Bernoulli trials). Source: Authors

see x events ("successes") happening in one unit of time is given by the $\mathrm{B}(n, p)$ probability mass function at x, where p is the probability of success in each Bernoulli trial (that is, the probability of an event happening in precisely one of the n time sub-intervals). Just as (8.18) suggests, we thus have for $X \sim \mathrm{B}(n, p)$ that

$$\mathbb{P}(X = x) = \binom{n}{x} \mathbb{P}(\text{“}x \text{ successes”}) \mathbb{P}(\text{“}n - x \text{ failures”}) = \binom{n}{x} p^x (1-p)^{n-x}.$$

Expanding the binomial coefficient $\binom{n}{x}$ and writing $p^x = (np)^x/n^x$, as well as $p = (np)/p$, we obtain that

$$\mathbb{P}(X = x) = \binom{n}{x} p^x (1-p)^{n-x} = \frac{n(n-1)\cdots(n-x+1)}{x!} p^x (1-p)^{n-x}$$
$$= \frac{n(n-1)\cdots(n-x+1)}{n^x} \frac{(np)^x}{x!} \frac{(1-(np/n))^n}{(1-p)^x}. \tag{8.21}$$

If we let the number n of division points of the unit time interval tend to infinity (so $n \to \infty$), then the probability p of seeing an event occurring in one of these smaller and smaller time sub-intervals tends to 0 (so $p \downarrow 0$). As we expect to see np events per unit of time, we must also have that $np \to \lambda$ (think of the email example; if $n = 60$, so that we are dividing an hour into minutes, the probability of receiving an email in one minute is $p = 5/60 = 1/12$, so we expect to see $np = 5 = \lambda$ emails per hour; note that we more formally introduce the concept of "expectation" in Section 8.4.1 later). So if we let $n \to \infty$ and $p \downarrow 0$ such that $np \to \lambda$, then the three fractions in (8.21) converge as follows. For the first, splitting the fraction into a product of fractions and dividing the numerator and denominator of each by n leads to

$$\frac{n(n-1)\cdots(n-x+1)}{n^x} = \frac{n(n-1)\cdots(n-x+1)}{n \times n \times \cdots \times n}$$
$$= \frac{n}{n} \frac{n-1}{n} \cdots \frac{n-x+1}{n}$$
$$= 1 \times \frac{1-1/n}{1} \cdots \frac{1+(-x+1)/n}{1}$$

from which it is immediate that each fraction (and thus the product) converges to 1. For the second fraction in (8.21), it follows from $np \to \lambda$ that $((np)^x/x!) \to (\lambda^x/x!)$. And for the third fraction in (8.21), the denominator converges to 1 (since $p \downarrow 0$). By Lemma 8.27, the numerator $(1-(np)/n)^n$ converges to $e^{-\lambda}$. We thus obtain that (8.21) converges to

$$f(x) = \begin{cases} (\lambda^x/x!)e^{-\lambda}, & x \in \mathbb{N}_0 = \{0, 1, 2, \dots\}, \\ 0, & \text{otherwise}. \end{cases} \tag{8.22}$$

Note that $x! = x(x-1) \times \cdots \times 2 \times 1$ denotes the *factorial* of the non-negative integer x (with $0! = 1$). And, as already encountered in Section 8.3.1, $e = 2.7182\dots$ denotes Euler's number. Clearly, f is non-negative. Knowing the series representation of the exponential function, $e^z = \sum_{k=0}^{\infty}(z^k/k!)$, we can verify that $\sum_{x=0}^{\infty} f(x) = e^{-\lambda} \sum_{x=0}^{\infty}(\lambda^x/x!) = e^{-\lambda}e^{\lambda} = e^0 = 1$, so that f is indeed a valid probability mass function, namely that of the *Poisson distribution*. Utilizing the Poisson distribution to approximate $\mathrm{B}(n, p)$ is known as the *Poisson limit theorem*. For practical purposes, the theorem is often written as $\mathbb{P}(X = x) \approx e^{-np}(np)^x/x!$ for x large and p small and referred to as the *law of rare events*. The latter name indicates that the probability of success p of $\mathrm{B}(n, p)$ is small; see also our discussion relating to extreme value theory in Section 9.2.

The approximation of a $\mathrm{B}(n, p)$ distribution by a Poisson distribution with parameter λ holds if $n \to \infty$, $p \downarrow 0$ and $np \to \lambda$. In practice one typically uses $n \geq 20$, $p \leq 1/20$ and $\lambda = np$ as a rule of thumb.

Example 8.29 (Poisson distribution)**.** The *Poisson distribution*, denoted by $X \sim \text{Poi}(\lambda)$ for a parameter $\lambda > 0$, is the distribution with probability mass function (8.22). Like the Poisson limit theorem, the Poisson distribution is named after Siméon Denis Poisson (1781–1840); see Figure 8.20.

Figure 8.20 Left: Siméon Denis Poisson (1781–1840). Right: Not to be confused with (or pronounced as) "poison". Sources: Wikipedia (left) and Enrico Chavez (right)

Similarly to Figure 8.17, Figure 8.21 shows the graphs of the probability mass function f and the corresponding distribution function F for different parameters $\lambda > 0$.

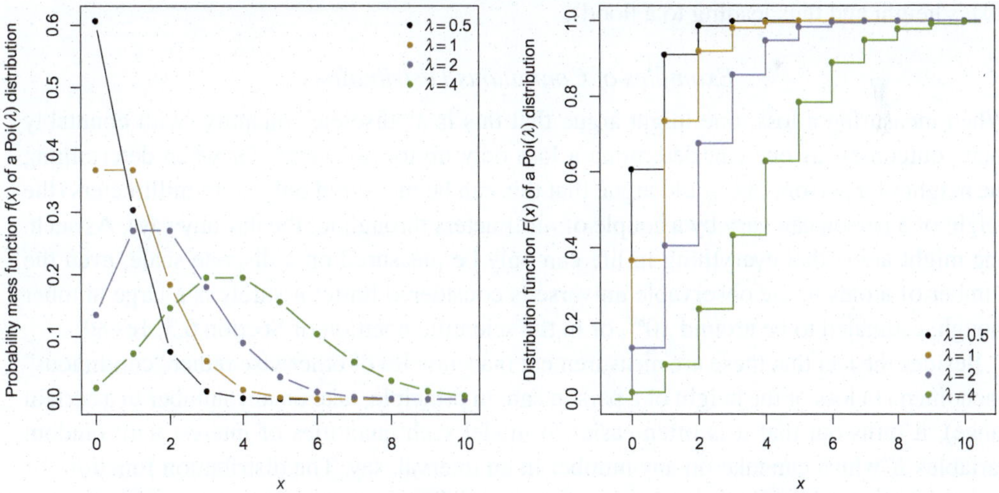

Figure 8.21 Probability mass functions (left) and distribution functions (right) of $\text{Poi}(\lambda)$ for $\lambda \in \{1/2, 1, 2, 4\}$. The lines connecting the dots in the left plot and the vertical lines in the right plot serve to visually distinguish the curves. Source: Authors

Coming back to the Poisson limit theorem of Derivation 8.28, the left-hand side of Figure 8.22 shows the probability mass function $\text{B}(n = 20, p)$ (solid lines) with that of $\text{Poi}(\lambda = np)$ (dashed lines) for $p \in \{0.01, 0.02, \ldots, 0.05\}$. The right-hand side of this figure shows the

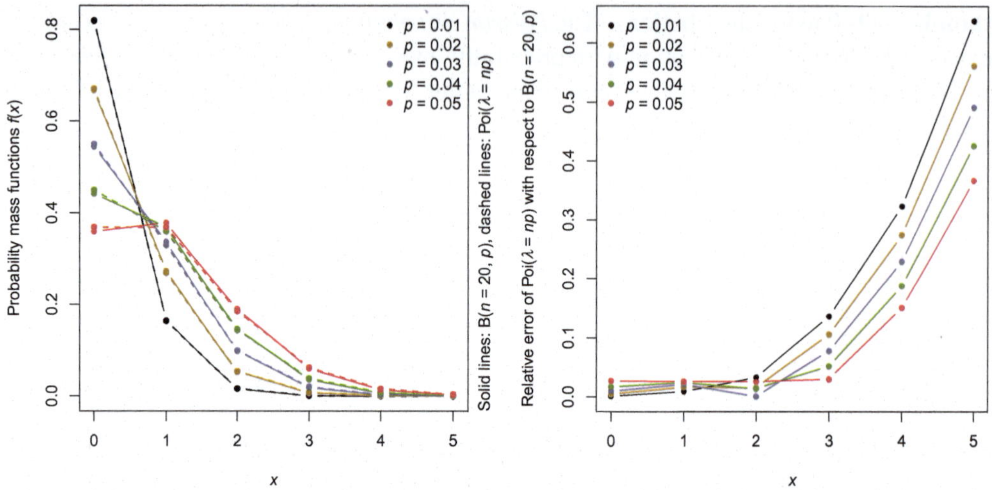

Figure 8.22 Left: Probability mass function of $\mathrm{B}(n = 20, p)$ (solid lines) with those of $\mathrm{Poi}(\lambda = np)$ (dashed lines) for $p \in \{0.01, 0.02, \ldots, 0.05\}$. Right: The corresponding relative errors. Source: Authors

respective relative errors (if $f_{\mathrm{B}(n,p)}$ and $f_{\mathrm{Poi}(\lambda)}$ denote the respective probability mass functions, their relative error is plotted as $|f_{\mathrm{Poi}(\lambda)}(x) - f_{\mathrm{B}(n,p)}(x)|/f_{\mathrm{B}(n,p)}(x)$).

We will revisit the Poisson distribution in Section 9.4 when we model the number of exceedances over a high threshold over time (think of extreme water heights exceeding a dike's height and thus leading to a flood).

Examples of Continuous Distributions

When measuring a loss, one might argue that this is a "discrete" quantity (with countably many outcomes) as one can determine a loss only up to, say, cents. Or when determining the height of a person, one could argue that this can be measured only up to millimeters (the height of a person can vary by a couple of millimeters throughout the day anyway). As such, one might argue that everything in life can only be measured on a discrete scale, even the number of atoms in the observable universe is considered finite; it surely is a large number though, estimated to be around 10^{80}, or in the scientific notation of Section 6.3, 1e+80.

Besides the fact that these are measurement inaccuracies of otherwise rather "continuous" phenomena (a loss or the height of a person can, in principle, take on any number in a certain range), it turns out that it is often easier to model such quantities of interest with random variables X which can take on any number in an interval, say. The distribution functions of such random variables X are continuous, and essentially all important continuous distribution functions allow for characterization via a function f akin to probability mass functions for discrete distribution functions. In the remaining part of this section we focus on such continuous distribution functions; the corresponding f are known as "densities".

Before being able to more formally introduce the "density" f of a distribution function F, we need another tool from high school mathematics, namely that of an integral. You may remember that the integral of a non-negative function is the area under its graph. Here is an example.

Example 8.30 (Pareto distribution)**.** For a parameter $\theta > 0$, consider $F(x) = 1 - 1/(1+x)^\theta$, $x \ge 0$. This is a distribution function; for $\theta = 1$ we have already seen its graph as the center panel of Figure 8.11. A random variable $X \sim F$ is said to follow a *Pareto distribution*, denoted by $X \sim \mathrm{Par}(\theta)$. The Pareto distribution is named after the economist Vilfredo Pareto (1848–1923); see Figure 8.23 (jointly with a cartoon). While at the University of Lausanne in 1896,

Figure 8.23 Left: Vilfredo Pareto (1848–1923). Right: The Pareto principle. Sources: Wikimedia Commons (left) and Enrico Chavez (right)

he showed that 80% of the land in Italy was owned by 20% of the population. This 80/20 ratio also applies to many other types of outcomes and is more generally known as the *Pareto principle* or 80/20 *rule*. According to the Pareto principle, 80% of consequences or outcomes come from 20% of the causes or inputs; see Figure 8.23 for a cartoonist's interpretation. Note that the 80/20 rule can also be formulated in terms of the Pareto distribution $\mathrm{Par}(\theta)$ for $\theta = 1.4$; see Embrechts et al. (1997, Definition 8.2.2, Table 8.2.5) for more details.

The distribution function F of a Pareto distribution has the property that there exists a function f such that $F(x)$ can be interpreted as the area under the graph of f to the left of x. As so often, we can use mathematical language to make this statement precise. Here it is in terms of an integral:

$$F(x) = \int_{-\infty}^{x} f(z)\,\mathrm{d}z, \quad x \in \mathbb{R} \tag{8.23}$$

(read: "F of x is the integral of f of z from minus infinity to x"). For $\mathrm{Par}(\theta)$, the function f (the "density" of the Pareto distribution) is

$$f(x) = \begin{cases} \theta(1+x)^{-\theta-1}, & x > 0, \\ 0, & x \le 0, \end{cases} \tag{8.24}$$

so it is a *power law*. As in the discrete case, we call the set of x such that $f(x) > 0$ the *support* of f (or F or $X \sim F$), which is $(0, \infty)$ for Pareto distributions. For f as in (8.24), we can indeed verify that (8.23) holds:

$$\int_{-\infty}^{x} f(z)\,\mathrm{d}z = \int_{0}^{x} f(z)\,\mathrm{d}z = \int_{0}^{x} \theta(1+z)^{-\theta-1}\,\mathrm{d}z = \left[-(1+z)^{-\theta}\right]_{0}^{x}$$
$$= -(1+x)^{-\theta} - (-1) = 1 - (1+x)^{-\theta}.$$

The correctness of this integration can also be verified by differentiating F to obtain f, so $F'(x) = (1-(1+x)^{-\theta})' = -(-\theta)(1+x)^{-\theta-1} = \theta(1+x)^{-\theta-1}$; see also Remark 8.32 (2) below.

The representation (8.23) is somewhat reminiscent of adding up the probability mass function to obtain $F(x)$ for a discrete F; see Example 8.25. Indeed, an integral can be viewed as the limit of such sums. If this limit exists such that writing an integral as in (8.23) makes sense, we call f *integrable*. Figure 8.24 shows the graphs of the distribution function F and its corresponding density f for different parameters θ.

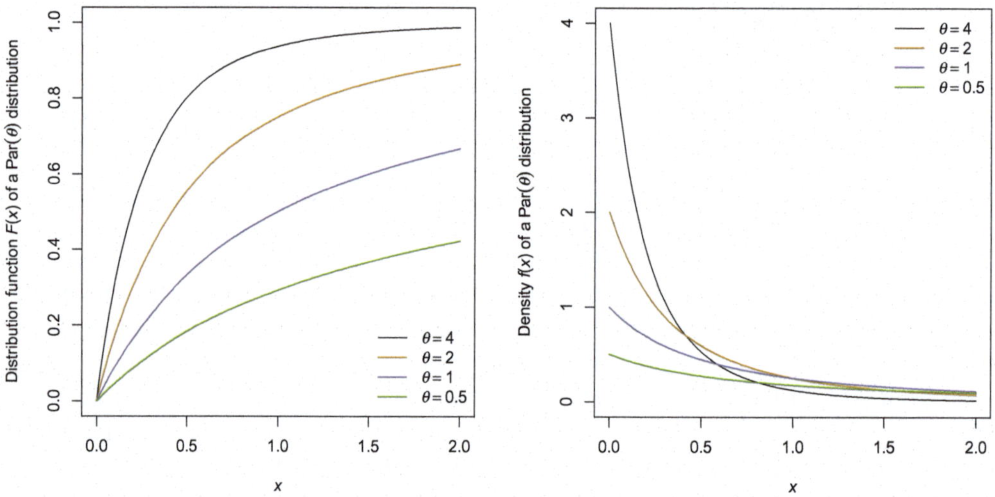

Figure 8.24 Left: Par(θ) distribution functions for $\theta \in \{4, 2, 1, 1/2\}$. Right: The corresponding densities. Source: Authors

Overall, the Pareto distribution is a very nice example of a continuous distribution to keep in mind, as most quantities of interest are easily calculated for it and as it has important properties for risk management purposes; see, for example, Sections 8.4.2 and 8.4.3. In Chapter 9 we will encounter a generalized version of the Pareto distribution with two parameters. This distribution will turn out to be of the utmost importance for the modeling of extreme events.

We can now more formally introduce the concept of a density f of a distribution.

Concept 8.31 (Density). If the distribution function F of a random variable X can be written as

$$F(x) = \int_{-\infty}^{x} f(z)\,\mathrm{d}z, \quad x \in \mathbb{R}, \tag{8.25}$$

for a function f which is non-negative, integrable and which satisfies

$$\int_{-\infty}^{\infty} f(z)\,\mathrm{d}z = 1, \tag{8.26}$$

then X and F are called *(absolutely) continuous* and f is called the *density* of X or F; see the left-hand side of Figure 8.25 for an interpretation (the area under the density f up to 4 equals $F(4) = 0.8$).

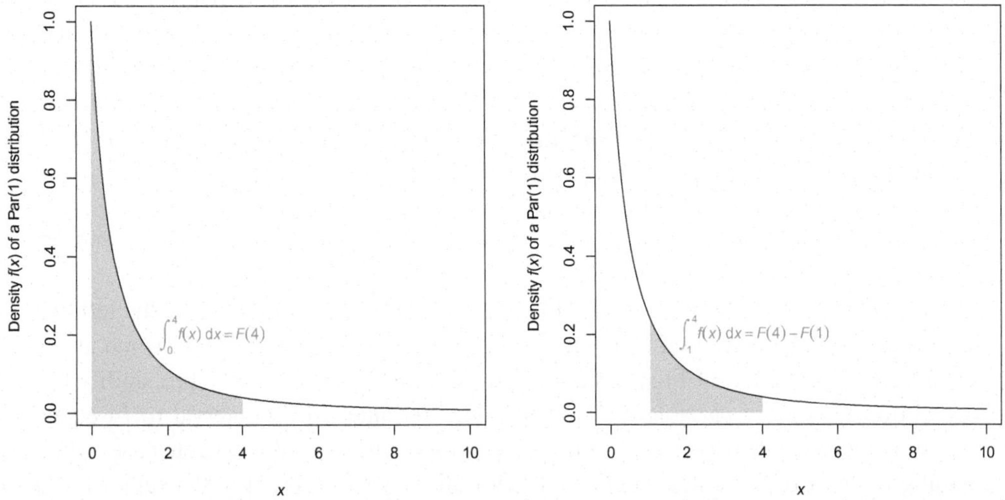

Figure 8.25 Left: The gray area under the density f of a Par(1) distribution from 0 to 4 equals $F(4) = 0.8$. Right: The same but from 1 to 4 equals $F(4) - F(1) = 0.8 - 0.5 = 0.3$. Source: Authors

Several remarks are in order here.

Remark 8.32.

(1) The right-hand side of (8.25) is the area under the graph of the density f (so it is the area between the x-axis and the graph of f) to the left of x.
(2) As already seen in Example 8.30, the density f can be obtained by differentiating the distribution function F.
(3) The density is one of the central objects in statistics. Given some data, viewed as realizations of random variables $X_1, X_2, \ldots, X_n$ from an unknown distribution function F, one often assumes that F is of some parametric form, say, F is the distribution function of a Pareto distribution Par(θ) but we do not know θ. The goal is then to *estimate* θ from $X_1, X_2, \ldots, X_n$, that is, to approximate the value of θ based on $X_1, X_2, \ldots, X_n$ such that the distribution function of the resulting Par(θ) distribution "best" represents the unknown F. Different ideas about what "best" means lead to different methods of finding this "best" θ. A function which maps the random variables $X_1, X_2, \ldots, X_n$ to θ is called an *estimator* and its realization (a number based on realizations $x_1, x_2, \ldots, x_n$ of $X_1, X_2, \ldots, X_n$) is an *estimate*. An estimator or estimate is typically denoted by $\hat{\theta}_n$, the hat notation indicating an estimator or estimate and the subscript n denoting the sample

size; we already encountered the estimate $\hat{F}_n$ for the empirical distribution function when it was introduced; see Concept 8.16. There we estimated a whole distribution function instead of just a parameter of a distribution. Using the $\mathrm{Par}(\hat{\theta}_n)$ distribution then best represents (the realizations of) $X_1, X_2, \ldots, X_n$ in the sense in which we constructed $\hat{\theta}_n$. One of the most popular estimators is the *maximum likelihood estimator*. It utilizes the density (here: the Pareto density) to return the value of θ under which obtaining the given data is "most likely" among all distributions considered (here: $\mathrm{Par}(\theta)$ for all $\theta > 0$). Being based on the density of the distribution of interest, the popular maximum likelihood estimator $\hat{\theta}_n$ of a distributional parameter θ is one reason for the importance of densities in statistics. There are many more statistical applications that utilize the density. And the same is true for discrete distributions with a probability mass function.

(4) An important property following from the central formula (8.11) and representation (8.25) is that if $X \sim F$ has density f then, for all $-\infty < a < b < \infty$,

$$\mathbb{P}(a < X \leq b) = F(b) - F(a) = \int_{-\infty}^{b} f(z)\,\mathrm{d}z - \int_{-\infty}^{a} f(z)\,\mathrm{d}z = \int_{a}^{b} f(z)\,\mathrm{d}z, \qquad (8.27)$$

where we use the geometric property that the area under the graph of f to the left of b minus the area under the graph of f to the left of a is the area under f between a and b; see the right-hand side of Figure 8.25 for an illustration. In words, the probability that a continuous random variable $X \sim F$ will take on values in the interval $(a, b]$ equals the area under the density f of F over $(a, b]$. This is similar to the discrete case, where adding the probability mass function $f(x)$ for all values $x \in (a, b]$ in the support of f provides us with $\mathbb{P}(a < X \leq b)$. In particular, summing a probability mass function f over all values in its support results in all the probability mass being summed up, and so 1, the corresponding condition for densities is (8.26).

(5) One typically considers densities f only for values of x in their support; for a $\mathrm{Par}(\theta)$ distribution, for example, $x > 0$. We see from (8.27) why this is sufficient: if $f(x) = 0$ for all $x \in (a, b]$ then $\int_a^b f(z)\,\mathrm{d}z = 0$ and thus $\mathbb{P}(a < X \leq b) = 0$. In words, regions where f is 0 correspond to regions where X does not take on values (or, equivalently, where F is flat, as we have already learned), and so regions outside the support of f are simply not relevant to consider. As such, one would typically simply call $f(x) = \theta(1 + x)^{-\theta-1}$, $x > 0$, the density of the $\mathrm{Par}(\theta)$ distribution, instead of the more complicated (8.24); recall the phrase "A good mathematician is a lazy mathematician" from Morris (2016) and Figure 8.10.

(6) Again let $X \sim F$ with density f. For some real value x and a small quantity $\delta > 0$, consider $a = x - \delta$ and $b = x$ in (8.27):

$$F(x) - F(x - \delta) = \int_{x-\delta}^{x} f(z)\,\mathrm{d}z. \qquad (8.28)$$

If we let δ go to 0 then the right-hand side converges to 0 (the area under the graph of f gets smaller and smaller) and the left-hand side converges to $F(x) - F(x-)$. As we already know from (8.12), the latter difference is $\mathbb{P}(X = x)$, so we obtain that $\mathbb{P}(X = x) = 0$ for any x; because of this property, we can also interpret the integral in (8.27) as $\mathbb{P}(a \leq X \leq b)$, $\mathbb{P}(a \leq X < b)$ or $\mathbb{P}(a < X < b)$. We already established the fact that $\mathbb{P}(X = x) = 0$ for continuous distribution functions in Section 8.3.1. The reason why here we mention this again is that densities are often misinterpreted (which is reminiscent of the many trap doors one could easily fall through; see Section 8.1). The probability $\mathbb{P}(X = x)$ is in general not equal to the density f at x. For example, for the continuous Pareto distribution, we know that $\mathbb{P}(X = x) = 0$ for all x, yet the density $f(x) = \theta(1 + x)^{-\theta-1}$ of $X \sim \text{Par}(\theta)$ is positive for all $x > 0$. Note that this is unlike the discrete case, where $\mathbb{P}(X = x)$ is indeed the probability mass function at x. So how can we interpret a density f at a particular value x if not as the probability $\mathbb{P}(X = x)$?

To motivate the answer to this question, consider the right-hand side of (8.28), which is the area under the graph of f over the interval $(x-\delta, x]$. For sufficiently small $\delta > 0$, f can be considered as approximately equal to $f(x)$ over the small interval $(x - \delta, x]$. The area over $(x - \delta, x]$ is thus approximately equal to a rectangle with length $x - (x - \delta) = \delta$ (the length of the rectangle) and height $f(x)$ (the height of the rectangle), so

$$F(x) - F(x - \delta) = \int_{x-\delta}^{x} f(z)\,\mathrm{d}z \approx \delta f(x).$$

If we divide both sides by δ we obtain

$$f(x) \approx \frac{F(x) - F(x - \delta)}{\delta}$$

and so $f(x)$ is approximately $\mathbb{P}(X \in (x - \delta, x])/\delta$. The right way to interpret $f(x)$ is therefore not as $\mathbb{P}(X = x)$ but as $\mathbb{P}(x - \delta < X \leq x)/\delta$ for small $\delta > 0$, and so as the probability that X will fall in a small interval around x relative to the length of the interval.

We now cover several more examples of continuous distributions with densities which play important roles in the modeling of risks.

Example 8.33 (Exponential distribution). A random variable X follows an *exponential distribution* with parameter $\lambda > 0$, denoted by $X \sim \text{Exp}(\lambda)$, if it has distribution function

$$F(x) = 1 - \exp(-\lambda x), \quad x \geq 0.$$

As mentioned in Remark 8.32 (2), the density of F can be obtained by differentiation. You may recall that the exponential function $\exp(x)$ has the property that its derivative is $\exp(x)$ itself (geometrically speaking, the slope of the graph of $\exp(x)$ at any point is the function value $\exp(x)$). Using this property one obtains that the exponential distribution function F has density

$$f(x) = \lambda \exp(-\lambda x), \quad x > 0.$$

Figure 8.26 shows the graphs of the distribution function F and the corresponding density f for different parameters λ. Remark 8.40 later addresses an important property of exponential distributions.

Remark (Animals in the zoo of distributions and importance of the concept). There are many more 'animals' in the 'zoo of distributions' that we could consider; see Figure 8.27 for some of them. Many of these distributions (and many more not listed in this figure) are important for modeling random phenomena in general and for modeling risks in particular. Having an available repertoire of different distributions is important for statistical modeling, as otherwise the *law of the instrument* applies; see, for example, Maslow (1966):

If the only tool you have is a hammer, it is tempting to treat everything as if it were a nail.

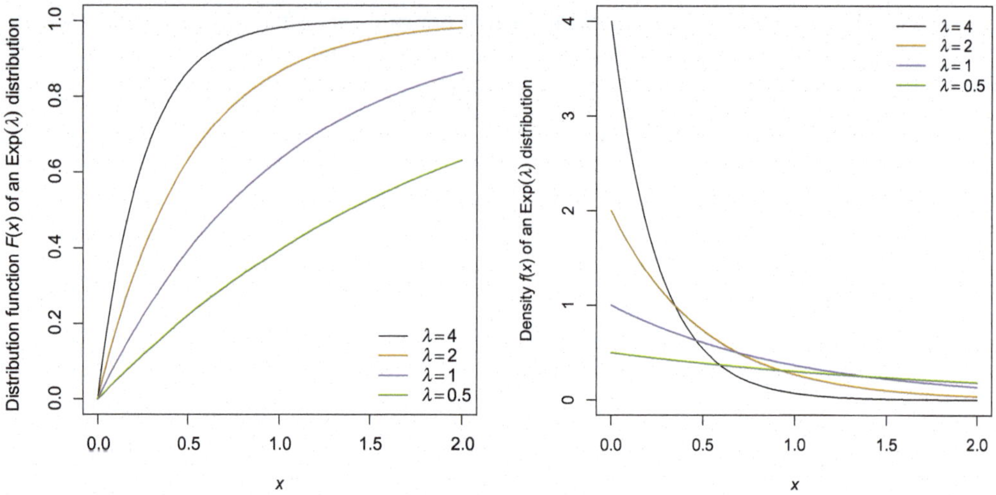

Figure 8.26 Left: Exp(λ) distributions for $\lambda \in \{4, 2, 1, 1/2\}$. Right: The corresponding densities. Source: Authors

Even more important is the following. We will frequently encounter new distributions in what follows and in Chapter 9. The details (the form of the distribution function, probability mass function or density, etc.) of every such 'animal' in the 'zoo' are not nearly as important for us as knowing what they enable us to model or the types of questions they enable us to answer. What is important for us to know is the general concept of a distribution function F and how its properties translate to properties of $X \sim F$. And, the other way around, it is important to be familiar with the general concept of a random variable X and how its properties translate to properties of its distribution function F; see Section 8.3.1. This helps us to understand how risks can be modeled. Beyond that guidance, look at Figure 8.27 from a distance and imagine coloring it, and it may start to look as some piece of abstract art or a highly complicated subway map, as for instance the subway map of Tokyo.

For the Pareto and the exponential distribution one has $f(x) = 0$ for $x < 0$, so random variables X from these distributions do not attain negative values almost surely ($\mathbb{P}(X \leq 0) = 0$). The following distribution is an example of a distribution whose support is the real numbers $\mathbb{R}$ (so, for which $f(x) > 0$ for all $x \in \mathbb{R}$). In particular, a random variable X from such a distribution can take on positive and negative values.

Example 8.34 (Normal distribution)**.** The *normal distribution*, denoted by $\mathrm{N}(\mu, \sigma^2)$, with parameters $\mu \in \mathbb{R}$ and $\sigma > 0$, is defined by the density

$$f(x) = \frac{1}{\sqrt{2\pi\sigma^2}} \exp\left(-\frac{1}{2}\left(\frac{x-\mu}{\sigma}\right)^2\right), \quad x \in \mathbb{R}. \tag{8.29}$$

Because of the shape of its graph, the density (8.29) is also known as the *bell curve*. The normal distribution is of such importance to statistics that its density (8.29) was used on the former 10 Deutsche Mark banknote before the euro was introduced; see Figure 8.28. Other examples where mathematicians appear on banknotes include Bohr (Denmark),

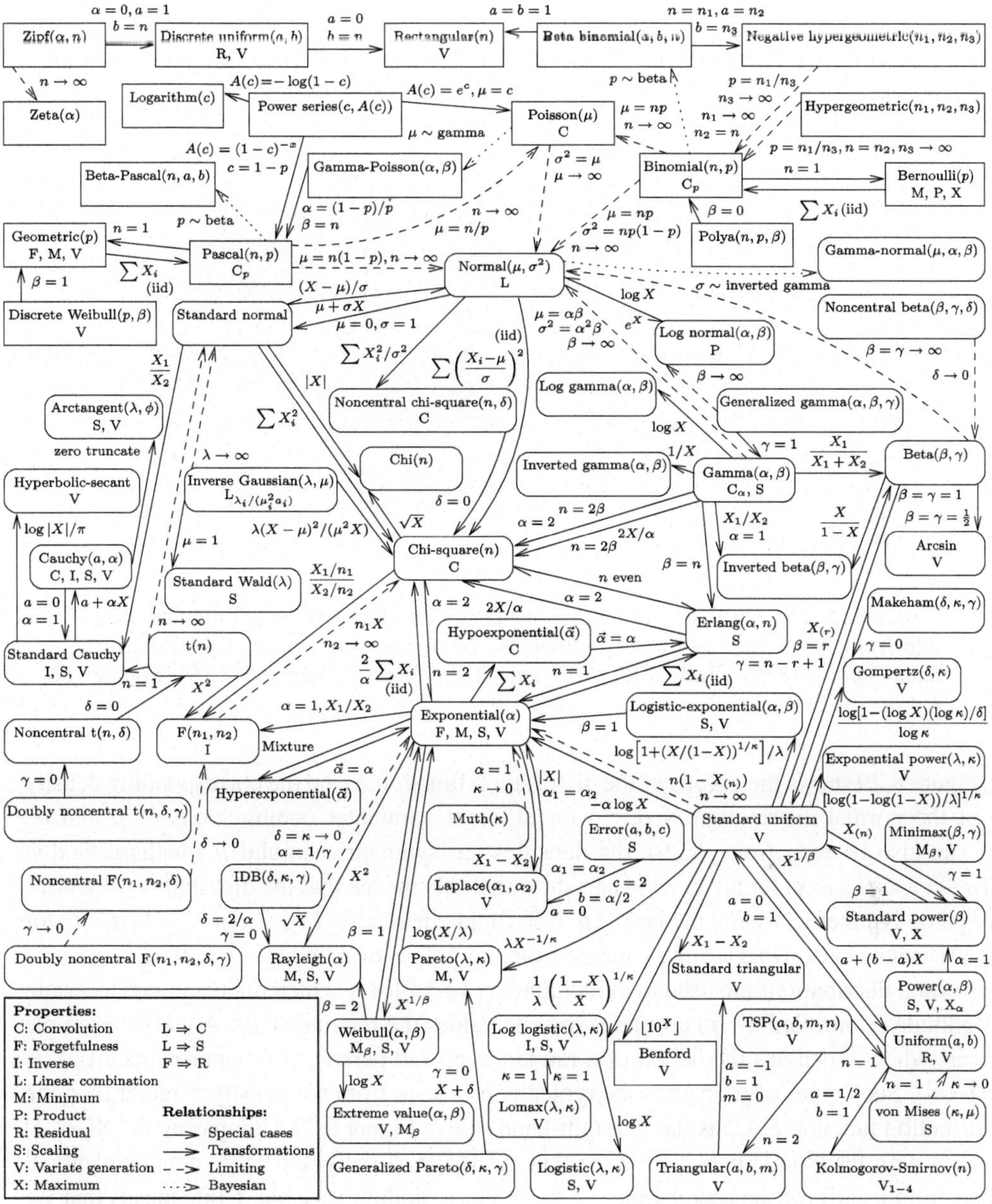

Figure 8.27 A selection of 'animals' in the 'zoo of distributions', to be enjoyed from a distance. Source: Leemis and McQueston (2008), © American Statistical Association, reprinted with permission of Taylor & Francis Ltd.

Descartes (France), Einstein (Israel), Euler (Switzerland), Galileo (Italy), Newton (United Kingdom) and Pascal (France); see Bui (2010). A most recent addition is Alan Turing on the new 50 pound note in the United Kingdom. Alan Turing (1912–1954) stood at the cradle of computer science and cryptography. He will for always be remembered as the mathematician who broke the Enigma code in World War II. We briefly met Turing in Section 7.4 in the context of the movie "The Imitation Game"; the film tells the dramatic story of his life. Needless to say, we wish more mathematical concepts found their way into people's wallets and pockets.

Figure 8.28 The obverse of the former 10 Deutsche Mark banknote officially available from April 16, 1991 to December 31, 2001, showing Johann Carl Friedrich Gauss (1777–1855) jointly with the graph of the $\mathrm{N}(\mu,\sigma^2)$ density (8.29). Source: Wikimedia Commons

Figure 8.29 shows the graphs of the distribution function F and the corresponding density f of the normal distribution $\mathrm{N}(\mu,\sigma^2)$ for different parameter combinations of μ and σ. As can also be seen from (8.29), the densities are symmetric around μ (mathematically: $f(\mu - x) = f(\mu + x)$ for all x). One can show that this means that the distribution functions are point-symmetric around the point $(\mu, 1/2)$ (mathematically: $F(\mu - x) = 1 - F(\mu + x)$ for all x), which is also what we can confirm from the graphs of F in Figure 8.29.

Normal distributions are used to model random variables X which fluctuate with a certain magnitude (controlled by σ) around a specific value (the parameter μ). A classic example occurs when X denotes the height of a randomly picked person; then, approximately, $X \sim \mathrm{N}(\mu,\sigma^2)$. Such types of properties are typically best seen from the density f rather than the distribution function F. Consider the right-hand side of Figure 8.29. Comparing the $\mathrm{N}(-2,1)$ density with the $\mathrm{N}(0,1)$ density, we see that the shape of the graph of f remains identical apart from a shift (of the center from -2 to 0). For a random variable X this means that we expect to see it fluctuating around -2 if $X \sim \mathrm{N}(-2,1)$ and around 0 if $X \sim \mathrm{N}(0,1)$. On the other hand, when comparing $\mathrm{N}(0,1)$ with $\mathrm{N}(0,(1/2)^2)$ and $\mathrm{N}(0,2^2)$, we see that the graphs of all three densities f share the same center ($\mu = 0$) but the smaller the value of σ, the more the curves are concentrated around μ. What does this mean for $X \sim \mathrm{N}(\mu,\sigma^2)$? Consider as an example the interval $(2,\infty)$. Comparing the densities on this interval we see that the area under the graph of the density of $\mathrm{N}(0,2^2)$ is larger than that of $\mathrm{N}(0,1)$, and the area under the graph of the density of $\mathrm{N}(0,1)$ is larger than that of $\mathrm{N}(0,(1/2)^2)$. So, the probability $\mathbb{P}(X > 2)$ becomes smaller and smaller, the smaller is σ. As the densities are symmetric around μ, the same can be seen for the interval $(-\infty,-2)$. Therefore small values $X < -2$ and large values $X > 2$ are less likely, the smaller is σ. To summarize, σ controls how much $X \sim \mathrm{N}(\mu,\sigma^2)$

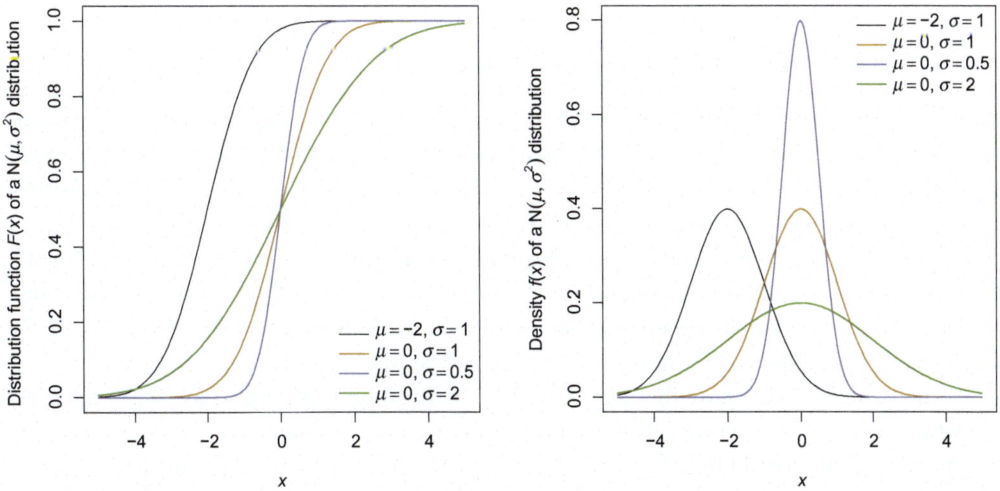

Figure 8.29 Left: $N(\mu,\sigma^2)$ distributions for $(\mu,\sigma) \in \{(-2,1),(0,1),(0,1/2),(0,2)\}$. Right: The corresponding densities. Source: Authors

fluctuates around μ; the smaller is σ, the less fluctuation. In Section 8.4.1 we will interpret μ as the "mean" and σ^2 as the "variance" (and σ as the "standard deviation") of $X \sim N(\mu,\sigma^2)$; see Example 8.44 (1).

For $\mu = 0$ and $\sigma = 1$, one speaks of the *standard normal distribution*. Its density is typically denoted by ϕ (instead of f) and its distribution function by Φ (instead of F); once more, we meet several Greek letters. The density ϕ is given by

$$\phi(x) = \frac{1}{\sqrt{2\pi}} \exp\left(\frac{-x^2}{2}\right), \quad x \in \mathbb{R}. \tag{8.30}$$

The distribution function Φ of $N(0,1)$ (and thus also that of $N(\mu,\sigma^2)$) is not available in analytical form. All one can say is that Φ can be obtained from ϕ by integration via (8.25). However, Φ can be approximated numerically with high precision and such approximations can be found in any piece of statistical software. We thus treat Φ as "given".

Derivation 8.35 (Evaluating the normal distribution function)**.** We said that Φ is available in statistical software. But how then can we evaluate the distribution function F of a general $N(\mu,\sigma^2)$ distribution?

Similarly to what we did in Examples 8.13 and 8.23, we consider a stochastic representation to answer this question. As two random variables X, Z instead of just one random variable will now be of interest, we distinguish the distribution functions by a subscript, writing $X \sim F_X$ and $Z \sim F_Z$. For $\mu \in \mathbb{R}$ and $\sigma > 0$, consider

$$X = \mu + \sigma Z; \tag{8.31}$$

the random variable X is a *linear transformation* of the random variable Z, and one also says that X and Z are of the same *type*, in this case; see also Section 9.3.1. What is the distribution function F_X of X in terms of F_Z? The stochastic representation (8.31) is sufficiently simple to answer this question in one line:

$$
\begin{aligned}
F_X(x) &= \mathbb{P}(X \le x) = \mathbb{P}(\mu + \sigma Z \le x) = \mathbb{P}(Z \le (x-\mu)/\sigma) \\
&= F_Z((x-\mu)/\sigma), \quad x \in \mathbb{R}.
\end{aligned} \tag{8.32}
$$

Let us assume for a moment that any $X \sim \mathrm{N}(\mu, \sigma^2)$ allows for the stochastic representation (8.31) in terms of $Z \sim \Phi$; then we can already answer our initial question at this point. By (8.32), we know how to evaluate F_X (the distribution function of $\mathrm{N}(\mu, \sigma^2)$) in terms of Φ (the distribution function of $\mathrm{N}(0,1)$). This has the major advantage that Φ is a single function which does not depend on any parameters. As such, one can go into great detail to approximate this function satisfactorily in software, independently of the choice of μ and σ. To find directly a numerical approximation of the distribution function of $\mathrm{N}(\mu, \sigma^2)$ which works equally well for all possible $\mu \in \mathbb{R}$ and $\sigma > 0$ would be a much harder task.

We now have left to show that any $X \sim \mathrm{N}(\mu, \sigma^2)$ can be represented as (8.31) for $Z \sim \Phi$ (which we assumed above). If F_Z has density f_Z, we can obtain the density of X by differentiating its distribution function F_X using (8.32). You may first want to recall the chain rule for calculating derivatives of composite functions. With this rule, we obtain from (8.32):

$$
f_X(x) = F'_X(x) = \frac{F'_Z((x-\mu)/\sigma)}{\sigma} = \frac{f_Z((x-\mu)/\sigma)}{\sigma}, \tag{8.33}
$$

so if X, Z are related via (8.31), then the densities are related via (8.33). As densities uniquely characterize distributions, the converse statement is also true: if densities f_X, f_Z are related via (8.33), then the corresponding random variables X, Z are related via (8.31). With $f_Z = \phi$, we can plug $(x-\mu)/\sigma$ into (8.30) and multiply the expression by $1/\sigma$ to check that we obtain (8.29) and that the density of $\mathrm{N}(\mu, \sigma^2)$ and that of $\mathrm{N}(0,1)$ are related to each other via (8.33) and thus X and Z are related to each other via (8.31).

This result is remarkable. By using densities we established that $Z \sim \mathrm{N}(0,1)$ if and only if $X = \mu + \sigma Z \sim \mathrm{N}(\mu, \sigma^2)$, without having analytically tractable expressions for F_Z or F_X. Relying on the uniqueness of densities, not shown here, we have moreover established that the densities f_Z (of F_Z) and f_X (of F_X) are related to each other as in (8.33) if and only if $X = \mu + \sigma Z$. Note that this result holds irrespective of whether the two distributions are normal! Establishing such connections, patterns or laws is what the core of mathematics is about.

To quantify the certainty or significance of scientific results, one often reads about the five- or six-sigma rule. The following example covers what this means.

Example 8.36 (n-sigma rule)**.** On October 8, 2013, it was announced that François Englert and Peter Higgs had received the Nobel Prize in Physics:

> [...] for the theoretical discovery of a mechanism that contributes to our understanding of the origin of the mass of subatomic particles, and which recently was confirmed through the discovery of the predicted fundamental particle, by the ATLAS and CMS experiments at CERN's Large Hadron Collider.

The *Conseil Européen pour la Recherche Nucléaire (CERN)* is the European Organization for Nuclear Research, located in Geneva, Switzerland; we will meet CERN again in Chapter 12 regarding the development of the World Wide Web. The Large Hadron Collider (LHC) is a large-scale particle accelerator, frequently called the "largest machine in the world", a 27 km long tunnel at about 175 m depth, built by over 10 000 scientists and engineers between

1998 and 2008. In it, particles (mostly protons) are accelerated to nearly the speed of light in order to initiate high-energy collisions. The resulting decay products (the *decay signature*) of such collisions are registered by detectors such as ATLAS and CMS, mentioned in the above quote, and the resulting data are analyzed with the goal of investigating known or new elementary particles. The long-term goal is to verify or be able to improve the *Standard Model* of particle physics, which is the theory describing three of the four fundamental forces, the electromagnetic, weak and strong interactions (without gravity), and includes a classification of all known elementary particles.

On March 14, 2013, CERN announced that physicists are now confident they have discovered a new particle, the Higgs boson (also known as the God particle), which had been theorized in a two-page paper (Higgs, 1964) to explain why matter has mass; see CERN (2013) and Heilprin (2013). Our wording here ("are now confident they have discovered" as opposed to "have discovered") is from the latter and stresses that the discovery of a new particle is affected by uncertainty. Producing Higgs bosons by particle collisions is a rare event (of the order of 1 in 10 billion at the LHC) and the particles have a mean lifetime of about 1.56×10^{-22} seconds, so there is extremely little time for measuring and detecting them.

The discovery of a new particle is ultimately determined via a hypothesis which is rejected (or not) on the basis of the observed data from the collisions. The hypothesis was that there is no Higgs boson in the Standard Model (so that, if true, the observed effects that hinted at the existence of a Higgs boson are due only to random events, or perhaps other new particles). To be able to reject this hypothesis and thus report the discovery of the Higgs boson, particle physicists require the *n-sigma rule* to hold with $n = 5$. According to this rule, the chance that the observed decay signatures are truly due to random events in the Standard Model has to be less than $\bar{\Phi}(n) = 1 - \Phi(n)$. If the hypothesis is indeed true and no Higgs boson exists, the n-sigma rule limits the chance of rejecting the hypothesis although it is true (a *false positive*).

Note that $\bar{\Phi}(n) = 1 - \Phi(n) = 1 - \mathbb{P}(Z \le n) = \mathbb{P}(Z > n)$ for $Z \sim \mathrm{N}(0, 1)$, so $\bar{\Phi}(n)$ is an exceedance probability of Z over the threshold n. The threshold n is of course simpler to report than a small probability. For example, for $n = 5$ it is easier to say the Higgs boson was discovered "at a five-sigma level of certainty" (or *confidence* as one also says) than to say "with a probability of a false positive of only 0.0000002866516" or "with a probability of a false positive of only 1 in 3 488 556". To put this in perspective, this is roughly equal to the chance of obtaining only "heads" (so a *run* of "heads") when tossing a fair coin 22 times, or throwing 8 sixes in a row when rolling a fair die 8 times. As a fraction, this is about the same as the ratio of three-quarters of a second to a month.

The n-sigma rule is also used to report the significance of results in the social sciences ($n = 2$), empirical sciences ($n = 3$) and quality management ($n = 4.5$ or even $n = 6$). Figure 8.30 shows the graph of $\bar{\Phi}(n)$ for $n \in \{1, 1.5, 2, \dots, 6\}$, with the secondary y-axis on the right showing the corresponding number of obtaining only "heads" when tossing a fair coin.

8.3.3 From Distribution Functions to Conditional Distribution Functions

One of the questions we raised in the beginning of Section 8.1 was about the probability that a 40 year old person will eventually reach the age of 85. Such a question is naturally a

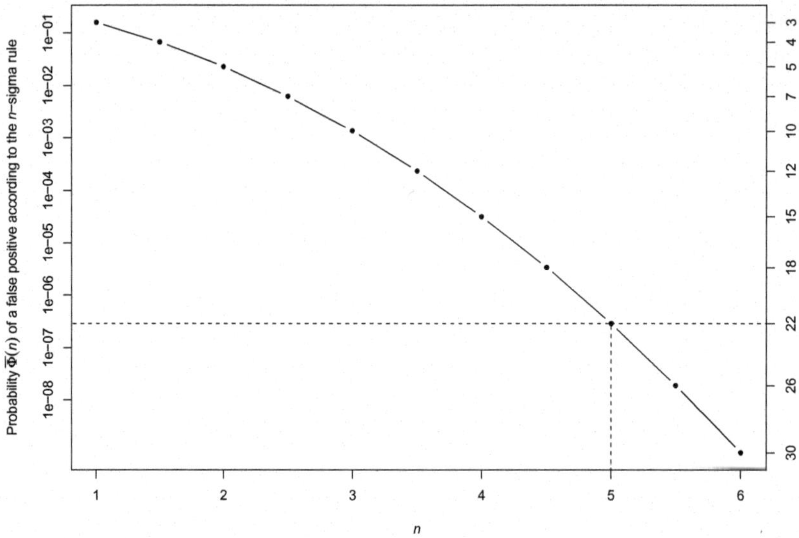

Figure 8.30 Plot of the probability $\bar{\Phi}(n)$ of a false positive according to the n-sigma rule versus n, including, as secondary y-axis on the right, the corresponding number of "only heads" obtained when tossing a fair coin. The case $n = 5$ is indicated by a dashed line. Source: Authors

conditional one: we are given that the person has already reached the age of 40, and are now asking for the probability that this person also reaches age 85. If X denotes the (random) duration of human life in a well-specified population, this conditional probability is

$$\mathbb{P}(X > 85 \mid X > 40)$$

(read: "the probability of X exceeding 85 given that X exceeds 40"). Or suppose we are interested in the probability of a sea-level exceedance of the 1953 flood mark of 4.20 m at Neeltje Jans (Figure 1.6) given that the water level has already exceeded the safety level of 3 m at which the sluices of the Oosterscheldekering close. If X denotes the corresponding maximal daily sea-level height in meters, we are asking for the probability $\mathbb{P}(X > 4.20 \mid X > 3)$. Another example comes from reinsurance (the insurance of one insurance company by another). Say that an insurer covers all loss claims over the next year less than or equal to a threshold u of one million USD. For losses exceeding this threshold, the insurer buys insurance from a reinsurer. From the reinsurer's perspective, the distribution of the losses exceeding the threshold u is the loss distribution of interest. If X denotes the next year's loss in the same monetary unit as u, then the loss distribution function of interest is $\mathbb{P}(X \leq x \mid X > u)$, $x \geq u$. This leads to the following concept.

Concept 8.37 (Conditional distribution function)**.** For a random variable $X \sim F$ and an event B with $\mathbb{P}(B) > 0$, the function $F_B(x) = \mathbb{P}(X \leq x \mid B)$ is a distribution function known as the *conditional distribution function* of X given B.

Conditional distributions and probabilities allow us to address the *what if* type of questions. The following example covers the prominent case where B is the event that X lies in an interval $(a, b]$, in which case $F_{(a,b]}(x) = \mathbb{P}(X \leq x \mid X \in (a, b])$.

Example 8.38 (Conditional distribution with interval constraint)**.** Let $-\infty < a < b < \infty$ and let $X \sim F$ be a random variable with $\mathbb{P}(X \in (a,b]) > 0$. By (8.2), the conditional distribution function $F_{(a,b]}(x) = \mathbb{P}(X \le x \mid X \in (a,b])$ is

$$F_{(a,b]}(x) = \frac{\mathbb{P}(X \le x,\ X \in (a,b])}{\mathbb{P}(X \in (a,b])} = \frac{\mathbb{P}(X \le x,\ a < X \le b)}{\mathbb{P}(a < X \le b)}. \tag{8.34}$$

For $x \in [a,b]$, the numerator is the probability of the event that $X \le x$ and that $a < X \le b$, that is, the probability of the event that $a < X \le x$. This probability and the denominator can be expressed in terms of the distribution function F of X via (8.11) as

$$F_{(a,b]}(x) = \frac{\mathbb{P}(a < X \le x)}{\mathbb{P}(a < X \le b)} = \frac{F(x) - F(a)}{F(b) - F(a)}, \quad x \in [a,b]. \tag{8.35}$$

Note that for $x < a$ we have $F_{(a,b]}(x) = 0$ (since $\mathbb{P}(a < X \le x) = 0$), and for $x > b$ we have $F_{(a,b]}(x) = 1$ (since $\mathbb{P}(X \le x,\ X \in (a,b]) = \mathbb{P}(a < X \le b)$, so $F_{(a,b]}(x) = 1$ follows from (8.34)). In other words, $F_{(a,b]}$ is a linear transformation of F.

The three defining properties of a distribution function from Section 8.3.1 are easily verified for $F_{(a,b]}$, as given in (8.35), so the conditional distribution function of X given that $X \in (a,b]$ is indeed a proper distribution function. A random variable from $F_{(a,b]}$ is typically written as

$$X \mid X \in (a,b]$$

(read: "X given that X lies in the interval from a to b"), a rather intuitive notation for (conditional) random variables.

If F has density f, then the conditional distribution function $F_{(a,b]}$ has density

$$f_{(a,b]}(x) = F'_{(a,b]}(x) = \frac{f(x)}{F(b) - F(a)}, \quad x \in (a,b); \tag{8.36}$$

outside (a,b), we have that $f_{(a,b]}$ is 0. This allows for the interpretation that the density $f_{(a,b]}$ is the density f, scaled to be a density on (a,b). One can argue similarly in the case where F is discrete with probability mass function f.

Later in Example 8.67 and especially in Chapter 9, we will consider the special case of Example 8.38 where $a = u$ for some threshold u and b tends to ∞. The limit formula corresponding to (8.35) is then

$$F_{(u,\infty)}(x) = \frac{F(x) - F(u)}{1 - F(u)}, \quad x \ge u. \tag{8.37}$$

This is a valid distribution function for all u such that $F(u) < 1$. In terms of our concept of survival functions, the denominator in (8.37) can also be written as $\bar{F}(u)$. The conditional distribution function $F_{(u,\infty)}$ starts at $x = u$ (for $x < u$ it is 0). In practice one typically has $u > 0$ and one is looking for an adequate parametric distribution function as a model for $F_{(u,\infty)}$. Most parametric distribution functions of positive random variables start at 0, for example $\mathrm{Par}(\theta)$ or $\mathrm{Exp}(\lambda)$. To avoid having to shift these parametric distribution functions to start at u for modeling $F_{(u,\infty)}$, one typically shifts $F_{(u,\infty)}$ to start at 0 by considering $F_{(u,\infty)}(x+u)$, $x \ge 0$. For the distribution function $F_{(u,\infty)}(x+u)$ one can then conveniently use known parametric distribution functions as a model. Since this shifted conditional distribution function $F_{(u,\infty)}(x+u)$ plays a major role in Chapter 9, it deserves to be introduced here.

Concept 8.39 (Excess distribution function)**.** For $X \sim F$, the *excess distribution function* over the threshold u is given by

$$F_u(x) = F_{(u,\infty)}(x+u) = \mathbb{P}(X \le x+u \mid X > u) = \frac{F(x+u) - F(u)}{1 - F(u)}, \quad x \ge 0. \tag{8.38}$$

Since $\mathbb{P}(X \le x+u \mid X > u) = \mathbb{P}(X - u \le x \mid X > u)$, the excess distribution function F_u can be interpreted as the distribution function of the *excess* $X - u$ over the threshold u, given that $X > u$.

The following example shows an important property of the exponential distribution in terms of its excess distribution function.

Example 8.40 (Memorylessness of the exponential distribution)**.** Let $X \sim \text{Exp}(\lambda)$ with distribution function $F(x) = 1 - \exp(-\lambda x)$, $x \ge 0$. By (8.38), for all $u \ge 0$,

$$\begin{aligned} F_u(x) = F_{(u,\infty)}(x+u) &= \frac{F(x+u) - F(u)}{1 - F(u)} = \frac{1 - e^{-\lambda(x+u)} - (1 - e^{-\lambda u})}{1 - (1 - e^{-\lambda u})} \\ &= \frac{e^{-\lambda u} - e^{-\lambda(x+u)}}{e^{-\lambda u}} = \frac{e^{-\lambda u}(1 - e^{-\lambda x})}{e^{-\lambda u}} = 1 - e^{-\lambda x} = F(x), \quad x \ge 0. \end{aligned} \tag{8.39}$$

So for $\text{Exp}(\lambda)$ distributions, the excess distribution function F_u is simply F itself. Writing $F_u(x) = F(x)$ in terms of probabilities, we obtain from (8.39) that

$$\mathbb{P}(X \le x+u \mid X > u) = \mathbb{P}(X \le x).$$

In terms of survival functions, we thus have

$$\mathbb{P}(X > x+u \mid X > u) = \mathbb{P}(X > x). \tag{8.40}$$

This property of the exponential distribution is known as the *memoryless property* and one can show that the exponential distribution is the only continuous distribution with this property.

Remark 8.41 (Memoryless property for the exponential distribution)**.** In Derivation 8.28 we motivated the $\text{Poi}(\lambda)$ distribution as the distribution of the number of events if these events occur at a constant rate $\lambda > 0$ per unit of time and independently of the time that the last event occurred. In such a set-up, the time gap between two consecutive events is known as the *waiting time* or *interarrival time*. This time gap itself is a random variable and one can show that its distribution is $\text{Exp}(\lambda)$; there is thus a connection between the discrete $\text{Poi}(\lambda)$ and the continuous $\text{Exp}(\lambda)$. With this interpretation of $\text{Exp}(\lambda)$ we can interpret the memoryless property. The probability $\mathbb{P}(X > x+u \mid X > u)$ is the probability that we have to wait more than $x+u$ units of time for the next event to happen, given that we have already waited for u units. And by the memoryless property (8.40), this probability equals the probability $\mathbb{P}(X > x)$, the probability that we have to wait more than x units of time. Therefore, the next event's occurrence does not memorize (hence the name) the fact that we have already waited u units of time for it to happen (so, informally, there is no related increase in 'statistical pressure' for the next event to happen).

You may feel a bit uneasy about the somewhat hidden nature of time in the above discussion. A proper treatment would need the introduction of a time-indexed family of (Poisson) random variables, referred to as a (Poisson) stochastic process. We will refrain from doing so and very much hope that the intuition given above suffices.

8.4 From Distribution Functions to Summary Statistics

In this section we think about how we can summarize important properties of a distribution function F or its random variable X in terms of a few numbers, known as *summary statistics*. We start with those summary statistics that are important mainly for general statistical applications and then cover summary statistics that more specifically focus on summarizing risks, so-called *risk measures*.

8.4.1 Mean and Variance

When you roll a fair die once, what is the value you expect to obtain? You will see one of the numbers $1, 2, \dots, 6$, each with probability $1/6$. Computing the weighted sum $1 \times (1/6) + 2 \times (1/6) + \cdots + 6 \times (1/6)$ (with summands of the form "outcome times probability of that outcome") leads to 3.5. Although 3.5 is not printed on any of the sides of a die and thus will never appear itself as a number, it represents a central value (or *location*) of the (random) rolled number X or of its discrete uniform distribution $\mathrm{U}(\{1, 2, \dots, 6\})$. The idea readily extends to discrete distributions with probability mass function f and support $\{x_1, x_2, \dots\}$, in which case

$$\mathbb{E}(X) = \sum_{k=1}^{\infty} x_k f(x_k) \tag{8.41}$$

is the *mean* or *expectation* of X, that is, a weighted sum with summands of the form "outcome times probability of that outcome". We see that the more probability mass f puts at x_k, the more weight x_k has in this sum, and the more the value x_k contributes to the mean $\mathbb{E}(X)$ of X. In the case where f puts probability mass 1 on, say, x_1 (and thus probability mass 0 on all other values), we simply obtain $\mathbb{E}(X) = x_1$, which is indeed what we would expect to see in this case (think of a die with all sides containing the number 1).

For a continuous random variable $X \sim F$ with density f, the mean of X is

$$\mathbb{E}(X) = \int_{-\infty}^{\infty} x f(x)\,\mathrm{d}x. \tag{8.42}$$

Knowing that integrals are limits of sums, (8.42) looks similar to (8.41) in the sense that we sum over all x and the summands are of the form $x f(x)$. We are tempted to finish the last sentence by saying "... of the form 'outcome times probability of that outcome,'" but recall from Remark 8.32 (6) that $f(x)$ is not $\mathbb{P}(X = x)$ in the continuous case. A more precise interpretation is readily obtained by invoking a small interval $(x - \delta, x]$ as we did in Remark 8.32 (6): the larger f is over this interval, the more weight the values in $(x - \delta, x]$ have in this integral, so the more these values contribute to the mean $\mathbb{E}(X)$ of X.

The mean of a random variable X with a mixed-type distribution function can also be defined. Although its form (that is, the type of integral used) is irrelevant to us, it subsumes the discrete and continuous case in a single formula for the mean of X, which allows for easier reference.

Concept 8.42 (Mean, expectation)**.** If it exists, the *mean* or *expectation* of $X \sim F$ (or F) is

$$\mathbb{E}(X) = \int_{-\infty}^{\infty} x \,\mathrm{d}F(x). \tag{8.43}$$

If F has probability mass function f, then $\mathbb{E}(X)$ is given by (8.41), and if F has density f, then $\mathbb{E}(X)$ is given by (8.42).

As hinted at in Concept 8.42, there are distributions for which a mean does not exist, so for which the integral (8.43) does not make mathematical sense in an unambiguous way. This can occur when the distribution of X puts its probability mass further and further out in both tails, and thus when $F(x)$ converges too slowly to 0 for $x \to -\infty$ or when it converges too slowly to 1 for $x \to \infty$, or indeed both. For distributions whose support is the positive real line (the most important distributions for modeling non-negative losses in risk management applications), it can happen that $\mathbb{E}(X) = \infty$, but at least the mean can be computed and thus exists. This is the case, for example, for the $\mathrm{Par}(\theta)$ distribution with $\theta \in (0, 1]$; see Example 8.44 (2) below. However, most applications (but not all) do require a finite mean.

The condition of a finite mean of any random variable X can be expressed with the help of the *absolute value* $|X|$ of X (where $|x|$ equals x if $x \geq 0$ and $-x$ if $x < 0$), which is a non-negative random variable. The random variable $|X|$ now necessarily has mean either $\mathbb{E}(|X|) < \infty$ (in which case the mean $\mathbb{E}(X)$ of X is finite) or $\mathbb{E}(|X|) = \infty$. In other words, the condition for any random variable $X \sim F$ to have a finite mean can be expressed as $\mathbb{E}(|X|) < \infty$, the condition that the mean of the absolute value of X is finite. We could now revisit the previously introduced distributions and check whether their means are finite, but there are more important aspects for us to consider than how to calculate integrals.

We can simply view $\mathbb{E}(|X|)$ as the mean of the new random variable $|X|$ constructed from X. In the same way we can view $\mathbb{E}(h(X))$ as the mean of the new random variable $h(X)$ constructed from X and a function h. Apart from $h(x) = |x|$, are there other functions h that are of particular interest? Indeed, for example, the functions $h(x) = x^k$, $k \in \mathbb{N}$, which give rise to the concept of moments. The *kth moment* of X (or F) is $\mathbb{E}(X^k)$, the first moment being the mean of X as already discussed. Another important special case of the function h is $h(x) = (x - \mathbb{E}(X))^2$, in which case we obtain the notion of the variance.

Concept 8.43 (Variance, standard deviation)**.** If it exists, the *variance* of $X \sim F$ (or F) is

$$\mathrm{var}(X) = \mathbb{E}\big((X - \mathbb{E}(X))^2\big); \tag{8.44}$$

one can show that $\mathrm{var}(X) = \mathbb{E}(X^2) - (\mathbb{E}(X))^2$, which is often useful in calculations. The *standard deviation* of X (or F) is $\mathrm{sd}(X) = \sqrt{\mathrm{var}(X)}$.

The variance is the mean squared deviation of X from its mean $\mathbb{E}(X)$, so $\mathrm{var}(X)$ quantifies how much X fluctuates around $\mathbb{E}(X)$. If $\mathbb{E}(X)$ is a summary statistic of the location of X, as we have already discussed, then $\mathrm{var}(X)$ and $\mathrm{sd}(X)$ are summary statistics of the *scale* of X. This can also be seen from the square function $h(x) = (x - \mathbb{E}(X))^2$, which is 0 if and only if $x = \mathbb{E}(X)$. One can show that this implies that var(X) in (8.44) is 0 (corresponding to no fluctuation) if and only if $X = \mathbb{E}(X)$ almost surely (that is, if and only if the random variable X is constant; indeed X does not fluctuate then). The standard deviation sd(X) is also a measure of scale. In comparison with var(X) (which contains the squared deviation of X from $\mathbb{E}(X)$), the standard deviation sd(X) measures the deviation of X from $\mathbb{E}(X)$ on the original scale of X. For example, if X is associated with some unit, say meters, then sd(X) is a summary statistic of the scale in meters, whereas var(X) is associated with square meters. Both measures of scale have their advantages and are widely used in statistics and risk management.

Example 8.44 (Mean and variance of normal and Pareto distributions).

(1) Calculating (8.42) and (8.44) for the normal density as given in (8.29), one can show that a random variable $X \sim \mathrm{N}(\mu, \sigma^2)$ has mean $\mathbb{E}(X) = \mu$ and variance $\mathrm{var}(X) = \sigma^2$. The two parameters μ, σ^2 of a normal distribution thus control the distribution's location (via μ) and scale (via σ^2). As such, the parameters of normal distributions allow for an easy interpretation. They can also be easily estimated from data. These are some of the many reasons for the popularity of this distribution.

(2) If $X \sim \mathrm{Par}(\theta)$, we know from Example 8.30 that the density f of X is $f(x) = \theta(1+x)^{-\theta-1}$ for $x > 0$, and 0 for $x \leq 0$. For $\theta > 1$, utilizing (8.42) and applying integration by parts, we obtain

$$\begin{aligned}\mathbb{E}(X) &= \int_{-\infty}^{\infty} x f(x)\,\mathrm{d}x = \int_{0}^{\infty} x\theta(1+x)^{-\theta-1}\,\mathrm{d}x \\ &= \left[-x(1+x)^{-\theta}\right]_0^{\infty} - \int_0^{\infty} (-1)(1+x)^{-\theta}\,\mathrm{d}x \\ &= 0 - \left[\frac{1}{\theta-1}(1+x)^{-\theta+1}\right]_0^{\infty} = \frac{1}{\theta-1};\end{aligned}$$

for $\theta \in (0,1]$, one has $\mathbb{E}(X) = \infty$. For $\theta > 2$, applying integration by parts twice leads to $\mathbb{E}(X^2) = 2/((\theta-1)(\theta-2))$. Via $\mathrm{var}(X) = \mathbb{E}(X^2) - (\mathbb{E}(X))^2$ we find $\mathrm{var}(X) = \theta/((\theta-1)^2(\theta-2))$; for $\theta \in (0,2]$, one has $\mathrm{var}(X) = \infty$.

We now return to our discussion about the existence and finiteness of the mean of a random variable. Since $(X - \mathbb{E}(X))^2$ is a non-negative random variable, the variance var(X) of X is either finite or infinite but always exists. One can show that $\mathrm{var}(X) < \infty$ occurs if and only if the second moment of X is finite, so that $\mathbb{E}(X^2) < \infty$. As the mean of X also enters (8.44) one should ask: if $\mathbb{E}(X^2) < \infty$, do we then also have $\mathbb{E}(|X|) < \infty$? This is indeed the case. Moreover, one can show that if $\mathbb{E}(|X|^k) < \infty$, then $\mathbb{E}(|X|^l) < \infty$ for all $1 \leq l < k$. In words, the finiteness of higher moments implies the finiteness of lower moments. Although these results seem technical, they have important consequences for risk management applications. For extreme risks X (such as the ones we more precisely call "heavy-tailed," as those of Section 8.4.3 below), it is not uncommon that $\mathrm{var}(X) = \infty$ but $\mathbb{E}(|X|) < \infty$; Par(θ) distributions for $\theta \in (1,2]$ provide such examples. But if even $\mathbb{E}(|X|) = \infty$ (and thus also $\mathrm{var}(X) = \infty$), this is a sign of an even more extreme risk X.

A certainly oversimplified but nonetheless interesting example is the following. Insurance contracts are based on the idea of matching the mean premiums paid by the insured with the mean loss the insurer has to pay in regard of the contract. Extreme losses with infinite mean are not insurable any longer in this sense as the finite premiums paid produce a finite mean which cannot be matched with an infinite-mean loss.

A drastic (but hopefully convincing) example concerns the use of nuclear energy to solve mankind's energy problems. What to do with the nuclear waste? One could put it on a rocket to shoot it to the Moon (or beyond). But what happens if the rocket explodes and creates a nuclear fallout over Earth? How about buying insurance for such a rocket? Well, who would be willing to step in to insure the *what if* loss due to a rocket explosion, an event of non-zero probability but with disastrous impact and of virtual infinite loss (as we only have one Earth, we should not destroy it!) and thus an infinite-mean loss. Perhaps you may find this story a bit far-fetched, but is it? There are more examples out there where extreme heavy-tailedness occurs. These include the modeling of operational risk in finance, cyber risk overall and indeed losses in the realm of climate change. In all these examples, whenever a careful statistical analysis points in the direction of infinite-mean models, and such analyses do, then the risk manager involved has to take a methodological step back. In such cases, (s)he is confronted with a problem where standard economic tools and techniques such as cost-benefit analysis, as well as risk diversification, are not functioning the way we learned for less heavy-tailed risks. Of course, in reality an infinite-mean risk does not exist; we have therefore chosen our wording above very carefully as "points in the direction of infinite-mean models".

If you are interested in these issues, we strongly advise you to read about the discussion between the environmental economists Martin Weitzman (1942–2019) and Nobel Memorial Prize winner William Nordhaus around Weitzman's dismal theorem; see, for example, Weitzman (2009) and Nordhaus (2009). From a methodological point of view, the breakdown of diversification benefits in markets exposed to very heavy-tailed risks is discussed in Ibragimov and Walden (2007). In the end, the non-insurability of such risks becomes a major issue and invariably leads to the question: "Have we as a human species pushed our luck too far?" Nowhere can this debate be seen more acutely than in the realm of climate change.

As a random variable, $X \mid X \in (a, b]$ (see Section 8.3.3) also has a mean, the *conditional mean* or *conditional expectation* of X given that $X \in (a, b]$. The conditional mean is denoted by $\mathbb{E}(X \mid X \in (a, b])$ and is a summary statistic of the location of X given that $X \in (a, b]$. Equivalently, it is the mean of the conditional distribution function $F_{(a,b]}$ from (8.35) and can thus be calculated via (8.42) based on (8.36). As already mentioned, in Chapter 9 we will frequently consider the excess distribution function $F_u(x) = F_{(u,\infty)}(x + u)$ and its mean $\mathbb{E}(X - u \mid X > u)$. When viewed as a function of the threshold u, the latter conditional mean is known as the mean excess function. Again we refer to Chapter 9 for its importance and use.

Concept 8.45 (Mean excess function). If the mean of $X \sim F$ exists, the *mean excess function* is the function

$$e_F(u) = \mathbb{E}(X - u \mid X > u). \tag{8.45}$$

Equivalently, $e_F(u)$ is the mean of F_u as a function of u.

8.4.2 An Example from Financial Regulation: Value-at-Risk

When risks are communicated in the daily news or when we are confronted with the consequences of a particular disaster, the reporting typically does not mention the distribution function F of some underlying risk random variable X but, rather, a summary statistic in the form of a number, a *risk measure*. In the realm of catastrophic events, we often learn that a particular disaster is for instance a one-in-so-many-years event, an example of a *return period*. For instance, the Delta Report in Section 1.3 communicated new dike heights through such return periods. Once a type of risk measure has been agreed upon, for example a return period, one needs to specify a required safety level (the "so-many"), for example, 1 in 10 000 years. From this input one should be able to calculate the corresponding safety quantity, in this case the necessary dike height. At this point, stochastic modeling enters. First of all, one has to be clear about the threat X against which precautionary safety measures have to be taken. In case of the dikes, X could be the yearly maximal height of a storm surge at Neeltje Jans. Clearly, this needs a very careful, multidisciplinarily supported, precise definition of "storm surge". Measurements (data) have to be taken and a model for the distribution function F of X has to be proposed and estimated. Here statistics enters, together with possible physical and environmental models for storm surges. From an estimated model for F we can then obtain the resulting estimated dike height and also quantify the various model uncertainties underlying the reported solution. This process is very much the one followed in the famous van Dantzig report of Section 1.3. We will return to the dike example in Chapter 9.

In this subsection we shall formalize the above approach through the important notion of a quantile function as risk measure, and we will offer several applications. In doing so, we go to the heart of the understanding and communication of risk. We start with a rather remarkable example from banking and insurance regulation, where the above thinking has become hard-wired in international law.

The risk of a portfolio, the financial result of a business unit or of a whole company, depends on a diverse set of input parameters. These range from the quality of the managers involved (indeed, one hopes for this) to the numerous factors that drive the ambient economic environment. It is clear that randomness does play a role. A clear example of the latter is the demise of Barings Bank brought upon by a rogue trader Nick Leeson in Singapore (see Leeson, 2015); besides numerous internal factors that contributed to the downfall of the bank, a final, relevant, contributing random event turned out to be the Kobe earthquake of January 17, 1995. Several further examples were encountered in Chapter 3. For the purpose of our discussion, we assume that this financial or business outcome can be modeled by a random variable $X \sim F$. In doing so, we take a one-period view on risk. In the case of banking, typically this is one day, two weeks or a year. We are writing about the early 1990s, when the world of banking experienced a dramatic growth in the complexity and volume of so-called derivative products, which make bank-internal as well as regulatory oversight a highly challenging task. This situation led to the introduction of a particularly popular risk measure we already mentioned in Chapter 3, the value-at-risk (VaR).

In order to get a better feeling for the risk embedded in the so-called trading book of the bank, around 1990 the then CEO of J. P. Morgan, Dennis Weatherstone (1930–2008), requested a one-page market risk report, say, on a daily basis within 15 minutes after market closure; hence the name the *4:15 Weatherstone report*. Through this report, VaR was born.

This risk measure was going to revolutionize quantitative risk management, reporting and communication in banking worldwide. It is no coincidence that the bible on the topic, Jorion (2006), in its original 1996 edition carries as a subtitle *The New Benchmark for Controlling Derivatives Risk*. One of the architects of VaR was Till Guldimann, who first helped develop VaR as a risk measure within J. P. Morgan and later promoted it outside the firm. The methodology underlying VaR met with considerable interest from industry and regulators alike; it was made public as a technical document in 1994 in the form of a framework named *RiskMetrics*. See Morgan and Reuters (1996), Guldimann (2000) and Holton (2014, Section 1.9.5) for more historical details. In 1996, VaR entered the so-called *Basel Accords*, worldwide guidelines on banking regulation issued by the Basel Committee on Banking Supervision. Effective as of 1998, an amendment to the first Basel Accord (known as Basel I) detailed the use of VaR as a risk measure for market risk. Following on from these guidelines, national regulators made VaR-based risk management a legal requirement for larger international banks; see McNeil et al. (2015, Chapter 1) for further details. So, calculating a risk measure like VaR transformed from something "nice to have" to a "legal requirement to calculate and communicate".

What is VaR? We now return to our earlier discussion and let $X \sim F$ be a risk defined over a fixed time horizon. In words (the precise mathematical definition is given in Concept 8.46), the *value-at-risk* of X at *(confidence) level* $\alpha \in (0,1]$ (typically taken close to 1) is the threshold x (say, in USD) that is exceeded by our risk position X with probability at most $1-\alpha$ (typically small, thus corresponding to a rare event). For example, X may stand for the negative *profit-and-loss* (P&L) of the current trading book of a bank, projected as it stands today into the future. In this case one typically considers two trading weeks (10 days); here, clearly, randomness enters. The negative P&L of our portfolio is a function of the change in value of the various investment constituents. In this case, the VaR of X at level $\alpha = 0.95$ is that amount x in USD which is exceeded with probability at most 0.05 by the end of that period of 10 consecutive trading days. VaR thus provides a threshold that is expected to be violated only infrequently, where the exact meaning of "infrequently" is controlled by the level α. In the parlance of *if* versus *what if*, VaR is very clearly an *if* measure of risk; it measures loss frequency, not the more important *what if*, namely, the severity. VaR contains no information whatsoever on the loss severity beyond VaR. We will come back to this important point later.

In the language of mathematics, we can express VaR in a more compact form.

Concept 8.46 (Value-at-risk). The *value-at-risk* of $X \sim F$ at *(confidence) level* $\alpha \in (0,1)$ is

$$\mathrm{VaR}_\alpha(X) = \inf\{x \in \mathbb{R} : F(x) \geq \alpha\} \tag{8.46}$$

(read: "the value-at-risk of X at level alpha is the smallest x such that F of x is at least as large as alpha"). The "inf" stands for the "infimum", which, for our purposes, can be interpreted as "minimum".

In Gaivoronski and Pflug (2005), the authors introduced the notation "V@R" for value-at-risk. It is a pity that this suggestion did not catch on, as indeed the notation follows the pronunciation perfectly, and one would have further avoided the occasional clash with "Var" or "var" for the statistically important concept of variance. We would also like to add that there is of course much more to a bank-internal VaR calculation than 'just' (8.46).

Remark (Interpretation)**.** Since $F(x) \geq \alpha$ is equivalent to $\mathbb{P}(X > x) = 1 - F(x) \leq 1 - \alpha$, we obtain from (8.46) the interpretation that $\mathrm{VaR}_\alpha(X)$ is the smallest x such that the probability $\mathbb{P}(X > x)$ of X exceeding x is at most $1 - \alpha$.

8.4.3 The Concept of Quantile Functions

Viewing the right-hand side of (8.46) as a function of α, a function well known to probabilists and statisticians arises, namely the quantile function of a distribution function F.

Concept 8.47 (Quantile function, quantile)**.** For a distribution function F, its *quantile function* is defined by

$$F^{-1}(y) = \inf\{x \in \mathbb{R} : F(x) \geq y\}, \quad y \in (0, 1), \tag{8.47}$$

(read: "the quantile function of F at y is the smallest x such that F of x is at least as large as y"). For $y \in (0, 1)$, $F^{-1}(y)$ is also known as the *y-quantile* of F. If F is strictly increasing and continuous, one has $F^{-1}(F(x)) = x$, $x \in \mathbb{R}$, and $F(F^{-1}(y)) = y$, $y \in (0, 1)$.

Comparing (8.46) with (8.47) we see that $\mathrm{VaR}_\alpha(X) = F^{-1}(\alpha)$, so VaR at level α is simply the α-quantile of F. Quantiles are frequently encountered in everyday life. For example, you have probably heard of the *median* (the 0.5-quantile or 50%-quantile; it represents the 'middle' value of data points and is thus a measure of location similar to the mean) or of the *kth percentile* (the $(k/100)$-quantile or k%-quantile, used to score sports or exam results; for example, everyone scoring above the 95th percentile is scoring in the top 5% of all participants).

Figure 8.31 shows the graphs of the quantile functions $F_1^{-1}, F_2^{-1}, F_3^{-1}$ corresponding to the three distribution functions F_1, F_2, F_3 shown in Figure 8.11. Comparing F_2^{-1} in the center

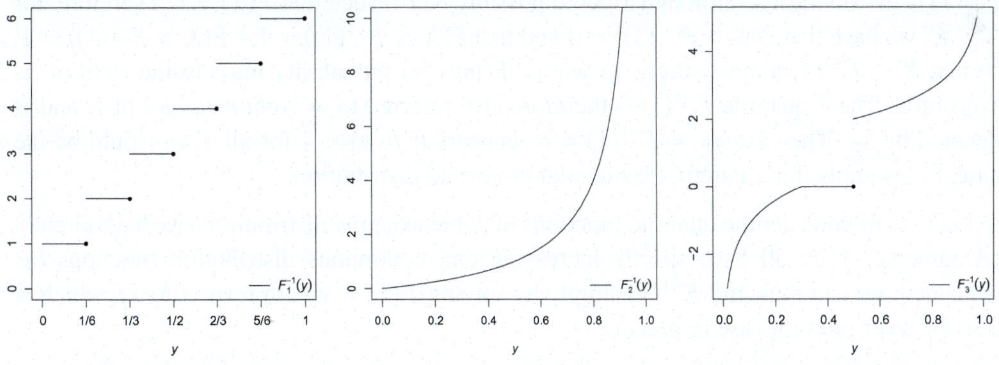

Figure 8.31 Quantile function F_1^{-1} (left), F_2^{-1} (center) and F_3^{-1} (right) corresponding to the distribution functions F_1, F_2 and F_3 shown in Figure 8.11. Source: Authors

panel of Figure 8.31 with its corresponding distribution function F_2 in the center panel of Figure 8.11, one can check that one graph is obtained from the other by a reflection. In a Cartesian coordinate system (where one unit of length on the x-axis equals one unit of length on the y-axis), the axis of reflection is the line $y = x$, the line through the origin with slope 45°. This is also how the square root $\sqrt{x}$ is related to its inverse the quadratic function x^2 or how the natural logarithm $\log(x)$ is related to its inverse the exponential function $\exp(x)$ or e^x. We can thus interpret the quantile function F^{-1} as the inverse of the distribution function F (hence the notation F^{-1}). Although a bit harder to spot, this also applies to F_1^{-1} and F_3^{-1} in Figure 8.31. To this end, note that the reflection of a flat part of a distribution function F becomes a jump of the quantile function F^{-1}. And the reflection of the jump-part of a distribution function F becomes a flat part of the quantile function F^{-1}. Viewing quantile functions as inverses of distribution functions is also not too far-fetched to imagine from the defining formula (8.47). Since, for strictly increasing and continuous F, for every $y \in (0, 1]$ there is a unique x such that $F(x) = y$, calculating (8.47) is equivalent to solving $F(x) = y$ with respect to x, and so the y-quantile $F^{-1}(y)$ of F is simply the inverse of the distribution function F at y. However, using the solution of $F(x) = y$ with respect to x as a definition of the y-quantile would allow us to have quantile functions only for strictly increasing and continuous F such as F_2 in the center panel of Figure 8.11, but not F_1 or F_3. For example, for F_3 and $y = 0.4$, there is no x such that $F_3(x) = y$. And for $y = 0.5$, there are infinitely many x such that $F_3(x) = y$ (namely, all $x \in [0, 2]$). For this reason, using $F(x) = y$ to define the y-quantile of F is too restrictive. One way out of this difficulty is to replace $F(x) = y$ by $F(x) \geq y$ and to choose as y-quantile of F the smallest x such that this inequality is fulfilled, which is exactly what (8.47) does.

Viewing quantile functions as inverses of distribution functions provides us with the intuition that a quantile function F^{-1} indeed uniquely characterizes a distribution function F – if you know one, you know the other. A quantile function F^{-1} therefore also characterizes the random variable X of F; exactly how it does this and exactly how useful this result is will become clear in Section 8.5 below.

Remark 8.48 (Right endpoint). With the convention that $\inf \emptyset = \infty$, the definition of the quantile function $F^{-1}(y)$ of a distribution function F also makes sense at $y = 1$. By definition, $F^{-1}(1)$ is the smallest x (although x could possibly be ∞) such that $F(x) = 1$. Therefore, for $X \sim F$, we have that $\mathbb{P}(X > F^{-1}(1)) = 0$ and that $\mathbb{P}(X \leq F^{-1}(1)) = 1 - \mathbb{P}(X > F^{-1}(1)) = 1$, so that $X \leq F^{-1}(1)$ almost surely. In words, F puts no probability mass to the right of its 1-quantile. The 1-quantile $F^{-1}(1)$ is therefore also referred to as *right endpoint* of F and is denoted by x_F. The case $x_F = F^{-1}(1) = \infty$ appears if $F(x) < 1$ for all x, as would be the case, for example, for a Pareto, exponential or normal distribution.

Let us now consider the quantile functions of some example distributions we have already encountered. They all have strictly increasing and continuous distribution functions (as explained, we can thus find $F^{-1}(y)$ simply by solving $F(x) = y$ with respect to x), which is also the most relevant case in practice.

Example 8.49 (Quantile functions of Pareto, exponential and normal distributions).

(1) If $X \sim \mathrm{Par}(\theta)$, we know from Example 8.30 that the distribution function F of X is $F(x) = 1 - 1/(1+x)^\theta$, $x \geq 0$. This is a strictly increasing and continuous distribution function, so we can solve $F(x) = y$ with respect to x to obtain $F^{-1}(y)$. Since $1 - 1/(1+x)^\theta = y$ if and only if $(1+x)^{-\theta} = 1 - y$, so that $1 + x = (1-y)^{-1/\theta}$ and thus $x = (1-y)^{-1/\theta} - 1$, we obtain that $F^{-1}(y) = (1-y)^{-1/\theta} - 1$, $y \in (0, 1)$.
(2) If $X \sim \mathrm{Exp}(\lambda)$, we know from Example 8.33 that the distribution function F of X is $F(x) = 1 - \exp(-\lambda x)$, $x \geq 0$. Following the same steps as for the Pareto case, we obtain that $F^{-1}(y) = -\log(1-y)/\lambda$, $y \in (0, 1)$.

(3) If $X \sim \mathrm{N}(\mu, \sigma^2)$, we know from (8.32) that $F(x) = \Phi((x-\mu)/\sigma)$, $x \in \mathbb{R}$; recall that Φ denotes the $\mathrm{N}(0,1)$ distribution function. This is a strictly increasing and continuous distribution function, so, as before, we can solve $F(x) = y$ with respect to x to obtain $F^{-1}(y)$. Since $\Phi((x-\mu)/\sigma) = y$ if and only if $(x-\mu)/\sigma = \Phi^{-1}(y)$, so that $x = \mu + \sigma\Phi^{-1}(y)$, we obtain that $F^{-1}(y) = \mu + \sigma\Phi^{-1}(y)$, $y \in (0,1)$. Like Φ, Φ^{-1} is not analytically tractable but can be evaluated numerically with statistical software.

From Example 8.34 we know that $F(\mu - x) = 1 - F(\mu + x)$ for all $x \in \mathbb{R}$. Equivalently, $\mu - x = F^{-1}(1 - F(\mu + x))$ for all $x \in \mathbb{R}$. Letting $p = 1 - F(\mu + x)$ and noting that then $x = F^{-1}(1-p) - \mu$, we obtain $2\mu - F^{-1}(1-p) = F^{-1}(p)$ for all $p \in (0,1)$. This is the point symmetry property of normal distributions in terms of their quantile functions. In particular, if $\mu = 0$ and $\sigma = 1$ then $-\Phi^{-1}(1-p) = \Phi^{-1}(p)$, $p \in (0,1)$, a property to which we will return in Section 8.6.3.

Figure 8.32 shows the distribution function F and corresponding quantile function F^{-1} of $\mathrm{Par}(1)$, $\mathrm{Exp}(1)$ and $\mathrm{N}(0,1)$ distributions. We can visually confirm the relationship between F and F^{-1} by a reflection of their graphs on the given axes.

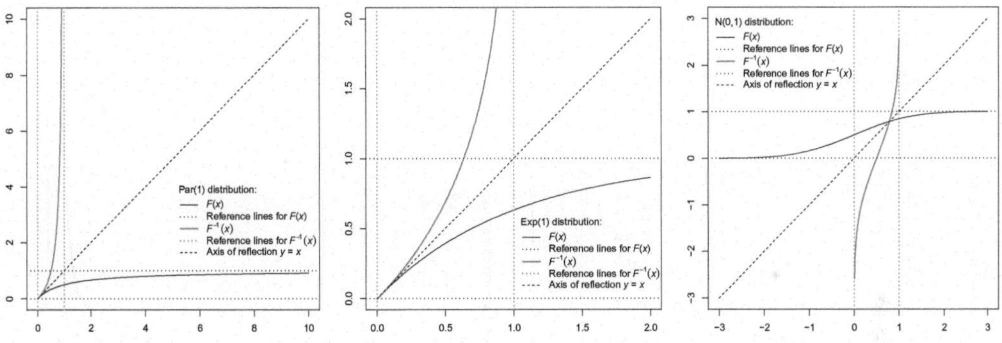

Figure 8.32 Distribution function F and corresponding quantile function F^{-1} for the $\mathrm{Par}(1)$ (left), $\mathrm{Exp}(1)$ (center) and $\mathrm{N}(0,1)$ (right) distributions. Source: Authors

The following example addresses the quantile function of an empirical distribution, which is an instance of a discrete distribution. Though the concept in the example below remains straightforward, the jumps make the notation a bit cumbersome. If the various indices start dancing on the page, feel free to immediately go to Example 8.51.

Example 8.50 (Quantile functions of empirical distributions). In short, the quantile function $\hat{F}_n^{-1}$ of an empirical distribution function $\hat{F}_n$ is called *empirical quantile function*. But how can we construct or think of $\hat{F}_n^{-1}$?

Recall from Concept 8.16 the concept of an empirical distribution function. With the convention that $x_{(n+1)} = \infty$, the empirical distribution function $\hat{F}_n$ based on the data points $x_1, x_2, \dots, x_n$ satisfies $\hat{F}_n(x) = k/n$ for $x \in [x_{(k)}, x_{(k+1)})$ and $k = 1, 2, \dots, n$; see Remark 8.17. Thus, for values $y \in ((k-1)/n, k/n]$, one must have that $\hat{F}_n^{-1}(y) = x_{(k)}$. These things are easiest to verify by drawing the graph of the step function $\hat{F}_n$ and choosing particular values for k. For example for $k = 1$, the y values considered are $y \in (0, 1/n]$, so the first time $\hat{F}_n$ is at least as large as such a y value occurs when $\hat{F}_n$ jumps from 0 to $1/n$, which happens at the smallest data point, so $\hat{F}_n^{-1}(y) = x_{(1)}$ and thus indeed $x_{(k)}$ for $k = 1$. Describing the empirical quantile function $\hat{F}_n^{-1}(y)$ in terms of which of the bins $(0, 1/n]$, for $k = 1$, $(1/n, 2/n]$, for $k = 2$, and so on, y falls into is inconvenient. However, we can describe "$\hat{F}_n^{-1}(y) = x_{(k)}$ for $y \in ((k-1)/n, k/n]$" more explicitly. The condition $y \in ((k-1)/n, k/n]$ is equivalent to $ny \in (k-1, k]$ or $\lceil ny \rceil = k$, where $\lceil x \rceil$ is the *ceiling function* of x (read: "the ceiling of x"). Using $k = \lceil ny \rceil$ in $x_{(k)}$, we obtain the compact expression

$$\hat{F}_n^{-1}(y) = x_{(\lceil ny \rceil)}, \quad y \in (0, 1), \tag{8.48}$$

for the empirical quantile function $\hat{F}_n^{-1}$ based on the data points $x_1, x_2, \ldots, x_n$.

We can equally well define the quantile functions of mixed-type distributions; the following example contains such a case.

Example 8.51 (Quantile functions of disaster data)**.** We now revisit the disaster data of Example 8.18. The left-hand side of Figure 8.33 shows the empirical quantile function of the yearly number of worldwide deaths due to natural disasters since 1900; the latter was displayed on the left-hand side of Figure 8.14. For a given level $y \in (0, 1)$ we can read off

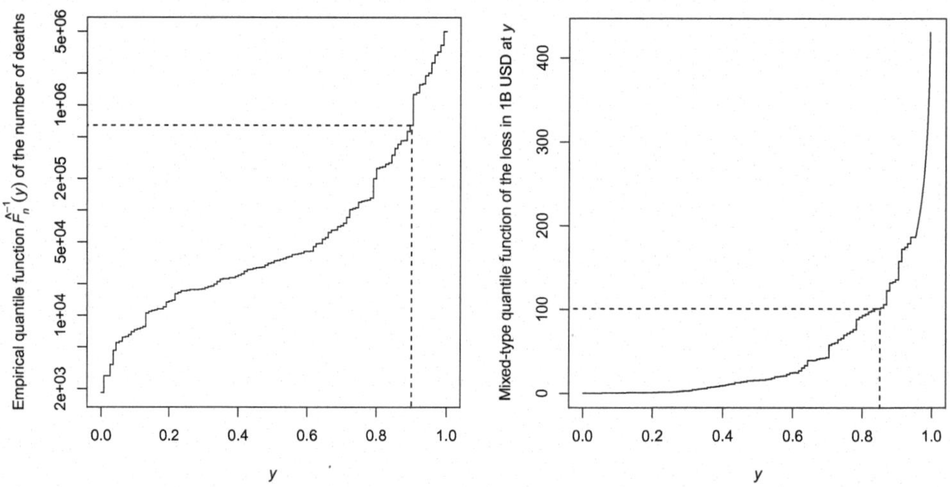

Figure 8.33 Left: Empirical quantile function of the yearly number of worldwide deaths due to natural disasters since 1900. Right: Quantile function of the empirical distribution function with continuous-tail model of the yearly loss, in one billion USD, due to natural disasters since 1900. The dashed lines indicate $y = 0.9$ (left) and $y = 0.85$ (right). Source: Authors

the corresponding number of deaths; for example, the 0.9 quantile is roughly 500 000, so the disasters with the largest 10% of deaths had a toll of at least 500 000 people.

The right-hand side of Figure 8.33 shows the quantile function corresponding to the distribution function displayed on the right-hand side of Figure 8.14. For a given level $y \in (0, 1)$ we can read off the corresponding loss in damage in one billion USD. For example, at level 0.85 we see a value of roughly 100, so the disasters with the largest 15% of losses had a loss of at least 100 billion USD, that is 100 000 000 000 USD. In relation to the 7.9 billion people on Earth at the time of writing, this amounts to costs of at least 12.66 USD per person (which many could not afford, and do not overlook the "at least").

Of course, conditional distributions also have quantile functions.

Example 8.52 (Quantile functions of conditional distributions)**.** Consider the conditional distribution function $F_{(a,b]}$ from Example 8.38 again. By (8.35), $F_{(a,b]}(x) = (F(x) - F(a))/(F(b) - F(a))$, $x \in [a, b]$. By definition, we thus obtain the quantile function $F^{-1}_{(a,b]}$ of $F_{(a,b]}$ via

$$\begin{aligned} F^{-1}_{(a,b]}(y) &= \inf\{x \in \mathbb{R} : F_{(a,b]}(x) \geq y\} = \inf\left\{x \in \mathbb{R} : \frac{F(x) - F(a)}{F(b) - F(a)} \geq y\right\} \\ &= \inf\{x \in \mathbb{R} : F(x) \geq F(a) + (F(b) - F(a))y\}. \end{aligned}$$

The latter expression is, again by definition, the quantile function of F evaluated at $F(a) + (F(b) - F(a))y$. We thus obtain that

$$F^{-1}_{(a,b]}(y) = F^{-1}\big(F(a) + (F(b) - F(a))y\big), \quad y \in (0, 1). \tag{8.49}$$

Feel free to follow the same steps for the excess distribution function from (8.38) to see that its quantile function is

$$F_u^{-1}(y) = F^{-1}\big(F(u) + (1 - F(u))y\big) - u, \quad y \in (0, 1).$$

Already in Section 8.3.1 we introduced the left and right "tails" of a distribution function F as the x values for which $F(x)$ is close to 0 or close to 1, respectively. With the notion of quantile functions, we can now make the concept of the tails of a distribution function more precise.

Concept 8.53 (Tails of a distribution)**.** The *left tail* (or *lower tail*) of a distribution function F refers to the values $x < F^{-1}(y)$ for small $y \in (0, 1)$. The *right tail* (or *upper tail*) of a distribution function F refers to the values $x > F^{-1}(y)$ for large $y \in (0, 1)$.

To see that this concept of, say, the right tail is equivalent to what we have informally described as "values for which $F(x)$ is close to 1", let us consider a strictly increasing and continuous F, so that $F(F^{-1}(y)) = y$; see Concept 8.47. If $x > F^{-1}(y)$, applying F to both sides of this inequality, we obtain $F(x) > F(F^{-1}(y)) = y$. For sufficiently large $y \in (0, 1)$, we thus see that the right tail of a distribution function F indeed corresponds to the region of x values where $F(x)$ is close to 1, with the added advantage that the quantile function allows us to precisely control how far we are into the tail of F. For example, if $y = 0.99$ then $F^{-1}(0.99)$ is the value x that is exceeded by $X \sim F$ with probability 0.01. The tails of a distribution function F thus contain the information about the extreme realizations of $X \sim F$ and are therefore a central object of interest in risk management practice.

The usefulness of describing tail events in terms of quantile functions becomes clear when we are comparing different random variables. For example, a "large" loss X can refer to one larger than 3.0902 if $X \sim \mathrm{N}(0, 1)$, larger than 6.9078 if $X \sim \mathrm{Exp}(1)$ and larger than 999 if $X \sim \mathrm{Par}(1)$. Each value corresponds to the 0.999-quantile of the respective distribution, so if we had spoken of three losses that exceed their 0.999-quantiles we would have had a better intuition what this means for the respective losses, namely an extreme value that is exceeded with probability 1 in 1000. Quantiles thus allow us to more easily compare different random losses. Also, quantiles allow us to describe a *return level*, a threshold that is exceeded with a given probability. For example, as we just saw, the 0.999-quantile describes a threshold that is exceeded with probability $1/1000$. In other words, if we had realizations of $X_1, X_2, \dots, X_{1000} \overset{\text{iid}}{\sim} F$, then $F^{-1}(0.999)$ is the threshold that is expected to be exceeded by (at most) one of these 1000 realizations; if F is continuous, one can omit the "at most". We therefore also see that return levels and return periods are connected, the former describing a threshold that is exceeded with a probability determined by the latter and the latter describing a frequency at which a threshold determined by the former is exceeded.

The idea of using $X \sim \mathrm{N}(\mu, \sigma^2)$ to model fluctuations around μ of magnitude σ seems tempting for the modeling of risks: the larger σ, the larger the fluctuations, thus the higher the uncertainty and thus the higher the risk. Now that we have defined the tail of a distribution, we can think about when using $\mathrm{N}(\mu, \sigma^2)$ to model losses is a good idea (or, rather, when it is *not*). For simplicity, we consider the standard normal distribution $\mathrm{N}(0, 1)$ with density (8.30). As we can see from the graph of this density (the brown curve on the right-hand side of Figure 8.29) and also from the mathematical form in (8.30), it converges to 0 rather fast for large values x, because of the exponential factor $\exp(-x^2/2)$ in (8.30). Because of this factor, one also speaks of *exponentially fast* convergence to 0 and says that normal distributions have *light tails* (or *thin tails*), hinting at the fact that there is not much probability mass distributed further out in the tails. Another example of a light-tailed distribution is the exponential distribution $\mathrm{Exp}(1)$; its density contains the factor $\exp(-x)$, another example of an exponentially fast convergence to 0 in the (here, right) tail. In contrast, the $\mathrm{Par}(1)$ distribution has density $f(x) = 1/(1 + x)^2$, which converges to 0 for $x \to \infty$ like the power function $1/x^2$. Power functions do not converge to 0 as quickly as $\exp(-x^2/2)$ or $\exp(-x)$. Pareto and other distributions with such "power tail" behavior are thus said to have *heavy tails*. Suppose a loss X has a heavy-tailed distribution function F_{true} that is unknown to us and that we want to model. If we model this true, but unknown, distribution function F_{true} with a light-tailed distribution function F then this leads to an underestimation of risk, since large losses exceeding a certain threshold (so-called *tail losses*) are less likely to appear if $X \sim F$ than if $X \sim F_{\text{true}}$.

This can become a problem for a bank, say, if X models the losses over the next year and suddenly the actual losses exceed the company's *regulatory capital*, that is, the capital reserve that a bank is required to hold, by the Basel Accords, to account for such losses. Studying the tails of distributions and implications on the corresponding random variables is thus of utmost importance when choosing an adequate model for a risk $X \sim F_{\text{true}}$. We have already encountered in Section 3.5 the fact that $\mathrm{N}(\mu, \sigma^2)$ often does *not* provide an adequate risk model. There is also the more complicated case of *multivariate distributions* involving more than one random variable; see (3.1), which describes a certain probability involving two random variables in terms of an underlying bivariate normal distribution. In this book, we focus on *univariate distributions*, the distributions of single random variables (or sequences of independent random variables). We now present another distribution which addresses exactly this weakness of the exponentially fast converging density of $\mathrm{N}(\mu, \sigma^2)$ by essentially replacing the exponential function in the density with a power function. As a result, it allows one to model heavier tails than those of normal distributions and is therefore typically a much more adequate model F for F_{true}. Multivariate extensions are not far around the corner.

Example 8.54 (t distribution)**.** Student's t distribution is named after William Sealy Gosset (1876–1937); see Figure 8.34. Gosset studied chemistry and mathematics and then worked

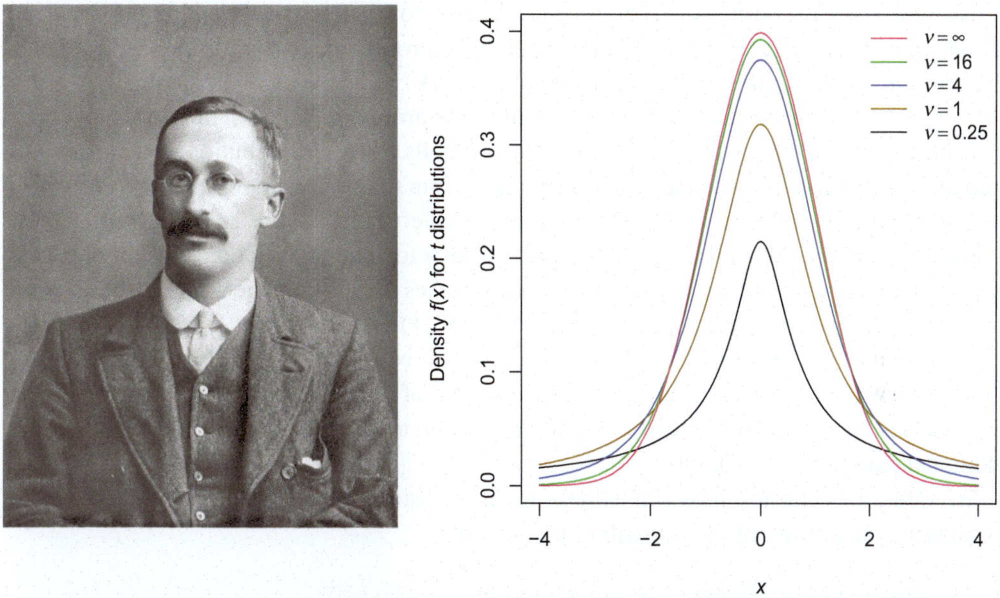

Figure 8.34 Left: William Sealy Gosset (1876–1937) also known as "Student". Right: t_ν densities for $\nu \in \{1/4, 1, 4, 16, \infty\}$. Sources: Wikipedia (left) and authors (right)

for the Dublin brewery Arthur Guinness & Son as Head Experimental Brewer. Motivated by his work at the brewery, Gosset made contributions to problems relating to small sample sizes, for example. As the published work of other scientists working at the brewery revealed company secrets, the company banned its employees from publishing. After arguing that his work would be of no benefit to other brewers, Gosset was allowed to publish under a pseudonym, and he chose "Student" (two other employees published statistical work under the pseudonyms "Sophister" and "Mathetes", for example).

A random variable X follows a *(Student's) t distribution*, denoted by $X \sim t_\nu(\mu, \sigma^2)$ ($t_\nu(0,1)$ being denoted simply by t_ν), if X admits the stochastic representation

$$X = \mu + \sigma\sqrt{W}Z, \tag{8.50}$$

where $Z \sim \mathrm{N}(0,1)$, W is a non-negative random variable (following a distribution we have not introduced but which depends on the *degrees of freedom* parameter $\nu > 0$) and W and Z are independent. The stochastic representation (8.50) is not far from that of a normal distribution, see (8.31) (use $Z \sim \mathrm{N}(0,1)$ in the latter), so the only additional quantity that (8.50) involves is $\sqrt{W}$, a random variable which influences the shape of the resulting t distribution. As for normal distributions, the distribution and quantile functions of t distributions are not available analytically, but they are numerically in software.

Similarly to (but different from) normal distributions, the parameters μ, σ^2 of t distributions are location and scale parameters, respectively. What is different is that μ is not necessarily the mean $\mathbb{E}(X)$ of $X \sim t_\nu(\mu, \sigma^2)$ and σ^2 is no longer the variance $\mathrm{var}(X)$. If $\mathbb{E}(|X|) < \infty$ (which one can show to be the case if and only if $\nu > 1$), then indeed $\mathbb{E}(X) = \mu$. But if $\mathrm{var}(X) < \infty$ (which one can show to be the case if and only if $\nu > 2$), then $\mathrm{var}(X)$ is $\sigma^2\nu/(\nu - 2)$.

In contrast with the normal distribution, the t distribution depends on the additional degrees of freedom parameter $\nu > 0$, which allows one to control the heaviness of the tails of the t distribution. For $\nu > 2$, this is already visible from $\mathrm{var}(X) = \sigma^2\nu/(\nu-2)$: the smaller is $\nu > 2$, the larger is the variance and thus the fluctuations of X around μ, which in turn leads to more probability mass being pushed out further into both tails of the t distribution in comparison with the normal distribution. This can also be seen from the stochastic representation (8.50) by studying the distribution of $\sqrt{W}$. The random variable $\sqrt{W}$ helps, for small ν, to push more probability mass into the tails of the t distribution so as to obtain heavier tails in comparison with those of the normal distribution. This effect vanishes for $\nu \to \infty$ when $\sqrt{W}$ can be shown to converge to the constant 1, and thus (8.50) and (8.31) are equivalent. For this reason we allow t distributions to have the degrees of freedom $\nu = \infty$ (so $t_\infty(\mu, \sigma^2) = \mathrm{N}(\mu, \sigma^2)$), and we can thus view the normal distribution as a special case of the t distribution. In other words, t distributions are more *flexible* (especially in the tails) than normal distributions owing to the additional parameter $\nu \in (0, \infty]$.

The effect of ν on the heavy-tailedness of the t distribution is most easily seen when considering the density of the t distribution, given by

$$f(x) = \frac{\Gamma((\nu+1)/2)}{\Gamma(\nu/2)\sqrt{\nu\pi\sigma^2}}\left(1 + \frac{((x-\mu)/\sigma)^2}{\nu}\right)^{-\frac{\nu+1}{2}}, \quad x \in \mathbb{R}; \tag{8.51}$$

see the right-hand side of Figure 8.34 for t_ν densities for different ν. The density (8.51) looks complicated (recall that it needs to integrate to 1, see (8.26), and thus involves a daunting-looking fraction of gamma functions and π). However, the most important takeaway is that there is a power function of the form $(1 + x^2/\nu)^{-(\nu+1)/2}$, which, depending on ν, creates heavier tails than the exponential function $\exp(-x^2/2)$ present in the normal density. To see this, consider the right-hand side of Figure 8.34, look at the interval $(3, 4]$ and compare the density values for different degrees of freedom parameters ν. We see that the smallest area (or least probability mass) under any of the curves is that below the red curve, which corresponds to $\nu = \infty$ and thus to the standard normal density ϕ. The left-hand side of Figure 8.35 shows more of the right tail. The curves are easier to distinguish in logarithmic scale; see the right-hand side of Figure 8.35. We can now more clearly see that the t_ν densities are all ordered according to ν. The smaller the value of ν, the larger the t density $f(x)$ and so the greater the concentration of probability mass in the tail of the t_ν distribution (making extremely large values more likely to appear). Note that, owing to the symmetry of $f(x)$ around $x = 0$, the same argument applies to the left tail and extremely small values, of course.

Because of these important properties, we would have preferred to see the t density (8.51) on a banknote – perhaps the 20 Deutsche Mark banknote, or, more importantly, a banknote of a currency that is still in use today. We could immediately come up with mathematical results from the realm of risk management important enough to warrant being printed on higher banknotes, too.

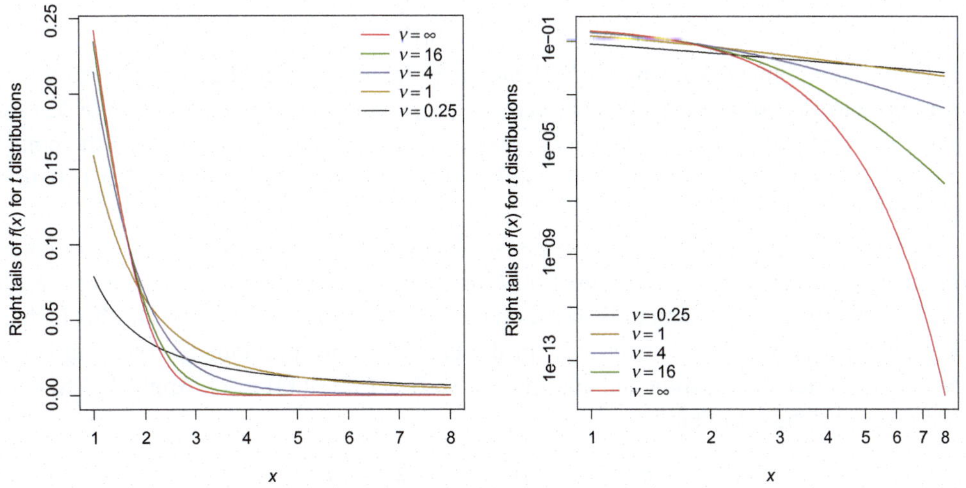

Figure 8.35 Left: t_ν densities for $\nu \in \{1/4, 1, 4, 16, \infty\}$ plotted in their right tail. Right: The same plot in logarithmic scale. Source: Authors

The dramatic difference that moving from a light-tailed normal distribution to a more heavy-tailed t distribution can make in assessing risk is also highlighted in Jorion (2008) in the context of the LTCM hedge fund disaster. Although, as mentioned before, we will not be discussing multivariate distributions, we could also revisit Section 3.5 at this point. Had a multivariate t distribution instead of a multivariate normal distribution been used as a model for pricing CDOs, then (3.1) adapted to a multivariate t distribution would have most likely only hurt, not killed, Wall Street; see also Hofert and Scherer (2011).

8.4.4 Quantile–Quantile Plots

Given n realized losses $x_1, x_2, \ldots, x_n$ one often needs to find an appropriate model that can produce such losses (recall our discussion on modeling F_{true} by some F above). Since in practice the true underlying distribution function F_{true} is unknown to us, how can we decide whether a hypothesized distribution function F (such as a t distribution function) produces realizations that "resemble" the data points $x_1, x_2, \ldots, x_n$ of F_{true} and thus whether F is an appropriate model based on those data points?

Conceptually, we should compare F with F_{true}. As we do not know F_{true}, we use the empirical distribution function $\hat{F}_n$ as an estimate of F_{true}; as already mentioned in Remark 8.17, $\hat{F}_n$ contains all the information about the n data points. This means that if n is sufficiently large, we should compare F with $\hat{F}_n$. This can be done in many ways, for example through a simple plot. But instead of plotting $\hat{F}_n(x)$ and $F(x)$ as functions of x and comparing two curves (thus comparing the two sets of points $(x, \hat{F}_n(x))$ and $(x, F(x))$), we could plot the single set of points $(\hat{F}_n(x), F(x))$ for specific x values. Reasonable x values are $x_1, x_2, \ldots, x_n$ as those are the interesting points, the ones where $\hat{F}_n$ jumps. To this end we use the order statistics $x_{(1)} \le x_{(2)} \le \cdots \le x_{(n)}$, the ordered data points. We then know from Remark 8.17 that $\hat{F}_n(x_{(k)}) = k/n$, $k = 1, 2, \ldots, n$, so that $\hat{F}_n$ jumps to k/n at the kth-smallest data point. We could then plot $(\hat{F}_n(x_{(k)}), F(x_{(k)}))$, $k = 1, 2, \ldots, n$, which is the set of points

$$(k/n, F(x_{(k)})), \quad k = 1,2,\dots,n; \tag{8.52}$$

the plot of these points is known as a *P–P plot* (P–P standing for "probability–probability"). If F is close to $\hat{F}_n$ and thus F_{true}, then $F(x_{(k)}) \approx \hat{F}_n(x_{(k)}) = k/n$ and so the points in (8.52) are roughly $(k/n, k/n)$, $k = 1,2,\dots,n$. This is easy to check visually as it means that the points should roughly lie on the line $y = x$ if n is sufficiently large and the hypothesized distribution function F is close to F_{true}.

Although P–P plots could be used to assess whether the given losses $x_1, x_2, \dots, x_n$ come from the hypothesized distribution function F and thus whether F is a good model for F_{true}, a more popular visual assessment is based on quantile functions. The reason for this is that, in particular, departures of F from F_{true} in the tails are more pronounced. For simplicity, suppose F is strictly increasing and continuous (so that $F^{-1}(F(x)) = x$). Apply F^{-1} to both coordinates in (8.52) to obtain

$$(F^{-1}(p_k), x_{(k)}), \quad k = 1,2,\dots,n, \tag{8.53}$$

where $p_k = (k - 1/2)/n$, $k = 1,2,\dots,n$; the slight shift from arguments k/n to $(k - 1/2)/n$ in the x-coordinate avoids the evaluation of 1-quantiles, which are ∞ if $x_F = \infty$. This leads to the concept of a Q–Q plot.

Concept 8.55 (Q–Q plot)**.** The plot of the points (8.53) is known as *Q–Q plot* (Q–Q standing for "quantile–quantile"). The x-coordinates of (8.53) (the quantiles under the hypothesized distribution function F) are referred to as *theoretical quantiles* and the y-coordinates (the sorted data) as *sample quantiles.*

The interpretation of a Q–Q plot remains the same as that for a P–P plot: if n is sufficiently large and the hypothesized distribution function F is close to F_{true}, then the points (8.53) lie on a straight line. Any departure from linearity can be interpreted as a departure of the hypothesized F from the true, but unknown, F_{true}.

Example 8.56 (Are Netflix negative log-returns normal?)**.** The left-hand side of Figure 8.36 shows the price S_t of one Netflix share in USD on day t from January 2, 2018 ($t = 0$), to December 30, 2021 ($t = 1006$). The plot on the right-hand side shows the *negative logarithmic returns (−log-returns)*

$$X_t = -\log\left(\frac{S_t}{S_{t-1}}\right), \quad t = 1,2,\dots,n, \tag{8.54}$$

for the sample size $n = 1006$ that we have here. For small $|x|$, we can use $-\log(1 + x) \approx -x$ to see from (8.54) that $X_t = -\log(1 + (S_t/S_{t-1}) - 1) \approx -((S_t/S_{t-1}) - 1) = -(S_t - S_{t-1})/S_{t-1}$, so that X_t is the (sign-adjusted) relative difference between S_{t-1} and S_t; the sign adjustment is made such that losses correspond to positive values of X_t, which are the main focus in risk management. For financial data like those in this example, the −log-returns are typically the starting point for finding an adequate model from a risk management perspective, which then implies a model for the price S_t of one Netflix share. Such models are used to project forward the price process of one Netflix share in order to make, for example, predictions based on these simulations; see also Section 8.5 and especially Section 9.5.4. The largest −log-return displayed on the right-hand side of Figure 8.36 is 0.1181, on March 16, 2020, which, by (8.54), corresponds to $\exp(-0.1181) = S_t/S_{t-1}$, so that $S_t = 0.8886 S_{t-1}$, which

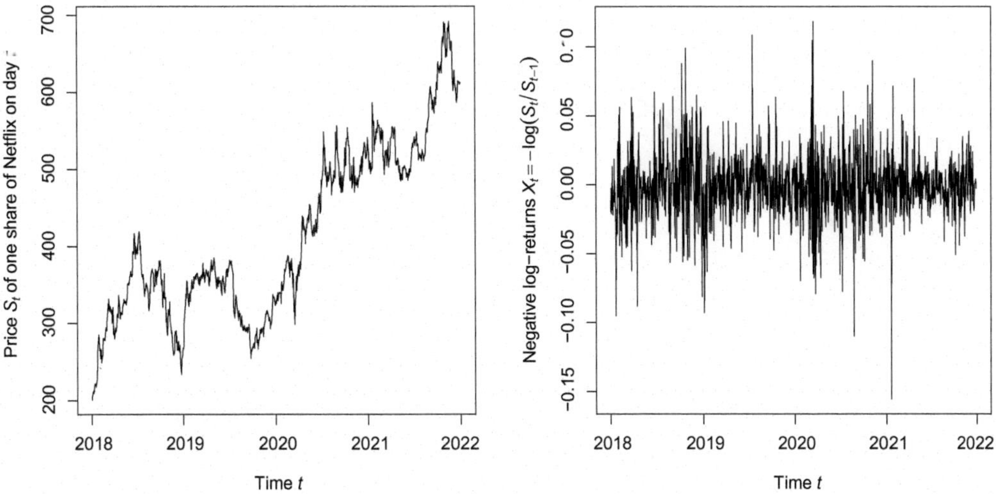

Figure 8.36 Left: Daily price of one Netflix share over the period January 2, 2018 until December 30, 2021. Right: The corresponding –log-returns. Source: Authors

corresponds to a price drop by $0.1114 = 11.14\%$ on that day in comparison with the day before. The larger the −log-return on a day, the larger the price drop that day.

For obvious reasons, the graph on the right-hand side of Figure 8.36 is also referred to as the heartbeat of the trader. The up-and-down behavior seen in this figure is rather typical of most financial data. In more technical jargon, it is often compared with white noise, known from signal processing. As we shall see, this observation leads very naturally to some statistical models.

A simple model for the fluctuation of X_t around a value μ and of a particular strength σ is the normal distribution. By Example 8.44 (1), a random variable $X \sim \mathrm{N}(\mu, \sigma^2)$ has mean $\mathbb{E}(X) = \mu$ and variance $\mathrm{var}(X) = \sigma^2$, and μ and σ can be estimated from data. If we thus estimate μ and σ from the given realizations of X_t we obtain the normal distribution $\mathrm{N}(-0.0011, 0.0006)$. The question is then whether this normal distribution indeed represents the true, but unknown, distribution (F_{true}) well, that is, whether $\mathrm{N}(-0.0011, 0.0006)$ can generate realizations like those we see on the right-hand side of Figure 8.36. To assess this question, the left-hand side of Figure 8.37 shows the corresponding Q–Q plot. As in (8.53), the points shown have x-coordinates given by the quantile function of $\mathrm{N}(-0.0011, 0.0006)$ evaluated at $p_k = (k - 1/2)/n$ and y-coordinates given by the ordered −log-returns (which are the $x_{(k)}$, $k = 1, 2, \ldots, n$, in (8.53)). As indicated by the departure of the points from the reference line $y = x$, the conclusion is that the −log-returns do not follow a normal distribution. The situation looks much better for the heavier-tailed t distribution with estimated degrees of freedom parameter ν equal to 3.69; see the right-hand side of Figure 8.37. In Section 9.5.4 we will show that, for a more detailed analysis of these data, we will have to dig a bit deeper into the toolkit of stochastic models.

It is important to consider the type of departure from the reference line in the Q–Q plot on the left-hand side of Figure 8.37 and to see what we can learn from this type of departure. Consider the theoretical quantiles (along the x-axis) greater than 0.05. Those

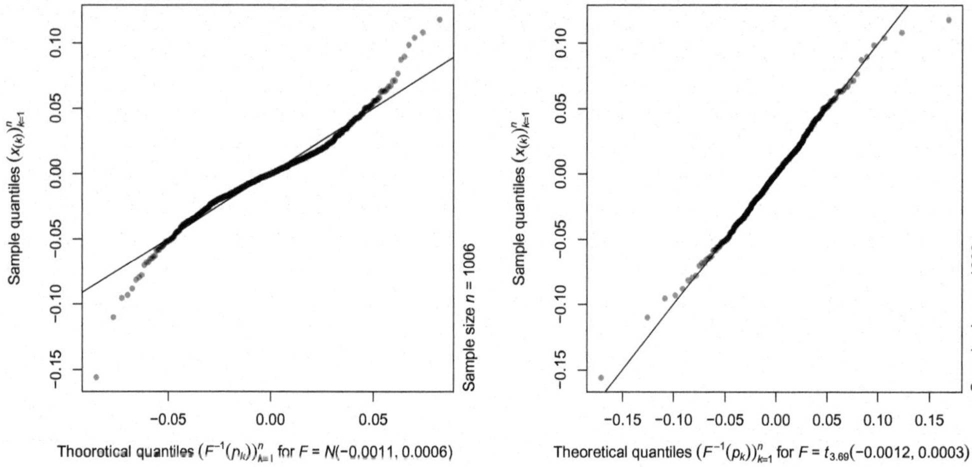

Figure 8.37 Q–Q plot of the −log-returns versus the estimated N(−0.0011, 0.0006) distribution (left) and versus a $t_{3.69}(-0.0012, 0.0003)$ distribution (right). Source: Authors

points correspond to large positive values of −log-returns, which means large losses of the Netflix share price. For theoretical quantiles beyond 0.05, we see not only a departure from the reference line in the direction of the y-axis, but also values *larger* than those of the reference line. This means that the corresponding −log-return values (the tail losses) are indeed *larger* than such losses would be under the hypothesized F (here, N(−0.0011, 0.0006)). This suggests that the underlying unknown F_{true} that generated these tail losses is heavier-tailed than the hypothesized F. And this means that if we used F as a model for the −log-returns, F would not produce sufficiently large tail losses like those we see in the data. As a result, F underestimates the risk of tail losses. Also, the value-at-risk VaR_α computed for $X \sim F$ for sufficiently large α would then not be as large as for $X \sim F_{\text{true}}$, which means that the computed regulatory capital computed from VaR_α would be too small to account for future losses appropriately. We thus again see that we would indeed have underestimated the risk. This line of thought is of utmost importance when teaching quantitative risk management courses and is frequently asked to be reproduced as an answer in exams on the topic, not because of the Q–Q plot itself, but because there are tools like Q–Q plots (and others not discussed here) that can help to determine whether a particular model for (tail) losses or loss distributions is appropriate.

If you rotate the Q–Q plot on the left-hand side of Figure 8.37 by 90° counterclockwise, you'll see that the points form a shape resembling an "Ƨ", an *inverted S shape*, instead of a line. As said above, the interpretation is that our hypothesized model F is lighter-tailed than what is suggested by F_{true}, and that translates into an underestimation of risk. If you see an *S shape* (so, a not inverted S), then, with a similar line of thought, your hypothesized F is heavier-tailed than F_{true}, which would lead to an overestimation of risk.

As a final remark, note that Q–Q plots are not always easy to interpret. For example, how far would a departure from the reference line need to be in order for F to be considered a non-adequate model for F_{true}? It needs experience to properly interpret Q–Q plots and the answer to this question also depends on the sample size n of the available data, for example. One can also include confidence intervals in Q–Q plots to help with the decision of whether a departure from the reference line is significant. Such confidence intervals can be obtained from a bootstrap procedure, for example; see Section 8.6.3. We do not dive into such details of Q–Q plots, but the interpretability of Q–Q plots provides us with another reason to consider confidence intervals in Section 8.6.3. A very simple approach for developing a gut feeling for Q–Q plots (and many other concepts) is to simulate them. For example, we could pick an F_{true} simulate data from this distribution and then create a Q–Q plot with F_{true} (and thus the correct distribution) as the hypothesized distribution to see and learn what typical departures under the correct model look like. This provides another motivation for what we will focus on in Section 8.5.

There is another important point we learn from Example 8.56. Just because we can estimate, to several digits after the decimal point, the best-fitting distribution ("best-fitting" in a sense we have not discussed; see also Remark 8.32 (3)) to given data does not mean this estimated distribution is actually appropriate for the given data, by any means! All that we can conclude is that we have found the best-fitting distribution in a certain parametric family of distributions (such as $\mathrm{N}(\mu, \sigma^2)$ for all $\mu \in \mathbb{R}$ and $\sigma > 0$); whether this distribution is appropriate requires careful inspection (with tools like Q–Q plots and others). Also, in Remark 9.12 and Section 9.5.4 we will meet more sophisticated models than a fitted normal or t distribution to address features of the Netflix data that these distributions cannot pick up.

8.4.5 From Value-at-Risk to Expected Shortfall

When we introduced VaR, we said that it is an *if* (frequency-based) risk measure, not a *what if* (severity-based) risk measure. For example, if $X \sim F_X$ and $Y \sim F_Y$ are two losses, then, for an α value of interest, the corresponding $\mathrm{VaR}_\alpha(X)$ and $\mathrm{VaR}_\alpha(Y)$ can very well be equal but F_X and F_Y can largely differ in their right tails beyond these VaR values. This is so because VaR_α is merely specifying is a single point, so that there is probability mass of (at most) $1 - \alpha$ beyond VaR_α, but VaR_α does not control how that probability mass is distributed over the values in the right tail beyond VaR_α. This probability mass could be located close to VaR_α, in the case of a light-tailed distribution like $\mathrm{N}(\mu, \sigma^2)$, or far out in the right tail for a heavy-tailed distribution like $\mathrm{Par}(\theta)$. To take the whole tail of a distribution into account, and thus to obtain a *what if* risk measure, one idea is to average out VaR_u over all levels $u \in (\alpha, 1)$. As we have just explained, no single fixed u takes the shape of the tail of the distribution beyond VaR_u into account, but if we average VaR_u over *all* $u \in (\alpha, 1)$, the whole tail of the distribution *is* taken into account. Since this average possibly contains uncountably many different values, the average does not use a sum but an integral. This leads to the concept of expected shortfall, a popular risk measure in practice.

Concept 8.57 (Expected shortfall)**.** *Expected shortfall (ES)* of $X \sim F$ at *(confidence) level* $\alpha \in (0, 1)$ is

$$\mathrm{ES}_\alpha(X) = \frac{1}{1-\alpha} \int_\alpha^1 \mathrm{VaR}_u(X)\,\mathrm{d}u. \tag{8.55}$$

In order for this integral to be finite, $F(x)$ is not allowed to be extremely heavy-tailed (a typical assumption is $\mathbb{E}(|X|) < \infty$, so the mean of F to be finite).

If the underlying distribution function F of X is continuous, which in most models and applications it is, then one can show that $\mathrm{ES}_\alpha(X) = \mathbb{E}(X|X > \mathrm{VaR}_\alpha(X))$, leading to various names for expected shortfall available in the literature, such as *conditional value-at-risk* (CVaR), *average value-at-risk* (AVaR) or *expected tail loss* (ETL). In the medical, reliability and insurance literature, where X often stands for a random variable modeling a lifetime, one frequently encounters the *mean residual life* of X, that is $\mathbb{E}(X - u \mid X > u)$ for some (excess-of-loss) threshold u; for $u = \mathrm{VaR}_\alpha(X)$ this equals $\mathrm{ES}_\alpha(X) - \mathrm{VaR}_\alpha(X)$.

In Example 8.56, X was the −log-return of the price of one Netflix share, which was an example of a negative profit-and-loss. When the focus is on modeling profits rather than losses (which is often the case in financial rather than risk management applications), one also frequently finds the (positive) profit-and-loss being considered. For example, we could have worked with the logarithmic (instead of the negative logarithmic) returns of the Netflix share price in Example 8.56. In this case, large losses are to be found in the left tail of the distribution of X rather than the right tail, so the respective literature considers small levels α for (suitably adapted) risk measures such as value-at-risk or expected shortfall. We will not further consider this approach (for us, losses are in the right tail of the considered distributions), but we felt obliged to quickly mention it in passing as one needs to be careful not to confuse the left and the right tail when computing risk measures in practice.

When banks and insurers first encountered expected shortfall as a *what if* risk measure, one concern emerged from a comparison with the *if* risk measure VaR_α. The concern was that for the same level α, ES_α would lead to unreasonably larger computed regulatory capital than VaR_α. This issue has been addressed by, for example, considering different levels for both risk measures, such as $\mathrm{VaR}_{0.99}$ versus $\mathrm{ES}_{0.975}$. Further skepticism has concerned the widely perceived larger difficulty in communication effort, at the level of top management, of ES versus VaR. This issue has resolved itself gradually over time.

Returning to the first concern, as an average of VaR_u for $u \in (\alpha, 1)$ it is clear that ES_α is at least as large as VaR_α, but the question is how much larger ES_α can be. The following example investigates this question for (normal and) t distributions. You may already have noticed that we regularly use language from banking regulation such as value-at-risk and regulatory capital. However, as we explained, VaR is 'just' a quantile and its applied interpretation very much depends on the context of the risk X to be modeled. So "regulatory capital" may then, for example, stand for "dike height" or "life time" depending on the context. The results below should hence be understood in this broader interpretation.

Example 8.58 (ES versus VaR for t distributions)**.** For a loss $X \sim F$, one can study the *shortfall-to-quantile ratio*

$$\frac{\mathrm{ES}_\alpha(X)}{\mathrm{VaR}_\alpha(X)} \tag{8.56}$$

to quantify how regulatory capital calculations with the risk measure ES_α differ from those with VaR_α. The left-hand side of Figure 8.38 shows the shortfall-to-quantile ratio (8.56) as a function of $\alpha \in [0.99, 1)$ (typical values of α are in the range $[0.95, 1)$ in banking practice) for $X \sim \mathrm{N}(0, 1)$. As is fully explained in the literature on quantitative risk management, the

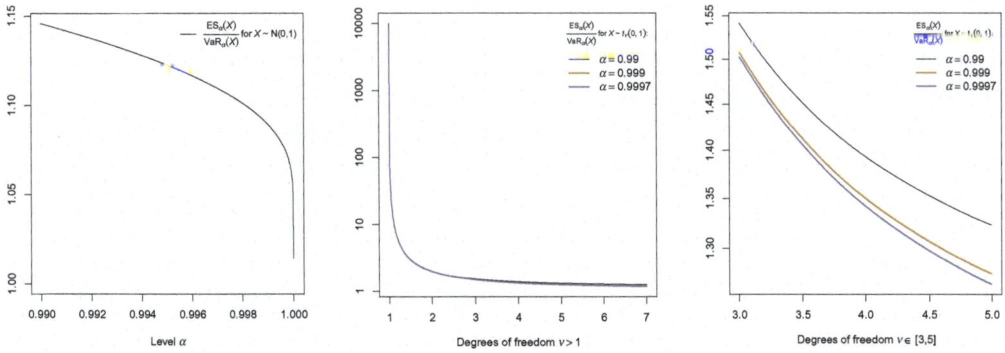

Figure 8.38 Shortfall-to-quantile ratio $\mathrm{ES}_\alpha(X)/\mathrm{VaR}_\alpha(X)$ as a function of $\alpha \in [0.99, 1)$ for $X \sim \mathrm{N}(0,1)$ (left); as a function of $\nu > 1$ for $X \sim t_\nu$ and $\alpha \in \{0.99, 0.999, 0.9997\}$ (center); and zoomed in for $\nu \in [3,5]$ (right). Source: Authors

expected shortfall enjoys several important properties over value-at-risk. An obvious one we have already mentioned is that, as a *what if* risk measure, ES_α measures risk beyond the *if* measure VaR_α. As a consequence, banks have become more and more aware of expected shortfall as a regulatory risk measure, especially for market risk portfolios. Their initial reaction however was "we have checked the influence of moving from value-at-risk to expected shortfall, and it does not make much of a difference." The latter was indeed true in a world driven by Gaussian models; for example, for $X \sim \mathrm{N}(0,1)$, where we see at most an increase of 15% when using ES_α instead of VaR_α, and that increase gets smaller the larger is α. However, one needs to be careful about the underlying assumptions! This maximal increase by 15% is only valid for $X \sim \mathrm{N}(0,1)$.

The center plot of Figure 8.38 shows what happens if we use a more heavy-tailed t_ν distribution for X. For all considered α we barely see a difference in the curves; plotting the difference between the curves would indeed be more meaningful but this is not the important takeaway here. What is important is that the smaller the degrees of freedom ν, hence the heavier the tails of the t_ν distribution function, the larger the shortfall-to-quantile ratio (8.56); note that $\nu > 1$ is the requirement for the integral in (8.55) to be finite, which is why we only consider $\nu > 1$. Moreover, note that the y-axis is given in logarithmic scale as otherwise the shortfall-to-quantile ratio explodes for small $\nu > 1$. The range of ν close to 1 completely dominates the plot. So *if* your random loss X over a future time period is $\mathrm{N}(0,1)$ distributed, then using VaR_α or ES_α to assess risk does not make a huge difference (the set-up of the left-hand side of Figure 8.38), but *what if* X is distributed with a heavier right tail, for example $X \sim t_\nu$ with small $\nu > 1$ (the set-up in the center plot of Figure 8.38)? Then the resulting regulatory capital can differ drastically in size depending on the risk measure used, and one needs to carefully determine the degree of heavy-tailedness in this case because small differences in ν can have a large impact on the assessment of risk.

To give concrete numbers, consider X to be the $-$log-return of the price of one Netflix share in USD, as in Example 8.56. We estimate the parameters ν, μ, σ of a $t_\nu(\mu, \sigma^2)$ distribution from the $-$log-return data and compute the shortfall-to-quantile ratio for this X; as an example, we take $\alpha = 0.999$ according to the Basel Accords for modeling credit and operational risk over a one-year time horizon. It is 1.3895, so the *what if* risk measure $\mathrm{ES}_\alpha(X)$ yields a 38.95% larger value than the *if* risk measure $\mathrm{VaR}_\alpha(X)$ in this case. The estimated degrees of freedom parameter ν was 3.69, clearly hinting at the heavy-tailedness of X.

One often finds degrees of freedom parameters $\nu \in [3,5]$ in financial data; see the right-hand side of Figure 8.38, which shows the center plot zoomed in to $\nu \in [3,5]$. If the t_ν distribution for X is an appropriate model, this leads to an increase between 27.50% ($\nu = 5$) and 50.86% ($\nu = 3$) when assessing financial risk with the *what if* risk measure $\mathrm{ES}_\alpha(X)$ instead of the *if* risk measure $\mathrm{VaR}_\alpha(X)$ at the same level $\alpha = 0.999$. Eventually the financial industry and the regulators have become more and more convinced of the superiority of expected shortfall as a regulatory risk measure, especially for market risk; see, for example, McNeil et al. (2015, Sections 1.3.1, 2.3.4 and 2.3.5) for a full discussion of the story.

8.5 From Quantile Functions to Sampling

An alternative title for this section could be "An introduction to stochastic simulation"; that is indeed what we provide in this section. Simulation methodology is of great importance not only in the world of risk, but also far beyond. It is therefore important that we introduce some basic facts on this topic. If you skip this section for now, do return to it at some later time. For the present, you should at least read the following three bullet points as well as Remark 8.68.

As we mentioned in Section 8.4.3, the quantile function F^{-1} uniquely characterizes a distribution function F and thus $X \sim F$. There is a more direct connection between F^{-1} and $X \sim F$ that we can establish. It turns out that F^{-1} can be utilized to derive a stochastic representation for X. This stochastic representation can be used to simulate realizations of X on a computer. This is useful in many ways, for example:

- For a fixed number n of realizations we can get results much faster than through manual experiments (if the latter are at all possible).
- We can choose a much larger sample size n. For example, rolling a fair die can be done in the order of hundreds of millions of rolls per second which allows us to look beyond what we could possibly do manually.
- Most importantly (and we will come back to this point multiple times), simulated realizations allow us to investigate, derive or confirm statistical properties and limit results that we would otherwise not be able to confirm or derive theoretically, whether because the necessary theoretical concepts are out of our reach or because there is no theory known yet. The latter case becomes relevant if the hopes and dreams of a beautiful mathematical theory are shattered by a violation of one or more of its assumptions in a practical application. More realistic assumptions often lead to significantly more complex and thus less analytically tractable models. In this case one has to rely on simulations to obtain sufficiently accurate results (for example, to be able to answer questions of the *what if* type).

We have already met a number of distributions and explained how they can be used to model random phenomena. More will come. One of the simplest continuous distributions is the uniform distribution. This distribution is the key ingredient for establishing a stochastic representation of any random variable with distribution function F in terms of the quantile function F^{-1} of F. This stochastic representation can be exploited for generating realizations of $X \sim F$ on a computer, as we will see later.

Example 8.59 (Uniform distribution)**.** A random variable X follows a *uniform distribution*, denoted by $X \sim \mathrm{U}(a, b)$ for parameters $-\infty < a < b < \infty$, if it has distribution function

$$F(x) = \begin{cases} 0, & x < a, \\ \dfrac{x-a}{b-a}, & x \in [a, b), \\ 1, & x \geq b. \end{cases}$$

Characteristic of a uniform distribution is that its distribution function is linear (a line with slope $1/(b-a)$ from a to b). By differentiation we can easily obtain the density of $\mathrm{U}(a, b)$, namely $f(x) = 1/(b-a)$, $x \in (a, b)$, so f is constant over (a, b). The mean and variance of $X \sim \mathrm{U}(a, b)$ can be calculated to be $(a+b)/2$ (the midpoint between a and b) and $(b-a)^2/12$ (note that the longer the interval from a to b, the larger the variance), respectively. By solving $F(x) = y$ with respect to y, the quantile function F^{-1} is seen to be $F^{-1}(y) = a + (b-a)y$, $y \in (0, 1)$, so F^{-1}, like F, is linear – which is not surprising given our interpretation of the graph of F^{-1} as a reflection of the graph of F; see Section 8.4.3.

Invoking the central formula (8.11), we see that for $X \sim \mathrm{U}(a, b)$ and $0 < c < b - a$,

$$\begin{aligned}\mathbb{P}(X \in (x-c, x]) &= \mathbb{P}(x - c < X \leq x) = F(x) - F(x-c) \\ &= \frac{x}{b-a} - \frac{x-c}{b-a} = \frac{c}{b-a}\end{aligned}$$

for all $x \in [a + c, b]$, so the probability that X falls in an interval of length c depends only on the length of the interval (the value of c) and not where this interval is located inside $[a, b]$; this is also clear from the constancy of the density and the interpretation that the area under the density over an interval is the probability that X falls in this interval. In other words, the probability that X falls in an interval of fixed length is the same *uniformly* (hence the name of this distribution) over all locations of this interval; see also Example 8.61 below. Hence this distribution provides the model for drawing a number at random in the interval $[a, b]$.

The special case $a = 0$, $b = 1$ of $\mathrm{U}(a, b)$ is known as the *standard uniform distribution*, denoted by $U \sim \mathrm{U}(0, 1)$; note that a random variable from $\mathrm{U}(0, 1)$ is typically denoted by U instead of X. The distribution function of $\mathrm{U}(0, 1)$ is simply $F(x) = x$, $x \in [0, 1]$, and its density is $f(x) = 1$, $x \in (0, 1)$.

The importance of the standard uniform distribution $\mathrm{U}(0, 1)$ comes from the following result, an example of a rather general but extremely useful stochastic representation.

Concept 8.60 (Quantile transformation, inversion method, sample, sampling)**.** Let F be a distribution function and $U \sim \mathrm{U}(0, 1)$; then

$$F^{-1}(U) \sim F. \tag{8.57}$$

This result is known as *quantile transformation*. For the purpose of generating realizations $x_1, x_2, \ldots, x_n$ of $X \sim F$, it is also known as the *inversion method*. Realizations $x_1, x_2, \ldots, x_n$ from $X \sim F$ are also called a *sample* from X (or F) and the process of generating such realizations is known as *sampling*.

The correctness of (8.57) can easily be verified if F is strictly increasing and continuous (so that $F(F^{-1}(y)) = y$) since then

$$\mathbb{P}(F^{-1}(U) \le x) = \mathbb{P}(F(F^{-1}(U)) \le F(x)) = \mathbb{P}(U \le F(x)) = F(x), \quad x \in \mathbb{R}; \tag{8.58}$$

note that this result can also be justified for general F. The beauty of (8.57) is that we obtain a general stochastic representation for any $X \sim F$ in terms of $U \sim \mathrm{U}(0,1)$ via

$$X = F^{-1}(U). \tag{8.59}$$

An *algorithm* is an unambiguous, finite, set of instructions (typically executed on a computer) for solving a problem. An algorithm or its implementation in computer software to produce realizations $x_1, x_2, \dots, x_n$ of $X \sim F$ is known as a *(pseudo-)random number generator*. One pseudo-random number generator for $\mathrm{U}(0,1)$ is *L'Ecuyer's combined multiple-recursive generator*; see L'Ecuyer (1999). If we utilize it to sample $u_1, u_2, \dots, u_n$ from $U \sim \mathrm{U}(0,1)$, then, by (8.59), we know how to sample $X \sim F$, since we can simply convert the sample $u_1, u_2, \dots, u_n$ from $U \sim \mathrm{U}(0,1)$ to a sample $x_1, x_2, \dots, x_n$ from $X \sim F$ via

$$x_i = F^{-1}(u_i), \quad i = 1, 2, \dots, n. \tag{8.60}$$

Let us again stress the importance of distinguishing between random variables (as in (8.59)) and their realizations (as in (8.60)); see also, for example, Section 8.2. For example, confusing the random variable U with a realization u in the derivation (8.58) would make no sense, as the event $F^{-1}(u) \le x$ either holds or does not hold, so $\mathbb{P}(F^{-1}(u) \le x)$ is 0 or 1 but not $F(x)$, in general. Similarly nonsensical would be to try to plot $F^{-1}(U_i)$, $i = 1, 2, \dots, n$, for random variables $U_i \overset{\text{iid}}{\sim} \mathrm{U}(0,1)$ instead of their realizations, which would be the case if the realizations u_i in (8.60) were confused with respective random variables. We suspect that we are not the only teachers who frequently face students complaining about having received reduced marks in an exam because U has been written as u or u written as U. Note that, for pedagogical reasons, we still prefer to write capital letters in the y-axis labels of the plots in Figure 8.39, for example, even if plots always show realizations; we could have added "Realizations of..." instead, but, you know, "A good mathematician is a lazy mathematician".

In what follows we look at various example distributions that we have already encountered and plot the corresponding samples (the realizations $x_1, x_2, \dots, x_n$ from (8.60)) against their indices to link the shapes of these plots to properties of the respective distributions. Studying such a plot can be a very useful way to learn about properties of a distribution, as this allows us to more directly develop a gut feeling for how a random variable X from this distribution behaves, via its realizations rather than via the analytical function F (or its probability mass function or density f, for example).

Example 8.61 (Sampling continuous uniform distributions)**.** Figure 8.39 shows plots of samples of size 1000 from $X \sim \mathrm{U}(1/4, 2/3)$ (left) and from $U \sim \mathrm{U}(0,1)$ (center) and the same realizations of $U \sim \mathrm{U}(0,1)$ with an interval $(x-c, x]$ highlighted (right). For $X \sim \mathrm{U}(a,b)$, $F^{-1}(y) = a + (b-a)y$ and so the inversion method (8.59) implies that $X \sim \mathrm{U}(a,b)$ has the stochastic representation $X = a + (b-a)U$ for $U \sim \mathrm{U}(0,1)$. This is how we generated the realizations of X (based on realizations of $U \sim \mathrm{U}(0,1)$) shown on the left-hand side of Figure 8.39. The $\mathrm{U}(0,1)$ realizations used, see the center panel of this figure, were generated with the function `runif()` in the statistical software R (which uses a specific algorithm, not further

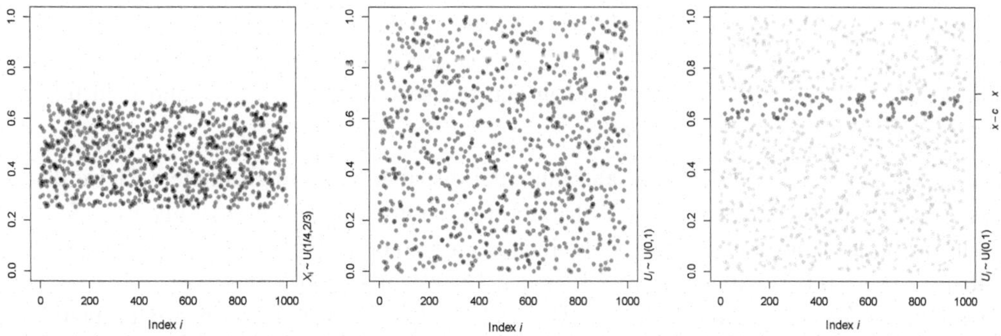

Figure 8.39 1000 realizations of $X \sim \mathrm{U}(1/4, 2/3)$ (left), of $U \sim \mathrm{U}(0,1)$ (center) and the same realizations of U with an interval $(x-c, x]$ highlighted (right). Source: Authors

discussed here, in order to mimic the complete randomness that we see in the center panel of Figure 8.39).

On the right-hand side of Figure 8.39 we see an interval $(x-c, x]$ (of the same sample as that shown in the center) that has been highlighted. As in Example 8.59, the probability that $U \sim \mathrm{U}(0,1)$ falls in $(x-c, x]$ is $\mathbb{P}(x-c < U \leq x) = x-(x-c) = c$ for every $x \in [c, 1]$, so it remains the same no matter where we shift this interval in $[0, 1]$; as already mentioned, this is what the word "uniform" captures. And this behavior is also visible from the realizations on the right-hand side of Figure 8.39, as the number of highlighted points in the interval $(x-c, x]$ remains roughly the same if we move this interval in the direction of the y-axis.

For other distributions we will discuss, this interpretation is also helpful. Shifting an interval along the y-axis and determining the fraction of points in the interval in comparison to all the samples shown provides us with an estimate of the probability that a random variable from this distribution will fall in that interval.

Not all random number generators utilize (8.59) since F^{-1} is not always numerically tractable. There are other sampling algorithms known, too. But, conceptually, we can think of the stochastic representation (8.59) as a random number generator from any distribution function F of interest given that we have a random number generator of $U \sim \mathrm{U}(0,1)$ (such as `runif()` in the statistical software R or similar functions in many other pieces of software). Also, (8.59) can sometimes be useful for sampling $X \sim F$ even if F^{-1} is not numerically tractable. For example, if

$$X = \sum_{i=1}^{n} X_i \quad \text{for} \quad X_1, X_2, \ldots, X_n \overset{\text{iid}}{\sim} G$$

for some distribution function G (see Example 8.13 (1)), then, often, neither F nor F^{-1} is tractable. But if the quantile function G^{-1} of G is tractable, then we can use (8.59) n times (namely for each summand) to obtain the stochastic representation

$$X = \sum_{i=1}^{n} G^{-1}(U_i) \quad \text{for} \quad U_1, U_2, \ldots, U_n \overset{\text{iid}}{\sim} \mathrm{U}(0,1). \tag{8.61}$$

If we want to sample m realizations of X, we would generate mn realizations from $U \sim \mathrm{U}(0,1)$, say $u_{k1}, u_{k2}, \ldots, u_{kn}$ for $k = 1,2,\ldots,m$ (the realizations u are labeled with two indices here, the first running from 1 to m and the second from 1 to n), and then build $x_k = \sum_{i=1}^{n} G^{-1}(u_{ki})$ for $k = 1,2,\ldots,m$.

Example 8.62 (Sampling Bernoulli and binomial distributions). In Example 8.21 we discussed the Bernoulli distribution. The distribution function F of $X \sim \mathrm{B}(1,p)$ is given by (8.15) and its quantile function F^{-1} is easily obtained from F to be $F^{-1}(y) = I_{(1-p,1)}(y)$, $y \in (0,1)$, so that $F^{-1}(y) = 0$ if $y \in (0, 1-p]$ and $F^{-1}(y) = 1$ if $y \in (1-p, 1)$.

Utilizing (8.59), we see that $F^{-1}(U) = I_{(1-p,1)}(U) = I_{\{U>1-p\}}$ is a stochastic representation of $X \sim \mathrm{B}(1,p)$ which we can use to sample X. Since $\mathbb{P}(1-U < x) = \mathbb{P}(U > 1-x) = 1-\mathbb{P}(U \le 1-x) = 1-(1-x) = x$, $x \in [0,1]$, we see that $1-U$ has the same distribution function as U, so $1-U \sim \mathrm{U}(0,1)$. Thus we also have the slightly simplified stochastic representation

$$X = I_{\{U<p\}} \tag{8.62}$$

for $X \sim \mathrm{B}(1,p)$. This is rather intuitive, as for $X \sim \mathrm{B}(1,p)$ we expect X to be 1 with probability p and 0 with probability $1-p$, which is exactly what (8.62) represents in terms of $U \sim \mathrm{U}(0,1)$.

Using this stochastic representation and (8.41) we can also derive the mean of $X \sim \mathrm{B}(1,p)$. Using that the indicator $I_{\{U<p\}}$ is 1 if and only if $U < p$, and that $U \sim \mathrm{U}(0,1)$, we obtain $\mathbb{E}(X) = 0 \times \mathbb{P}(X=0) + 1 \times \mathbb{P}(X=1) = \mathbb{P}(X=1) = \mathbb{P}(I_{\{U<p\}} = 1) = \mathbb{P}(U<p) = p$; with the same logic one can show that the probability $\mathbb{P}(A)$ of any event A is the expectation of the indicator I_A of that event, so probabilities are means of indicators. One can also show that $\mathrm{var}(X) = p(1-p)$.

The stochastic representation (8.62) can also be used for sampling $\mathrm{B}(1,p)$. If $u_1, u_2, \ldots, u_n$ denote realizations of $U \sim \mathrm{U}(0,1)$, then, for every $i = 1,2,\ldots,n$, set $x_i = 1$ if $u_i < p$ and $x_i = 0$ otherwise. We can simulate independent Bernoulli trials such as coin tosses in this way. To this end, choose $p = 1/2$ for a fair coin, and choose $p \neq 1/2$ for a biased coin (something not easily done with a real coin, another advantage of simulation).

We can also use the stochastic representation (8.62) to sample a binomial distribution $\mathrm{B}(n,p)$ via (8.61), where G^{-1} is the quantile function of $\mathrm{B}(1,p)$, so $G^{-1}(y) = I_{(1-p,1)}(y)$, $y \in (0,1)$. Or, if you cannot justify that all Bernoulli trials have the same success probability just replace G^{-1} in (8.61) by another quantile function G_i^{-1}. As long as all involved quantile functions G_i^{-1} are tractable, one can sample from the corresponding X in (8.61) just as easily. Both this and the example of a biased coin are, in fact, examples in the spirit of the third bullet point in the beginning of this section.

Example 8.63 (Sampling discrete uniform distributions). Consider the discrete uniform distribution $\mathrm{U}(\{x_1, x_2, \ldots, x_n\})$, which puts probability mass $1/n$ on each of the data points $x_1, x_2, \ldots, x_n$; see Example 8.20. Its distribution function is that of the empirical distribution function $\hat{F}_n$ based on the data points $x_1, x_2, \ldots, x_n$, and so (8.48) provides the corresponding quantile function. We can thus apply the inversion method (8.59) for sampling $X \sim \mathrm{U}(\{x_1, x_2, \ldots, x_n\})$, which provides us with the stochastic representation

$$X = \hat{F}_n^{-1}(U) = x_{\lceil nU \rceil} \tag{8.63}$$

for X in terms of $U \sim \mathrm{U}(0,1)$; note that in comparison with (8.48) we can omit the sorting of the data here since each data point is picked with the same probability as we explain next.

What is important here is the intuition behind this result. If $U \sim \mathrm{U}(0,1)$, then nU gives us a value uniformly on $(0,n]$ (so $nU \sim \mathrm{U}(0,n)$). And the ceiling function moves the probability mass $1/n$ concentrated on each interval of the form $(k-1,k]$, for $k = 1,2,\dots,n$, to the next integer, namely k. In other words, $\lceil nU \rceil$ returns each of the values $1,2,\dots,n$, with probability $1/n$, so $\lceil nU \rceil \sim \mathrm{U}(\{1,2,\dots,n\})$. Since we need to return the values $x_1,x_2,\dots,x_n$ (instead of the numbers $1,2,\dots,n$), each with probability $1/n$, we can simply use the realizations of $\lceil nU \rceil$ to index the data points $x_1,x_2,\dots,x_n$, thus returning each of the values $x_1,x_2,\dots,x_n$ with probability $1/n$. And this is exactly what (8.63) does.

The stochastic representation (8.63) can be used to simulate the rolls of a fair die (taking $x_i = i$, $i = 1,2,\dots,6$), for example; see the left-hand side of Figure 8.40. It can also be

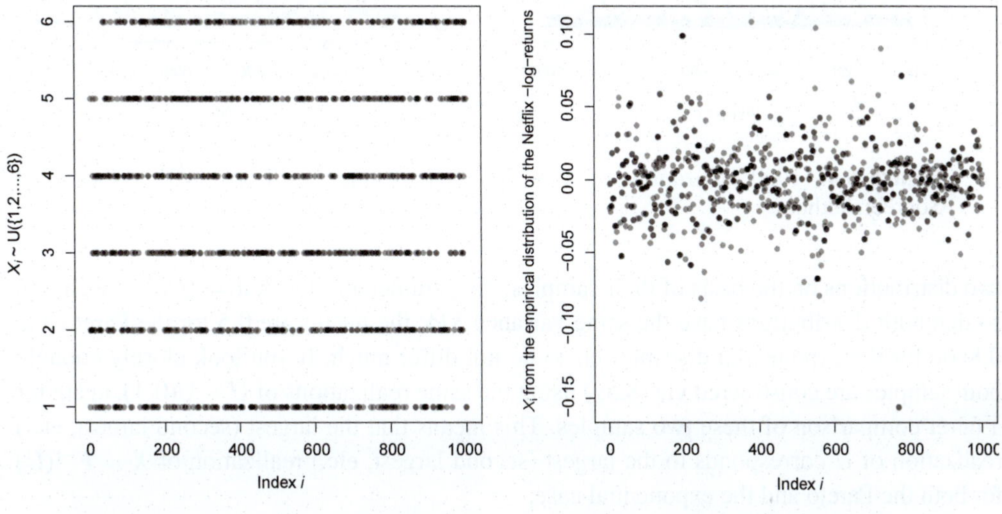

Figure 8.40 Left: 1000 realizations of $X \sim \mathrm{U}(\{1,2,\dots,6\})$. Right: Same for the empirical distribution of the −log-returns of the Netflix data. Source: Authors

used to sample from an empirical distribution based on the data points $x_1,x_2,\dots,x_n$. As an example, the right-hand side of Figure 8.40 shows a sample of size 1000 from the empirical distribution based on the −log-returns of the price of one Netflix share from Example 8.56.

Example 8.64 (Sampling Pareto, exponential and geometric distributions)**.** The quantile functions of the $\mathrm{Par}(\theta)$ and $\mathrm{Exp}(\lambda)$ distributions are given in Example 8.49. It is thus straightforward to apply the inversion method (8.59) to sample from these distributions; see Figure 8.41 for samples. By moving an interval of the form $(x-c,x]$ along the direction of the y-axis as we have done in Example 8.61, we see from the right-hand side of Figure 8.41 that the points are less dense for larger x. This is not surprising given that the $\mathrm{Exp}(\lambda)$ density $f(x) = \lambda\exp(-\lambda x)$, $x > 0$, is decreasing in x. Since the Pareto density is also decreasing we see a similar behavior on the left-hand side of Figure 8.41. The largest realization for the Pareto distribution is much further away from the bulk of the data than for the exponential distribution (note the different y-axis scales). This comes from the fact that the Pareto distribution is heavy-tailed whereas the exponential distribution is light-tailed.

Note that $\mathrm{Exp}(\lambda)$ has mean $1/\lambda$ and variance $1/\lambda^2$, and that $\mathrm{Par}(\theta)$ has mean $1/(\theta-1)$ (if $\theta > 1$) and variance $\theta/((\theta-1)^2(\theta-2))$ (if $\theta > 2$). To obtain a fair comparison between the

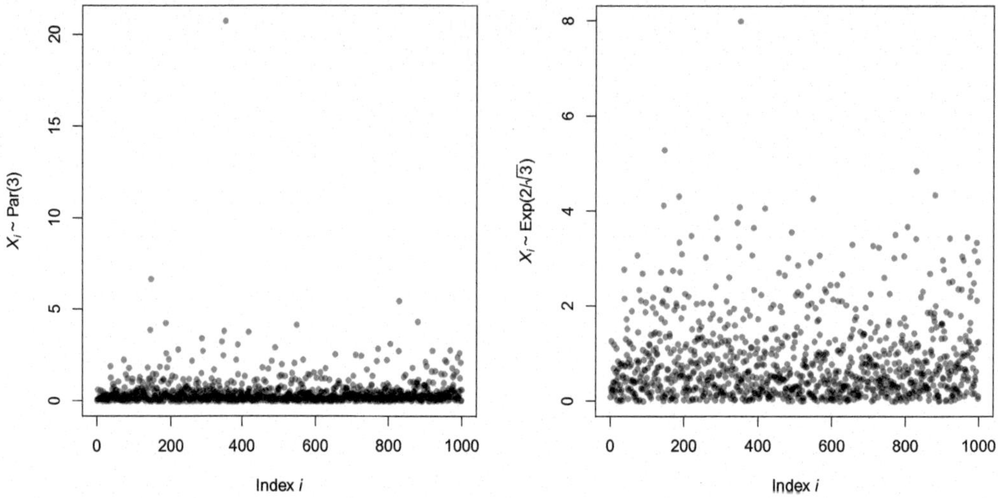

Figure 8.41 1000 realizations of $X \sim$ Par(3) (left) and $X \sim$ Exp($2/\sqrt{3}$) (right). Source: Authors

two distributions on the basis of their samples, the parameters ($\theta = 3$, $\lambda = 2/\sqrt{3}$) are chosen so that both distributions have the same variance 3/4; the means are 0.5 for the Pareto and 0.8660 for the exponential distribution, so do not differ much. If you look closely enough, both samples are constructed via (8.59) using the same realizations of $U \sim$ U(0, 1), again for a fairer comparison of these two samples. This means that the largest (second largest, etc.) realization of U corresponds to the largest (second largest, etc.) realization of $X = F^{-1}(U)$ for both the Pareto and the exponential case.

Similarly to how we obtained the stochastic representation $\lceil nU \rceil \sim U(\{1, 2, \dots, n\})$ from $U \sim$ U(0, 1), one can obtain a stochastic representation for a Geo(p) distribution from an exponential distribution: if $X \sim$ Exp($-\log(1 - p)$), then $\lceil X \rceil \sim$ Geo(p). This provides a convenient way to sample geometric distributions. Owing to this result, one can also view the geometric distribution as a discrete analog of the continuous exponential distribution. Note that this connection is missing in Figure 8.27. Finding the two distributions in the figure in the first place is somewhat reminiscent of the children's puzzle books "Where's Wally?" (known as "Where's Waldo?" in North America) by Martin Handford.

Example 8.65 (Sampling normal and t distributions). Figure 8.42 shows a sample of size 1000 from the N(0, 1) distribution (left), the N(2, 1/4) distribution (center) and the $t_3(2, 1/12)$ distribution (right). All three samples were constructed from the same 1000 realizations of $U \sim$ U(0, 1) (and, in fact, the same realizations of $U \sim$ U(0, 1) that have been used in all previous plot examples).

For the first sample (from N(0, 1)), the inversion method (8.59) provides us with $X = \Phi^{-1}(U)$, which can be evaluated by an implementation of the standard normal quantile function Φ^{-1} available in statistical software. For the second sample (from N(2, 1/4)), the inversion method provides us with $X = 2 + (1/2)\Phi^{-1}(U)$ (simply a linear transformation of the sample from $\Phi^{-1}(U)$); see also Example 8.49 (3). By comparison of the left-hand side and the center plots in Figure 8.42, we see that by shifting μ from 0 (left) to 2 (center), the

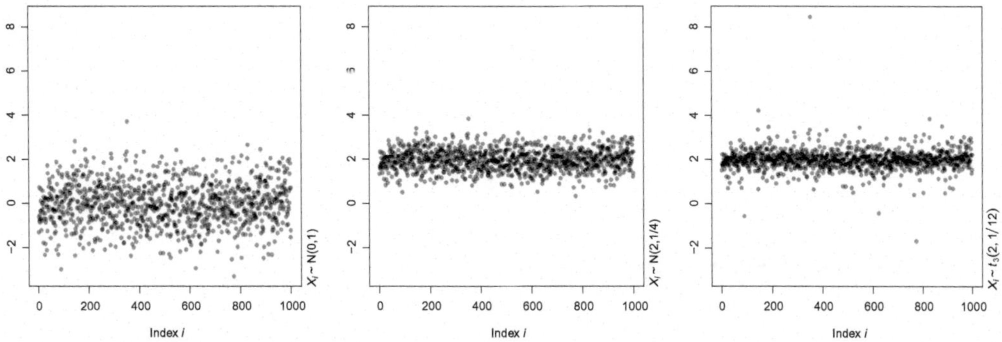

Figure 8.42 1000 realizations of $X \sim \mathrm{N}(0,1)$ (left), $X \sim \mathrm{N}(2,1/4)$ (center) and $X \sim t_3(2,1/12)$ (right). Source: Authors

whole sample is shifted upwards by 2 and now fluctuates around 2 (center) instead of 0 (left) as expected for the location parameter μ. At the same time, the scale parameter σ shifts from 1 (left) to 1/2 (center), which leads to a smaller dispersion (or a higher concentration) of the sample around μ.

For the third sample (from $t_3(2,1/12)$) on the right-hand side of Figure 8.42, the inversion method can be shown to give $X = 2+\sqrt{1/12}F_{t_3}^{-1}(U)$, where F_{t_3} denotes the quantile function of the t_3 distribution. The choice of μ and σ for the $t_\nu(\mu,\sigma^2)$ distribution is such that both the mean (μ; finite for $\nu > 1$) and the variance ($\sigma^2\nu/(\nu-2)$; finite for $\nu > 2$) of the $t_3(2,1/12)$ distribution are equal to the mean, 2, and variance, 1/4, of the $\mathrm{N}(2,1/4)$ distribution in the center panel of Figure 8.42. Similarly to the comparison of the Pareto and exponential sample, in Example 8.64, we see that the largest values of the heavier-tailed t distribution are located much further away from the bulk of the data (so, in the tail) than those of the normal distribution; again, this can be explained by the tails of the $t_3(2,1/12)$ distribution being heavier than those of the $\mathrm{N}(2,1/4)$ distribution.

Example 8.66 (Sampling mixed-type distributions)**.** The left-hand side of Figure 8.43 shows a sample of size 1000 from the mixed-type distribution with distribution function as given on the right-hand side of Figure 8.11. The right-hand side of Figure 8.43 shows a sample of size 1000 from the empirical distribution with continuous tail of the yearly disaster loss data in one billion USD, as given on the right-hand side of Figure 8.14. We used the inversion method (8.59) for sampling from these two rather exotic animals in the zoo of distributions. Also, as before, we used the same realizations of $U \sim \mathrm{U}(0,1)$.

The shape of the sample cloud on the left-hand side of Figure 8.43 is especially interesting to study, as it again helps us to develop a gut feeling for the connection between the properties of F and the behavior of realizations of $X \sim F$. The empty interval for y-axis values in $(0,2]$ stems from the flat part of the mixed-type distribution function F over this interval; see the right-hand side of Figure 8.11. Recall from Section 8.3.1 that a flat part of F over an interval $(a,b]$ means that $X \sim F$ almost surely does not take on values in $(a,b]$ – and we indeed see no realizations of X in $(0,2]$. Also recall that if F has a jump in x of height p, then $X \sim F$ takes on the value x with probability p. The darker line at the y-axis value 0 comes from all realizations that are equal to 0, which constitutes about 20% of them, as one can numerically

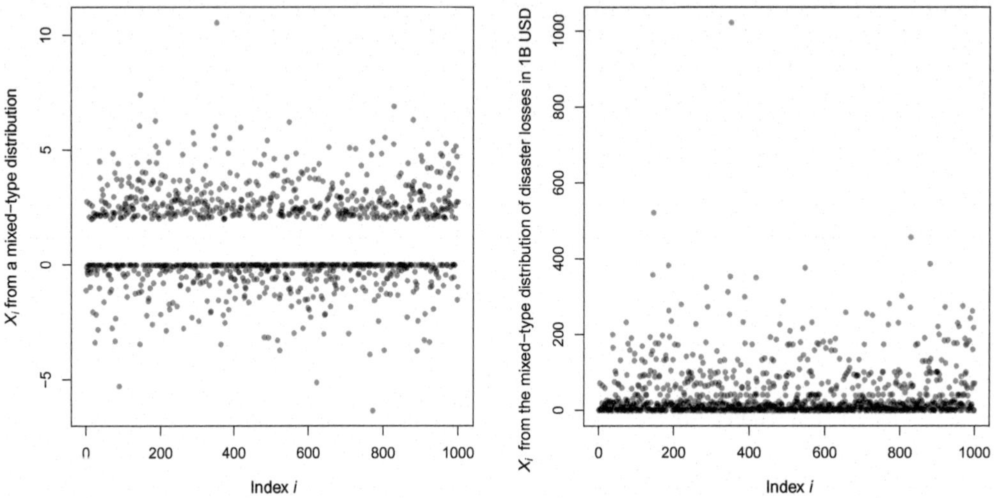

Figure 8.43 Left: 1000 realizations from the mixed-type distribution with distribution function shown on the right-hand side of Figure 8.11. Right: Same from the empirical distribution with continuous tail shown on the right-hand side of Figure 8.14. Source: Authors

determine – and the jump height of F at 0 is indeed $F(0) - F(0-) = 0.2$, so 20%. Although real data rarely look like this sample, it is a pedagogically meaningful example as it allows to "see" properties of F in terms of realizations of X, an indispensable skill when selecting an (or multiple) appropriate distribution(s) for real data.

Example 8.67 (Sampling conditional distributions)**.** The quantile function $F^{-1}_{(a,b]}$ of the conditional distribution function $F_{(a,b]}$ of a distribution function F is given by (8.49). It is thus straightforward to apply the inversion method for sampling from $F_{(a,b]}$ if both F and F^{-1} are tractable. Figure 8.44 shows a sample of size 1000 from $F_{(1,5]}$ when F is the $\mathrm{Par}(3)$ distribution function (left) and one from $F_{(-1,\infty)}$ when F is the $\mathrm{N}(0,1)$ distribution function (right). If we compare these two samples with the "unconditional" samples on the left-hand sides of Figures 8.41 and 8.42, we see a "zoom-in effect". The distribution of the conditional samples $X \mid X \in (a,b]$ in Figure 8.44 is of the same type as the distribution of the "unconditional" samples when one zooms in on the y-axis interval with lower endpoint a and upper endpoint b.

In Example 8.61 we have already seen such a zoom-in effect for a y-axis interval in the context of the continuous uniform distribution. And, indeed, for $U \sim \mathrm{U}(0,1)$ one can interpret $X \sim \mathrm{U}(1/4, 2/3)$ presented there as $U \mid U \in (1/4, 2/3]$; see the zoom-in effect on the left-hand side of Figure 8.39 in comparison with the plot in the center. This can also be verified through the lens of quantile functions, for example. If F is the standard uniform distribution function (so that $F(x) = x$, $x \in [0,1]$ and $F^{-1}(y) = y$, $y \in (0,1)$), we obtain from (8.49) that $F^{-1}_{(a,b]}(y) = a + (b-a)y$, which is indeed the quantile function of the $\mathrm{U}(a,b)$ distribution that we have already used in Example 8.61.

Remark 8.68. There are many important aspects to be added to the topic of stochastic simulation, even at this introductory level. We want to highlight the following:

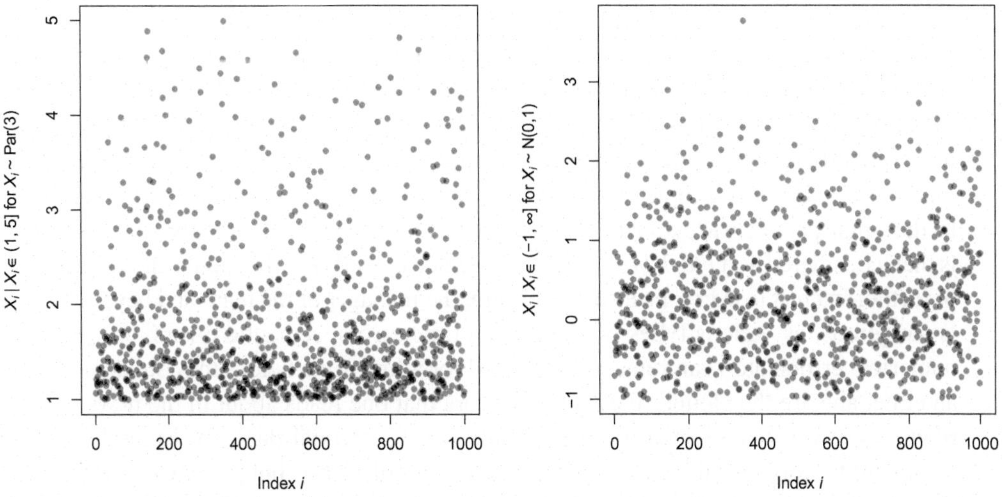

Figure 8.44 1000 realizations of $X \mid X \in (1,5]$ for $X \sim \text{Par}(3)$ (left) and $X \mid X > -1$ for $X \sim \text{N}(0,1)$ (right). Source: Authors

(1) The sample size matters. Although there are applications where studying a small sample size can be of interest, if we had shown only a handful of realizations for each of the distributions in this section, it would not have been possible to distinguish them nor to interpret each correctly in the sense we discussed in Example 8.61. Statistically important properties of distributions are revealed from their realizations only if we have a sufficiently large sample size; otherwise we would not get reliable simulation results but, rather, fluctuating answers depending on the actual random numbers used; see also (2) below. For example, an $\text{N}(\mu, \sigma^2)$ sample of size 10 does not allow one to see the meaning of its two parameters μ, σ, so that, every time we generate another sample of size 10, estimating the two parameters from such small samples can lead to quite different values.

(2) As shown in the center panel of Figure 8.39, random number generators of U(0, 1) produce sequences of numbers that mimic complete randomness. The numbers themselves are deterministic, which is not a contradiction to the fact that they mimic randomness when plotted (or measured with other metrics). What appears to be random is the starting point in those sequences, which is determined from the current time, the process identifier (the number assigned to the software with which you are generating the random numbers by the underlying operating system) and sometimes also movements you create by using your touchpad, mouse, keyboard, etc. In the software, one can fix the starting point (the so-called *seed*), which makes the numbers that are generated from this starting point *reproducible*, but they still mimic randomness. Reproducibility is an important aspect of a simulation, as one wants to be able to replicate results or also to make them more comparable. This we did when we used a sample of size 1000 of realizations of $U \sim \text{U}(0,1)$ in the examples above, as mentioned in Example 8.65. Again, this does not mean we destroyed the mimicking of randomness (each plot we considered revealed the statistical properties we were expecting to see); it just made the randomness reproducible since we had fixed the random number generator's starting value, the seed.

(3) Realizations of random variables are used as input to stochastic models which themselves can serve as input to stochastic simulation studies. Such simulation studies are crucial not only in practice to find solutions to complex problems, but also in mathematical and stochastic research itself in order to confirm new results, test their robustness to changes in the underlying assumptions, investigate conjectures, etc. Simulation studies are often implemented by those people who have the best training in *programming* and sometimes by the single person in a team with the best programming skills. Frequently, such software is then considered as a *black box*; only the interface (specifying the inputs and outputs) is of further interest to everyone else in the team, the manager of the team and beyond. This is dangerous, as various types of errors (run-time errors, semantic errors, user errors, numerical errors, etc.) can remain undetected for a long time before they suddenly have an effect and are thus found; the types of errors that one reads about in the news media often concern software security and include stolen personal data, for example. In an ideal world, people would not consider software as a black box but rather everyone in the team, including the manager, would at least have a good look inside the box, proofread it and verify what is done. Implementing a piece of software may be hard, but typically proofreading it to verify that the right quantities are computed typically is not. Similarly to "talking to the guys in the boiler room", which we encountered in Section 2.4, "talking to the guys in the computer room" comes to mind here. We have already encountered the relevance of the latter in Section 3.3 when we discussed the explosive growth of markets for mortgage-backed securities in the run-up to the 2007–2008 financial crisis.

One type of risk and the resulting errors that we want to further stress here is absolutely crucial and haunts every modeler: namely *model risk*, the risk of using the wrong model, for example due to insufficient data which results in incorrect model assumptions. If a risk $X \sim F_{\text{true}}$ is simulated by sampling from $X \sim F$ but F does not well represent F_{true} (for example if F is normal but F_{true} is more heavy-tailed, as for the Netflix data in Example 8.56), then we will not get accurate results for F_{true}, irrespectively of the sample size. As such, stochastic simulations have to be conducted with care and are not a free lunch and panacea for solving the world's problems. In computer science, this is summarized under the phrase *garbage in, garbage out*, so putting garbage (such as flawed assumptions or wrong data) into a model produces garbage as output (such as flawed predictions or non-representative output data), independently of the sample size used or the number of digits chosen to represent the output. Revisit Chris Rogers' quote in Section 3.6 and find a connection.

8.6 From Sampling to Limit Theorems and Beyond

Now that we have seen how to sample from various distributions, we can simulate various events, for example, whether we are rolling a 6 with a fair die or sustaining losses due to high water-mark exceedances. Viewed as realizations of iid random variables $X_1, X_2, \ldots \overset{\text{iid}}{\sim} F$, we can then investigate the simulated data to study the distributions of the quantities of interest. Examples of these include sums S_n, sample means $\bar{X}_n$ and maxima M_n; see Example 8.13 for the notation S_n, $\bar{X}_n$ and M_n. It is mathematically non-trivial to show that, under certain assumptions on F, quantities such as S_n, $\bar{X}_n$ and M_n, suitably location-scale transformed, converge (in some sense to be specified later) for $n \to \infty$ to random variables. The distribution functions of such *limit random variables* are known as *limit distributions* and the corresponding mathematical convergence results are known as *limit theorems*. Limit distributions are important as they provide approximations to the distributions of S_n, $\bar{X}_n$ or M_n even though we may not know the underlying distribution function F.

In this section we want to highlight some limit theorems and their limit distributions. This is a more strenuous hike, so watch your step! To make this more enjoyable, we often utilize simulation, which is in the spirit of the third bullet point at the beginning of Section 8.5. Also, as a disclaimer, we aim at striking a balance between intuition and mathematical accuracy, which means that not all results we cover are 100% intuitive, nor are the arguments we provide to help you understand them necessarily 100% mathematically accurate. For example, we will use plots of simulated data to convince ourselves of the aforementioned convergence even though one cannot decide convergence from finitely many values. In our opinion, such sacrifices are well worth it.

8.6.1 The Strong Law of Large Numbers

The *strong law of large numbers* (SLLN) is a limit theorem that addresses the convergence of sample means $\bar{X}_n$ of iid random variables to a constant (a "degenerate" random variable, if you like). As we already hinted at in Section 7.2 under the name "law of large numbers" (LLN), it is a major result in probability.

Theorem 8.69 (Strong law of large numbers). *If $X_1, X_2, \dots$ is a sequence of iid random variables with finite mean $\mu = \mathbb{E}(X_1)$, then*

$$\bar{X}_n \overset{a.s.}{\to} \mu \quad (n \to \infty). \tag{8.64}$$

In words, the sample average of iid random variables with mean μ converges almost surely to μ for n tending to infinity.

Some remarks are in order.

(1) The superscript "a.s." above the convergence arrow in (8.64) indicates that this convergence is *almost sure convergence*. This type of convergence means that the convergence holds with probability 1. This means that, for almost all realizations ω, $\bar{X}_n(\omega)$ is eventually (that is, for sufficiently large n) close to μ. With a slight abuse of mathematical accuracy, on occasion we will drop the "almost all" and write "for every". Also, for practical applications, we abbreviate (8.64) by

$$\bar{X}_n \underset{n \text{ large}}{\approx} \mu. \tag{8.65}$$

(2) The SLLN goes back to Kolmogorov. The "S" for "strong" hints at the fact that there may be another form of the LLN; this is indeed the case. It is called the *weak law of large numbers* (WLLN), which goes back to Jacob Bernoulli, who proved it for Bernoulli random variables. Its full generality was obtained by Alexandr Khinchin (1894–1959), one of the most significant contributors to the Soviet school of probability theory. For our purposes, we need not go into the mathematical difference between the WLLN and the SLLN and on occasion will just speak of the LLN; the WLLN has a weaker statement

under more restrictive assumptions but is much easier to prove. Its colloquial formulation as "sample averages converge to their population mean" often suffices for our purposes.

Proving the SLLN mathematically requires some heavy machinery, but (with our disclaimer at the beginning of Section 8.6 in mind), we can easily consider simulations to convince ourselves that this result holds; again, consider the third bullet point at the beginning of Section 8.5 in this regard.

Example 8.70 (SLLN for the probability of rolling a 6). As mentioned before, the event that we roll a 6 when rolling a fair die can be modeled as $X \sim \mathrm{B}(1, 1/6)$, with stochastic representation $X = I_{\{U<1/6\}}$ for $U \sim \mathrm{U}(0,1)$; see (8.62). The repetition of such rolls can thus be modeled as $X_1, X_2, \ldots \overset{\text{iid}}{\sim} \mathrm{B}(1, 1/6)$, with stochastic representations $X_i = I_{\{U_i<1/6\}}$ for $U_1, U_2, \ldots \overset{\text{iid}}{\sim} \mathrm{U}(0,1)$. Simulating realizations of $X_1, X_2, \ldots, X_m$ and then calculating $\bar{X}_n = (X_1 + X_2 + \cdots + X_n)/n$ for $n = 1, 2, \ldots, m$ allows us to plot $\bar{X}_n$ as a function of n (a so-called *sample path*) and thus visually confirm the convergence (8.64). The left-hand side of Figure 8.45 shows $\bar{X}_n$, $n = 1, 2, \ldots, 5000$; to this end we chose $m = 5000$ and generated one

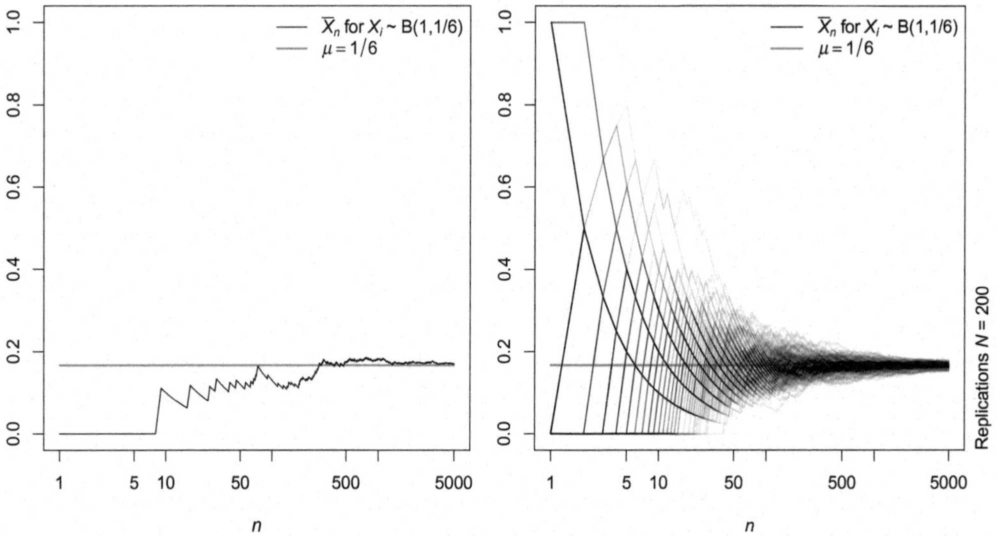

Figure 8.45 Left: The sample mean $\bar{X}_n$ as a function of n can be seen converging to the mean $\mu = 1/6 \approx 0.1667$, thus visualizing the SLLN. Right: $N = 200$ replications of realizations of $\bar{X}_n$ as a function of n. Source: Authors

realization from each of $X_1, X_2, \ldots, X_{5000}$. The first eight realizations were all 0 and the ninth was the first time we obtained 1, indicating that the first 6 was rolled on the ninth roll. We see that already around $n = 200$, $\bar{X}_n$ is close to the true mean μ, which is indicated by a horizontal line; in Example 8.62 we calculated the mean of $X \sim \mathrm{B}(1, p)$ as p, so $\mu = 1/6 \approx 0.1667$ here. You may want to look again at the Swiss postage stamp of Jacob Bernoulli and the SLLN (there referred to as LLN) in Figure 7.3; recall that the sample path on the stamp is an artistic impression.

The sample path of $\bar{X}_n$ ($\bar{X}_n$ as a function of n) could still be close to μ by chance, so we cannot yet be convinced that every such sample path is eventually close to μ, as it should be according to the SLLN. To this end, the right-hand side of Figure 8.45 shows $N = 200$ such sample paths. These N sample paths are based on N independent replications of $X_1, X_2, \ldots, X_{5000}$; we can label them $X_{1,b}, X_{2,b}, \ldots, X_{5000,b}$ for $b = 1, 2, \ldots, 200$. We thus obtain $N = 200$ replications of $\bar{X}_n$, that is, $\bar{X}_{n,1}, \bar{X}_{n,2}, \ldots, \bar{X}_{n,200}$, and, for each $b = 1, 2, \ldots, 200$, the right-hand side of Figure 8.45 shows $\bar{X}_{n,b}$ plotted against $n = 1, 2, \ldots, 5000$. Some of the 200 sample paths of $\bar{X}_{n,b}$ are already fairly close to μ for smaller n, some only for larger n, but, eventually, all 200 sample paths of $\bar{X}_{n,b}$ are close to μ.

As mentioned in our disclaimer, such plots do not constitute a proper mathematical proof, of course. However, they are often helpful to convince one that a mathematical result holds or that it fails to hold if certain assumptions are violated. Not infrequently, analyses of simulated data reveal a deeper mathematical property, which can then lead to a conjecture and eventually may turn into a mathematical proof. We cannot imagine the types of conjectures and results geniuses such as Jacob Bernoulli, Leonhard Euler or Carl Friedrich Gauss (and many more) would have come up with had they had the computational tools we have now at our disposal.

Strictly speaking, Example 8.70 has convinced us only that the SLLN holds for sequences of iid Bernoulli random variables. Let us therefore consider another example.

Example 8.71 (SLLN for Pareto losses). Figure 8.46 shows $N = 200$ paths of $\bar{X}_n$ for $X_1, X_2, \ldots \overset{\text{iid}}{\sim} \text{Par}(\theta)$, with $\theta = 3$ resulting in $\mu = \mathbb{E}(X_1) = 1/2$ (left), with $\theta = 1.1$ resulting in $\mu = \mathbb{E}(X_1) = 10$ (center), and with $\theta = 1/2$ (right). In comparison with the plot on the

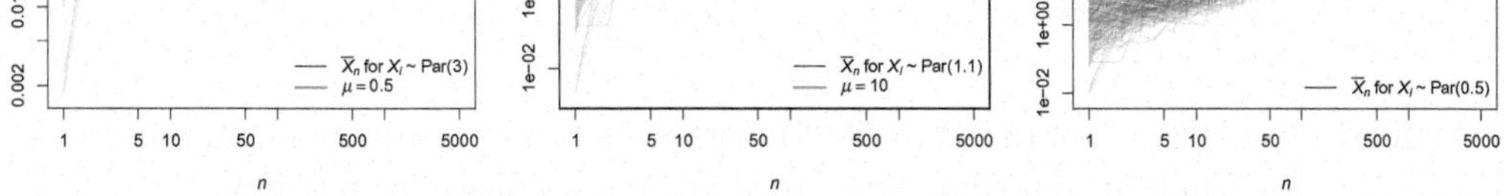

Figure 8.46 Sample mean $\bar{X}_n$ as a function of n for a sequence of iid random variables $X_1, X_2, \ldots$ from a Par(3) distribution (left), a Par(1.1) distribution (center) and a Par(1/2) distribution (right). Source: Authors

left, the paths of $\bar{X}_n$ in the center show a slower convergence towards μ overall. This is not surprising since the smaller the value of θ, the more heavy-tailed the Pareto distribution, and so the sample means $\bar{X}_n$ fluctuate more and thus take longer to settle around μ. Still, if we increased n we would see a similar behavior to that in the plot on the left. In the plot on the right, however, there is no μ included as the mean is ∞ (so is not finite). In this case, Embrechts et al. (1997, Section 2.1) noted that the SLLN still remains valid in the sense that (8.64) does hold for $\mu = \infty$, but this applies here only because our random variables are all non-negative. And indeed, we see from the plot that the paths of $\bar{X}_n$ seem to tend to infinity and not settle around any finite value.

As power-tail (hence Pareto-type) behavior is ubiquitous in risk management, the above comments are highly relevant. The consequences will resurface once we introduce extreme value theory in Chapter 9. In particular, the slow convergence for heavier-tailed models will come to haunt the risk modeler.

In Derivation 8.26 and Remark 8.32 (3) we briefly talked about estimating parameters. We now come back to this statistical task.

Remark 8.72 (The power of using the SLLN to construct estimators). The convergence (8.64) hints at the fact that the sample mean $\bar{X}_n$ is an estimator of the true underlying mean μ, that is, the mean $\mathbb{E}(X_1)$ of one of the underlying random variables. This is how we can (and in fact did in Example 8.56) estimate the location parameter μ of a distribution.

Since the variance σ^2 is also a mean (see (8.44)), an immediate estimator of the variance derived via (8.64) is $(1/n)\sum_{i=1}^n (X_i - \bar{X}_n)^2$. When computing its mean $\mathbb{E}(\hat{\sigma}_n^2)$ it turns out that one ends up with $\sigma^2(n-1)/n$, which is why one typically uses $\hat{\sigma}_n^2 = (1/(n-1))\sum_{i=1}^n (X_i - \bar{X}_n)^2$ as an estimator of σ^2. The estimator $\hat{\sigma}_n^2$, also known as the *sample variance*, indeed has σ^2 as its mean. An estimator of the standard deviation σ is the *sample standard deviation* $\hat{\sigma}_n$, the square root of $\hat{\sigma}_n^2$.

Parameters of distributions can often also be estimated in this way. For example, for $X \sim \text{Exp}(\lambda)$ we know that $\mathbb{E}(X) = 1/\lambda$, so solving this equation for the parameter λ gives $\lambda = 1/\mathbb{E}(X)$, and thus $1/\bar{X}_n$ is an estimator of λ based on the given data, $X_1, X_2, \dots, X_n \overset{\text{iid}}{\sim} \text{Exp}(\lambda)$. Similarly, one can derive $1 + 1/\bar{X}_n$ as an estimator of θ from $X_1, X_2, \dots, X_n \overset{\text{iid}}{\sim} \text{Par}(\theta)$.

Also, probabilities can be estimated in this way, as we can write them as expectations. For example, the exceedance probability $\mathbb{P}(Y > y)$ of a random variable Y over the threshold y is $\mathbb{E}(I_{\{Y>y\}})$, which is of the form $\mathbb{E}(X)$ for $X = I_{\{Y>y\}}$. Therefore, $\mathbb{P}(Y > y)$ can be estimated by the sample mean $\bar{X}_n = \frac{1}{n}\sum_{i=1}^n X_i$ for $X_i = I_{\{Y_i>y\}}$, $i = 1, 2, \dots, n$, based on data from $Y_1, Y_2, \dots, Y_n$. Using (8.65), we can thus write, more compactly,

$$\mathbb{P}(Y > y) = \mathbb{E}(I_{\{Y>y\}}) \underset{n \text{ large}}{\approx} \frac{1}{n}\sum_{i=1}^n I_{\{Y_i>y\}}.$$

Similarly, we obtain from (8.64) or (8.65) that the empirical distribution function based on $X_1, X_2, \dots \overset{\text{iid}}{\sim} F$ indeed approximates F. To see this, just replace the iid random variables $X_1, X_2, \dots$ in the SLLN by the (again iid) indicators $I_{\{X_1 \le x\}}, I_{\{X_2 \le x\}}, \dots$ We thus obtain for every $x \in \mathbb{R}$ that

$$\hat{F}_n(x) = \frac{1}{n}\sum_{i=1}^n I_{\{X_i \le x\}} \underset{n \text{ large}}{\approx} \mathbb{E}(I_{\{X_1 \le x\}}) = \mathbb{P}(X_1 \le x) = F(x),$$

so $\hat{F}_n(x)$ indeed approximates $F(x)$. Note that the SLLN applies here irrespectively of the existence of the mean of $X_1, X_2, \dots$, since we apply it to the indicators $I_{\{X_1 \le x\}}, I_{\{X_2 \le x\}}, \dots$, thus to Bernoulli random variables with finite mean. One can mathematically derive an even stronger result, namely that the maximal distance between $\hat{F}_n(x)$ and $F(x)$ over all real x converges to 0 for $n \to \infty$; this is known as the *Glivenko–Cantelli theorem*.

To cut a long story short, every quantity that we can write as a mean can be estimated by the corresponding sample mean on the basis of the given data. We omit a further discussion of whether the above SLLN-based estimators are 'good', whether they can fail, or when there are 'better' estimators available – one needs to ask a well-trained statistician for advice. We just want to give an idea of how one can construct an estimator (or its realization, the actual number, the estimate) from data.

The said data is sometimes obtained by sampling (with a large sample size), with the purpose of approximating a mean via (8.65). This is an extremely powerful computational tool in statistics with a wide range of applications, known under the name *Monte Carlo method*. The name refers to the Monte Carlo Casino in Monaco.

In Example 8.71 we learned from the right-hand side of Figure 8.46 that, just because we can calculate $\bar{X}_n$ from iid $X_1, X_2, \dots, X_n$, this does not imply that the mean $\mu = \mathbb{E}(X_1)$ (or the variance, or another moment) is finite. Although the realizations of $X_1, X_2, \dots, X_n \overset{\text{iid}}{\sim} \text{Par}(1/2)$ are always finite numbers, and so are the corresponding realizations of $\bar{X}_n$, we have $\mu = \mathbb{E}(X_1) = \infty$ in this example. Engineers or data scientists (who frequently learn to work with data only) often fall into the trap of assuming that all the moments of an unknown distribution to be estimated are finite. And this is so even in the case where a more sophisticated parameter estimator, such as the maximum likelihood estimator mentioned in Remark 8.32 (3), leads to an estimated distribution which clearly indicates that not all moments are finite. Underestimating the heavy-tailedness of a distribution can lead to an underestimation of risk. Similar observations were made when we compared the *t* distribution with the normal distribution in Example 8.54 (see also Example 8.58) or in regard to Q–Q plots in Example 8.56 for the Netflix data. Clearly, model risk is a problem in such cases.

8.6.2 The Central Limit Theorem

Again let us consider $X_1, X_2, \dots \overset{\text{iid}}{\sim} F$, but this time we will also assume that the variance $\sigma^2 = \text{var}(X_1)$ of the iid random variables is finite. This was the case in the set-up underlying the right-hand side of Figure 8.45 or the left-hand side of Figure 8.46. There we saw that, as n grows, $\bar{X}_n$ fluctuates less and less around μ. So, the distribution of $\bar{X}_n$ depends on the underlying F and on n. If we know F, we can at least hope to determine some properties of $\bar{X}_n$ analytically, for example one can show that $\text{var}(\bar{X}_n) = \sigma^2/n$, which converges to 0 for $n \to \infty$ as long as we know that $\sigma^2 < \infty$ and so indeed explains the behavior of $\bar{X}_n$ that we saw on the right-hand side of Figure 8.45 or the left-hand side of Figure 8.46.

The problem is that typically we do not know F (so neither do we know σ^2, for example). But perhaps we have at least good reason to believe that $\sigma^2 < \infty$ and that $X_1, X_2, \dots$ are iid from F. If, under these assumptions, we knew (at least approximately) the distribution of $\bar{X}_n$, we could quantify how fast $\bar{X}_n$ converges to μ in terms of the variance of $\bar{X}_n$. Or, of course, we could use this approximate distribution of $\bar{X}_n$ to compute the exceedance probabilities of $\bar{X}_n$ over thresholds, etc. We would also immediately get an approximate distribution of the sum $S_n = \sum_{i=1}^n X_i = n\bar{X}_n$, which we could use, for large enough n, to model the distribution of a total loss (again, under the assumption that all losses $X_1, X_2, \dots$ are iid and have finite variance). Knowing an approximate distribution of $\bar{X}_n$ would open the door to a large variety of different applications, even though we do not know the underlying F. Well, good news, one does know such an approximate or limit distribution under the iid and finite-variance assumption, which we cover next.

With $\bar{X}_n$ converging to μ as in (8.64) one has that $\bar{X}_n - \mu$ converges to 0 and so also

$$\frac{\bar{X}_n - \mu}{\sigma} \xrightarrow{\text{a.s.}} 0 \quad (n \to \infty);$$

as before, we interpret this as $(\bar{X}_n - \mu)/\sigma \approx 0$ for large n. But this does not give us a useful limit distribution yet, as the convergence is to 0 (it is a degenerate random variable whose distribution function is a jump of height 1 at 0).

The idea of the *central limit theorem* (CLT) is to multiply the standardized term $(\bar{X}_n - \mu)/\sigma$ by a scaling factor depending on n, in such a way that the limit no longer collapses to 0, but to a non-degenerate random variable. This sounds intuitively convincing, but we have not formally specified yet what we mean by "convergence to a random variable". The type of convergence we mean is *convergence in distribution*, which means that the distribution function of the scaled random variable $(\bar{X}_n - \mu)/\sigma$ converges to the distribution function of a non-degenerate limit random variable at all points where the limit distribution function is continuous. The scaling factor turns out to be $\sqrt{n}$ and the limit random variable turns out to be standard normal.

Theorem 8.73 (Central limit theorem). *If $X_1, X_2, \dots$ is a sequence of iid random variables with mean $\mu = \mathbb{E}(X_1)$ and variance $\sigma^2 = \mathrm{var}(X_1) < \infty$, then*

$$\sqrt{n}\frac{\bar{X}_n - \mu}{\sigma} \xrightarrow{d} \mathrm{N}(0, 1) \quad (n \to \infty). \tag{8.66}$$

In words, properly standardized sample means of iid random variables with finite variance converge in distribution to a standard normal, for n tending to infinity.

Several remarks are in order.

Remark 8.74.

(1) The "d" above the convergence arrow in (8.66) indicates that the type of convergence is convergence in distribution, so the distribution function of $\sqrt{n}(\bar{X}_n - \mu)/\sigma$ converges to that of $\mathrm{N}(0, 1)$, namely Φ, at all points where Φ is continuous, which simply means for all $x \in \mathbb{R}$. So for sufficiently large n, the random variable $\sqrt{n}(\bar{X}_n - \mu)/\sigma$ is approximately $\mathrm{N}(0, 1)$-distributed. A more practical interpretation of (8.66) is thus

$$\sqrt{n}\frac{\bar{X}_n - \mu}{\sigma} \underset{n \text{ large}}{\overset{\text{approx.}}{\sim}} \mathrm{N}(0, 1). \tag{8.67}$$

Note that, when we discuss convergence in distribution, it does not matter whether we write the abbreviating symbol for the distribution (here, $\mathrm{N}(0, 1)$), the distribution function (here, Φ) or its random variable (often denoted by Z) as the limit.

(2) The left-hand side of (8.67) is of the form $a\bar{X}_n + b$ for $a = \sqrt{n}/\sigma$ and $b = -\sqrt{n}\mu/\sigma$, so it is a linear transformation of $\bar{X}_n$. We have already established in Derivation 8.35 that a linear transformation of a normally distributed random variable is normally distributed, and we know how the parameters of the two normal distributions are related to each other: if $Z = a\bar{X}_n + b \sim \mathrm{N}(0, 1)$, then $d + cZ \sim \mathrm{N}(d, c^2)$. With $c = 1/a$ and $d = -b/a$ we

see that $(-b/a)+(1/a)(a\bar{X}_n+b) \sim \mathrm{N}(-b/a, 1/a^2)$, so $\bar{X}_n \sim \mathrm{N}(-b/a, 1/a^2) = \mathrm{N}(\mu, \sigma^2/n)$. As $a\bar{X}_n + b$ is only approximately normal, we have the approximation

$$\bar{X}_n \underset{n \text{ large}}{\overset{\text{approx.}}{\sim}} \mathrm{N}(\mu, \sigma^2/n), \tag{8.68}$$

which is equivalent to (8.67). In practical terms, we obtain a formula for the calculation of probabilities that involves averages, for example $\mathbb{P}(\bar{X}_n \le x) \approx \Phi(\sqrt{n}(x-\mu)/\sigma)$, where Φ is the standard normal distribution function, which can be calculated. In the same way we can also bring the factor $1/n$ from the left- to the right-hand side and obtain

$$S_n \underset{n \text{ large}}{\overset{\text{approx.}}{\sim}} \mathrm{N}(n\mu, n\sigma^2). \tag{8.69}$$

The equivalent of (8.66) in terms of S_n instead of X_n is obtained in the same way, by taking out a factor $1/n$ from the numerator in (8.66). We then obtain the convergence

$$\frac{S_n - n\mu}{\sqrt{n}\sigma} \overset{\text{d}}{\longrightarrow} \mathrm{N}(0,1) \quad (n \to \infty). \tag{8.70}$$

The left-hand side is still a linear transformation of S_n. We will come back to this in Section 9.3.1 when we are motivating an analog result to the CLT for extremes M_n instead of sums S_n.

Needless to say, (8.68) and (8.69) are very helpful in providing approximate distributions of the means and sums which frequently appear in applications. This is one reason (perhaps *the* reason) why the normal distribution is so important even though its density (8.29) may seem rather involved and its distribution function is not even available analytically. We wish we had a 10 euro banknote for every time a student asked us whether the normal distribution is really that important to know by heart in an upcoming exam.

(3) Why is the CLT so useful? One might wonder why we cannot just simulate realizations of $\bar{X}_n$ and then use the empirical distribution function of the simulated realizations of $\bar{X}_n$ as an approximate distribution of $\bar{X}_n$. The problem is that typically we do not know F and so cannot accurately sample the ingredients $X_1, X_2, \dots$ of $\bar{X}_n$. Knowing F is of course trivial in the case where we are interested in whether we will roll a 6 with a fair die. However, for modeling losses due to high water-mark exceedances, for example, we would not know F. One would then need to statistically estimate and check F or rely on expert opinion, but in both cases there is model risk involved; recall here our point about "garbage in, garbage out" from Remark 8.68 (3). The CLT is important, hence *central*, because it provides us with a limit distribution for $\bar{X}_n$ under little information about F. If we have $X_1, X_2, \dots \overset{\text{iid}}{\sim} F$ and if $\sigma^2 = \mathrm{var}(X_1) < \infty$, then, by (8.68), the CLT provides us with the approximate $\mathrm{N}(\mu, \sigma^2/n)$ distribution for $\bar{X}_n$, and F only enters this normal distribution through its mean μ and variance σ^2, both of which we can easily estimate as mentioned in Remark 8.72. The CLT is extremely powerful. Not only can the normal distribution be used to directly model $\bar{X}_n$, but it can also serve as a benchmark against which to compare other models for $\bar{X}_n$. Such other models could be specified by an expert, for example, which could lead to a much more well-rounded picture of the distribution of $\bar{X}_n$ or quantities derived from it. But in the very general applicability of the CLT lies a potential danger for the ubiquitous use of the normal distribution.

Throughout the book, we encounter many examples of this 'over-reliance on the bell curve', especially in the realm of risk modeling.

(4) The CLT with its limit normal distribution even extends beyond means and sums, for example to transforms of means with functions g satisfying some (rather weak) conditions (g differentiable, g' continuous, $g'(\mu) \neq 0$). For such g, one can show that

$$\sqrt{n}\frac{g(\bar{X}_n) - g(\mu)}{\sigma g'(\mu)} \xrightarrow{\text{d}} \text{N}(0, 1) \quad (n \to \infty);$$

compare this with (8.66). This result is known as the *delta method* and it allows the asymptotic normality of $\bar{X}_n$ to be extended to transformations of $\bar{X}_n$ (for example, $\exp(\bar{X}_n)$). This is especially powerful for constructing confidence intervals for quantities of interest of the form $g(\mu)$; we will come back to that in Section 8.6.3.

Remark (A historical comment)**.** On November 12, 2008, the Department of Statistics at the University of Chicago celebrated the 275th birthday of the normal distribution. As we can read in Daw and Pearson (1972), the authors refer to a note written on November 12, 1733, by Abraham de Moivre (1667–1754). This note is the earliest historical account of the formula for the normal distribution. The statistician and chemist William John ("Jack") Youden (1900–1971) immortalized the ubiquitous nature of the normal distribution in every aspect of society through a bell-shaped calligram; see Figure 8.47.

THE
NORMAL
LAW OF ERROR
STANDS OUT IN THE
EXPERIENCE OF MANKIND
AS ONE OF THE BROADEST
GENERALIZATIONS OF NATURAL
PHILOSOPHY. IT SERVES AS THE
GUIDING INSTRUMENT IN RESEARCHES
IN THE PHYSICAL AND SOCIAL SCIENCES AND
IN MEDICINE, AGRICULTURE AND ENGINEERING.
IT IS AN INDISPENSABLE TOOL FOR THE ANALYSIS AND THE
INTERPRETATION OF THE BASIC DATA OBTAINED BY OBSERVATION AND EXPERIMENT.

Figure 8.47 Calligram of William John Youden presenting the shape of the normal distribution. Source: Authors

A wonderful mechanical visualization of the CLT can be found in the *Galton board*, named after Sir Francis Galton (1822–1911). The left-hand side of Figure 8.48 shows a sketch of a Galton board, a vertical board with n rows, the kth row having k pegs. Through the top of the board, beads are dropped to make left or right turns at one of the pegs in each row, ending up in $n+1$ buckets labeled 0 to n. As there are $\binom{n}{k}$ possible left–right paths through the pegs that end up in bucket number k, it is more likely for beads to end up in buckets around $n/2$ than in bucket 0 or bucket n. In fact, denote the probability of a bead turning right by p and turning left by $1 - p$ (with a perfect board and perfect beads we have $p = 1 - p = 1/2$). Then the probability of a bead ending up in bucket k is $\binom{n}{k}p^k(1-p)^{n-k}$, the probability mass function of a $\text{B}(n, p)$ distribution. If $X_i = I_{\{\text{"bead turns right"}\}}$ denotes the indicator that a bead turns right, then $S_n = \sum_{i=1}^n X_i$ denotes the bucket number in which the bead lands. As a bead turns left or right at a particular peg independently of its turns at other pegs, and since $X_1, X_2, \ldots, X_n$ are iid with finite variance, we know from (8.69) that an approximate distribution of S_n can also be given by the CLT. With $\mu = \mathbb{E}(X_1) = p = 1/2$ and $\sigma^2 = \text{var}(X_1) = p(1-p) = 1/4$,

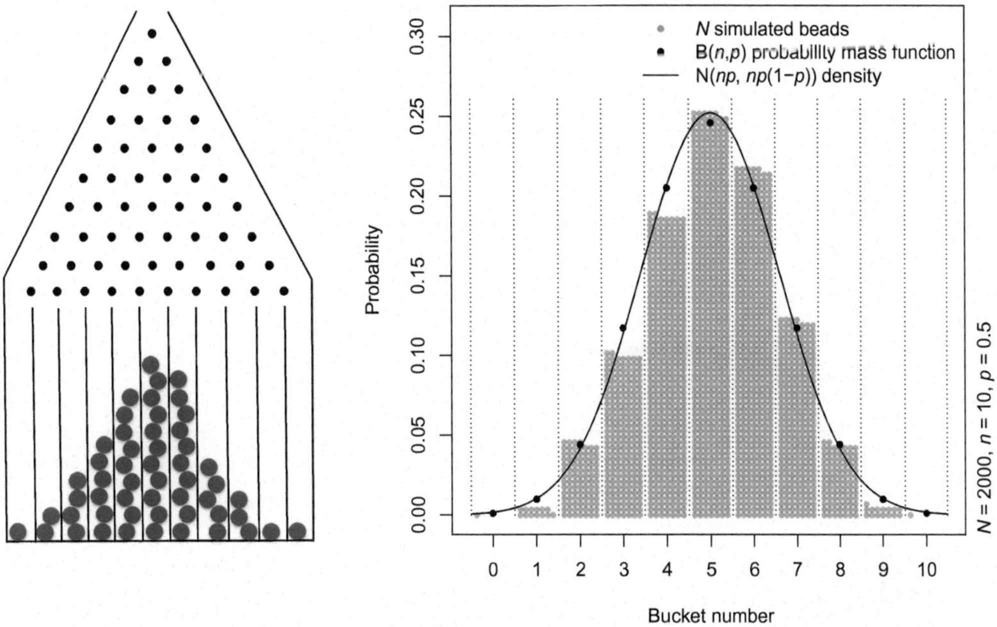

Figure 8.48 Left: Sketch of a Galton board with pegs and beads. Right: N simulated beads together with the $\mathrm{B}(n, p)$ probability mass function and $\mathrm{N}(np, np(1-p))$ density. Source: Authors

we obtain $S_n \underset{n \text{ large}}{\overset{\text{approx.}}{\sim}} \mathrm{N}(np, np(1-p))$. Mathematically one can indeed show that for sufficiently large n and with x close to np, the $\mathrm{B}(n, p)$ probability mass function at x approximates the $\mathrm{N}(np, np(1-p))$ density at x, a result known as *de Moivre–Laplace theorem*. The right-hand side of Figure 8.48 shows, for $n = 10$ and $p = 1/2$, the $\mathrm{B}(n, p)$ probability mass function overlaid with the $\mathrm{N}(np, np(1-p))$ density, as well as $N = 2000$ simulated beads in the $n + 1$ buckets, which seems convincing!

In Derivation 8.28 we already covered the Poisson limit theorem, which provided us with an approximation of $\mathrm{B}(n, p)$ by $\mathrm{Poi}(\lambda)$. Here we now saw how the $\mathrm{B}(n, p)$ distribution can be approximated by a normal. For this approximation to work well one typically uses $n \geq 20$, $np \geq 10$ and $n(1-p) \geq 10$ as a rule of thumb.

Example 8.75 (CLT for the probability of rolling a 6)**.** The left-hand side of Figure 8.49 shows the plot from the right-hand side of Figure 8.45, discussed in Example 8.70, with a vertical line at $n = 500$ included. We fix this n and select the $N = 200$ replications of $\bar{X}_n$ to visually confirm the CLT. Let these $N = 200$ replications of $\bar{X}_n$ be denoted by $\bar{X}_{n,1}, \bar{X}_{n,2}, \ldots, \bar{X}_{n,N}$. As each $\bar{X}_{n,b}$ is based on $X_{1,b}, X_{2,b}, \ldots, X_{n,b} \overset{\text{iid}}{\sim} \mathrm{B}(1, 1/6)$, we have $\mu = \mathbb{E}(X_{1,1}) = 1/6$ and $\sigma = \sqrt{\mathrm{var}(X_{1,1})} = \sqrt{(1/6)(1-1/6)} = \sqrt{5}/6$. We then standardize $\bar{X}_{n,1}, \bar{X}_{n,2}, \ldots, \bar{X}_{n,N}$ as in the CLT by forming the sample

$$Y_b = \sqrt{n}\frac{\bar{X}_{n,b} - \mu}{\sigma}, \quad b = 1, 2, \ldots, N. \tag{8.71}$$

According to the CLT, the sample $Y_1, Y_2, \ldots, Y_N$ should be approximately $\mathrm{N}(0, 1)$-distributed. The right-hand side of Figure 8.49 shows a plot of the sample $Y_1, Y_2, \ldots, Y_N$ together with realizations of $Z_1, Z_2, \ldots, Z_N \overset{\text{iid}}{\sim} \mathrm{N}(0, 1)$. Both samples show the same structure (for example,

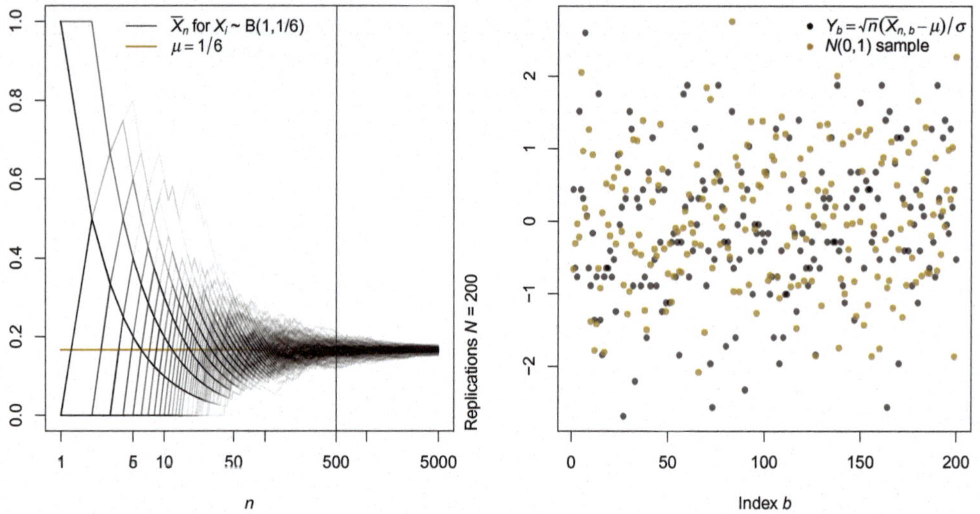

Figure 8.49 Left: $N = 200$ replications of realizations of $\bar{X}_n$ as a function of n. Right: A sample $Y_1, Y_2, \ldots, Y_N$ of standardized sample means as in (8.71) overlaid with N iid realizations from $\mathrm{N}(0, 1)$. Source: Authors

fluctuations around 0 of similar magnitude) in agreement with the CLT. There are also other ways to visually confirm the CLT, for example with a Q–Q plot or by comparing a density estimator with the $\mathrm{N}(0, 1)$ density.

As already mentioned in Remark 8.74 (3), the CLT provides us with a limit model, given by a normal distribution, for sample means and sums when there is little information about the common distribution function F of the underlying iid random variables $X_1, X_2, \ldots$ For the CLT to apply, all we need to know is that F has finite variance. Of course the devil is in the detail here. If $X_1, X_2, \ldots$ are independent, if they have a common distribution function F, if this F has finite variance and if n is sufficiently large, *then* the CLT holds. Those are a lot of *if* statements, and the *what if* question (in the sense of "what if any of these assumptions is violated?") is not far around the corner. Similarly to what we already mentioned at the end of Example 8.6 on the Monty Hall problem, it is often helpful to consider extreme cases to see what happens. So, to convince ourselves about what can go wrong, we could ask what happens if an assumption is violated to an extreme degree. For example, what if the sample size is only $n = 1$? The fact that $n = 1$ does not present enough data to be able to tell whether a sample mean or sum is approximately normally distributed should be pretty clear: both the mean and the sum are just X_1 and thus are distributed according to F, which can be far from a normal distribution. Another extreme violation would occur if all random variables were the same, which would be a strong form of dependence and thus a violation of the independence assumption. In this case $\bar{X}_n$ is simply $X_1 \sim F$ again and the sum is $S_n = nX_1$ with distribution function $F(x/n)$, a scaled version of F. As before, such distributions can be far from a normal. Such thought experiments can help one to understand why assumptions underlying mathematical results are necessary and not just "academic". Again model risk comes to mind here. Of course, it becomes more difficult to judge when assumptions are only mildly violated. One cannot give a general answer in such cases as it becomes relevant to know how the respective assumption is violated and whether that matters for the modeling purpose in mind. Once more we suggest asking a well-trained statistician, who can assess what must be taken into account and what options there are on the modeling side. As already mentioned before, simulations can also provide answers to *what if* questions and can allow one, for example, to study the effects of changing certain assumptions or parameters in an isolated manner while keeping others fixed.

8.6.3 Confidence Intervals

Roughly, a confidence interval (CI) is a random interval (an interval with random endpoints) that contains an unknown quantity of interest, with a given, typically large, probability. In Figure 2.6 of Section 2.3 we have already seen realizations of confidence intervals for estimates of a quantity of interest, in this case the probability of at least one O-ring failure. We now cover confidence intervals in more detail. Before starting on this stage of our hike, we would like to mention that, when it comes to the public communication of risk, being aware of the importance and usefulness of CIs is absolutely crucial; see Remark 8.77 (1) below.

Concept 8.76 (Confidence interval (CI)). Let $X_1, X_2, \dots, X_n \overset{\text{iid}}{\sim} F_{\text{true}}$ for some unknown distribution function F_{true} and let θ be a quantity of interest that depends on F_{true}, for example a mean, probability or parameter of F_{true}. A $(1-\alpha)$-*confidence interval (CI)* for θ is a random interval I_n with endpoints depending on $X_1, X_2, \dots, X_n$ such that

$$\mathbb{P}(\theta \in I_n) \geq 1 - \alpha. \tag{8.72}$$

Here, $\alpha \in (0,1)$ is known as the *significance level* and $\mathbb{P}(\theta \in I_n)$ as the *coverage probability* of the CI. One generally aims for coverage probabilities of precisely $1-\alpha$, but if that is not possible then the coverage probability should be at least $1-\alpha$, hence the inequality in (8.72). A typical choice is $\alpha = 0.05$, resulting in a 95%-CI.

Some remarks are in order now. How this seemingly abstract concept of CIs can be used in practice will become clear through the examples discussed below and later in the book.

Remark 8.77.

(1) Confidence intervals serve the purpose of quantifying the uncertainty when estimating a quantity of interest θ. They are thus important for assessing how much we can trust (or be confident in) an estimate of θ. Confidence intervals are therefore an important tool for communicating risk. As you can imagine, besides the assumptions relied upon in the underlying statistical analysis, the uncertainty quantification of estimates becomes a central issue, if not *the* central issue, in complex modeling problems such as the global average temperature increase in the context of climate change, which we mentioned as the last question in the first paragraph in Section 8.1. Other examples are the reporting of the reproduction numbers R_0 and R_e in the case of COVID-19 (see Section 6.3) and the communication of a necessary dike height for the Dutch coastal areas, as discussed in Chapter 1. Concerning the latter example, recall the second quote from Kruizinga and Lewis (2018) in Section 1.3 concerning model uncertainty.

(2) A realization of a CI I_n is an interval which either contains or does not contain the quantity of interest θ. If we could generate realizations of the random interval I_n independently of each other N times for large N, then roughly $(1-\alpha)N$ of the realized CIs would contain the unknown θ. We often just speak of a $(1-\alpha)$-CI and do not further specify whether the theoretical quantity (the CI estimator, an interval with random endpoints) or the empirical quantity (the CI estimate, an interval with real endpoints) is meant; this should become clear from the context. As for other estimators or estimates, the theoretical quantity in terms of random variables is used to study mathematical properties, and the empirical quantity providing actual numbers is computed, applied and reported in practice.

(3) For specific α, one often writes $1-\alpha$ as a percentage, $(1-\alpha)100\%$; for example, for $\alpha = 0.05$ one speaks of a 95%-CI (as we already did in Section 2.3). One typically considers 99%-, 95%- or 90%-CIs, with 95%-CIs being predominantly used; in fact, just speaking of a CI alone is typically interpreted as a 95%-CI. Note that a coverage probability of, say, 99.99% would rarely make sense in practice as the corresponding $(1-\alpha)$-CI would typically be far too wide to be useful.

(4) The significance level α (for example $\alpha = 0.05$, rather close to 0) of a CI should not be confused with the (confidence) level α (for example $\alpha = 0.99$, rather close to 1) of a risk measure such as VaR_α or ES_α. This is one of several reasons why we prefer to speak of the "level" α rather than the "confidence level" when considering risk measures such as VaR_α or ES_α.

(5) In this book, we concentrate on one-dimensional quantities of interest θ. The notion of a CI can easily be extended to include multi-dimensional quantities; one then speaks of *confidence regions*. In the infinite-dimensional case, when one for instance is estimating curves, one refers to *confidence bands*.

We now turn to an explicit example for constructing a CI as defined through (8.72). It should help our understanding of how the somewhat theoretical Concept 8.76 can actually be made operational in practice.

Example 8.78 (CI for $\theta = \mu$ for iid $\mathrm{N}(\mu, \sigma^2)$ data)**.** In this example, let $X_1, X_2, \dots, X_n \overset{\text{iid}}{\sim} F_{\text{true}}$, where F_{true} is the distribution function of $\mathrm{N}(\theta, \sigma^2)$ for a known $\sigma > 0$ and the mean θ of F_{true} is the unknown quantity of interest.

For the construction of a CI, we need a quantity that involves θ (the quantity of interest) and whose distribution we know (to ensure we obtain the correct coverage probability $1-\alpha$, see (8.72)). Finding this quantity is typically rather difficult. One usually starts with an estimator for θ; here we take $\hat{\theta}_n = \bar{X}_n$ as motivated by the SLLN. What we can exploit in this example is the fact that if $X_1, X_2, \dots, X_n \overset{\text{iid}}{\sim} \mathrm{N}(\theta, \sigma^2)$, then $\hat{\theta}_n = \bar{X}_n \sim \mathrm{N}(\theta, \sigma^2/n)$; we have not introduced tools to show this property in this book, nonetheless it is an important property of normal distributions. A location-scale transform then leads to

$$\sqrt{n}\left(\frac{\hat{\theta}_n - \theta}{\sigma}\right) \sim \mathrm{N}(0, 1). \tag{8.73}$$

Now you see that we have a quantity (on the left) that involves θ and whose distribution (the one on the right) we know. In terms of the $\mathrm{N}(0, 1)$ distribution and quantile functions, we can now construct a CI with coverage probability $1-\alpha$. For notational simplicity, let $z_p = \Phi^{-1}(p)$, $p \in (0,1)$, and recall from Example 8.49 (3) that $z_p = \Phi^{-1}(p) = -\Phi^{-1}(1-p) = -z_{1-p}$, $p \in (0,1)$, and hence, for $p = \alpha/2$, that $-z_{1-\alpha/2} = z_{\alpha/2}$. Because of (8.73), we know that

$$\mathbb{P}\left(-z_{1-\alpha/2} \le \sqrt{n}\frac{\hat{\theta}_n - \theta}{\sigma} \le z_{1-\alpha/2}\right) = \Phi(z_{1-\alpha/2}) - \Phi(-z_{1-\alpha/2})$$
$$= \Phi(z_{1-\alpha/2}) - \Phi(z_{\alpha/2}) = 1 - \frac{\alpha}{2} - \frac{\alpha}{2} = 1 - \alpha,$$

so we can already see two things, the correct coverage probability $1-\alpha$ and an interval with endpoints $-z_{1-\alpha/2}$ and $z_{1-\alpha/2}$ around the pivotal quantity $\sqrt{n}(\hat{\theta}_n - \theta)/\sigma$. The rest is easy: we just reorder the terms in the argument of $\mathbb{P}$ to see the (confidence) interval around θ. Since $\sqrt{n}(\hat{\theta}_n - \theta)/\sigma \ge -z_{1-\alpha/2}$ if and only if $\theta \le \hat{\theta}_n + z_{1-\alpha/2}\sigma/\sqrt{n}$ and since $\sqrt{n}(\hat{\theta}_n - \theta)/\sigma \le z_{1-\alpha/2}$ if and only if $\theta \ge \hat{\theta}_n - z_{1-\alpha/2}\sigma/\sqrt{n}$, we can write the above probability as

$$\mathbb{P}\left(\hat{\theta}_n - z_{1-\alpha/2}\frac{\sigma}{\sqrt{n}} \le \theta \le \hat{\theta}_n + z_{1-\alpha/2}\frac{\sigma}{\sqrt{n}}\right) = 1 - \alpha.$$

Hence $\mathbb{P}(\theta \in I_n) = 1 - \alpha$ for the $(1-\alpha)$-CI

$$I_n = \left[\hat{\theta}_n - z_{1-\alpha/2}\frac{\sigma}{\sqrt{n}},\ \hat{\theta}_n + z_{1-\alpha/2}\frac{\sigma}{\sqrt{n}}\right].$$

You might (still) wonder how useful this example is, given that (8.73) is probably rather specific to this setup of our example here and that we treated σ as known. And you are right. However, note that even if $X_1, X_2, \dots, X_n$ are not iid normal but iid with finite second moment (a much weaker condition), we know that (8.73) holds approximately for large n according to the CLT. And σ can be estimated. We will come back to these points later.

We now turn to the question of how we can construct CIs more generally. To this end, we present two approaches which are also applied throughout the book (with possible modifications for the specific situation at hand). Although the end of this hike is near, please take your time; there is no gain in rushing to the summit and risking falling off the last cliff.

Asymptotic Confidence Intervals

For the first approach, consider the problem of estimating $\mu = \mathbb{P}(\text{"rolling a 6"})$; this μ is indeed a mean since $\mathbb{P}(A) = \mathbb{E}(I_A)$ as we mentioned in Example 8.62. We thus construct a CI for the quantity of interest θ, which is μ here. Of course, for rolling a 6 with a fair die we know the underlying F_{true}, it is the distribution function of $\mathrm{B}(1,\mu)$ for $\mu = 1/6$, but in order to stay realistic we will not make use of this knowledge since we normally do not know F_{true}. In this chapter, this example serves only as a toy example, but one could imagine playing an online game involving a die and suspecting that this die is loaded. The true, but unknown, F_{true} is then still Bernoulli but with unknown probability μ. If the computed 95%-CI for μ does not contain $1/6$, then this is a strong indication that the die is indeed loaded.

As in Example 8.78, we use $\hat{\theta}_n = \bar{X}_n$ as an estimator of $\theta = \mu$, where $X_i = I_{\{i\text{th roll is a 6}\}}$, so that X_i indicates whether the ith roll is a 6. As already mentioned in Example 8.78, we now utilize the CLT to derive a $(1-\alpha)$-CI for θ based on the estimator $\hat{\theta}_n$ and the pivotal quantity $\sqrt{n}(\hat{\theta}_n - \theta)/\sigma$, which, by the CLT, is approximately N(0, 1)-distributed. With a similar calculation as in Example 8.78 but now based on the CLT, we obtain that

$$\mathbb{P}\left(-z_{1-\alpha/2} \leq \sqrt{n}\frac{\hat{\theta}_n - \theta}{\sigma} \leq z_{1-\alpha/2}\right) \underset{n \text{ large}}{\overset{\text{CLT}}{\approx}} \Phi(z_{1-\alpha/2}) - \Phi(-z_{1-\alpha/2}) = 1 - \alpha.$$

On the right we have the correct coverage probability $1-\alpha$, and on the left we have a quantity which contains the unknown parameter θ for which we want to construct a CI. Solving the inequalities in the argument of $\mathbb{P}$ with respect to θ leads to

$$\mathbb{P}\left(\hat{\theta}_n - z_{1-\alpha/2}\frac{\sigma}{\sqrt{n}} \leq \theta \leq \hat{\theta}_n + z_{1-\alpha/2}\frac{\sigma}{\sqrt{n}}\right) \underset{n \text{ large}}{\overset{\text{CLT}}{\approx}} 1 - \alpha.$$

We can now read off the interval

$$\left[\hat{\theta}_n - z_{1-\alpha/2}\frac{\sigma}{\sqrt{n}}, \hat{\theta}_n + z_{1-\alpha/2}\frac{\sigma}{\sqrt{n}}\right] \tag{8.74}$$

about θ and we know it has the correct coverage probability $1 - \alpha$. From $X_1, X_2, \ldots, X_n$ (or their realizations) we can calculate $\hat{\theta}_n = \bar{X}_n$, since n is known and α is given (for example $\alpha = 0.05$). But (8.74) still contains the unknown σ, the standard deviation of each of the random variables $X_1, X_2, \ldots, X_n$ from F_{true}, so (8.74) is not a CI just yet.

In Example 8.78 we assumed σ to be known. And from Example 8.62 we know that $\sigma = \sqrt{(1/6)(1-1/6)} = \sqrt{5}/6$ here, but we stated above that we will not make use of this knowledge. What can we do, then? We can estimate σ, namely by estimating the standard deviation of the data $X_1, X_2, \ldots, X_n$. If $\hat{\sigma}_n$ denotes the sample standard deviation, see Remark 8.72, then

$$I_n = \left[\hat{\theta}_n - z_{1-\alpha/2}\frac{\hat{\sigma}_n}{\sqrt{n}}, \hat{\theta}_n + z_{1-\alpha/2}\frac{\hat{\sigma}_n}{\sqrt{n}}\right] \tag{8.75}$$

is, for sufficiently large n (so that the CLT applies and also $\hat{\sigma}_n$ well approximates σ), a $(1-\alpha)$-CI for θ for which we can compute all ingredients. For small n, a statistical friend may reveal how, in this case, Student's t distribution enters into the construction of an appropriate CI.

As the interval (8.75) is derived from a limit theorem for $n \to \infty$ (the CLT), it is known as an *asymptotic* $(1-\alpha)$*-CI*. The quantity $\hat{\sigma}_n/\sqrt{n}$ is known as the *standard error (SE)*. Since, for $\alpha = 0.05$, $z_{1-\alpha/2} \approx 1.9600$, the asymptotic 95%-CI is often reported in the form "estimate ± 1.96 SE" (or, a bit less accurately, but easier to remember, "estimate ± 2 SE"). In Figure 2.5 of Section 2.3 we saw asymptotic 95%-CIs (one for each temperature t) when analyzing the O-ring failure probability in the Challenger disaster.

There are various powerful extensions of the construction of asymptotic CIs. We briefly mentioned the delta method in Section 8.6.2. It allows one to construct asymptotic CIs for transformed quantities of interest $g(\theta)$ from the estimator $g(\hat{\theta}_n)$ of $g(\theta)$ if $\hat{\theta}_n$ is asymptotically normally distributed.

The construction of asymptotic CIs seems to involve some handwaving, given that no one has told us what a "sufficiently large n" is or why the default significance level chosen is $\alpha = 0.05$, for example. Statisticians are sometimes blamed by other mathematicians for such apparent lack of precision. At the end of the day, statistics can provide guidance or answers (in the form of estimates) to problems that otherwise cannot be solved. The corresponding uncertainty can often be quantified in the form of CIs. Statistical software provides implementations of such solutions that can be applied by non-statisticians. To see theorems, algorithms and their implementation so directly applied to societal problems is a joy that perhaps not many other branches of mathematics can spark to a similar degree.

When viewed as an estimator, the asymptotic CI (8.75) is a random interval which contains the quantity $\theta = \mu$ with probability $1 - \alpha$. This cries for a check by simulation to convince us that our derivation is indeed correct. We compute, say, N realizations of (8.75) (and so N asymptotic $(1 - \alpha)$-CI estimates) and then check whether roughly $(1 - \alpha)N$ of them contain θ; to this end we can even recycle the sample paths we have already simulated and considered earlier.

Example 8.79 (Asymptotic confidence intervals for the probability of rolling a 6)**.** The left-hand side of Figure 8.50 contains the same $N = 200$ sample paths as shown on the right-hand side of Figure 8.45 but the plot now includes, for each $n \geq 25$, the asymptotic 95%-CI estimate I_n from (8.75) for $\theta = \mu = \mathbb{P}(\text{"rolling a 6"})$, computed using data from the first of the $N = 200$ sample paths. The lower and the upper endpoints of the CIs I_n

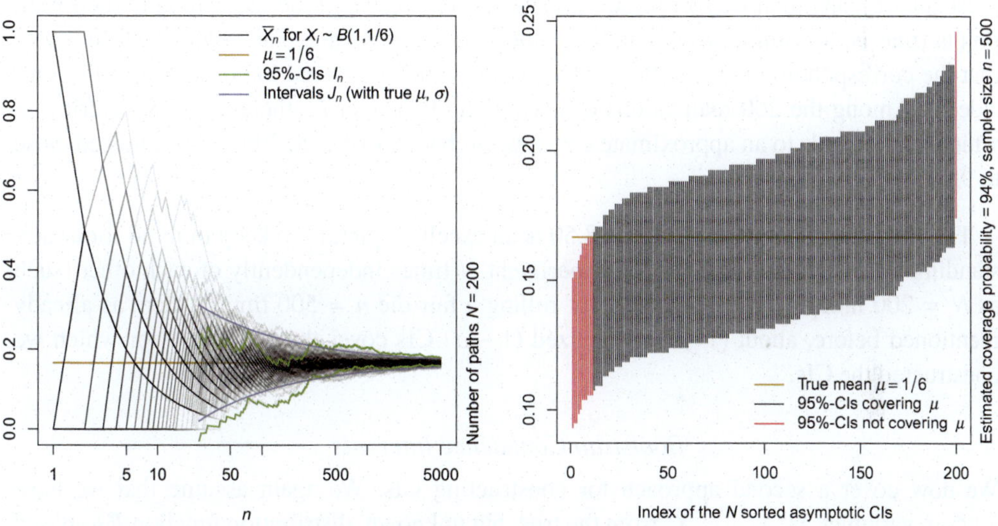

Figure 8.50 Left: Sample paths of $\bar{X}_n$ with the true mean $\mu = 1/6 \approx 0.1667$, asymptotic 95%-CI estimates I_n from (8.75) and intervals J_n for all $n \geq 25$. Right: Plot of the $N = 200$ realizations of asymptotic 95%-CIs for μ as vertical bars (for $n = 500$), sorted according to their midpoint and colored according to whether they contain μ. Source: Authors

are connected across different n for better visibility, which results in two paths of endpoints across different n. For small n, the endpoints of I_n fluctuate; we even see negative values of the lower endpoints of I_n for the smallest n. Obviously, n is not "sufficiently large" in these cases. Now let us consider large n. As already mentioned, in this toy example we know both the true mean μ (it is $1/6$) and the standard deviation σ (it is $\sqrt{5}/6$). We also include in the plot on the left-hand side of Figure 8.50 the endpoints of the interval $J_n = [\theta - z_{1-\alpha/2}\sigma/\sqrt{n},\ \theta + z_{1-\alpha/2}\sigma/\sqrt{n}] = [\mu - z_{1-\alpha/2}\sigma/\sqrt{n},\ \mu + z_{1-\alpha/2}\sigma/\sqrt{n}]$ for $n \geq 25$. The interval J_n is simply the CI I_n from (8.75) with the estimators $\hat{\theta}_n$ and $\hat{\sigma}_n$ replaced by the corresponding true quantities $\theta = \mu = 1/6$ and $\sigma = \sqrt{5}/6$, which we know here. Of course, in practice it is entirely unrealistic to have J_n available, but by including this interval in the plot in our case here we can visually confirm that I_n from (8.75) indeed estimates J_n for sufficiently large n, as expected by the SLLN.

We also see from the left-hand side of Figure 8.50 that the sample paths of $\bar{X}_n$ converge towards μ roughly at the same speed as the endpoints of J_n (and thus those of I_n, but it is more visible for J_n) converge towards μ. This comes from the term $1/\sqrt{n}$, which is known as the *speed of convergence* of the estimator $\bar{X}_n$ towards the quantity μ it estimates. The interpretation of this speed of convergence is that a hundred times larger sample size $100n$ leads to a $1/\sqrt{100n} = 1/(10\sqrt{n})$ more accurate estimate, the factor $1/10$ indicating one more digit of accuracy of the estimate.

The plot on the right-hand side of Figure 8.50 assesses the coverage probability of asymptotic 95%-CIs. We proceed as in Example 8.75 and fix $n = 500$; see the left-hand side of Figure 8.49. For each of the $N = 200$ realizations of $\bar{X}_n$ (which is $\hat{\theta}_n$ in the present case), we compute a realization of the asymptotic 95%-CI I_n for the mean μ (which is θ in the present case), and then we sort these estimates into increasing order according to their midpoints (that is, according to $\bar{X}_n$) and plot them as vertical bars. The color black indicates that the corresponding realized 95%-CI I_n covers μ; the color red indicates that μ lies outside I_n. Among the 200 realizations of asymptotic 95%-CIs I_n, there are 188 covering μ, which corresponds to an approximate coverage probability of $188/200 = 94\%$, indeed close to 95%.

The right-hand side plot of Figure 8.50 is an excellent picture to keep in mind for understanding CIs. If we can replicate an experiment N times independently of each other (such as $N = 200$ independent replications of rolling a fair die $n = 500$ times), then, as already mentioned before, about $(1 - \alpha)N$ realized $(1 - \alpha)$-CIs cover the quantity θ for which we constructed the CIs.

Bootstrap Confidence Intervals

We now cover a second approach for constructing CIs. We again assume that we have random variables $X_1, X_2, \dots, X_n$ from the true, but unknown, distribution function F_{true} based on which we construct an estimator $\hat{\theta}_n$ of a quantity of interest θ that depends on F_{true}. As a toy example we again consider the problem of estimating $\mu = \mathbb{P}(\text{"rolling a 6"})$ as we did earlier, so $\theta = \mu$. If we *knew* F_{true} (as we do in this toy example), we could generate B samples $X_{1,b}, X_{2,b}, \dots, X_{n,b}$, $b = 1, 2, \dots, B$, of size n each and compute the corresponding B estimates $\hat{\theta}_{n,1}, \hat{\theta}_{n,2}, \dots, \hat{\theta}_{n,B}$. In our toy example, we would have $\hat{\theta}_{n,b} = \bar{X}_{n,b} = \frac{1}{n}\sum_{i=1}^{n} X_{i,b}$ for $b = 1, 2, \dots, B$. Each of these B estimates is constructed from the same underlying distribution function F_{true} as $\hat{\theta}_n$ and with the same sample size n. The B estimates $\hat{\theta}_{n,1}, \hat{\theta}_{n,2}, \dots, \hat{\theta}_{n,B}$ are thus a sample from the same distribution as $\hat{\theta}_n$, which we denote by $F_{\hat{\theta}_n}$. Now that we have a whole sample $\hat{\theta}_{n,1}, \hat{\theta}_{n,2}, \dots, \hat{\theta}_{n,B}$ instead of just a single estimate $\hat{\theta}_n$, we can use this sample to study the behavior of $\hat{\theta}_n$, and so $F_{\hat{\theta}_n}$. For example, the empirical distribution function $\hat{F}_{\hat{\theta}_n,B}$ of $\hat{\theta}_{n,1}, \hat{\theta}_{n,2}, \dots, \hat{\theta}_{n,B}$ can be used to study the distribution of the estimator $\hat{\theta}_n$. Or $\hat{\theta}_{n,1}, \hat{\theta}_{n,2}, \dots, \hat{\theta}_{n,B}$ can be used to approximate the variance of the estimator $\hat{\theta}_n$. What would interest us most would be using $\hat{\theta}_{n,1}, \hat{\theta}_{n,2}, \dots, \hat{\theta}_{n,B}$ to construct a CI for θ. We only have one problem: typically we do not know F_{true}, so we cannot draw from this distribution, which produced the only sample $X_1, X_2, \ldots, X_n$ we have. Therefore, we cannot produce $\hat{\theta}_{n,1}, \hat{\theta}_{n,2}, \ldots, \hat{\theta}_{n,B}$ from $F_{\hat{\theta}_n}$. Time to call for help from Bootstrapman; see Figure 8.51.

Figure 8.51 Bootstrapman to the rescue (if F_{true} is unknown). Source: Enrico Chavez

The main idea of the *(non-parametric) bootstrap*, introduced by Bradley Efron (1979), is to replace the unknown F_{true} by its empirical equivalent $\hat{F}_n$ based on $X_1, X_2, \ldots, X_n$, which we know from Remark 8.72 approximates F_{true} well for sufficiently large n. Since we know $\hat{F}_n$ and how to sample from it (see Example 8.63), namely by sampling from $X_1, X_2, \ldots, X_n$ at random with replacement (see (8.63)), this is conceptually (but unfortunately not mathematically) an extremely convenient idea. The bootstrap proceeds by generating, for sufficiently large $B \in \mathbb{N}$, samples $X_{1,b}, X_{2,b}, \ldots, X_{n,b}$, $b = 1, 2, \ldots, B$, so-called *bootstrap samples*. In the same way as the estimate $\hat{\theta}_n$ is computed from $X_1, X_2, \ldots, X_n$, we then compute, for each $b = 1, 2, \ldots, B$, the estimate $\hat{\theta}_{n,b}$ from the bth bootstrap sample $X_{1,b}, X_{2,b}, \ldots, X_{n,b}$. The resulting *bootstrap estimates* $\hat{\theta}_{n,1}, \hat{\theta}_{n,2}, \ldots, \hat{\theta}_{n,B}$ can then be used to investigate the behavior of the estimator $\hat{\theta}_n$ of θ, and so $F_{\hat{\theta}_n}$. For example, the empirical distribution function $\hat{F}_{\hat{\theta}_n,B}$ of the bootstrap estimates $\hat{\theta}_{n,1}, \hat{\theta}_{n,2}, \ldots, \hat{\theta}_{n,B}$ is used as an approximation to the unknown distribution function $F_{\hat{\theta}_n}$ of $\hat{\theta}_n$. Or the variance of the bootstrap estimates is used as an approximation to the variance of $\hat{\theta}_n$ in order to measure the statistical variability of the estimator $\hat{\theta}_n$ and thus its precision. Being able to generate 'more' (estimates) out of 'less' (data) sounds too good to be true, in fact like pulling yourself up in the air by pulling at your own bootstraps, the small loops still found near the top of the heel of some boots even nowadays; see Figure 8.52. This is where the name of the bootstrap is believed to have come from. Figure 8.52 shows Baron Münchhausen, known for his impossible achievements, see Raspe (1785), pulling himself (and even his horse) out of a mud pit by tugging on his own hair; although there is no direct

Figure 8.52 Left: Bootstrap at the top of the heel of a boot. Right: Baron Münchausen by Theodor Hosemann pulling himself out of a swamp by tugging on his own hair. Source: Wikimedia Commons

connection to the bootstrap, this explanation is also often invoked to informally explain the idea of the bootstrap (achieving the seemingly impossible to generate 'more' out of 'less').

So how can we apply the bootstrap to construct CIs? There are several methods for deriving CIs via the bootstrap; see, for example, Davison and Hinkley (1997). They rest on what one could call the fundamental principle of bootstrapping, which says that $\hat{\theta}_n - \theta$ is approximately distributed as $\hat{\theta}_{n,b} - \hat{\theta}_n$ (for each $b = 1, 2, \ldots, B$). In words, the true θ is to the estimate $\hat{\theta}_n$ as the estimate is to the bootstrap estimate $\hat{\theta}_{n,b}$ (in the sense as just described). Or, formulated in terms of the underlying distributions (and understood in the same sense as before), F_{true} is to $\hat{F}_n$ as $\hat{F}_n$ is to $\hat{F}_{n,b}$, where $\hat{F}_{n,b}$ is the empirical distribution function of the bth bootstrap sample. Exploiting this, we can give an (admittedly vague) mathematical derivation of one method for constructing CIs via the bootstrap. To this end, let B be sufficiently large and let $\hat{\theta}_{n,(1)} \le \hat{\theta}_{n,(2)} \le \cdots \le \hat{\theta}_{n,(B)}$ be the order statistics of the B bootstrap estimators $\hat{\theta}_{n,1}, \hat{\theta}_{n,2}, \ldots, \hat{\theta}_{n,B}$. For any $b \in \{1, 2, \ldots, B\}$, the probability that a particular bootstrap estimator $\hat{\theta}_{n,b}$ belongs to the most extreme $\alpha \times 100\%$ of all bootstrap estimates is about $1 - \alpha$, so

$$\mathbb{P}\big(\hat{\theta}_{n,(\lceil(\alpha/2)B\rceil)} \le \hat{\theta}_{n,b} \le \hat{\theta}_{n,(\lceil(1-\alpha/2)B\rceil)}\big) \approx 1 - \alpha.$$

Considering $\hat{\theta}_n$ as fixed (so, as conditional on the original sample $X_1, X_2, \ldots, X_n$) and subtracting $\hat{\theta}_n$, the probability on the left-hand side is

$$\mathbb{P}\big(\hat{\theta}_{n,(\lceil(\alpha/2)B\rceil)} - \hat{\theta}_n \le \hat{\theta}_{n,b} - \hat{\theta}_n \le \hat{\theta}_{n,(\lceil(1-\alpha/2)B\rceil)} - \hat{\theta}_n\big).$$

By the fundamental principle of bootstrapping, this probability is approximately

$$\mathbb{P}\big(\hat{\theta}_{n,(\lceil(\alpha/2)B\rceil)} - \hat{\theta}_n \le \hat{\theta}_n - \theta \le \hat{\theta}_{n,(\lceil(1-\alpha/2)B\rceil)} - \hat{\theta}_n\big).$$

Rearranging terms, we obtain

$$\mathbb{P}\big(2\hat{\theta}_n - \hat{\theta}_{n,(\lceil(1-\alpha/2)B\rceil)} \leq \theta \leq 2\hat{\theta}_n - \hat{\theta}_{n,(\lceil(\alpha/2)B\rceil)}\big).$$

The interval

$$I_n = \big[2\hat{\theta}_n - \hat{\theta}_{n,(\lceil(1-\alpha/2)B\rceil)},\ 2\hat{\theta}_n - \hat{\theta}_{n,(\lceil(\alpha/2)B\rceil)}\big] \tag{8.76}$$

is the so-called *basic bootstrap* $(1-\alpha)$*-CI*. If the bootstrap estimates $\hat{\theta}_{n,1}, \hat{\theta}_{n,2}, \ldots, \hat{\theta}_{n,B}$ are symmetrically distributed around $\hat{\theta}_n$ (so that $\hat{\theta}_n - \hat{\theta}_{n,(\lceil(1-\alpha/2)B\rceil)} \approx \hat{\theta}_{n,(\lceil(\alpha/2)B\rceil)} - \hat{\theta}_n$ and $\hat{\theta}_n - \hat{\theta}_{n,(\lceil(\alpha/2)B\rceil)} \approx \hat{\theta}_{n,(\lceil(1-\alpha/2)B\rceil)} - \hat{\theta}_n$), then (8.76) is approximated by

$$I_n = \big[\hat{\theta}_{n,(\lceil(\alpha/2)B\rceil)},\ \hat{\theta}_{n,(\lceil(1-\alpha/2)B\rceil)}\big], \tag{8.77}$$

the so-called *percentile bootstrap* $(1-\alpha)$*-CI*. These are by far not the only bootstrap CIs; rather, they are the most basic ones. More sophisticated are studentized bootstrap CIs, where each bootstrap sample is again bootstrapped to estimate the standard deviation of each bootstrap estimate. These estimated standard deviations then enter a quantity from which one can derive CIs. Or there are bias-corrected and accelerated (BCa) bootstrap CIs, which are similar to percentile bootstrap CIs but with modified probabilities $\alpha/2$ and $1-\alpha/2$. Both studentized bootstrap CIs and BCa bootstrap CIs have the theoretical advantage that, under some conditions, the difference between the desired and the actual coverage probability is of lower order. However, this comes at the cost of a typically substantially increased run time.

There are debates about when one should or should not use a certain kind of bootstrap CI; see also Hesterberg (2015). The accuracy of a bootstrap method depends on whether its underlying mathematical assumptions are fulfilled and on how well the fundamental principle of bootstrapping holds in a particular situation. If F_{true} were identical to $\hat{F}_n$, then $\hat{\theta}_n$ and the bootstrap estimators would come from F_{true}. In this case the fundamental principle of bootstrapping holds. In general, however, the bootstrap estimates $\hat{\theta}_{n,1}, \hat{\theta}_{n,2}, \ldots, \hat{\theta}_{n,B}$ do not pull additional data from the true underlying but unknown F_{true}, but rather from the empirical distribution function $\hat{F}_n$ of the given realizations $X_1, X_2, \ldots, X_n$. This is the price one has to pay for being able to create 'more' (estimates) out of 'less' (data).

Again in the spirit of the third bullet point in the beginning of Section 8.5, we say it is time for a simulation!

Example 8.80 (Bootstrap confidence intervals for the probability of rolling a 6)**.** Figure 8.53 shows plots similar to those in Figure 8.50 but now for percentile bootstrap CIs instead of asymptotic CIs, with a bootstrap sample size of $B = 1000$. The plot on the left-hand side of Figure 8.53 shows the same $N = 200$ paths of $\bar{X}_n$ as before, so those paths are based on replications from the true underlying distribution $\mathrm{B}(1, 1/6)$. In contrast, the percentile bootstrap 95%-CIs plotted for each $n \geq 10$ (which, as before, are connected across different n for better visibility) are based only on the first of the N paths of $\bar{X}_n$, that shown on the left-hand side of Figure 8.45. So, the plotted percentile bootstrap CIs are obtained under the realistic scenario that we do not know the underlying true distribution function F_{true} (here, $\mathrm{B}(1, 1/6)$) and only have one sample of size n available from F_{true}. As for asymptotic CIs, the percentile bootstrap CIs fluctuate for small n and become more stable for larger n, so the larger the sample size n of the data we have, the better.

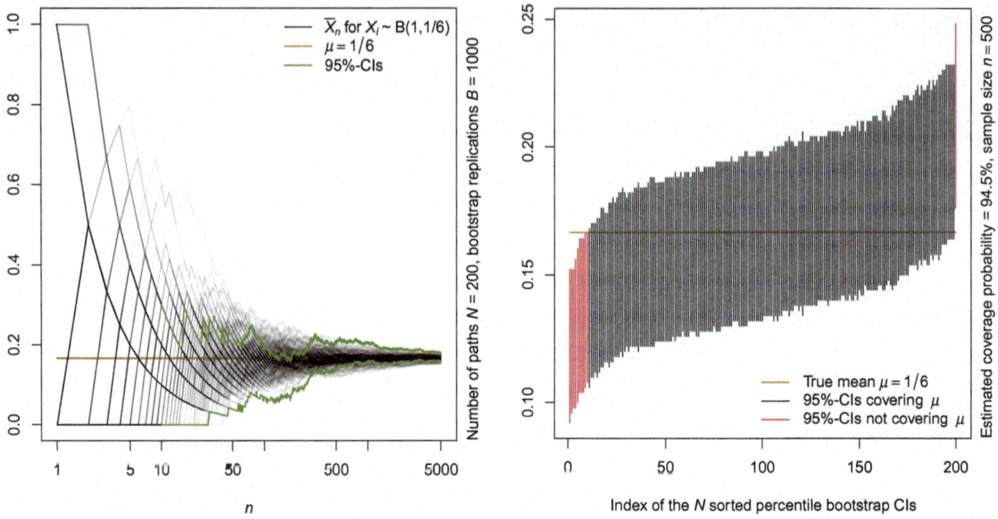

Figure 8.53 Left: Sample paths of $\bar{X}_n$ with true mean $\mu = 1/6 \approx 0.1667$ and percentile bootstrap 95%-CIs (8.77) (based on the data of the first path of $\bar{X}_n$) for all $n \geq 10$. Right: Plot of the $N = 200$ percentile bootstrap 95%-CIs for μ as vertical bars (each CI is computed from $B = 1000$ bootstrap samples and for $n = 500$), sorted according to their midpoint and colored according to whether they contain μ. Source: Authors

For the right-hand side of Figure 8.53, we fix $n = 500$ as in Example 8.79. From a single sample $X_1, X_2, \dots, X_n$ the bootstrap provides us with only one $(1 - \alpha)$-CI for μ. To assess whether the coverage probability of μ is indeed roughly $1 - \alpha$ as it should be, we need to simulate, say N times, percentile bootstrap $(1 - \alpha)$-CIs independently of each other. As on the right-hand side of Figure 8.50, we recycle the already simulated data for the $N = 200$ paths to this end. We thus obtain 200 independently simulated percentile bootstrap CIs. In short, each of the $N = 200$ percentile bootstrap CIs is computed from $B = 1000$ bootstrap samples of a single sample of size $n = 500$ from $\mathrm{B}(1, 1/6)$. From a computational point of view, one speaks of a *nested simulation* in this case, since this simulation is running over N replications (for computing N percentile bootstrap CIs in order to assess the coverage probability), where each replication entails a simulation running over B replications (for computing one percentile bootstrap CI). As one can count from the right-hand side plot in Figure 8.53, 190 of the $N = 200$ percentile bootstrap 95%-CIs contain the true mean μ, which indeed gives an estimated coverage probability of 95%.

Example 8.81 (Mean longest run of heads in a sequence of coin tosses)**.** If we toss a fair coin n times, what is the largest number of consecutive tosses giving heads that we expect to see, that is, the mean length of the longest run of heads?

What seems to be an innocent question is rather non-trivial to solve mathematically; see Binswanger and Embrechts (1994) and references therein. In the spirit of the third bullet point at the beginning of Section 8.5, we provide a solution by simulation. We simulate $N = 500$ coin tosses of length n ranging from 10 to 1000 and determine the longest run of heads in each sequence. The sample mean serves as estimate of the mean longest run of heads for each n. Figure 8.54 shows the estimated mean longest run of heads in n coin tosses as a function of n, including percentile bootstrap 95%-CIs based on the bootstrap sample size $B = 1000$. Note that the asymptotic 95%-CIs are essentially indistinguishable from the percentile bootstrap 95%-CIs in this case.

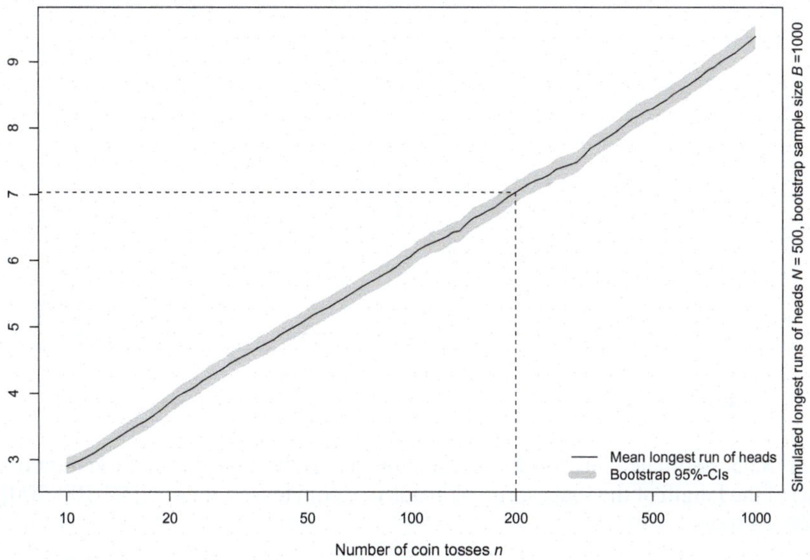

Figure 8.54 Estimated mean length of the longest run of heads in n coin tosses (n ranging from 10 to 1000, the case $n = 200$ being indicated by dashed lines; for each n, $N = 500$ coin tosses are simulated), together with percentile bootstrap 95%-CIs (with bootstrap sample size $B = 1000$) for this mean. Source: Authors

We see from Figure 8.54 that for $n = 200$ coin tosses we expect the mean length of the longest run of heads to be about 7, perhaps a rather unexpectedly large value. This can be turned into an excellent classroom experiment. Divide a group of students into two parts and ask each student in the first group to write down, without giving it any thought, a sequence of $n = 200$ realistic-looking coin tosses on a card (write down 0 for tails and 1 for heads), with their name (but not the group number) on the back. Ask each student in the second group to do the same, but to generate the 200 tosses of 0s and 1s by throwing an actual coin. Then collect the cards of all participating students and shuffle them. Randomly pick one of the cards and claim to know, with high probability, the group to which the student belongs. As we know from Figure 8.54, we expect to see a run of 1s of length 7 on the card if the sequence of 0s and 1s were generated truly at random. However, asking students to write down a representative sequence of 0s and 1s typically leads to far too many switches between the two numbers, so far too few longer runs of each of the two outcomes. So, by determining the length of the longest sequence of 1s we should be able to identify, with high probability, the correct group.

How large is this probability of identifying the correct group? For this we need to study the distribution of the length of the longest run of heads in n coin tosses. Figure 8.55 shows the estimated probability mass function (left) and distribution function (right) of the length of the longest run of heads in n coin tosses, for $n \in \{50, 100, 200\}$; the probability mass function was estimated by the relative frequency of respective lengths of the longest run of heads based on 5000 replications and the distribution function was computed from the probability mass function, as we did for the geometric distribution in (8.19), for example. We can use the estimated distribution function for $n = 200$ to determine the probability of

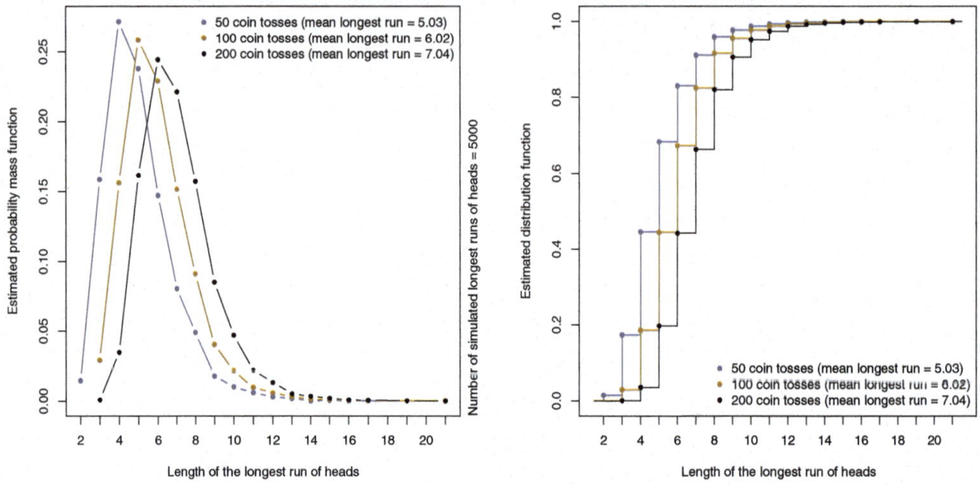

Figure 8.55 Estimated probability mass function (left) and distribution function (right) of the length of the longest run of heads in n coin tosses, for $n \in \{50, 100, 200\}$. Source: Authors

correctly identifying the student's group in the game. For example, if the longest run of 1s on the card is 4, then you know that the probability of seeing such a longest run or a shorter one in a truly randomly generated sequence is about 0.04, so the probability that you correctly identify the student as having written down that sequence manually is $1 - 0.04 = 0.96$ or 96%. If 5 is the length of the longest run of 1s on the card, the distribution function is about 0.2 (see the right-hand side plot of Figure 8.55), so the probability of correctly identifying the corresponding student as having written down that sequence manually is still about 80%.

You may at this point want to revisit the Swiss postage stamp of Jacob Bernoulli and the SLLN for coin tosses in Figure 7.3, the artist apparently having fallen into the same trap when drawing a sample path of $\bar{X}_n$ (the average in the postage stamp jumps below and above the mean 1/2 too often). Another application of this type can be found in high-frequency trading, where one is interested in downward and upward movements (reminiscent of coin tosses) of, say, a certain stock within a short period of time. Again, longer runs of, say, downward movements are not as unexpected as one might think.

A similar story is that of Peacock (2013), where Eloise Peacock tells the story of how in 2003 she beat an online roulette game, making 2000 USD in two days before being barred by the casino. A roulette table consists of 36 numbers and a zero; variations exist, for example, in the US version there are two zeros. The numbers alternate in color: 1 is red, 2 is black, 3 is red, etc. Peacock realized the online roulette game she played was roughly forcing the lower third of numbers (1–12), the middle third (13–24) and the upper third (25–36) each to appear with probability about 33% (with the occasional 0 appearing about 1% of the cases), and this over about 50 spins only (instead of, say, 1000 or more). It implies that there can be at most 33 spins (66%) with numbers from only two of the three dozens in 36; if there were 34 (and so 17 from each of the first two dozens), then there could be at most $50 - 34 = 16$ spins with numbers from the third dozen, which would underrepresent spins from the third dozen in comparison with the other two. Peacock waited until one of the three dozens had not been landed on for eight spins in a row, and then started betting on that dozen to show up, as there were only $33 - 8 = 25$ spins left until the software would make the ball land on the dozen she picked. If losses occurred among one of these 25 spins, she would simply bet more on the next spin in order to compensate for the losses; such strategies over 25 consecutive spins are no longer possible nowadays as (online) casinos have set much lower bet limits. To end this story, we quote from Peacock (2013): "When red has come up many times in a row, black is surely due its turn soon. That is the gambler's fallacy; it has ruined millions." It is perhaps interesting to know that in 1913 in a Monte Carlo casino, the ball landed on the same color 26 times in a row. Again, such events are not as rare as one might think, they happen with probability $2 \times (18/37)^{26} \approx 7.3 \times 10^{-9}$; the multiplication by 2 here comes from the fact that we do not specify a specific color. If a single roulette wheel were spun twice every minute, it would take about 65 years until we expect to see the same color 26 times; see also Figure 8.30 from which we learn that this is a 5.7-sigma event.

Returning to asymptotic CIs and the different variations of bootstrap CIs, the technical details or scenarios under which one method is preferred over another is not important for us in this book; if we encounter them, we thus only speak of asymptotic CIs or bootstrap CIs and omit further details. What is important for us to know is that a $(1 - \alpha)$-CI is a random interval which contains, with probability $1 - \alpha$, the true quantity of interest we are estimating and it allows us to quantify the estimation uncertainty of the said quantity.

In the communication of risk, one often reports relevant risk parameters, such as a reproduction number in the case of COVID-19, or an estimate of the return period of a certain catastrophic event. One should always mention CIs for these parameters. The notion of a CI entails parameter uncertainty both at the level of the statistical model used and at the level of the (always) restricted data available on which to base decisions. This important point is worth mentioning, as in the current hype of artificial intelligence and big data some may believe that such environments get rid of model and data uncertainty.

8.7 Lessons Learned

We have now reached the finish line of a very long chapter. As we suggested before the start of this strenuous hike, we very much hope that you took sufficient pauses along the way. You now are equipped with the basics of the mathematical language used in probability theory. A brief summary recalls what we have achieved. Starting from several historical examples, we introduced the concept of a probability space as developed by Kolmogorov in 1933. In that context, we defined independent events and conditional probabilities. When confronted with this axiomatic set-up for the first time, the less mathematically versed reader might ask "Why make it so difficult?" As we showed through examples, a clear and concise mathematical development helps in avoiding some of the paradoxical traps existing even at the level of easy examples involving chance experiments. We encouraged the drawing of tree diagrams for calculating simple probabilities. We quickly forgot about these mathematical necessities and went from the more abstract notion of a probability space to random variables and their distribution functions. From the latter we gave many examples, from discrete distribution functions (for example, binomial, Poisson) to continuous ones (for example, normal, Pareto). In that realm we then defined all concepts needed further and introduced some workhorses of probability theory: the SLLN and the CLT. In particular we recalled main summary statistics of distribution functions relevant for risk analyses, such as the mean, variance, quantiles (important for the introduction of return periods) and the risk measures value-at-risk and expected shortfall widely used in the financial industry. The various examples given should help to make the transition to some key concepts of mathematical statistics. An important task in the many case studies given throughout our book is the quantification of statistical uncertainty. We achieve this through the introduction of CIs. For the calculation of the latter we introduced you to asymptotic CIs (based on the CLT) and bootstrap CIs (based on resampling methodology). We now continue our hike with a second longer stretch, moving from stochastics on to extreme value theory; see Figure 8.56 for a cartoon interpretation.

Figure 8.56 From Gauss to Gumbel. Source: Enrico Chavez

9

The Modeling of Extreme Events

If you do not expect the unexpected, you will not find it;
for it is unfathomable and difficult.

Heraclitus of Ephesus (c. 535 to c. 475 BCE),
translated by Frank Van Dael

9.1 Heraclitus of Ephesus

Ephesus, or Efes in Turkish, is located at the Turkish coast, just outside present-day Selçuk in the province of İzmir. At Ephesus, around 550 BCE, the great Temple of Artemis was completed. It was considered one of the Seven Wonders of the Ancient World. At the same site, but built much later (c. 110 CE), you can find the remains of the famous Library of Celsus. You may wonder why we mention Heraclitus and indeed, why we quote him. Recall that we started Chapter 1 of this book with the 1953 great flood in The Netherlands and highlighted the important work done by Dutch scientists in the context of the Delta Committee. In particular, we discussed the so-called van Dantzig formula (1.1). In the present chapter, we will dig a little bit deeper in the understanding of that formula. In order to do so, we will rely on the concepts introduced in Chapter 8 and develop a better understanding of the stochastic modeling of extremes and rare events. A fundamental paper that we shall encounter along the way is by the Dutch probabilist Laurens de Haan. This paper (de Haan, 1990) contains an analysis of extreme storm-surge heights along the Dutch coast based on extreme value theory (EVT). It starts with the following quote from Theophrastus (c. 371 to c. 287 BCE) in which he refers to Heraclitus as follows: "The most beautiful order in the world is but a heap of random sweepings" (we thank Frank Van Dael for the translation). Theophrastus was the successor of Aristotle in the latter's school of philosophy. However, Heraclitus is known for other quotes relevant for our book. One of them can be translated, according to Harris (1994), as "It is in changing that things find repose". This quote is very often (somewhat erroneously) translated as "There is nothing permanent except change". Through this 'translation', we do come close to the epitaph on Jacob Bernoulli's tombstone "Although changed, I rise again the same"; see Figure 7.3. The quote by Heraclitus that no doubt is most often mentioned is "You can never step into the same river twice"; this at least is a version of a statement by Heraclitus as it was later referred to by Plato (c. 428 to c. 348 BCE). If you are interested in this quote, in Greek philosophy and in particular in Heraclitus, you should consult Graham (2021). In any case, make sure that next time you are in Rome, you visit the Vatican and admire the 1509–1511 masterpiece "The school of Athens" by Rafaello Sanzio da Urbino (1483–1520), better known as Raphael; see Figure 9.1. The 5 m by 7.7 m fresco depicts numerous Greek

Figure 9.1 "The school of Athens", a fresco by Raphael in the Apostolic Palace, Vatican City. Source: Wikimedia Commons

philosophers, mathematicians and scientists including Aristotle, Euclid, Ptolemy, Pythagoras (see also Section 7.1), Archimedes, Socrates and Diogenes. Interestingly, two philosophers are singled out to be represented by the great Renaissance artists Leonardo da Vinci and Michelangelo. In the top center, standing next to a younger Aristotle, Plato very much looks like Leonardo da Vinci whereas in the bottom center, sitting down and writing, a solitary Heraclitus no doubt resembles Michelangelo. We leave it to you to find a cameo appearance (self-portrait) of Raphael in the fresco.

We now return to the above epigraph by Heraclitus. As is often the case with quotes from antiquity, a modern-day interpretation is seldom unambiguous; the same is true in this case. William Harris (1926–2009) explains the quote via Einstein's quest for the understanding of the nature of light:

> Einstein is recorded as stating that his sole original interest was in the phenomenon of light, which on the basis of faith he felt to be supremely important. This unexpected insight, coupled with his lifelong dogged determination, is what brought him finally in the direction of his major conclusions about light. (Harris, 1994)

We can also view Andrew Wiles' quest towards the solution of Fermat's Last Theorem from Section 7.1 in the context of the above quote. The Nobel Prize in Physics 1921 was awarded to Albert Einstein "for his services to Theoretical Physics, and especially for his discovery of the law of the photoelectric effect." Interestingly, Einstein did not receive the Nobel Prize for his work on the theory of relativity; if you want to read more about why this is so, see Pais (1982). Niels Bohr, whom we encountered in Section 7.3 together with Albert Einstein in the context of their famous discussion on randomness and the universe, was awarded the 1922 Nobel Prize in Physics "for his services in the investigation of the structure of atoms and of the radiation emanating from them".

A somewhat broader interpretation of the quote by Heraclitus can guide us to the heart of EVT and the modeling of extremes and rare events. We have already encountered many examples of extreme events; more will follow. The question we turn to in this chapter is to what extent probability theory and statistics can be used to describe and possibly forecast such events. Often a rare, extreme, event is regarded as unexpected. For instance, the coronavirus pandemic, discussed in Chapter 6, found society at large fully unprepared. On the other hand, researchers worldwide did warn that such a catastrophic event was "around the corner". In that sense, science was "expecting the unexpected". To what extent the consequences of such an event can be sufficiently quantified (and hence becomes "fathomable") is what we want to address in this chapter. Certain rare events can be fully explained and measured. A typical example is given by the longest-sequence-of-runs in Example 8.81. Other extreme events, like 9/11, are difficult, if not impossible to predict and their consequences are very much "unfathomable". The theory (EVT) that we shall briefly discuss in this chapter will guide us to find the line separating fathomable from unfathomable, quantifiable from non-quantifiable. The final word in the quote by Heraclitus, "difficult", will surely be with us along the way. By virtue of the rareness of catastrophic events, a statistical analysis will always be fraught with a lack of data and hence be difficult to perform. Rather than developing the theory, for which many excellent textbooks exist, we shall concentrate on some examples and highlight within these examples the kind of questions that EVT can (or cannot) answer. Titus Flavius Clemens, also known as Clement of Alexandria (c. 150 to c. 215 CE) interpreted the words of Heraclitus as "you have to make a leap of faith in order to understand the truth". We will later see that within EVT, this "leap of faith" in a way corresponds to a fundamental scaling property (see Section 9.4) of the underlying mathematical model.

9.2 Some Early History

An *Act of God* is an accident or event resulting from natural causes without human intervention, and one that could not have been prevented by reasonable foresight or care. For example, insurance companies often consider a flood, a volcanic eruption, an earthquake or a tornado to be an Act of God. From this modern day interpretation, it is not difficult to understand that, historically, extreme (natural) events were for a long time viewed as due to divine intervention and their prevention at best belonged to the realm of superstition. A main aim of this chapter will be to understand, on the basis of EVT, such Acts of God better from a scientific point of view. In particular, how can we build resilience against the consequences of such events? For instance, the construction of a sufficiently high dike (resilience) corresponds to the Act of God of an extreme storm event. Of course, modern environmental and meteorological modeling and prediction has moved "God" further and further away from the "Act" of an extreme event such as a storm.

In Chapter 8, we reviewed the early days of probabilistic modeling, mainly coming from games of chance. When it comes to the modeling of extremes, very few early, historical, examples exist. One of the first such examples was reported in Gumbel (1958, Section 1.0.2), an important textbook to which we will return later. This example was formulated by Nicolaus I Bernoulli in 1709 and concerns the actuarial problem of calculating the mean duration of life of the last survivor among n men of equal age who all die within t years. Bernoulli reduced this question to the following: if n points lie at random on a straight line of length

t, calculate the mean largest distance from the origin; see Figure 9.2. The last survivor clearly corresponds to a problem on extremes, surely not averages. With the notation and very basic theory introduced in Chapter 8, this problem can be solved easily. Indeed, denote the time of death of person i, $i = 1,2,\dots,n$, by X_i and assume, as Nicolaus I Bernoulli did, that $X_1, X_2, \dots \overset{\text{iid}}{\sim} \mathrm{U}(0,t)$, so $X_1, X_2, \dots$ are independent and uniformly distributed on the interval $(0,t)$ with common density function $f(x) = (1/t)I_{\{0<x<t\}}$ (so that $f(x) = 1/t$ for $x \in (0,t)$ and $f(x) = 0$ otherwise). The time of death of the last surviving person is $M_n = \max\{X_1, X_2, \dots, X_n\}$, a random variable we already encountered in Example 8.13 (3). The problem then simply becomes that of calculating the mean $\mathbb{E}(M_n)$. In Example 8.23 we derived the distribution function F_{M_n} of M_n as the nth power of the distribution function of each of the random variables, so that $F_{M_n}(x) = (x/t)^n$, $x \in [0,t]$, and thus the density f_{M_n} of F_{M_n} is its derivative $f_{M_n}(x) = nx^{n-1}/t^n$, $x \in (0,t)$. Using this density in (8.42), we obtain that

$$\begin{aligned}\mathbb{E}(M_n) &= \int_{-\infty}^{\infty} x f_{M_n}(x)\,\mathrm{d}x = \int_0^t xn\frac{x^{n-1}}{t^n}\,\mathrm{d}x = \frac{n}{t^n}\int_0^t x^n\,\mathrm{d}x = \frac{n}{t^n}\left[\frac{x^{n+1}}{n+1}\right]_0^t \\ &= \frac{n}{t^n}\frac{t^{n+1}}{n+1} = \frac{n}{n+1}t,\end{aligned}$$

a result that is intuitively clear. Indeed, omitting the randomness (by putting n points equally spaced on an interval of length 1) leads to the distances $1/(n+1)$ between the points, so the largest is located at $n/(n+1)$; for an interval of length t multiply by t.

In Section 7.2 we have already met Nicolaus I Bernoulli, who published the book *Ars Conjectandi* by his uncle Jacob Bernoulli posthumously in 1713. At this point, you may want to revisit Figure 7.5 for the genealogical tree of the Bernoulli family. The above example of Nicolaus I Bernoulli of an extremal-type question is just what it is, an example. No doubt the first mathematical result in the realm of probability theory addressing rare events is to be found in Poisson's limit theorem of rare events; see Derivation 8.28. In words, it tells us that in n replications of a dichotomous experiment (that is, one with two possible outcomes, say, success or failure), with a small probability p of success, the distribution function of the total number of successes is approximated by a Poisson distribution with parameter $\lambda = np$. Siméon Denis Poisson, see Figure 8.20, is quoted as the scientific father of this result; it will play a fundamental role in our discussion of EVT. This role can be grasped easily. For instance, in Section 1.2 we discussed the construction of the Oosterscheldekering as part of the Delta Works. The sluices are closed as soon as the level of the sea water reaches 3 m above NAP; see the mark on the dam in Figure 1.6. Now think of daily maximal storm surges as generating dichotomous data: 'failure' corresponds to a level below 3 m, 'success' to a

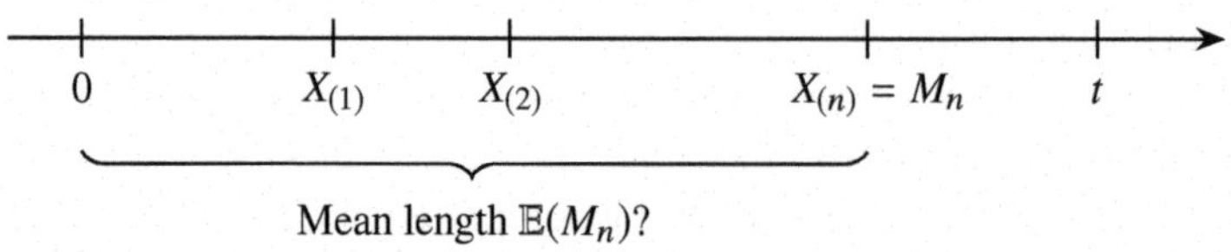

Figure 9.2 Sketch of the problem of Nicolaus I Bernoulli, which was to determine the mean lifetime $\mathbb{E}(M_n)$ of the last of n survivors if all n persons are of equal age and die within the next t years. Source: Authors

level above 3 m (see the last paragraph of Example 8.24 for an explanation for the choice of success/failure). The probability of success of a storm surge above 3 m is then a rare event and hence a Poisson-type limit theorem may be useful here. In fact it will be useful, but this will require some careful data analysis and the checking of necessary conditions for the theorem to hold; more on this in Section 9.4. Whereas Poisson's name will forever be linked to the above theorem, a first textbook treatment by Ladislaus Josephovich Bortkiewicz (1868–1931) (or von Bortkewitsch as he appears on the cover) proved to be very important; see Bortkiewicz (1898) and Figure 9.3. Its title *Das Gesetz der kleinen Zahlen* ("The Law of Small Numbers")

Figure 9.3 Left: Ladislaus Josephovich Bortkiewicz (1868–1931). Right: His book *Das Gesetz der kleinen Zahlen* ("The Law of Small Numbers"). Sources: Wikimedia Commons (left) and SUB Göttingen (right)

is telling and should be contrasted with Jacob Bernoulli's law of large numbers in the *Ars Conjectandi* (a special case of Theorem 8.69 for $B(1, p)$ random variables). Bortkiewicz's name is also closely linked to the famous dataset of the number of deaths due to horse-kicks in the Prussian army; see, for example, Preece et al. (1988).

From the early twentieth century, the stochastic modeling of extremes became more and more an important branch of probability theory and mathematical statistics. We shall meet the key contributors while introducing some of the main models from EVT. We will aim for a close interplay between methodological developments and concrete examples. We have already mentioned the important textbook by Gumbel. It is fair to say that this book has had a considerable influence on the field. It summarizes the knowledge on the modeling of extremes up to the mid twentieth century; of course, at the time, Gumbel was also standing on the shoulders of giants, the names of whom we will encounter soon. The book's numerous applications, mainly to engineering, paved the way for a wider acceptance of EVT techniques in the scientific community. It is no coincidence that, in his book, Gumbel dedicates a chapter to the important scientific work of Bortkiewicz. By now, many textbooks exist on the topic of EVT, which also has its own scientific journal, *Extremes*; see springer.com/journal/10687. Further, specialized conferences are devoted to the topic. From a broader point of view, the biennial Extreme Value Analysis (EVA) conferences bring together specialists from diverse backgrounds. At the end of this section, however, we want to briefly highlight the personal life of Emil Julius Gumbel (1891–1966); see Figure 9.4. As well as being an excellent scientist, Gumbel was a highly engaged citizen who used the

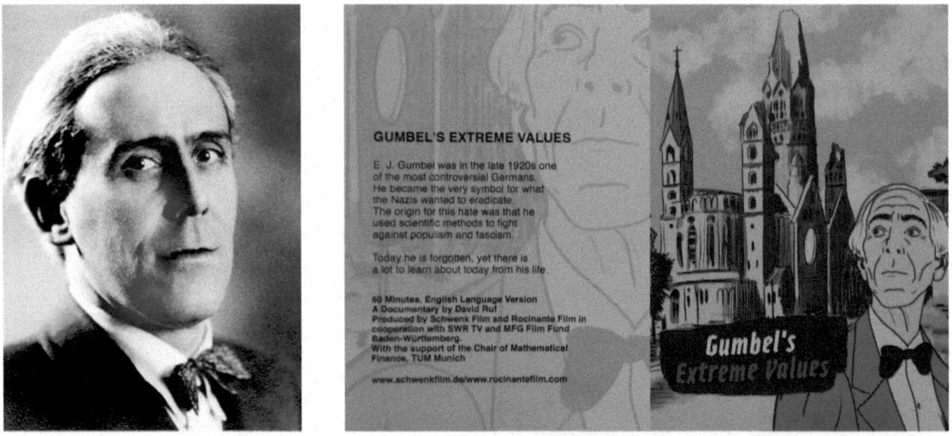

Figure 9.4 Left: Emil Julius Gumbel (1891–1966). Right: An announcement of a documentary about his life (right). Sources: University of Heidelberg (left) and Rocinante Film and David Ruf (right)

power of statistics to its full extent in order to warn German society in the 1920s to 1930s of the upcoming danger of National Socialism (usually referred to as Nazism, from the German "Nationalsozialismus"). As a German Jewish academic, he was forced out of his position at Heidelberg in 1932, after which he and his family first fled to France and then to the USA, where he eventually became a professor at Columbia University; an extensive microfilm collection of Gumbel's political writings is held there. For an overview, see Brenner (1990) and also Brenner (2002). From the latter book's cover we read:

> This biography chronicles the life of professor Emil J. Gumbel (1891–1966) one of Weimar Germany's foremost left-wing intellectuals. A pacifist, socialist, and human rights activist, Gumbel is best remembered for his exposés on political violence and politicized justice in Nazi Germany. A "one-man party", at the center of the most acrimonious political battle in Weimar academia, Gumbel stood alone among his peers courageously speaking out against the Nazis.

To read more about an excellent documentary on the life of Gumbel, see Ruf (2019), an advertisment for which is reproduced in Figure 9.4.

9.3 A Stochastic Theory of Extremes

9.3.1 The Block Maxima Theorem

In many of the examples discussed so far, and indeed several to come, we are faced with the necessity to make scientifically based statements on the occurrence of events that lie at the edge of, or even well beyond, the range of available data. As we discussed in the context of the central limit theorem (CLT; Theorem 8.73), the normal distribution plays a fundamental (indeed a *central*) role when we want to understand the random fluctuations around averages or sums of iid random variables. Extreme value theory, however, is very much concerned with the behavior of extremes, the maxima (or minima) of random variables. In this chapter, we will shift our attention from sums S_n (or averages $\bar{X}_n$) to maxima M_n; see Example 8.13 (3) or Example 8.23, where we have already encountered M_n. By sign reflection, the theory for minima is similar. The story of the 2007–2008 financial crisis, told in Chapter 3, highlights the fact that decisions based on the normal distribution typically lead to serious underestimation of the underlying risks whenever extreme market movements are involved; see also Sections 8.4.3 and 8.4.4. The preface of Embrechts et al. (1997) contains the following relevant quotes by leading researchers Richard Smith and Jonathan Tawn in the field of EVT. Smith says:

> There is always going to be an element of doubt, as one is extrapolating into areas one doesn't know about. But what EVT is doing is making the best use of whatever data you have about extreme phenomena.

And Tawn:

> The key message is that EVT cannot do magic – but it can do a whole lot better than empirical curve-fitting and guesswork [gormless guessing]. My answer to the sceptics is that if people aren't given well-founded methods like EVT, they'll just use dubious ones instead.

These quotes highlight the need for a theory adapted to the analysis of extremes.

Fully recognizing the relevance of the old proverb "The proof of the pudding is in the eating", later in the chapter we will together sample some EVT puddings and very much hope that you will like their taste. A particularly convincing example of EVT's success will be offered in Section 9.8, where we will tell the story of the MV Derbyshire (a ship). But before we can reach that summit, for now we will have to end our rest and resume our hike, this time for some serious climbing in EVT-land.

The methodological task is very similar to our discovery of the normal distribution in the CLT. A key question we want to answer is whether there are well-defined (limit) distribution functions (hence stochastic models) that play a role in the modeling of maxima of iid random variables similar to the central role played by the standard normal distribution N(0, 1) in modeling averages $\bar{X}_n$ and sums S_n. The answer is yes. It will be useful to keep the comparison with the CLT in mind and try to mimic its formulation, but now for the case of maxima. We have already presented in (8.70) the formulation

$$\frac{S_n - n\mu}{\sqrt{n}\sigma} \xrightarrow{\text{d}} \mathrm{N}(0,1) \quad (n \to \infty)$$

of the main statement of the CLT in terms of sums, where $\mu = \mathbb{E}(X_1)$ is the mean and $\sigma = \sqrt{\mathrm{var}(X_1)}$ is the standard deviation of X_1. Writing $a_n = \sqrt{n}\sigma$ and $b_n = n\mu$ we obtain the reformulation that, for iid random variables with finite variance, we can find normalizing constants $a_n > 0$ and b_n such that

$$\frac{S_n - b_n}{a_n} \xrightarrow{\text{d}} \mathrm{N}(0,1) \quad (n \to \infty). \tag{9.1}$$

Hence, standardization of the sum S_n through the mean and standard deviation leads to the appropriate limit. The real magic is that the CLT holds true *for all* underlying distribution functions F with finite variance. So we know an approximate distribution of properly location-scale- (b_n being the location, a_n the scale) transformed sums of (sufficiently many) iid random variables with finite variance even though we don't know F; see also the beginning of Section 8.6 and after Example 8.75, where we made this point. As a shorter reformulation, we say that any distribution function F with finite variance belongs to the *(sum) domain of attraction* of the normal distribution, written $F \in \mathrm{DA}(\mathrm{N}(0,1))$. Moreover, the convergence to the $\mathrm{N}(0,1)$ limit is comparably *fast* (the convergence rate is $1/\sqrt{n}$ irrespective of the particular F, which we have also seen in the left-hand side of Figure 8.50, for example). These two aspects (*for all* and *fast*) imply that the CLT is widely applicable and is one of the reasons why the normal distribution is a ubiquitous model in statistics. Although we can obtain a similar result for maxima (replacing S_n by M_n), unfortunately we lose both of these two advantages (*for all* and *fast*). This is the price one has to pay for moving from questions regarding the center of the data to questions relating to the edge of the data, or even beyond the available data. Already in our short discussion above and surely more so when we discuss the block maxima theorem (Theorem 9.3) below, we have to cut methodological corners. If you are interested in some more strenuous mountain climbing, you can consult Chapters 2 (for S_n) and 3 (for M_n) in Embrechts et al. (1997) for a full mathematical treatment. For an introduction to EVT within the realm of quantitative risk management (QRM), see McNeil et al. (2015), especially Chapters 5 and 16.

The basic methodological task of EVT is as follows. Given the maximum M_n of $X_1, X_2, \ldots, X_n \overset{\text{iid}}{\sim} F$, the question is whether one can find normalizing constants $c_n > 0$ and d_n such that, for a distribution function H of a (non-degenerate) random variable,

$$\frac{M_n - d_n}{c_n} \overset{\mathrm{d}}{\to} H \quad (n \to \infty). \tag{9.2}$$

Compare with (9.1) to see the similarity. In the case of (9.2) we say that F belongs to the *maximum domain of attraction* (MDA) of H and write $F \in \mathrm{MDA}(H)$. The task then becomes to characterize, for a given F, all limit distributions H together with the corresponding normalizing constants c_n and d_n and also, for a given possible limit distribution function H, to determine all $F \in \mathrm{MDA}(H)$. The functions H found as limit distributions in (9.2) will replace the normal distribution when it comes to answering practical questions of the rare-event type. Before we formulate the main result, the block maxima theorem, let us take a rest and consider two examples that make the above task and what $\mathrm{MDA}(H)$ means more concrete.

Example 9.1 (To which MDA does the exponential distribution belong?)**.** Suppose $F(x) = 1 - \exp(-\lambda x)$, $x \geq 0$, is the distribution function of the $\mathrm{Exp}(\lambda)$ distribution with parameter $\lambda > 0$; see Example 8.33. From Example 8.23 we immediately obtain that

$$\begin{aligned}\mathbb{P}((M_n - d_n)/c_n \leq x) &= \mathbb{P}(M_n \leq c_n x + d_n) = F^n(c_n x + d_n)\\ &= \big(1 - \exp(-\lambda(c_n x + d_n))\big)^n.\end{aligned}$$

So far, so good. We now have to find sequences $c_n > 0$ and d_n such that the right-hand side in the above displayed equation converges, for $n \to \infty$, to a (non-degenerate) distribution function H. Taking $c_n = 1/\lambda$ and $d_n = \log(n)/\lambda$ we obtain that $\mathbb{P}((M_n - d_n)/c_n \leq x) = (1 - e^{-x}/n)^n$, which, by Lemma 8.27 (for the special case of the constant sequence e^{-x}), converges for $n \to \infty$ to $\exp(-e^{-x})$ for $x \in \mathbb{R}$. As one can check from the three defining properties of a distribution function in Section 8.3.1, $\exp(-e^{-x})$ is indeed a proper distribution function on the real line, so we obtain $H(x) = \exp(-e^{-x})$, $x \in \mathbb{R}$, in this case. The distribution function H carries the obvious name *double exponential distribution* or *Gumbel distribution* and it is denoted by Λ (instead of H); see Figure 9.5 for a plot of Λ and its density Λ'. Drawing a distribution function is always a bit dull; it just increases from 0 to 1. Its density

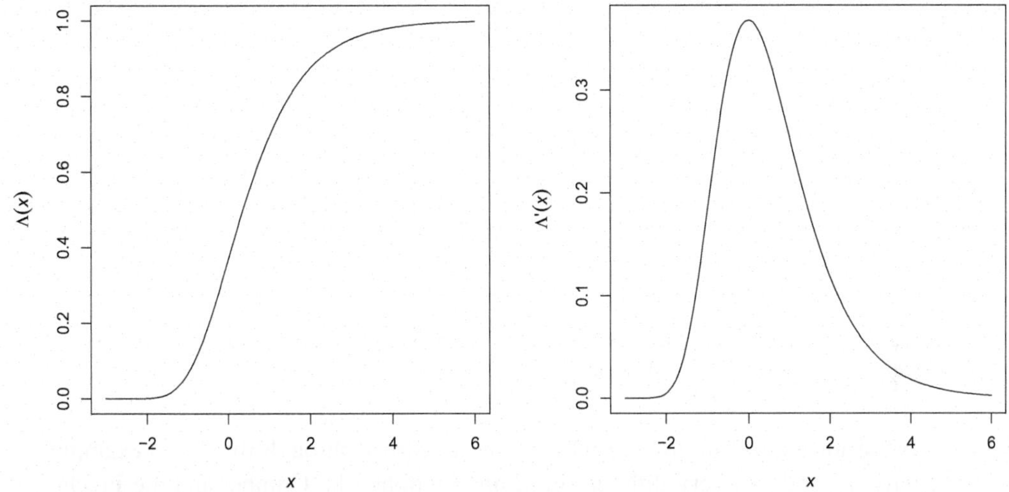

Figure 9.5 Distribution function (left) and corresponding density (right) of the Gumbel distribution. Source: Authors

is more exciting as it already characterizes practical aspects of the underlying problem. For instance, in the case of the Gumbel distribution, the density Λ' is skewed to the right; see also Figure 8.56. Thinking about models for extremes this should be intuitively clear. *Skewness*, as a form of asymmetry, should rule here, and indeed it does. The symmetric normal distribution, which we met in Example 8.34, does not play a role here. We have thus shown that the exponential distribution belongs to the MDA of the double exponential (Gumbel) distribution, or, with a slight abuse of notation, that $\mathrm{Exp}(\lambda) \in \mathrm{MDA}(\Lambda)$.

Example 9.2 (To which MDA does the Pareto distribution belong?). Suppose $F(x) = 1 - 1/(1+x)^\theta$, $x \geq 0$, is the distribution function of the $\mathrm{Par}(\theta)$ distribution with parameter $\theta > 0$; see Example 8.30. As in Example 9.1, we obtain from Example 8.23 that

$$\mathbb{P}((M_n - d_n)/c_n \leq x) = F^n(c_n x + d_n) = (1 - 1/(1 + c_n x + d_n)^\theta)^n.$$

Taking $c_n = n^{1/\theta}$ and $d_n = -1$ we obtain that $\mathbb{P}((M_n - d_n)/c_n \leq x) = (1 - x^{-\theta}/n)^n$, which, by Lemma 8.27, converges for $n \to \infty$ to $\exp(-x^{-\theta})$, for $x > 0$. As one can check from the three defining properties of a distribution function in Section 8.3.1, $\exp(-x^{-\theta})$ is indeed a proper distribution function on the positive real line, so we obtain $H(x) = \exp(-x^{-\theta})$, $x \geq 0$ (with $H(0) = 0$), in this case. The distribution function H is known under the name of the *Fréchet distribution* and it is denoted by Φ_θ (rather than H). Figure 9.6 shows a plot of Φ_θ and its density Φ'_θ for $\theta \in \{2, 1, 1/2, 1/4\}$; in this case, the skewness of the Fréchet limit model becomes even more clear. We have thus shown that the Pareto distribution belongs to the MDA of the Fréchet distribution, or, with a slight abuse of notation, that $\mathrm{Par}(\theta) \in \mathrm{MDA}(\Phi_\theta)$.

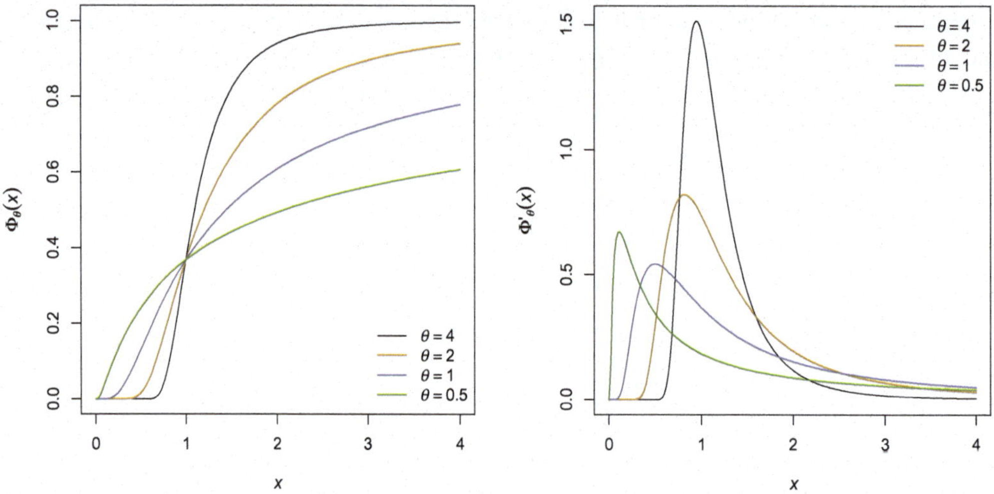

Figure 9.6 Distribution functions (left) and corresponding densities (right) of the Fréchet distribution. Source: Authors

Let us pause for a moment and see what we have learned through these two examples. First, we have derived two very different limit distributions (the Gumbel and the Fréchet) which, on its own, is already very different from the case of sums of iid data with finite variance, for which we would always end up with the normal as a limit distribution. Second, the normalizing constants in both examples are not the mean and standard deviation. This of course begs the question how to find these constants. We will not discuss this issue here; if you are interested, see the examples given in Embrechts et al. (1997, Tables 3.4.2–3.4.4). Third, the way in which we derived these results may look somewhat ad hoc insofar that other choices of normalizing constants could have possibly led to other limits. Let us immediately settle the latter issue: whereas we may fiddle a bit with the precise definition of c_n and d_n and also possibly obtain a somewhat different form of the two limit distributions, it turns out that the resulting limit distributions H can differ only by a location-scale (that is, a linear) transformation of their arguments. One also says that the possible limit distributions H are of the same *type*, so all possible limit distributions are equal up to their type. This is a remarkable result known as the *convergence to types theorem*; see Embrechts et al. (1997, Theorem A1.5). To be more specific, two distribution functions H_X and H_Y are of the same type if there exist $a > 0$ and b such that $H_Y(y) = H_X((y - b)/a)$ for all y; in this case the random variable $Y \sim H_Y$ then allows for a stochastic representation $Y = aX + b$ in terms of the random variable $X \sim H_X$. We have a similar situation for the CLT, where we can also easily obtain as a limit distribution any normal distribution $\mathrm{N}(\mu, \sigma^2)$; it does not have to be the standard normal $\mathrm{N}(0, 1)$. This is possible because we obtain from Derivation 8.35 that, for any μ and $\sigma > 0$, $\mathrm{N}(\mu, \sigma^2)$ and $\mathrm{N}(0, 1)$ are of the same type.

We are soon ready to climb our first serious mountain. For this ascent we need special equipment in the form of a new distribution, the *generalized extreme value distribution* (GEVD). This distribution function is given by the so-called Jenkinson–von Mises representation

$$H_{\xi,\mu,\sigma}(x) = \begin{cases} \exp(-(1+\xi(x-\mu)/\sigma)^{-1/\xi}), & \xi \neq 0, \\ \exp(-e^{-(x-\mu)/\sigma}), & \xi = 0, \end{cases} \tag{9.3}$$

for x such that $1 + \xi(x - \mu)/\sigma > 0$, where $\xi \in \mathbb{R}$ is the *shape* parameter, $\mu \in \mathbb{R}$ the *location* parameter and $\sigma > 0$ the *scale* parameter. Figure 9.7 shows a plot of H_ξ (the standardized version $H_{\xi,0,1}$, which is $H_{\xi,\mu,\sigma}$ for $\mu = 0$ and $\sigma = 1$) and its density h_ξ for $\xi \in \{-1, -1/2, 0, 1/2, 1\}$. Note that, for $\xi < 0$, the GEVD function has a finite right endpoint

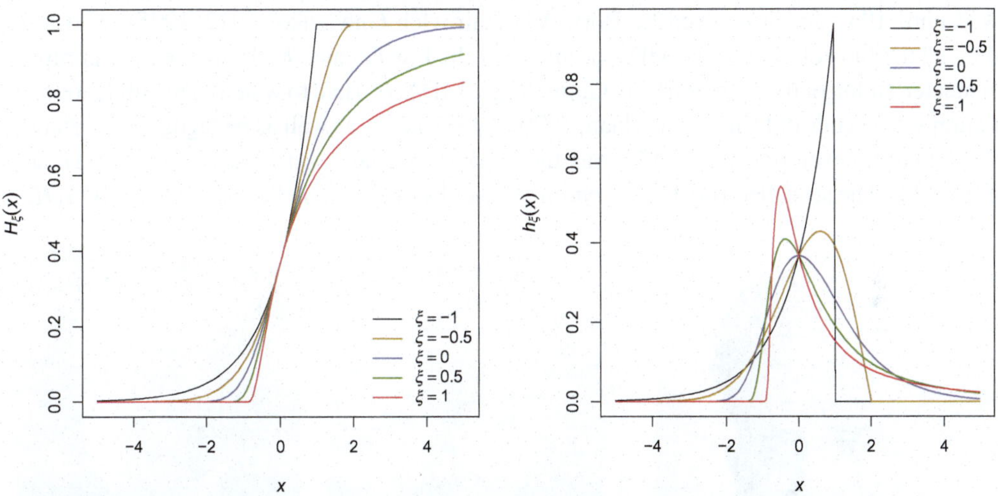

Figure 9.7 Distribution functions (left) and corresponding densities (right) of the GEVD. Source: Authors

$\mu - \sigma/\xi$, and for $\xi > 0$ the GEVD function is unbounded above. The mean of $X \sim H_{\xi,\mu,\sigma}$ exists only if $\xi < 1$ and the variance exists only if $\xi < 1/2$. In general, the larger the shape parameter ξ, the more heavy-tailed the distribution becomes; compare, for example, for different $\xi > 0$, the density values $h_\xi(x)$ for $x \in [2, 4]$ in the plot on the right-hand side of Figure 9.7. From these properties we already learn that the most important parameter of $H_{\xi,\mu,\sigma}$ is the shape parameter $\xi \in \mathbb{R}$. Also important to note is that, whereas μ and σ denote the location and scale parameters, in general they are not equal to the mean and standard deviation of $H_{\xi,\mu,\sigma}$.

The importance of the GEVD becomes clear from the *Fisher–Tippett–Gnedenko theorem.*

Theorem 9.3 (Fisher–Tippett–Gnedenko theorem, block maxima theorem)**.** *Given* $X_1, X_2, \ldots, X_n \overset{\text{iid}}{\sim} F$ *and* $M_n = \max\{X_1, X_2, \ldots, X_n\}$*, suppose that there exist constants* $c_n > 0$ *and* d_n *such that*

$$\frac{M_n - d_n}{c_n} \xrightarrow{d} H \quad (n \to \infty)$$

for a non-degenerate limit distribution H*, then* H *must be* $H_{\xi,\mu,\sigma}$ *for some* $\xi \in \mathbb{R}$*,* $\mu \in \mathbb{R}$ *and* $\sigma > 0$*. In other words, suppose that* $F \in \mathrm{MDA}(H)$ *for some non-degenerate distribution function* H*, then* H *must be a GEVD function.*

Theorem 9.3 goes under various names. We will refer to it as the *block maxima theorem* owing to its connection with the method of the same name that we will introduce in Section 9.3.2. In particular, it will become clear why the "block" appears in the block maxima theorem. Historically, the block maxima theorem goes back to Ronald Aylmer Fisher (1890–1962), the founding father of modern statistical science, and Leonard Henry Caleb Tippett (1902–1985). They formulated their result in 1928, obtaining the three limit distributions corresponding to the cases MDA(H_ξ) for $\xi < 0$ (the so-called *Weibull case*), $\xi = 0$ (the *Gumbel case*) and $\xi > 0$ (the *Fréchet case*). The publication of a rigorous proof had to wait until 1943 and was given by Boris Vladimirovich Gnedenko (1912–1995), who also derived the full solution of the MDA problem for the Fréchet and Weibull cases. A definitive characterization of the more intricate Gumbel case MDA(H_0) had to wait for the publication of Laurens de Haan's PhD thesis (de Haan, 1970). Of scientific, as well as pedagogic importance for the early development of EVT, was the work of Richard von Mises (1883–1953); see Figure 9.8. The so-called *von Mises conditions* yield sufficient conditions for $F \in$ MDA(H_ξ)

Figure 9.8 Left: Maurice René Fréchet (1878–1973). Center: Ernst Hjalmar Waloddi Weibull (1887–1979). Right: Richard von Mises (1883–1953). Sources: Wikimedia Commons

for $\xi < 0$, $\xi = 0$ and $\xi > 0$, respectively, in terms of the density f of F, assumed to exist; see Embrechts et al. (1997, Section 3.3). An excellent account of the important influence of Gnedenko to the field of probability in general and EVT in particular is given in Bingham (2013). We mention these publication dates as, too often, especially within quantitative risk management in finance and risk management in finance and economics, the analysis of risk-related questions is still today anchored all too often in a normal (Gaussian) world, neglecting about 100 years of scientific work going "beyond the normal".

Both the CLT and the block maxima theorem involve a convergence in distribution of location-scale transformed quantities of iid random variables. However, a major difference between the CLT and the block maxima theorem is that, in contrast with the former, in the latter we know the analytical form of the limit distribution, namely (9.3). For the moment we concentrate on the standard GEVD function H_ξ. The somewhat special parameterization of the GEVD in (9.3) is used to obtain $H_0 = \Lambda$ as the limit of H_ξ for $\xi \to 0$. Also, be aware of the fact that throughout the EVT literature different parameterizations occur, so always check for the one that is being used.

As already mentioned, there are three important subclasses of GEVDs, those corresponding to $\xi < 0$, $\xi = 0$ and $\xi > 0$; what can be said about their respective maximum domains of attraction (MDAs)? In Example 9.1 on maxima from an Exp(λ) distribution, we have already seen that $c_n = 1/\lambda$ and $d_n = \log(n)/\lambda$ lead to the double exponential distribution function $H_0 = \Lambda$, corresponding to the GEVD H_ξ with $\xi = 0$. Another example in MDA(H_0) is the normal distribution, thus we can regard it as the distribution of the maximum of a sample of iid N(μ, σ^2) random variables. When properly normalized by sequences $c_n > 0$ and d_n, this distribution can be approximated by a double exponential. The appropriate c_n and d_n can be found in Embrechts et al. (1997, p. 156). Whereas both the exponential and the normal distributions belong to MDA(H_0), the corresponding convergence rates to the double exponential limit (that is, how fast the corresponding distribution functions of $(M_n - d_n)/c_n$ converge to H_0) differ significantly: the convergence is very fast in the exponential case, but very slow for the normal case, another important difference between the CLT and the block maxima theorem. Figure 9.9 illustrates this in terms of the corresponding densities; to see the exact differences we could have plotted differences relative to the Gumbel density h_0, but the difference in the speed of convergence is already visible by just comparing the densities themselves. If F is the Exp(1) distribution function, already for $n = 1000$ the density of $(M_n - d_n)/c_n$

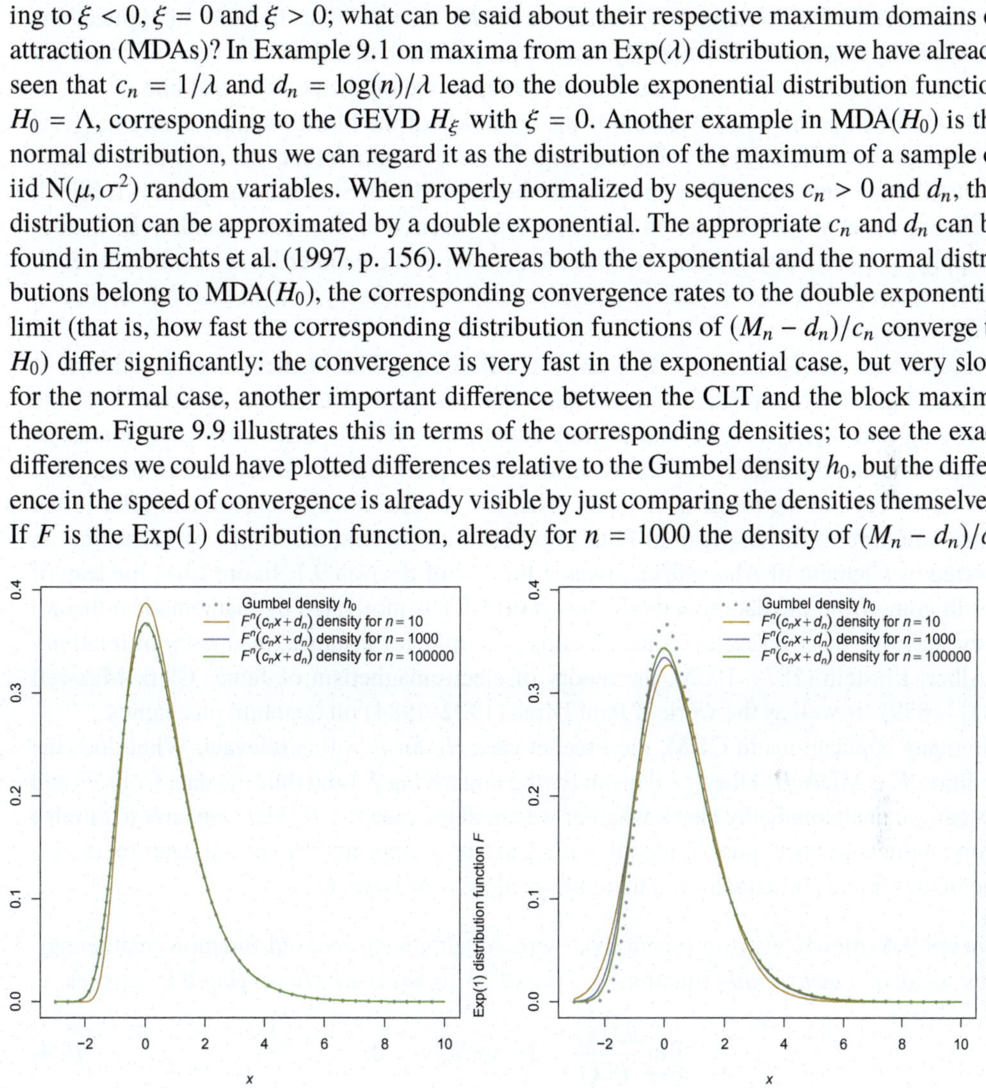

Figure 9.9 Density of the distribution function $F^n(c_n x + d_n)$ for $n \in \{10, 1000, 100\,000\}$ for F an Exp(1) distribution (left) and an N(1, 1) distribution (right), plotted with the Gumbel density h_0. Source: Authors

is indistinguishable from the Gumbel density h_0, whereas if F is the N(1, 1) distribution function we see that the density of $(M_n - d_n)/c_n$ is still quite far away from h_0 (for example, for negative evaluation points) even for $n = 100\,000$; note that Exp(1) and N(1, 1) have the same mean, 1, and variance, 1. In Example 9.2 we encountered a second important subfamily of GEVDs, the Fréchet distribution Φ_θ, which corresponds to H_ξ for parameters $\xi > 0$; note that $\Phi_\theta = H_{1/\theta,1,1/\theta}$. The distribution is linked to Maurice René Fréchet (1878–1973);

see Figure 9.8. And the final GEVD subfamily is H_ξ for $\xi < 0$, the so-called *Weibull distribution*, named after Ernst Hjalmar Waloddi Weibull (1887–1979); see also Figure 9.8. An example in this class is provided by the uniform distributions. In the context of EVT, note that the name Weibull distribution is reserved for H_ξ with $\xi < 0$ as in (9.3). In applications like reliability theory or survival analysis, the name Weibull distribution is typically reserved for a distribution function that is similar, but is supported on the positive real line.

An important point worth stressing is the condition of the block maxima theorem: "[...] suppose that $F \in \mathrm{MDA}(H)$ for some non-degenerate distribution function H". Most continuous distribution functions F belong to one of the three MDAs, so this assumption is fulfilled; see Embrechts et al. (1997, Section 3.1) for a brief discussion. However, standard EVT does not apply to discrete distributions such as the Poisson or the geometric; see Embrechts et al. (1997, Section 3.1) for a brief discussion. Hence a special theory (which exists) is needed in the discrete case.

Assume that $X_1, X_2, \dots, X_n \overset{\text{iid}}{\sim} F$ with $F \in \mathrm{MDA}(H_\xi)$. The mathematical notation $F \in \mathrm{MDA}(H_\xi)$ hides the important link between the unknown distribution function F of the underlying data $X_1, X_2, \dots, X_n$ and a limit model, the GEVD. After having estimated its three parameters, the GEVD will allow us to answer questions on the occurrence of extreme events even beyond the range of the data. This is a first instance where we can refer to the quote "you have to make a leap of faith in order to understand the truth", by Heraclitus as reported by Clement of Alexandria, given at the end of Section 9.1. In our case, the leap of faith in going from the data to a model based on EVT is motivated by mathematical theory. Many scientific discoveries are guided by this principle; for instance, the theory of relativity of Albert Einstein (1879–1955), the theory of electromagnetism of James Clerk Maxwell (1831–1879), as well as the work of Paul Dirac (1902–1984) on quantum mechanics.

In many applications in QRM, the Fréchet case H_ξ for $\xi > 0$ is relevant. What does the condition $F \in \mathrm{MDA}(H_\xi)$ for $\xi > 0$ mean for the underlying F (and thus the data from F) and how can we mathematically check whether we are in the case $\xi > 0$? Here we have to climb a second mountain, once more guided by Gnedenko and supported by special equipment, this time in the form of functions $L(x)$ that change slowly for large x.

Concept 9.4 (Slowly varying functions). A reasonably well-behaved function (mathematicians speak of a *measurable* function) $L : (0, \infty) \to (0, \infty)$ is *slowly varying* if it satisfies

$$\lim_{x \to \infty} \frac{L(tx)}{L(x)} = 1 \quad \text{for all } t > 0. \tag{9.4}$$

Interpreting (9.4) as "for all $t > 0$, $L(tx) \approx L(x)$ for large x", we do indeed see that slowly varying functions L are functions that change slowly. Examples include functions that asymptotically (for large arguments) behave like a constant or a logarithm. The mathematical definition (9.4) of a slowly varying L may look rather innocent; the methodological as well as practical impact, however, is considerable. Slowly varying functions are absolutely crucial throughout a vast range of mathematical sub-disciplines; see Bingham et al. (1987). The theory of slowly varying functions goes back to the Serbian mathematician Jovan Karamata (1902–1967) and his school within the University of Belgrade.

It was essentially the work of Karamata in the 1930s that allowed Gnedenko to give a full solution of the $\mathrm{MDA}(H_\xi)$ problem for $\xi \neq 0$. We focus on the important case $\xi > 0$ and formulate the following theorem, referred to as *Gnedenko's theorem*, as a precise mathematical result, going straight for the top of the mountain. However, we will then pause and allow you to admire the view and will describe what we have achieved.

Theorem 9.5 (Gnedenko's theorem; MDA of the Fréchet distribution). *Let F be a distribution function. Then* $F \in \text{MDA}(H_\xi)$ *for some* $\xi > 0$ *if and only if there exists a slowly varying function L such that, for* $x > 0$,

$$\bar{F}(x) = x^{-1/\xi} L(x).$$

If we briefly forget about the slowly varying function L multiplying $x^{-1/\xi}$ in Theorem 9.5, then the survival or tail function $\bar{F}(x) = 1 - F(x)$ behaves like a power function with so-called *tail index* $1/\xi$ (again be careful with the names given in references that you consult; the tail index is also referred to as the *Pareto exponent*, for example). Power laws play a truly fundamental role throughout the sciences, where their occurrence is ubiquitous. Famous historical examples include the growth of cities (Zipf's law), economics (the Pareto principle; see Example 8.30) and finance (Mandelbrot's scaling laws). In Wikipedia (2021c) you can find that "More than a hundred power-law distributions have been identified in physics (e.g., sandpile avalanches), biology (e.g., species extinction and body mass), and the social sciences (e.g., city sizes and income);" see for instance Andriano and McKelvey (2007).

One way of discovering power-law tail behavior in data is by plotting the empirical survival function $1 - \hat{F}_n(x)$ against x for large realizations x in log–log scale (with both axes in logarithmic scale). If the data support a power-tail behavior, then such a plot should exhibit a downward sloping linear trend. Instead of plotting $1 - \hat{F}_n(x)$ in log–log scale, directly plotting $\log(1 - \hat{F}_n(x))$ versus $\log(x)$ for large realizations x has one particular advantage here: the slope of the linear trend we see in the plot of the points $\big(\log(x), \log(1 - \hat{F}_n(x))\big)$ for large realizations x is the negative tail index $-1/\xi$, so reading off this slope allows us to roughly estimate the negative tail index. Figure 9.10 shows example plots for different distributions with parameter $\theta = 2$, namely a *Pareto type I* distribution function $F(x) = 1 - x^{-\theta}$, $x \geq 1$ (top left), a Par(2) (top right) distribution function and a distribution function $F(x) = 1 - x^{-\theta} \log(e - 1 + x)$, $x \geq 1$ (bottom left). The three corresponding slowly varying functions are $L(x) = 1$ (top left), $L(x) = (1 + 1/x)^{-1/\theta}$ (top right) and $L(x) = \log(e - 1 + x)$ (bottom left). Furthermore, the points $\big(\log(x), \log(1 - \hat{F}_n(x))\big)$ shown in each of these plots are computed for the largest 10% of 10 000 simulated realizations x of the three distribution functions (the largest realization x is omitted since $1 - \hat{F}_n(x) = 0$). The lines shown in the three plots have the negative tail index (so $-\theta = -2$) as slope and were determined in such a way that they go through $\big(\log(q), \log(1 - \hat{F}_n(q))\big)$, where q is the 90% quantile of the 10 000 realizations. We can clearly see that the slowly varying functions $L \neq 1$ in the plot on the top right and the bottom left of Figure 9.10 affect the linearity.

From Figure 9.10 we see how the presence of L can affect our ability to detect power-tail behavior. The presence of some unknown L is also the cause of a considerable amount of methodological headache, with consequences reaching into statistical estimation; see, for example, Embrechts et al. (1997, Example 4.1.12). The plot at the bottom right of Figure 9.10 shows the same kind of tail plot; however, it is now based on the –log-returns of the Netflix data from Example 8.56 (instead of simulated data). Again you see a departure from linearity, hinting at the presence of some unknown L. This is why EVT experts occasionally quote "A slowly varying function L can ruin a risk manager's life."

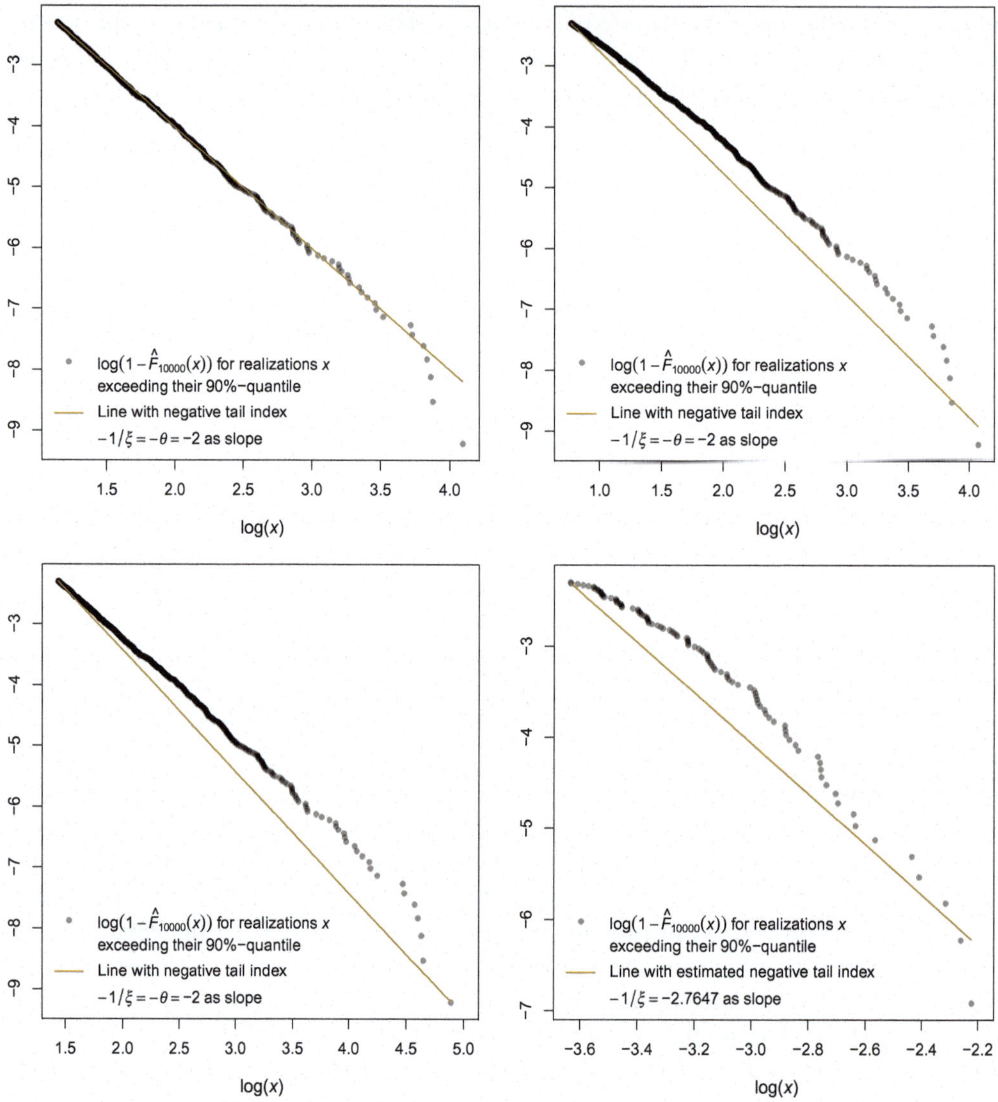

Figure 9.10 Plots of the points $(\log(x), \log(1 - \hat{F}_n(x)))$ together with lines with a negative tail index as slope for the largest 10% of 10 000 simulated realizations x, for a perfect power-tail distribution (top left) and under the presence of a slowly varying function L (top right and bottom left). The bottom right plot is based on the −log-returns of the Netflix data from Example 8.56. Source: Authors

Like the GEVD, the three subclasses $\mathrm{MDA}(H_\xi)$ for $\xi < 0$, $\xi = 0$ and $\xi > 0$ possess some distinct features mainly related to the support of the underlying data model $F \in \mathrm{MDA}(H_\xi)$. For the Weibull case ($\xi < 0$), the right endpoint x_F of F (see Remark 8.48) is finite, so $x_F < \infty$. It is not difficult to show that, for instance, the historical example from Nicolaus I Bernoulli in Section 9.2, from an EVT point of view, corresponds to $\xi = -1$, indeed $\mathrm{U}(0, t) \in \mathrm{MDA}(H_{-1})$. In that case $x_F = t$. For the Gumbel case ($\xi = 0$) we have $x_F \leq \infty$, so both cases, $x_F < \infty$

as well as $x_F = \infty$, may occur, although most standard models belong to the latter category. For the Fréchet case ($\xi > 0$), we always have $x_F = \infty$. This discussion further stresses the delicate underlying conditions that probabilistic models have to satisfy when using EVT. As we shall see in the next section, in a statistical analysis many of these underlying conditions remain hidden. They enter importantly, however, when questions of model uncertainty and robustness are addressed. One question that often surfaces concerns models with infinite support, that is, $x_F = \infty$. Surely all possible measurements are finite and even have a natural upper bound, so that from a practical point of view x_F must be finite. Nevertheless we use an infinite-support normal distribution throughout numerous applications on data that have an obviously finite upper endpoint. For example, think of biological measurements such as length and body mass. Here nobody would question the applicability of the normal. Talking about extremes, where we really want to move into the tails of the distribution, this issue becomes more acute. As a consequence, EVT-based models, like the Fréchet, occasionally are cut off at some high threshold or tapered, replacing the heavy upper tail by a thinner one such as an exponential.

Whereas such approaches may be useful, we advise you to reread the quotes by Richard Smith and Jonathan Tawn at the start of this section. The discussion becomes even more prevalent when an EVT-based analysis leads to an infinite-mean model: this happens in the Fréchet case when $\xi \geq 1$. For instance, for $X \sim \mathrm{Par}(1/2)$ where $\xi = 2$, you can check that $\mathbb{E}(X) = \infty$. Such examples do occur, for example in models for natural catastrophes, climate change, operational and cyber risk, or nuclear accidents. When a sound statistical analysis hints at such models, the risk manager involved has to take a step back and question whether standard economic techniques such as cost–benefit analysis and our understanding of risk aggregation and diversification are still applicable. A relevant example in the context of the economics of climate change was discussed towards the end of Section 8.4.1, when we mentioned Weitzman's dismal theorem. So, when applying EVT, it is always good to have an EVT specialist at hand, ready to assist, when necessary, with the final conclusions and reporting of a statistical analysis involving extreme events.

9.3.2 The Block Maxima Method

In the previous subsection we offered a first glimpse of EVT, a theory that offers probabilistic models that are potentially useful for statistical analyses of extreme events. In this subsection, we will explain how the block maxima theorem enables one to do just that; the term "block" used will then become clear. We start with an iid sample from an unknown distribution function F; we assume that F belongs to some $\mathrm{MDA}(H_\xi)$ so that we can apply the block maxima theorem. Recall that for data coming from a continuous distribution function, imposing an MDA condition is not really restrictive. Later we shall discuss weakening the iid assumption. From a statistical point of view, we are interested in estimating quantities such as a high quantile of F, an associated risk measure (for example, value-at-risk (VaR) or expected shortfall (ES)), the probability of a future record event, a return level for a given frequency of an extreme event or the return period of such an event. In an EVT set-up, a key task concerns the estimation of the all-important shape parameter ξ; from there, we can start answering the above questions. Through the basic setup of the block maxima theorem, estimation of the location-scale parameters μ and σ corresponds to the estimation of the normalizing sequences d_n and c_n, respectively. So we consider from the start that $F \in \text{MDA}(H_{\xi,\mu,\sigma})$. There are essentially two approaches concerning the statistical theory within EVT, the *block maxima method* (BMM) and the *peaks over threshold method* (POTM). The BMM has its origin mainly in applications in hydrology. The POTM has advanced to become the preferred method for the statistical analysis of one-dimensional extremes. We will discuss both. We start with the BMM; for the POTM, see Section 9.4.

You may find the notation a bit heavy going in this section. The figures, as well as the later examples, will however make the BMM set-up clear. As mentioned in Section 8.2, we will occasionally mix notation between data (for example, lower-case x) and model (upper-case X) according to our focus. Whenever truly necessary, we will adhere to notational correctness. The statistical analysis according to the BMM starts from $X_1, X_2, \ldots, X_{n_F} \overset{\text{iid}}{\sim} F$, where $F \in \text{MDA}(H_{\xi,\mu,\sigma})$; see the top left plot of Figure 9.11 for some stylized data. Though this is not important for our purpose here, the data were generated from a $t_{3.5}(6,1)$ distribution conditional on those values exceeding 2; see Example 8.54. Note that, for pedagogic reasons, we use the notation n_F for the (original) sample size here rather than the usual n, as we will use n for the number of samples that we gather to form a block. We partition the n_F data points $X_1, X_2, \ldots, X_{n_F}$ into n_b successive and disjoint blocks of (about) equal size n each (so, approximately, $n_F = n_\text{b} n$; in words, the sample size is the number of blocks times the block size):

$$\underbrace{X_1, X_2, \ldots, X_n}_{\text{block } 1}, \underbrace{X_{n+1}, X_{n+2}, \ldots, X_{2n}}_{\text{block } 2}, \ldots, \underbrace{X_{(n_\text{b}-1)n+1}, X_{(n_\text{b}-2)n+2}, \ldots, X_{n_F}}_{\text{block } n_\text{b}};$$

see the top right plot of Figure 9.11. In each block, we then determine the maximum, and so the *block maxima* $M_{n,1}, M_{n,2}, \ldots, M_{n,n_\text{b}}$; see the bottom left plot of Figure 9.11. In some applications, the block maxima are available from the start, for example if only yearly maxima of some hydrological variable such as river height are recorded. If not, then the block maxima are obtained from the original data $X_1, X_2, \ldots, X_{n_F}$ as just described. The latter is the case, for example, if data are observed over time, say, one observation per day over several years. In this case one often considers monthly block maxima, where the block size $n \in \{28, 29, 30, 31\}$ varies a bit depending on the length of the respective month.

The statistical modeler needs to determine n and n_b. An appropriate choice of the block size n (or, equivalently, the number of blocks n_b) is key. If the block size n is large enough, then the block maxima theorem allows us to approximate the distribution function of each $M_{n,i}$, $i = 1, 2, \ldots, n_\text{b}$, by the three-parameter GEVD $H_{\xi,\mu,\sigma}$. Because of the iid assumption for $X_1, X_2, \ldots, X_{n_F}$, the block maxima $M_{n,1}, M_{n,2}, \ldots, M_{n,n_\text{b}}$ are also iid, so we are in the standard statistical situation of an iid sample (namely $M_{n,1}, M_{n,2}, \ldots, M_{n,n_\text{b}}$) from a common distribution function (namely $H_{\xi,\mu,\sigma}$). The three unknown parameters ξ, μ, σ can then be estimated from the realizations $m_{n,1}, m_{n,2}, \ldots, m_{n,n_\text{b}}$ of $M_{n,1}, M_{n,2}, \ldots, M_{n,n_\text{b}}$, the actual block maxima data; see the bottom right plot of Figure 9.11. The fact that the density $h_{\xi,\mu,\sigma}$ of the GEVD $H_{\xi,\mu,\sigma}$ is known analytically helps in numerically computing the maximum likelihood estimator (see Remark 8.32 (3)). We omit the details here and just mention that such procedures are available in statistical software. We thus obtain estimates $\hat{\xi}_{n_\text{b}}, \hat{\mu}_{n_\text{b}}, \hat{\sigma}_{n_\text{b}}$ of ξ, μ, σ.

This leads us to the unavoidable Achilles heel of the BMM (as well as of any other EVT-based method), the so-called *bias–variance trade-off*. Ideally, both the block size n and the

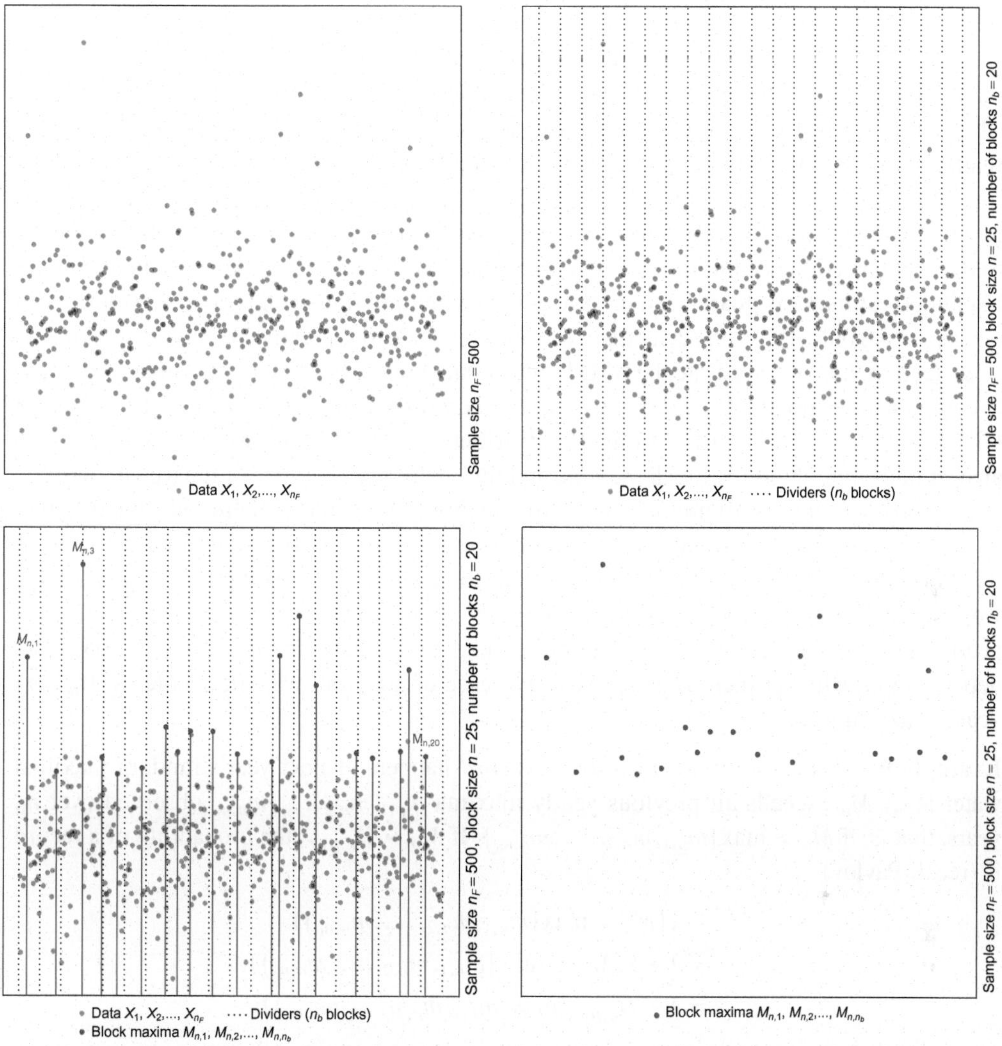

Figure 9.11 Stylized illustration of the BMM. The data (top left), with blocks of equal size (top right), with block maxima (bottom left) and with just the block maxima data used to fit the parameters of the GEVD (bottom right). Note that here we use capital letters in this stylized illustration without axis annotations, to highlight the construction of the random block maxima and their GEVD – in a real application these are the corresponding realized block maxima data. Source: Authors

number of blocks n_b are large; we need a sufficiently large block size n in order for each block maximum $M_{n,i}$ to be approximately distributed according to a GEVD by the block maxima theorem, but we also need a sufficiently large number n_b of block maxima for computing the estimator, for example the maximum likelihood estimator. However, in practice we have a fixed sample size $n_F = n_b n$, so increasing one of n_b or n inevitably decreases the other. If n is too small, $M_{n,i}$ does not follow a GEVD, which may lead to what is referred to as a *bias*. The technicalities are not important; just think of the extreme case $n = 1$ to convince yourself that the distribution of $M_{n,i}$ might be far from a GEVD: for $n = 1$, we have $M_{n,i} = X_i \sim F$ and F can indeed be far from a GEVD. On the other hand, if n_b is too small, then we have only a small number of block maxima for computing the maximum likelihood estimator. This can be shown to lead to a large variance of the estimators of ξ, μ, σ, and thus their realizations, the estimates $\hat{\xi}_{n_b}, \hat{\mu}_{n_b}, \hat{\sigma}_{n_b}$, are affected by a large uncertainty (wide confidence intervals (CIs)). An optimal choice of n (and thus n_b) has to be found and this often requires repetition of the BMM for different choices of n. We leave a discussion of the technical details to the specialized literature; for example, one should be aware of the fact that the maximum likelihood estimator is guaranteed to have good asymptotic statistical properties (for $n \to \infty$ and $n_b \to \infty$) only if $\xi > -1/2$, a range of values one indeed typically encounters in practice.

In Section 9.5.2 we illustrate the BMM for the Dutch sea-level data using standard software to compute the estimates $\hat{\xi}_{n_b}, \hat{\mu}_{n_b}, \hat{\sigma}_{n_b}$ of ξ, μ, σ and corresponding CIs for ξ, μ, σ. The resulting GEVD $H_{\hat{\xi}_{n_b}, \hat{\mu}_{n_b}, \hat{\sigma}_{n_b}}$ is then the estimated distribution function that we will work with to answer questions about the block maxima and thus extreme outcomes of the true underlying but unknown F, the distribution function in which we are actually interested. In particular, from $H_{\hat{\xi}_{n_b}, \hat{\mu}_{n_b}, \hat{\sigma}_{n_b}}$ we can obtain estimates of several relevant statistical quantities, three of which we now consider in more detail. In order to be specific in the applied language used, we assume for the discussion below that the n_b block maxima correspond to the successive yearly maxima of a specific risk, for example, a high water mark; in this case $n \in \{365, 366\}$, depending on whether there is a leap year. For ease of notation, we will denote by M_n next year's maximum.

Example 9.6 (Probability of a record)**.** A *record* happens if next year's random maximal water level M_n exceeds all previous yearly maxima $m_{n,1}, m_{n,2}, \ldots, m_{n,n_b}$ observed over n_b years, that is, if $M_n > \max\{m_{n,1}, m_{n,2}, \ldots, m_{n,n_b}\}$. If $M_n \sim H_{\xi,\mu,\sigma}$, the probability of a record is (recall Pavlov)

$$\begin{aligned}
&\mathbb{P}(M_n > \max\{m_{n,1}, m_{n,2}, \ldots, m_{n,n_b}\}) \\
&= 1 - \mathbb{P}(M_n \leq \max\{m_{n,1}, m_{n,2}, \ldots, m_{n,n_b}\}) \\
&= 1 - H_{\xi,\mu,\sigma}(\max\{m_{n,1}, m_{n,2}, \ldots, m_{n,n_b}\}).
\end{aligned}$$

Since, approximately, $M_n \sim H_{\hat{\xi}_{n_b}, \hat{\mu}_{n_b}, \hat{\sigma}_{n_b}}$, we obtain the estimated probability of a record as

$$1 - H_{\hat{\xi}_{n_b}, \hat{\mu}_{n_b}, \hat{\sigma}_{n_b}}(\max\{m_{n,1}, m_{n,2}, \ldots, m_{n,n_b}\}).$$

Note that all appearing quantities are known: the estimates $\hat{\xi}_{n_b}, \hat{\mu}_{n_b}, \hat{\sigma}_{n_b}$, the corresponding distribution function $H_{\hat{\xi}_{n_b}, \hat{\mu}_{n_b}, \hat{\sigma}_{n_b}}$ and the yearly maxima $m_{n,1}, m_{n,2}, \ldots, m_{n,n_b}$ from the previous n_b years.

Example 9.7 (Return level)**.** In Section 8.4.3, we have already encountered return levels and described them through quantiles. For block maxima, the *return level* $r_{n,p}$ is the level which M_n exceeds with *return probability* p. As such, $r_{n,p}$ satisfies $p = \mathbb{P}(M_n > r_{n,p}) = 1 - H_{\xi,\mu,\sigma}(r_{n,p})$, and thus

$$r_{n,p} = H^{-1}_{\xi,\mu,\sigma}(1 - p).$$

The estimated return level is thus

$$\hat{r}_{n,p} = H^{-1}_{\hat{\xi}_{n_b}, \hat{\mu}_{n_b}, \hat{\sigma}_{n_b}}(1 - p).$$

The (return) probability p is often specified as a return period (see also Section 8.4.2), thus as a one-in-so-many-years event. For example, the 100 year return level asks for $\hat{r}_{n,1/100}$, which is

$$\hat{r}_{n,0.01} = H^{-1}_{\hat{\xi}_{n_b},\hat{\mu}_{n_b},\hat{\sigma}_{n_b}}(0.99) = \hat{\mu}_{n_b} + \frac{\hat{\sigma}_{n_b}}{\hat{\xi}_{n_b}}\big((-\log(0.99))^{-\hat{\xi}_{n_b}} - 1\big);$$

the latter equality follows from (9.3) through inversion, assuming $\hat{\xi}_{n_b} \neq 0$ (which is typically the case).

Example 9.8 (Return period). In Section 8.4.2, we already encountered return periods; see also Example 9.7 where we just spoke of a 1 in 100 years event. For block maxima, the *return period* $k_{n,u}$ of an event $\{M_n > u\}$ with *return probability* $p = \mathbb{P}(M_n > u) = \bar{H}_{\xi,\mu,\sigma}(u)$ is

$$k_{n,u} = \frac{1}{p} = \frac{1}{\bar{H}_{\xi,\mu,\sigma}(u)},$$

which is the mean of a Geo(p) distribution, and so the expected number of Bernoulli trials with success probability p; see Example 8.25. In other words, the return period of $\{M_n > u\}$ is the smallest number of blocks of size n for which we expect to see one block exceeding u. For example, if u is a future high water mark and $n \in \{365, 366\}$ as before, then $k_{n,u}$ is the smallest number of years for which we expect to see a future yearly maximum that exceeds u. Plugging in estimated parameters, we obtain the estimated return period as

$$\hat{k}_{n,u} = \frac{1}{\bar{H}_{\hat{\xi}_{n_b},\hat{\mu}_{n_b},\hat{\sigma}_{n_b}}(u)}.$$

As should be clear, the return level $r_{n,p}$ and the return period $k_{n,u}$ are connected. One can check that $r_{n,1/k_{n,u}} = u$, that is, the return level of an event with return probability $1/k_{n,u}$ (specified through the return period $k_{n,u}$) is u. And $k_{n,r_{n,p}} = 1/p$, so the return period of an exceedance event over $r_{n,p}$ is $1/p$ (specified through the return probability p). Computing probabilities of records, return levels or return periods is an excellent exercise; see, for example, Hofert et al. (2020a, Exercise 5.21) and its solution (Hofert et al., 2020b, Exercise 5.21) in the context of stock market crashes.

All three calculations above yield explicit formulas as functions of parameter estimates based on block maxima data. Further, CIs can be computed using standard statistical software. We stress that the all-important underlying assumption is the iid property of the data. When this does not hold, extra modeling needs to be done before the BMM can be applied. We will return to this very important issue when analyzing Dutch sea-level data in Section 9.5.2.

9.4 Analyzing Extremes through Exceedances

9.4.1 The Peaks over Threshold Method and the Corresponding Theorem

First, a comment about language: a *threshold* u is a (typically large, positive) level above which we observe more extreme data. An *exceedance* is a data point X that lies above u (that peaks over the high threshold u, hence the name *peaks over threshold method*) and the corresponding *excess* is $E = X - u$ (the distance from the threshold u to the exceedance X).

In most cases, data are sampled through time. We thus refer to the time at which a data point X exceeds u as the *exceedance time*.

Figure 9.12 shows the same stylized data as in Figure 9.11 (top left), with a threshold u (top right), the corresponding excesses (bottom left) and just the excesses (bottom right). Figure 9.12 is the equivalent of Figure 9.11 but now for the POTM instead of the BMM.

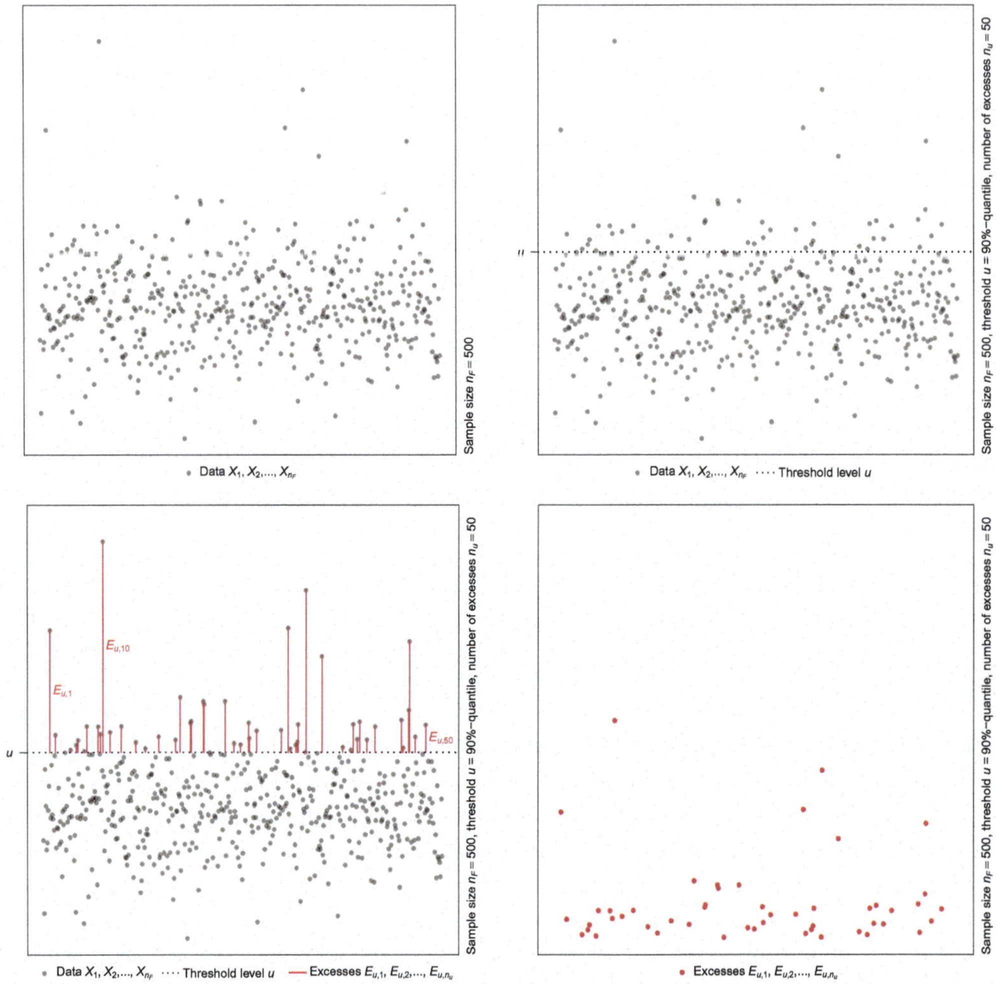

Figure 9.12 Stylized illustration of the POTM. The data (top left; the same as in the top left plot of Figure 9.11), with threshold u (top right), excesses (bottom left) and just the excess data used to fit the parameters of the GPD (bottom right). Source: Authors

The POTM analyzes extreme events using a mathematical model for exceedance times that are over a high threshold and their corresponding excesses. In what follows we describe the POTM steps in detail, including the relevant underlying mathematical results.

We start with a brief comment on the exceedance times. In Section 9.2, we promised to provide a link between the Poisson limit theorem (see Derivation 8.28) and the POTM.

This we can now do. If the data $X_1, X_2, \ldots, X_{n_F}$ (for comparability with the BMM we also use the notation n_F for the (original) sample size here to introduce the POTM) are iid with distribution function F, then the number $N_u = \sum_{i=1}^{n_F} I_{\{X_i>u\}}$ of exceedances of u has a binomial distribution with parameters n_F and $p = \mathbb{P}(X_1 > u) = 1-F(u)$ (so that $N_u \sim \mathrm{B}(n_F, 1-F(u))$); use (8.17) to convince yourself of the correctness of this statement (recall that $I_{\{X_i>u\}} = 1$ if $X_i > u$ and 0 if $X_i \le u$). Assume now that the threshold u is chosen as a function of n_F (that is, $u = u(n_F)$), so that $n_F(1-F(u(n_F)))$ converges to $\lambda > 0$ as $n_F \to \infty$; the latter is the condition "$np \to \lambda$" in Derivation 8.28. By the Poisson limit theorem, the distribution of N_u then converges to the Poisson distribution $\mathrm{Poi}(\lambda)$. So, if n_F and u are sufficiently large, then for iid data, the number of exceedances of u is Poisson distributed. As so often, and indeed also here, the devil lies in the tails, but more on that later.

Now let us turn to the excesses. As already hinted at in the caption of Figure 9.12 and, as in the case of the BMM (where we have the underlying GEVD; see (9.3)), the POTM also depends on a specific distribution function, namely a *generalized Pareto distribution* (GPD), given by

$$G_{\xi,\beta}(x) = \begin{cases} 1-(1+\xi x/\beta)^{-1/\xi}, & \xi \neq 0, \\ 1-\exp(-x/\beta), & \xi = 0, \end{cases} \tag{9.5}$$

where ξ is the *shape* parameter and β is the *scale* parameter. In order to guarantee that $G_{\xi,\beta}$ is a proper distribution function, we require that $\beta > 0$, and we also require that $x \ge 0$ if $\xi \ge 0$ and that $0 \le x \le -\beta/\xi$ if $\xi < 0$. Figure 9.13 shows a plot of $G_{\xi,\beta}$ and its density $g_{\xi,\beta}$ for $\xi \in \{-1, -1/2, 0, 1/2, 1\}$ and $\beta = 1$. As we shall see below, the parameter ξ corresponds

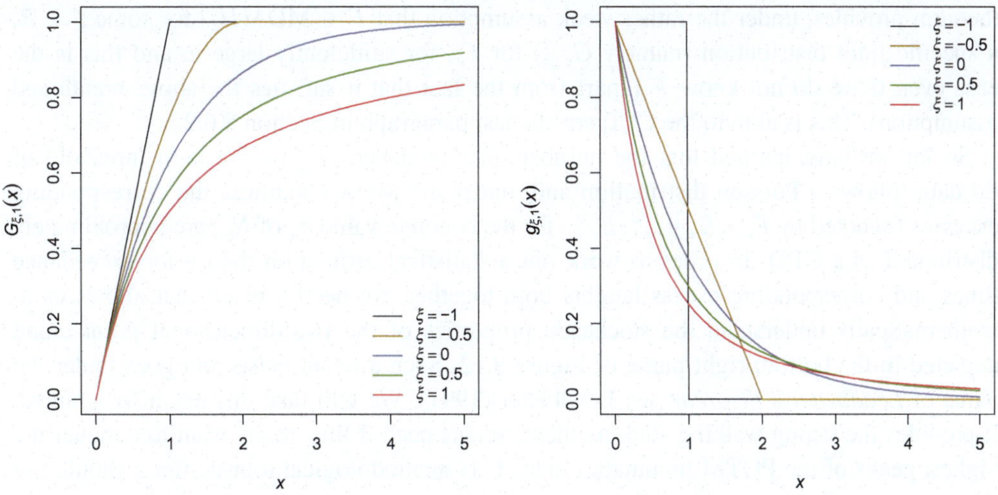

Figure 9.13 Distribution functions (left) and corresponding densities (right) of the GPD in (9.5) for $\beta = 1$. Source: Authors

to the shape parameter in the GEVD and its value is of utmost importance as it largely determines the shape of the GPD. The parameter $\beta > 0$ is a scale parameter. Note that, as in the case of the GEVD, we have an explicit analytical form of the GPD. The following key result underlying the POTM shows that the GPD enters as the distribution of excesses; recall from (8.39) that the excess distribution function is $F_u(x) = \mathbb{P}(X \le x + u \mid X > u)$.

Theorem 9.9 (Pickands–Balkema–de Haan theorem, threshold excess theorem)**.** *A distribution function F is in* MDA(H_ξ) *for some $\xi \in \mathbb{R}$ if and only if there exists a positive, measurable function $\beta(u)$ such that*

$$\lim_{u \to x_F} \sup_{0 \le x < x_F - u} |F_u(x) - G_{\xi,\beta(u)}(x)| = 0.$$

Historically, the threshold excess theorem goes back to a 1974 joint paper by Guus Balkema and Laurens de Haan and a paper published independently in 1975 by James Pickands III. As a consequence, it is typically referred to as the *Pickands–Balkema–de Haan Theorem*. Owing to its connection with the POTM, we will simply refer to it as *threshold excess theorem*. Occasionally, Theorem 9.9 is also called the second theorem of EVT, Theorem 9.3 being the first. It is worth stressing that Gumbel (1958, Section 2.2) has as a title "The Distribution of Exceedances". In that section, Gumbel mainly concentrates on the Poisson character of the exceedance times, not, however, on the excess distribution.

What does the limit in the threshold excess theorem even mean? The innermost part $|F_u(x) - G_{\xi,\beta(u)}(x)|$ is, for a fixed x, the distance between $F_u(x)$ and $G_{\xi,\beta(u)}(x)$, or the error when approximating $F_u(x)$ by $G_{\xi,\beta(u)}(x)$. And $\sup_{0 \le x < x_F - u}$ is a fancy mathematical way to express the maximal such distance for all evaluation points x. Together with $\lim_{u \to x_F}$, we can rephrase the limit in words as "for larger and larger thresholds u, the maximal distance between F_u and $G_{\xi,\beta(u)}$ converges to 0", in other words, for sufficiently large thresholds u, the excess distribution function F_u is well approximated by the GPD $G_{\xi,\beta(u)}$ from Equation 9.5. Now that we understand the threshold excess theorem, we can appreciate it together. The theorem provides, under the rather weak assumption that $F \in$ MDA(H_ξ) for some $\xi \in \mathbb{R}$, a specific limit distribution (namely $G_{\xi,\beta}$) for F_u, for sufficiently large u, and this is the case even if we do not know F (apart from the fact that it satisfies the above-mentioned assumption). This is akin to the CLT; see the last paragraph in Section 8.6.2.

So far we have learned that the number of exceedances N_u over a high threshold of iid data follows a Poisson distribution and, under the MDA condition, the corresponding excesses (denoted by $E_{u,1}, E_{u,2}, \dots, E_{u,n_u}$ for the concrete value n_u of N_u) are approximately distributed as a GPD. In order to work out a statistical estimation theory for exceedance times and corresponding excess heights both together, we need a result that allows us to mathematically understand the stochastic properties of the two-dimensional point cloud depicted in the bottom right panel of Figure 9.12. Such a result exists and goes under the name of *Leadbetter's theorem*; see Leadbetter (1991). We will take this result for granted. It provides the strong walking stick mathematicians need if they really want to conquer the highest peaks of the POTM mountain range. This methodological tour-de-force should not deter us from our more gentle path.

9.4.2 The Peaks over Threshold Tail and Quantile Estimators

On the basis of what we have learned so far we can derive, in a natural way, an estimator of the survival function $\bar{F}(x) = 1 - F(x)$ for $x \ge u$, the high threshold chosen for our POTM analysis. From there we can also obtain an estimator of $F^{-1}(y)$ for $y \in [F(u), 1)$. The derivation below is rather easy and we strongly advise you to walk through the calculations at your own pace. Doing so, you will enjoy the natural beauty not just of the peaks over threshold (POT) landscape, but also why the POT (tail or quantile) estimators we will discover together are also very natural from a mathematical point of view.

Derivation 9.10 (POT tail estimator, POT quantile estimator)**.** Suppose $X \sim F \in \mathrm{MDA}(H_\xi)$. If we use the definition of conditional probability (see (8.2)) and that the excess distribution function is $F_u(x) = (F(x+u)-F(u))/(1-F(u))$, $x \geq 0$ (see (8.38)), we arrive at the following identity, valid for all $x \geq u$:

$$\begin{aligned}\bar{F}(x) &= 1 - F(x) = (1 - F(u))\frac{1 - F(x)}{1 - F(u)} = \mathbb{P}(X > u)\,\mathbb{P}(X > x \mid X > u) \\ &= \mathbb{P}(X > u)\mathbb{P}(X - u > x - u \mid X > u) = \bar{F}(u)(1 - F_u(x - u)). \end{aligned} \tag{9.6}$$

From this identity we can immediately suggest a *plug-in estimator* for the survival function $\bar{F}(x)$ for $x \geq u$ (that is, an estimator for $\bar{F}(x)$ for x beyond the high threshold u) by plugging in estimators for the unknown parts $\bar{F}(u) = 1 - F(u)$ and $1 - F_u(x - u)$ on the right-hand side of (9.6). Since $\bar{F}(u) = \mathbb{P}(X > u)$ is the exceedance probability over u, we can simply estimate $\bar{F}(u)$ by the fraction of points $X_1, X_2, \dots, X_{n_F}$ above u, which equals n_u/n_F. For $1 - F_u(x - u)$, we can replace F_u by the GPD we get from the threshold excess theorem when we plug in estimates $\hat{\xi}_{n_u}, \hat{\beta}_{n_u}$ for the unknown GPD parameters ξ, β. Recall that a hat denotes an estimated quantity, to be computed from the data, and the index (sometimes) used denotes the number of observations from which the estimator is computed. Since the estimate $\hat{\xi}_{n_u}$ of the shape parameter ξ is typically non-zero, we obtain

$$1 - \hat{F}_u(x - u) = 1 - G_{\hat{\xi}_{n_u}, \hat{\beta}_{n_u}}(x - u) = \left(1 + \hat{\xi}_{n_u}\frac{x - u}{\hat{\beta}_{n_u}}\right)^{-1/\hat{\xi}_{n_u}},$$

where n_u is the number of excesses or exceedances (the concrete realization of N_u based on the realizations of $X_1, X_2, \dots, X_{n_F}$). This provides us with an estimator for the second factor $1 - F_u(x - u)$ in (9.6). Putting these two pieces of information together, we obtain the *POT tail estimator*

$$\hat{\bar{F}}(x) = \frac{n_u}{n_F}\left(1 + \hat{\xi}_{n_u}\frac{x - u}{\hat{\beta}_{n_u}}\right)^{-1/\hat{\xi}_{n_u}}, \quad x \geq u, \tag{9.7}$$

for $\bar{F}(x)$, $x \geq u$. Note that this estimator is valid only for $x \geq u$; also it should more correctly be denoted by $\hat{\bar{F}}_{n_u}(x)$, $x \geq u$.

From the POT tail estimator (9.7) we obtain that

$$\hat{F}(x) = 1 - \frac{n_u}{n_F}\left(1 + \hat{\xi}_{n_u}\frac{x - u}{\hat{\beta}_{n_u}}\right)^{-1/\hat{\xi}_{n_u}}, \quad x \geq u.$$

As mentioned before, note that the form of $\hat{F}(x)$ is valid only for $x \geq u$. Setting this equal to y and solving with respect to x leads to

$$\hat{F}^{-1}(y) = u + \frac{\hat{\beta}_{n_u}}{\hat{\xi}_{n_u}}\left(\left(\frac{1 - y}{n_u/n_F}\right)^{-\hat{\xi}_{n_u}} - 1\right), \quad y \in [1 - n_u/n_F, 1); \tag{9.8}$$

again, the range of allowed y-values is important here, as the form of $\hat{F}(x)$ that we used is valid only for $x \geq u$. What we have derived is an estimator of the quantile function of F at y for sufficiently large y. We call $\hat{F}^{-1}$ in (9.8) the *POT quantile estimator*.

Similarly to the beauty behind the interpretation of the threshold excess theorem, the power of the POT tail estimator (9.7) comes from the fact that we now have a model for $\bar{F}(x) = \mathbb{P}(X > x)$ for any $x \geq u$, in particular those x far beyond our largest data point, in a region where, for example, the empirical distribution function (see Section 8.3.1) based on the data points breaks down by being a constant, 1 (so the corresponding survival function is the constant 0) and thus is not useful from a modeling perspective. In contrast, we can use (9.7) and its quantile function (9.8) to answer probabilistic questions that we might have beyond our data. Of course, (9.7) might also not perfectly estimate the tail of F, but at least we have a model that's typically both reasonable (backed by theory) and useful (for answering practical questions involving the tail for $x \geq u$). At a high level, what we have done is to obtain estimates for our model ingredients from within the data (by considering threshold excesses for the threshold u within the range of the data to get estimates of ξ, β and also of $1 - F(u)$) and to express $\bar{F}(x)$ for $x \geq u$ in terms of these estimated ingredients. In this sense, EVT allows us to "look beyond the data", with care, however. This brings us back to the leap of faith statement of Heraclitus and Clement of Alexandria, quoted in Section 9.1.

We derived the POT tail estimator and the POT quantile estimator explicitly in order to show how some basic mathematical theory, in this case from the realm of EVT, leads to the desired result. It also shows how naturally the GPD enters in its formulation. Of course, mathematics is still used heavily in deriving the distributional properties of the estimators in order that we can obtain CIs, for example. And numerical optimization tools within a statistical software environment play an important role for finding the various parameter estimates. This has all been done and is ready to be used in specific applications, as we shall soon discover.

We now address how to compute return levels in the POTM.

Example 9.11 (Return level)**.** As in the BMM, once we have a distribution for the largest values, we can determine the return level as a quantile of the said distribution. In the POTM, these largest values are the exceedances over u and thus the *return level* $r_{u,p}$ is

$$r_{u,p} = F^{-1}(1-p), \quad p \in (0, \bar{F}(u)].$$

The estimated return level can thus be expressed in terms of the POT quantile estimator:

$$\hat{r}_{u,p} = u + \frac{\hat{\beta}_{n_u}}{\hat{\xi}_{n_u}}\left(\left(\frac{p}{n_u/n_F}\right)^{-\hat{\xi}_{n_u}} - 1\right), \quad p \in (0, n_u/n_F]. \tag{9.9}$$

9.4.3 Threshold Choice

In the POTM, the parameter estimates $\hat{\xi}_{n_u}, \hat{\beta}_{n_u}$ are typically obtained using maximum likelihood techniques based on the excess data $E_{u,1}, E_{u,2}, \ldots, E_{u,n_u}$ above the threshold u. For this to work well, we need u to be sufficiently large that we can safely apply the threshold excess theorem. At the same time a lower u yields more excess data (a larger n_u), which is needed to achieve a smaller variance of the estimators of ξ, β, and so a more precise estimation. We also need u to be not too high, so that n_u/n_F is a reasonably good estimator for $\bar{F}(u) = \mathbb{P}(X > u)$.

We find ourselves back at the bias–variance trade-off that we encountered in the BMM. For the POTM to succeed, an optimal choice of threshold u is called for, just as in the case of the BMM, where we needed optimality with respect to the block size n (or the number of blocks n_{b}). For the POTM also, no unique best solution exists. It is good practice to report the results of a POTM analysis across a range of different threshold values. We have now seen twice that the statistical analysis of extreme or rare events does not come for free; surely this was to be (or should have been) expected!

Let us dive a bit further into the ideas underlying the choice of the threshold u. To this end, we would like to mention one more mathematical result, a stability property for the class of GPDs that further stresses the relevance of the GPD within EVT. This stability property (with important applications in practice) allows us to first "go down into the data" and then "estimate the model parameters where sufficient data are available"; this is followed by an estimation of the relevant extremal quantities such as high quantiles which need "moving up in the data". To explain the stability property, assume that the excess distribution function F_u over the threshold u is exactly the GPD $G_{\xi,\beta}$ for some parameters ξ, β; according to the threshold excess theorem we will later apply this result in the case where F_u is only approximately a GPD. It then follows that the excess distribution function F_v above any higher threshold $v \geq u$ is $G_{\xi,\beta+\xi(v-u)}$, so it remains a GPD with the same shape parameter ξ (only the scale parameter changes); see McNeil et al. (2015, Lemma 5.22) for a proof in a similar spirit to that of Derivation 9.10.

This stability property in the shape parameter for high excess distributions forms the basis of an important graphical method that can be used to assist one in determining an optimal threshold above which a POTM analysis can be performed; see the discussion on the sample mean excess plot in McNeil et al. (2015, p. 151). Here the mean excess function e_F of Section 8.4.1 becomes relevant; see (8.45). The conditioning events in the definitions of F_u and e_F (so in $F_u(x) = \mathbb{P}(X \leq x + u \mid X > u)$ and $e_F(u) = \mathbb{E}(X - u \mid X > u)$ for $X \sim F$) highlight the fact that EVT, by definition, is mainly concerned with a *what if* approach to questions related to risk (literally, "what if $X > u$?"). It is not difficult to show that, for heavy-tailed distributions, $e_F(u)$ increases linearly in u, whereas for $X \sim \mathrm{Exp}(\lambda)$ it is constant (namely $e_F(u) = 1/\lambda$; see Remark 8.41 where we mentioned that $\mathrm{Exp}(\lambda)$ has no memory) and for light-tailed distributions (like the normal) it goes to 0. As we are mostly concerned with the case $\xi \geq 0$, the graphical method for choosing a threshold u then proceeds as follows. An estimator $\hat{e}_F(u)$ of $e_F(u)$ is plotted for a range of thresholds u and one chooses as optimal threshold the smallest u (to utilize most data) after which $\hat{e}_F(u)$ becomes linear (that is, where $\hat{e}_F(v)$ for $v \geq u$ looks roughly linear). Note that there are also other graphical tools that can assist in the choice of the threshold u, for example the estimated ξ as a function of the threshold choice can be plotted for various thresholds, and then a threshold can be chosen on the basis of the bias–variance trade-off.

In the discussion following Gnedenko's theorem we mentioned that power laws, hence distributions with power-tail behavior, are ubiquitous in practical applications. A simple comparison between risk estimates based on the (thin-tailed) normal and the (power-law, heavy-tailed) Pareto quickly opens the modeler's eyes to the typically considerable differences between the resulting estimates. The mean excess function e_F offers a magnifying glass to this phenomenon. The fact that, for a Pareto distribution (with finite mean), $e_F(u) = \mathbb{E}(X - u \mid X > u)$ *increases* with the threshold u should deprive a risk modeler, in a Pareto world, of sleep. However large the observed loss $X > u$, the expected excess loss above that u

becomes larger (linearly) as a function of u. By the way, this property characterizes power-law distributions. In contrast, the normal (Gaussian) world has hardly any risk left once we move further out in the tail (beyond the 6-sigma range, say). Generations of risk managers have been trained rather exclusively on the paradigm of normality; the normal distribution has become part of their risk-genetic pool. *Extreme value theory is certainly not a panacea for handling risk!* However, being aware of how the numbers change once you leave the familiarity and (perceived) safety offered by the normal assumption and enter reality, governed by power-law distributions, is definitely helpful. You may want to reread the quotes by Richard Smith and Jonathan Tawn at the start of Section 9.3.1.

9.4.4 BMM versus POTM

Before we discuss some applications of the BMM and the POTM, we want to highlight that the block maxima theorem and the threshold excess theorem are linked through the condition that F belongs to the MDA of a GEVD. This implies that for data satisfying the latter condition, both the BMM and the POTM can be used; the block maxima theorem and the threshold excess theorem share the same ξ. Both methods have their advantages and disadvantages. Which one to use very much depends on the data structure and the questions asked. It is fair to say that currently the POTM is more widely used than the BMM. For a scientifically based comparison of both methodologies, see Ferreira and de Haan (2015). These authors conclude as follows:

> From all these studies, some even with mixed views, the following two features seem dominant. First, POT [the POTM] is more efficient than BM [the BMM] in many circumstances, though needing, on average, a number of exceedances larger than the number of blocks. Secondly, POT and BM often have comparable performances, for example, for large sample sizes.

One way of deviating from the iid property is achieved by mathematicians within the broader context of *stationarity*. At the risk of oversimplification (something mathematicians always dislike; Albert Einstein famously wrote, essentially that “everything should be made as simple as possible, but no simpler”), we try an intuitive explanation. We can think of *stationary* data as data over which one can slide an observation window with a certain width and find that all relevant stochastic data properties remain the same from window to window, and for each chosen window width. Extreme value theory turns out to be still applicable if data are stationary, with the added conditions that observations far apart must be almost independent and also that the data must not show obvious clustering behavior. Keeping Einstein’s quote in mind, check the precise mathematical details in Embrechts et al. (1997, Section 4.4). For sea-level data, “sufficiently far apart independence” is typically violated when data show a long-term trend. The clustering of larger observations within a given storm event certainly poses a problem. For most weather-driven events (storms, heat waves, fires, sea surges, . . .) a key development that drives the EVT modeler out of the comfort zone of stationarity, let alone “iid-ness”, is climate change. We have addressed this issue on numerous occasions throughout the book. In Chapter 10 we will have a somewhat broader discussion on the topic of climate change and climate risk.

Non-stationarities in data can sometimes be handled by pre-analyzing the data, removing trends (*detrending*), removing clusters (*declustering*) or pre-modeling seasonal effects. Both the BMM and the POTM allow for their parameters to depend on other variables, so-called *covariates* (such as time), in order to address non-stationarities. In Section 9.5, we will briefly mention time-dependent GPD parameters in the context of the POTM.

9.4.5 An Historical Comment on Value-at-Risk (VaR)

All the above results were available and well understood by the mid 1980s. Also recall from Section 8.4.2 that the financial risk measure value-at-risk, VaR_α, was introduced for the banking industry in the early 1990s. A question to the industry and regulators was that, given that VaR_α is used and the fact that α is typically chosen close to 1, why not use an EVT-based VaR_α estimate rather than one based on the assumption of an underlying normal distribution? In the iid case, an EVT-based VaR_α estimator can easily be given in terms of the POT quantile estimator (9.8):

$$\widehat{\mathrm{VaR}}_\alpha = \hat{F}^{-1}(\alpha) = u + \frac{\hat{\beta}_{n_u}}{\hat{\xi}_{n_u}}\left(\left(\frac{1-\alpha}{n_u/n_F}\right)^{-\hat{\xi}_{n_u}} - 1\right), \quad \alpha \in [1 - n_u/n_F, 1).$$

Compare the form of this estimator with the one we obtain from the normal assumption (with estimates $\hat{\mu}, \hat{\sigma}$ of the normal parameters):

$$\widehat{\mathrm{VaR}}_\alpha = \hat{\mu} + \hat{\sigma}\Phi^{-1}(\alpha), \quad \alpha \in (0, 1).$$

At the time, the reaction of Wall Street was that the EVT-based formula would not be used as, in contrast with the volatility parameter σ (volatility is another name for standard deviation, typically used in finance), Wall Street has no feeling for the financial interpretation of these mysterious parameters ξ and β. We do understand this reaction but this is not an excuse for using statistical models and formulas which clearly fall short in times of need. That these times came we have unfortunately experienced repeatedly; see Chapter 3. We surely do not claim that an EVT-based VaR measure would have avoided these financial crises: far from it. We do however suggest that some EVT training for those working in banking (and insurance) would be time well spent. Through such training a better awareness of the difference between the *if* and the *what if* might become more tangible.

This all became much more pronounced when the possible replacement of value-at-risk by expected shortfall entered the regulatory discussion for the financial industry. Also, in this case EVT-based *what if* thinking early on led the way. In 1995, a two-day seminar on "A survival kit to quantile estimation" took place at a UBS Quant Workshop in Zurich. In that seminar, the basics of EVT were introduced, as discussed in the previous sections. It was clearly stated that the finance industry should be interested in the excess distribution function F_u and the corresponding mean excess function e_F. The latter corresponds trivially to the expected shortfall risk measure, which became famous through the seminal work of Artzner et al. (1999). An interesting early discussion with practitioners concerned the EVT property that, for $X \sim F \in \mathrm{MDA}(H_\xi)$, $0 < \xi < 1$, one has

$$\lim_{\alpha \uparrow 1} \frac{\mathrm{ES}_\alpha(X)}{\mathrm{VaR}_\alpha(X)} = \frac{1}{1-\xi} > 1;$$

so, for α near 1, it holds that $\mathrm{ES}_\alpha(X) \approx \mathrm{VaR}_\alpha(X)/(1-\xi)$ (with equality for $\xi = 0$). See also Example 8.58 and note that $t_\nu(\mu, \sigma^2)$ is in $\mathrm{MDA}(H_\xi)$ for $\xi = 1/\nu$. To prove this result in its full generality, one needs Gnedenko's theorem and Karamata's theory of slow variation. Recall that, in financial regulation, α is close to 1 so that the above limit reflects the difference in value between the two risk measures rather well. For market risk, it is often the case that $1/4 < \xi < 1/3$, so that the above limit lies between 1.3333 and 1.5. Changing from VaR_α (*if*) to ES_α (*what if*) thus yields a considerable regulatory capital increase of between 33% and 50%. Of course these are just stylized calculations; nevertheless they help in understanding the relevance of EVT-based thinking. Note that Artzner et al. (1999) goes well beyond these numerical values; it yields a set of axioms that any good risk measure ought to satisfy, the so-called *axioms of coherence*. These axioms are violated by VaR_α but not by ES_α.

9.5 EVT at Work

9.5.1 On Data

Thus far we have explained new statistical methodology mostly by using simulated data. Learning from simulated data is indeed important as it shows us that what is predicted in theory actually happens; see also the beginning of Section 8.5 in this regard. Furthermore, in some cases there simply is no other technique available. One finds an important example of this in so-called ensemble weather forecasting, which uses massive simulation models run under slightly varying initial conditions. The remarkable success of modern-day weather forecasting yields ample proof of the relevance of simulation technology; if you are interested in this topic in the context of weather forecasting, we can recommend Blum (2019).

Whereas, in the early days of statistical science, data were rare, currently a data deluge under the umbrella of "big data" is all over the media; one occasionally hears the comment "it is raining data". Various government-related institutes or non-profit organizations provide nice-looking websites with interfaces to download data. A paradise for researchers? Rather rarely. Whether data are available (accessible) or useful for your purpose (convenient) is an entirely different story. Even obtaining the correct historical closing price of Bear Stearns, shown in Figure 3.2, turned out to be challenging (the data available from a well-known public financial database turned out to be nonsensical), even though financial data are typically comparatively simple to obtain. For an historical example of playing the data-detective, see Remark 13.1 in the case of the famous Dutch tulip bubble.

In preparation for this chapter, we explored several possibilities in order to exemplify the use of EVT on real data. Our aim was to go beyond the well-trodden data examples one so often finds in academic publications. The goal we set ourselves was threefold. First, find data that fit well within the stories told in the book; second, the examples should have a strong pedagogical character regarding the possibilities and limitations of EVT; and third, the questions discussed should be sufficiently challenging from a modeling point of view. We had to put a fair amount of work into obtaining the data we present in this chapter, including conversion into formats convenient to work with (more on that later). Also, some data we wanted to analyze were inaccessible or our requests to the maintainers of the respective databases remained unanswered. This becomes a particular problem if you need synchronized data, say measurements at different locations but at the same time points, or anything else that is of a "multivariate" and potentially "high-dimensional" nature – a case we do not even consider in this book (the theoretical treatment of it is already more involved than what was presented in Chapters 8 and 9). In comparison with, say, 10 to 15 years ago, what has changed is that there are many more websites that, on the surface, provide access to data. However, a recurring theme is that those websites often only provide limited data access, and only a small number of variables of interest over only small time periods of interest. Having to manually click through all the settings, download and merge (even with software), say, 200 datasets is far from being fail-safe, besides being frustratingly time consuming. For weather data, this might be fine for a large number of users, as they perhaps only need a current weather forecast. But for statistical analyses in the context of research, typically one would hope for the whole dataset (or at least a large part of it). For a pre-internet view on the need for high-quality statistical data, see Andrews and Herzberg (1985).

The field of EVT meets both the world of big data and the world of small data: "big" because of possibly massive amounts of underlying data, but at the same time "small", as data availability in the range of extremes, where it actually matters, is by definition rare. The "leap of faith" that EVT offers builds a bridge between these two worlds; below we cross this bridge together in terms of several real-life example datasets. These examples come from the realm of natural catastrophes (sea-level and earthquake data) as well as finance (Netflix data). Later on, we add some further examples related to climate change (Chapter 10) and the seventeenth-century tulip bubble (Chapter 13). We will also invite you to participate in some of our detective work corresponding to finding and preparing the data for statistical analyses. By definition, discussions will be application-specific, though we strongly believe that, through them, you will become aware of the fact that they share with most serious statistical applications, that they do not start from a repository of nicely cleaned, iid, data conveniently saved in a ready-to-be-used file on your computer or in some cloud storage, to be analyzed by the push of the right button in a statistical software package.

Remark 9.12 (On the time series character of data and how to model it)**.** One important issue regarding real data is their time series character. A sequence of random variables indexed by time, say $X_{t_1}, X_{t_2}, \dots, X_{t_n}$ for time points $t_1 < t_2 < \dots < t_n$, is called a *time series*. Time series data are almost never independent (X_{t_k} typically depends on the previously observed $X_{t_1}, X_{t_2}, \dots, X_{t_{k-1}}$) and often not identically distributed. For example, consider the plot of the Netflix data on the right-hand side of Figure 8.36, in Example 8.56. Given past values, the variance of the −log-return changes can locally (in different periods of time) be small or large. Our focus in Example 8.56 was not the perfect modeling of the Netflix −log-returns, so we did not address the time dependence there, but it will become relevant in this chapter as well as in Chapter 10. Modeling time series data is statistically a fundamentally different situation from modeling iid data. Indeed, iid data $X_1, X_2, \dots, X_n$ can be viewed as observing the same phenomenon n times with outcomes independent of each other (like rolling a fair die n times; the order is irrelevant in this case). On the other hand, time series data $X_{t_1}, X_{t_2}, \dots, X_{t_n}$ typically arise from observing potentially different phenomena, each only once and typically depending on past observations (so that their order matters).

In what follows we deal with time series data $X_{t_1}, X_{t_2}, \dots, X_{t_n}$. In Section 9.5.4, for example, we decompose such data into two component parts, one component that takes care of the dependence over time and one component that models what is left after removing the dependence on time, that is, the *residuals*. If the model for the time component fits well, the residuals are typically iid, at which point we can apply everything we have learned so far about modeling iid data. For example, we can compute bootstrap CIs of quantities of interest by resampling the residuals. We apply this idea in Sections 9.5.4 and 10.2 for constructing bootstrap CIs; note that such CIs do not take into account the statistical uncertainty in estimating the model for the time component. The model for the original data $X_{t_1}, X_{t_2}, \ldots, X_{t_n}$ then conceptually consists of the model for the time component (for which we use standard models in Sections 9.5.4 and 10.2) and the model for the (iid) residuals. As such, the overall model for $X_{t_1}, X_{t_2}, \ldots, X_{t_n}$ consists of "just" one additional, albeit important, modeling layer on top of iid data.

In the following section on sea-level data, the situation is somewhat simpler. Despite climate change effects over longer time periods, the considered block maxima or peaks over the threshold do not show significant time dependence. We do however assess non-stationarity in various ways not reported. Our analyses will hence be based mainly on an iid assumption. If so, we can then resample the resulting data, estimate the GEVD parameters (for the BMM) or the GPD parameters (for the POTM), and obtain bootstrap CIs. Our main aim throughout the analyses of Section 9.5.2 is to show how the theoretical tools and results from EVT are capable of modeling sea-level data and can answer relevant risk-related questions.

9.5.2 Dutch Sea-Level Data

In Chapter 1 we discussed the 1953 great flood, which mainly impacted severely parts of Zeeland, South Holland and North Brabant in The Netherlands. In response to this disaster, the Dutch government initiated the Delta Works project, leading to a massive reinforcement of the coastal defenses of the affected areas. Two major dike constructions were the Maeslantkering (Figure 1.5) and the Oosterscheldekering (Figure 1.6). In this section, we consider sea-level data at Hoek van Holland, a Dutch town at the mouth of the shipping canal (New Waterway) linking the North Sea to the harbor of Rotterdam; see the right-hand side of Figure 1.1. There are several gauge stations along the western coast of The Netherlands. Our choice of Hoek van Holland is mainly based on its geographic location, its economic relevance for the harbour of Rotterdam and the fact that an analysis of its sea-level data occurs in many publications. Below, we will provide EVT-based answers to questions about the return level of high sea levels at Hoek van Holland, the basic safety question posed in the Delta Report (Deltacommissie, 1961). One such question concerns the construction of dikes withstanding a 1 in 10 000 years flood event.

In our discussion below, we neglect the engineering tasks, thus fully concentrating on the issue of statistical estimation and its uncertainty. In a more involved modeling approach than ours, careful consideration should be given to various factors influencing storm-surge heights. Obvious examples around a storm event are astronomical, tidal and meteorological information. At a longer time scale the influence of climate change will be crucial, as well as possible topographic changes to the sea floor. We will encounter the latter, in a rather surprising way, in our discussion in Section 9.5.1. Consequently, sea-level data show a lot of structure that goes well beyond the iid assumptions underlying standard EVT as discussed so far. In de Haan (1990) these factors and their potential influences are discussed and indeed modeled and compensated for when necessary.

Before we start with our statistical analysis, we would like to stress that lessons learned from the Delta Works have had considerable implications for coastal protection worldwide.

About the Data

In this subsection we describe the quest for the sea-level data at Hoek van Holland and the concrete questions we will answer based on those data. The main reason for sharing the data-gathering story in detail is to underscore our statement made before, namely that statistical data ready to be analyzed do not come free (in the sense of "time and effort invested").

The database Rijkswaterstaat (2021) provides an excellent way to access sea-level data across The Netherlands. Although one does not seem to be able to change the language of the website, one can quickly figure out how to get sea-level data from it. Go to the Rijkswaterstaat website (again see Rijkswaterstaat, 2021) and select the "Expert" tab. Then go to "Waterkwantiteit" and afterwards to "Waterhoogten" (translated as "water heights"). There you find "Waterhoogte berekend Oppervlaktewater t.o.v. Mean Sea Level in cm", "Waterhoogte berekend Oppervlaktewater t.o.v. Normaal Amsterdams Peil in cm", "Waterhoogte Oppervlaktewater t.o.v. Mean Sea Level in cm" and "Waterhoogte Oppervlaktewater t.o.v. Normaal Amsterdams Peil in cm". Be careful, because they all sound somewhat similar. We would like to work with the water heights above Normaal Amsterdams Peil (NAP; the standard we already mentioned in Section 1.1), but is it "Waterhoogte berekend Oppervlaktewater t.o.v. Normaal Amsterdams Peil in cm" or "Waterhoogte Oppervlaktewater t.o.v. Normaal Amsterdams Peil in cm"? Indeed, only the latter are observed data; the former are "berekend" (translated as "calculated") based on astronomical data; we thank a data scientist at Rijkswaterstaat for that clarification and for spotting that we had, initially, accidentally made the wrong choice! After selecting this option, choose "Download meer data" and "Huidige gebruiken". You can then pick a time period. After that, choose "Uit lijst" and go to the list of locations to pick "Hoek van Holland". If you then provide an email address, a request is started and, after approval, you will receive an email with a link to download the data. Easy, right? Well, not quite – although, rest assured, this is one of the cleanest and easiest-to-use databases on environmental data that we have seen.

The problem is that this email request will fail if you selected (roughly) more than a year of data. So you would need to repeat this for every single year of data – and we are interested in all those from 1888 (and, originally, also in more locations, actually). After a request sent via the contact form on the Rijkswaterstaat website, we were sent the data directly after a couple of days. However, a quick plot revealed that the highest water level (3.85 m above NAP at Hoek van Holland; see Section 1.1) was still not to be found in the data. Why? The reason is that the data from Rijkswaterstaat (2021) contain measurements on an equidistant grid; Before 1961 the measurements were at 02:40, 05:40, ..., 23:40 Central European Time (CET); from 1961 to the end of 1970 the measurements were at 02:00, 05:00, ..., 23:00; and from 1971 onward they have been recorded every full hour. And this grid does not contain the exact time point with the largest measurement, 3.85 m above NAP, which was observed on February 1, 1953 at 04:20 CET! As a data scientist from Rijkswaterstaat explained, the data downloadable from the website are only a subset of the data available in the underlying (but not publicly available) database DONAR, which stands for "Data Opslag Natte Rijkswaterstaat". After another request, we were sent the full DONAR dataset for Hoek van Holland from August 1, 1887 to December 31, 2019, measured in centimeters above NAP; in what follows we have converted all heights to meters, interpreted as meters above NAP. These are the data we work with below. Needless to say at this point, without the extensive help from Rijkswaterstaat, we would have been unable to access these data (or perhaps we would have worked with the downloadable data and thus potentially underestimated the risk of high sea levels).

All reported times of sea-level heights are measured with respect to CET, which is UTC+01:00 in terms of Coordinated Universal Time (UTC). And why is the time zone relevant? Because the measurements are not equally spaced and thus should be treated as having the correct time labels, so as a time series. We omit more details here, but data scientists would know that forcing the time zone CET on the data can lead to (here, more than a handful of) missing values, as European countries switch from CET to Central European Summer Time (CEST) on the last Sunday of March (since its (forced) introduction in 1940 and then again in 1977), which means on 02:00 CET clocks switch forward by one hour and thus all times in-between technically do not exist! There are more data-related traps of this sort than you might think; another one that we found and actually reported to Rijkswaterstaat to correct in their database is mentioned below.

Before starting any type of modeling, it is always a good idea to consider a plot of the raw data first in order to spot any distinct features or anomalies. The top left plot of Figure 9.14 shows a line plot (connecting consecutive data points by line segments) of the time series of all measurements from January 1, 1888 to December 31, 2019. Speaking of anomalies, in the DONAR dataset that Rijkswaterstaat sent to us, we immediately spotted from this line plot a measurement of −4.14 m on March 15, 1933 at 23:00 CET. When asking Rijkswaterstaat how such a low value could appear, it turned out that the correct value should have been −0.41 m; and this is of course what we show in the top left plot of Figure 9.14. So a line plot can be quite revealing in such cases. What the line plot over such a large time period is not revealing, however (not even with semi-transparent lines, for example), is what a plot of the raw data reveals, namely that tides affect the iid assumption of the data; see the top right plot of Figure 9.14. This effect can also be seen from the bottom left plot of Figure 9.14, which shows just the years 1951 and 1952. What this zoomed-in plot also reveals is that the winter months (in particular roughly the beginning of November to the end of February) face higher sea levels. Because of the daily tidal effects, one typically considers daily maxima as the raw data for modeling sea-level heights. They are shown in the bottom right plot of Figure 9.14 over the whole time period of available data and serve as the basis for the questions we set out to answer below.

There is more to say about sea-level data than just meets the eye in terms of the above comments on the graphical representations of data. Of course, from the start one knows that sea-level data consist of a so-called astronomical part, mainly driven by the gravitational forces of the Sun and the Moon, and a set-up ("opzet" in Dutch) mainly due to wind-driven factors. One typically has two high tides and two low tides during a so-called lunar day (about 24 h 50 min). The reason for the 50 min is that the Earth's rotation and the Moon's orbit go in the same direction so that a point on the Earth's surface (Hoek van Holland, say) is "chasing" a fixed point on the Moon's surface. Further, by nature, sea-level data are periodic across summer and winter periods since North Sea storms typically happen during the winter months, thus leading to rising sea levels. Finally, over longer periods, as in our case, we may discover trends in the data possibly due to greenhouse effects and climate change; all but the bottom left plot in Figure 9.14 visually hint at such a trend. Such effects need to be modeled.

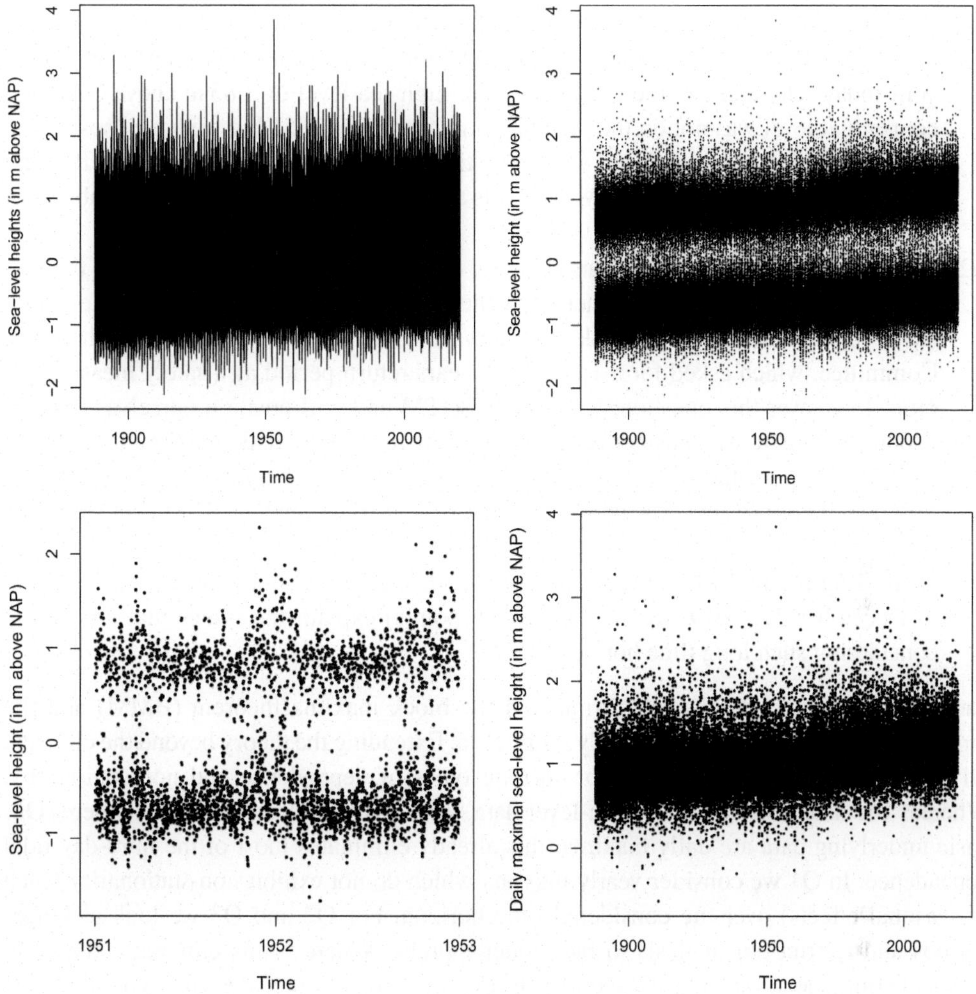

Figure 9.14 Top left: Line plot of the sea-level height measurements (in meters above NAP) as a time series from January 1, 1888 to December 31, 2019. Top right: Scatter plot of these measurements. Bottom left: The data zoomed in on the time period from January 1, 1951 to December 31, 1952. Bottom right: Plot of the daily maxima from January 1, 1888 to December 31, 2019. Source: Authors

In our discussion, we will address some, but not all, of these issues. For a more detailed analysis we will refer to the specialized literature on the topic.

Questions

In what follows, we aim at finding answers to the following questions:

Q1: On Saturday, February 1, 1953 at 04:20 CET the sea level at Hoek van Holland reached its maximal height of 3.85 m above NAP. Let us do the following thought experiment. Suppose it is December 31, 1952, a little more than one month before that catastrophic night. You have at your disposal realizations of daily maximum sea levels $X_1, X_2, \ldots, X_{n_F}$

for Hoek van Holland from January 1, 1888 to ('today') December 31, 1952. From those $n_F = 23\,741$ data points and the BMM (take one year as the block size), how would you 'today' characterize the risk of a future high sea level of 3.85 m above NAP? Of course that level was not on the planning horizon in late 1952 but we have chosen it to assess how likely such an event was at the time. Compute the probability that the next year's (1953's) maximum sea level exceeds all previous ones (that is, the probability of a record; see Example 9.6).

Q2: Recall the 1956 publication (van Dantzig, 1956) containing the formula for the calculation of the necessary dike height at the Dutch coast; see (1.1). What dike height (with corresponding 95% confidence interval) would you have suggested to the Delta Committee, which asked for a 1 in 10 000 years return period as a safety measure? In order to answer this question, we use the POTM and will position ourselves in 1956 ('today'). As data we consider realizations of the $n_F = 8297$ daily maximum sea levels $X_1, X_2, \ldots, X_{n_F}$ of the winter month observations from the beginning of November to the end of February, over the period January 1, 1888 to December 31, 1956 at Hoek van Holland.

Q3: Also using the POTM and the same type of data as in Q2 but from January 1, 1888 to December 31, 2019 ($n_F = 15\,872$ data points) explain the issues that arise from considering such long time horizons.

Remark 9.13 (Beyond iid)**.** We formulated the block maxima theorem (BMM) and the threshold excess theorem (POTM) only for iid data. Extending the theory beyond the classical iid context is possible and allows for specific types of dependence and non-stationarity. Whereas the latter is relevant for sea-level data, we do not discuss these results here. Our basic underlying data are daily maxima; this already eliminates most of the intra-day tidal dependence. In Q1 we consider yearly maxima which do not exhibit non-stationarity (such as seasonal effects) over the considered time horizon. For Q2 and Q3 we look at longer periods and restrict our analysis to the winter months. Whereas this eliminates the main seasonal component, the data still exhibit cluster effects during storm events. The latter can be addressed with appropriate declustering techniques; this we did. As the results obtained did not differ significantly, we decided to not go that extra mile in theory and analysis. At the level of our book, the EVT answers given to Q1–Q3 should be seen as a first illustration of the usefulness of that theory.

Q1: Block Maxima Method

In Section 9.3.2 we explained the details of the BMM. We now apply it to address Q1. Our initial data are thus daily maximal sea-level heights, in meters above NAP, from January 1, 1888 to December 31, 1952 at Hoek van Holland, denoted by $X_1, X_2, \ldots, X_{n_F}$ with $n_F = 23\,741$. The corresponding yearly maxima are $M_{n,1}, M_{n,2}, \ldots, M_{n,n_b}$ with $n_b = 65$.

The realizations of the yearly maxima are shown on the left-hand side of Figure 9.15. Motivated by the block maxima theorem, we fit a GEV distribution to the yearly maxima. The estimates $\hat{\xi}_{n_b}, \hat{\mu}_{n_b}, \hat{\sigma}_{n_b}$ of the GEVD parameters ξ, μ, σ and corresponding bootstrap 95%-CIs are $\hat{\xi}_{n_b} = -0.0857$ with 95%-CI $[-0.29, 0.11]$, $\hat{\mu}_{n_b} = 2.1697$ with 95%-CI $[2.08, 2.26]$ and $\hat{\sigma}_{n_b} = 0.2854$ with 95%-CI $[0.23, 0.36]$. Note that 0 is included in the 95%-CI for ξ, so there is no strong statistical evidence against $\xi = 0$, which results in a Gumbel distribution.

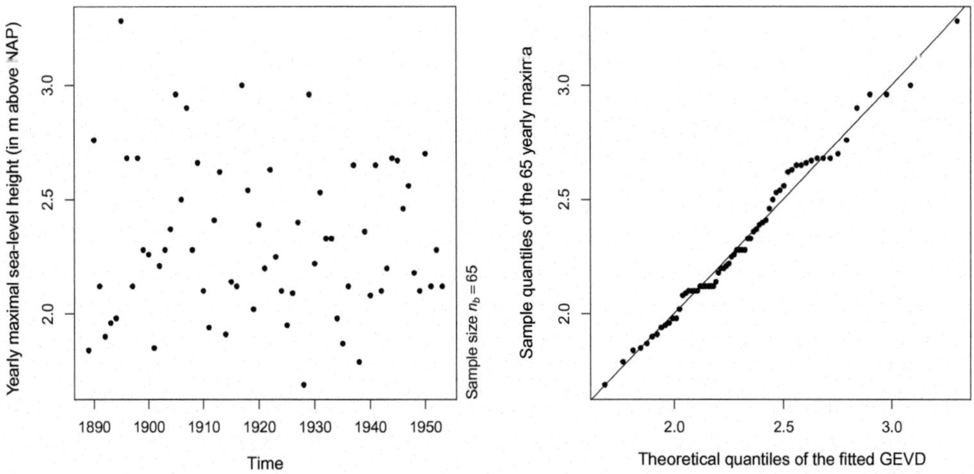

Figure 9.15 Q1: Left: Yearly maximal sea-level heights in meters from January 1, 1888 to December 31, 1952 at Hoek van Holland. Right: Q–Q plot against the fitted GEVD. Source: Authors

This choice also corresponds with the analysis of the Hoek van Holland data used in de Haan (1990).

Whether the GEVD $H_{\hat{\xi}_{n_b},\hat{\mu}_{n_b},\hat{\sigma}_{n_b}}$ with fitted parameters is indeed an adequate representation for the yearly maxima data as shown on the left-hand side of Figure 9.15 is an important next task to assess. To this end we utilize a Q–Q plot; see Section 8.4.4 where we introduced this graphical tool. The Q–Q plot of the yearly maxima against the fitted GEVD $H_{\hat{\xi}_{n_b},\hat{\mu}_{n_b},\hat{\sigma}_{n_b}}$ is shown on the right-hand side of Figure 9.15 and there is indeed no strong evidence to believe that $H_{\hat{\xi}_{n_b},\hat{\mu}_{n_b},\hat{\sigma}_{n_b}}$ is not an adequate model for the yearly maxima from 1888 to 1952. As a final remark on the BMM, the software we used to fit the GEVD parameters also allows these parameters to depend on covariates. We therefore also fitted a GEVD with parameters depending on the year as covariate but found no strong evidence against the hypothesis that the GEVD parameters are constant over the time period from 1888 to 1952.

Now that we have a model for yearly maxima we can use it to answer Q1. The probability of a record is computed as in Example 9.6 and we obtain 0.0088, so there is about a 1% chance of seeing a record in the year 1953 (which indeed happened). The corresponding bootstrap 95%-CI constructed as described in Remark 9.12 is [0, 0.0272], and so has an upper CI endpoint of about 3%.

Q2: Dike Height Estimate

We now turn to Q2. As storms at the location of Hoek van Holland typically happen over the winter months, we restrict our analysis to the months November until February from January 1, 1888 to December 31, 1956, which provides us with $n_F = 8297$ daily maxima $X_1, X_2, \ldots, X_{n_F}$. We proceed with the modeling according to the POTM. As we briefly discussed in Remark 9.13 above, we could have added a declustering step here; see de Haan (1990).

The first step within the POTM concerns the choice of a sufficiently large threshold u, that is, *sufficiently large* that we can apply the GPD approximation from the threshold excess theorem, see Theorem 9.9, but *not too large*, in order to end up with enough excesses for the estimation of the GPD parameters. As we briefly discussed when we introduced the POTM, the choice of the threshold u is important, and indeed a more detailed reporting should be made across various values of u. The work of de Haan (1990, Section 3) details how the data used in the Delta Report were selected. In particular, the winter months November to January were used, augmented with additional flood events outside this period (for example, the great flood from February 1, 1953). Even though the BMM was used as the modeling approach, the data were thresholded at 1.70 m above NAP (which corresponds to the 96.20% quantile of the data). Further (de Haan, 1990, Section 3), a certain subset of that thresholded, augmented, data was then picked based on the highest tides during severe storms. Our preliminary analysis came up with a slightly larger threshold value, $u = 1.83$ (so, 1.83 m above NAP), which corresponds to the 97.5% quantile of the data we are considering. This threshold u results in $n_u = 208$ excesses $E_{u,1}, E_{u,2}, \ldots, E_{u,n_u}$; in particular, note the much larger sample size $n_u = 208$ in comparison the value $n_b = 65$ for the BMM. The corresponding realized excesses are shown on the left-hand side of Figure 9.16. The estimates $\hat{\xi}_{n_u}, \hat{\beta}_{n_u}$ of

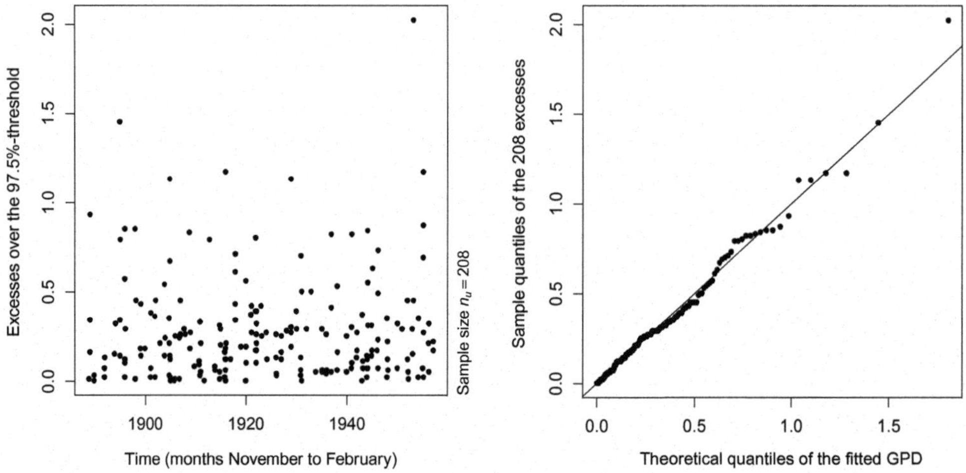

Figure 9.16 Q2: Left: Excesses in meters over the 97.5%-quantile as threshold u (1.83 m above NAP) for the daily maxima of the winter months from January 1, 1888 to December 31, 1956 at Hoek van Holland. Right: Q–Q plot against the fitted GPD. Source: Authors

the GPD parameters ξ, β and corresponding bootstrap 95%-CIs (computed as described in Remark 9.12) are $\hat{\xi}_{n_u} = 0.0395$ with 95%-CI $[-0.09, 0.26]$ and $\hat{\beta}_{n_u} = 0.2656$ with 95%-CI $[0.21, 0.32]$. In comparison with Q1, where $\hat{\xi}_{n_b} = -0.0857 < 0$, we now obtain a positive shape parameter estimate $\hat{\xi}_{n_u} = 0.0395 > 0$. Of course, the data used for the current analysis are not only different; they now include the extreme 1953 event. Note however that for both Q1 and Q2, 0 is contained in the 95%-CI of the shape parameter (of the GEVD in the BMM for Q1 and of the GPD in the POTM for Q2). As in the case of Q1, we now turn to the issue of assessing how well our model fits the data. The right-hand side of Figure 9.16 shows the Q–Q plot against the fitted GPD. There is no strong evidence to believe that the GPD with parameter estimates $\hat{\xi}_{n_u}, \hat{\beta}_{n_u}$ is not an adequate model for the excesses of the winter months from January 1, 1888 to December 31, 1956. We also fitted a GPD with parameters depending on the day of the excess as covariate, but found no strong evidence against the hypothesis that the GPD parameters are constant over time.

So, overall, the GPD model based on the POTM yields an adequate model for the excess sea levels over the threshold of 1.83 m above NAP at Hoek van Holland. We are thus ready to tackle the dike height problem. For this, we want to estimate the 1 in 10 000 years return level based on (9.9). Since we are considering only the four winter months (about 120 days) of a year, we have to calculate (9.9) for $p = 1/(10\,000 \times 120) \approx 8.3333 \times 10^{-7}$ (a 1 in 10 000 years event corresponds to a 1 in 120 000 winter-days event). Plugging in the estimated parameter values, we obtain a return level estimate $\hat{r}_{u,p}$ of 5.2057 (about 5.21 m above NAP) with corresponding 95%-CI [3.96, 11.76]. The estimate is close to the 5.13 m mentioned in the Delta Report (see p. 24 in the van Dantzig Report; de Haan (1990, p. 48) mentions 5.14 m), even though different data and the BMM were used in the analysis for the Delta Report. Taking into account uncertainty bounds for various input parameters in his formula, van Dantzig (1956) reported the value 6.73 m for Hoek van Holland; see also Section 1.3. This is well within the 95%-CI we obtained for a 1 in 10 000 years event. As we quoted in Section 1.3, van Dantzig himself found this too pessimistic and concluded that roughly 6.00 m was "a reasonable estimate of a sufficiently safe height". As we know, the Delta Committee in the end decided on 5.00 m above NAP.

For a more complete picture, we also considered larger values of the return probability p and plotted the estimated return level as a function of p. This *return level plot* is shown in Figure 9.17. It first and foremost shows the return level estimate $\hat{r}_{u,p}$ from (9.9) as a function of p (solid black line), together with bootstrap 95%-CIs for $r_{u,p}$ at each p (gray region), constructed as described in Remark 9.12. Additionally, the plot shows the decreasingly sorted exceedances of u (black dots) plotted against their empirical probabilities, thus against (a slightly adjusted version of) $1/n_F, 2/n_F, \ldots, n_u/n_F$. The intuition behind these points is that, purely empirically, an event of the height of the kth largest exceedance is expected with empirical probability k/n_F. That such "empirical thinking" breaks down rather soon (beyond the largest exceedance, to be more precise) is clear from Figure 9.17. In contrast, the return level estimate $\hat{r}_{u,p}$ can provide answers well beyond the data (that is, for return probabilities p smaller than $1/n_F = 1/8297$). Although we did not ask for it specifically as part of Q2, being at the end of 1956, you might wonder what the probability of a record is, in other words the probability $\mathbb{P}(X > 3.85)$ where X is a random daily maximum sea level in a winter month. This is precisely given by $\hat{\bar{F}}(3.85)$ and turns out to be 0.0032% with 95%-CI [0.0000, 0.0002]. Although this seems to be a small probability (note the percentage sign), let us think about how this translates into the probability of the event of seeing at least one such exceedance (which would then be a record) over the next n years. We need to scale to the time period used for modeling, so "the next n years" means the "the next $120n$ winter days". We thus obtain:

$$
\begin{aligned}
&\mathbb{P}(\text{"the event happens at least once in the next } n \text{ years"})\\
&= \mathbb{P}(\text{"the event happens at least once in the next } 120n \text{ winter days"})\\
&= 1 - \mathbb{P}(\text{"the event does not occur in any of the next } 120n \text{ winter days"})\\
&= 1 - \prod_{i=1}^{120n} \mathbb{P}(\text{"the event does not occur on winter day } i \text{ from now on"})\\
&= 1 - (1 - \hat{\bar{F}}(3.85))^{120n};
\end{aligned}
$$

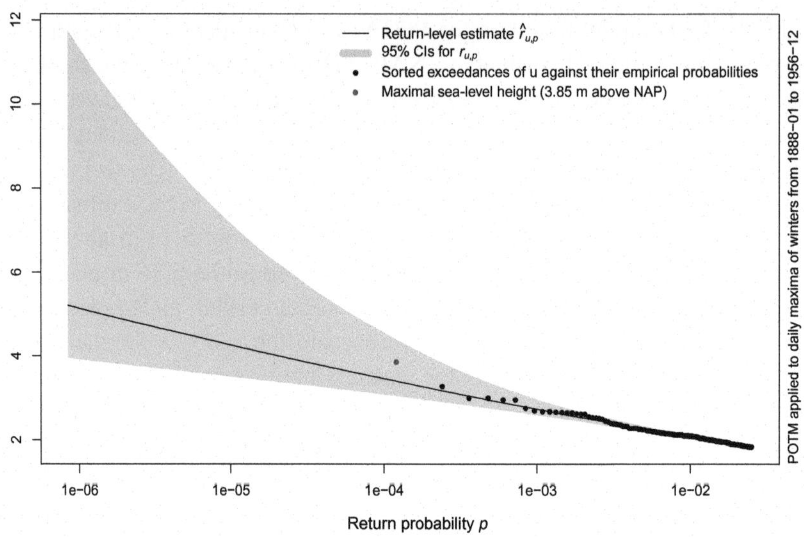

Figure 9.17 Q2: Plot of the return level estimate $\hat{r}_{u,p}$ from (9.9) as a function of p, together with bootstrap 95%-CIs for $r_{u,p}$ constructed as described in Remark 9.12. Included are the decreasingly sorted exceedances over the threshold $u = 1.83$ (97.5%-quantile of the daily maxima of the winter months from January 1, 1888 to December 31, 1956 at Hoek van Holland) plotted against their empirical probabilities. For a 1 in 10 000 years event (so that $p = 1/1\,200\,000$, the smallest p shown), we have $\hat{r}_{u,p} = 5.2057$ with 95%-CI [3.96, 11.76]. Source: Authors

see also Example 8.25 and Derivation 8.26, and note how we used the assumption of independence in going from the third to the fourth line and the assumption of an identical distribution from the fourth to the fifth. Given the rise in sea levels due to climate change (see also Q3), this assumption may yield a conservative estimate. Figure 9.18 shows this estimated probability of seeing at least one exceedance event of 3.85 m above NAP (that is, the probability of seeing at least one record) over the next n years as a function of n, including bootstrap 95%-CIs constructed as described in Remark 9.12. As an example, we obtain a probability of 0.3835% (with bootstrap 95%-CI [0, 0000, 0.0178]) of such a record happening over the next year and a probability of 32% (with bootstrap 95%-CI [0.0004, 0.8329]) of such a record happening over the next 100 years.

Q3: Future Events

Let us now consider Q3 from earlier in this subsection. The left-hand side of Figure 9.19 shows a plot of the $n_F = 15\,872$ available daily maxima in the winter months November to February of the years 1888 to 2019. In contrast with Q1 and Q2, over such a long time period the trend in the data becomes apparent and non-negligible. Just looking at the data, one also notices a possible shift upwards around 1965. If real, then in statistics one would refer to this phenomenon as a *change point* of the mean. A change-point detection procedure (we omit the details) found November 21, 1965 as change point in the trend of the daily maximal sea-level heights in the winter months from 1888 to 2019. The right-hand side

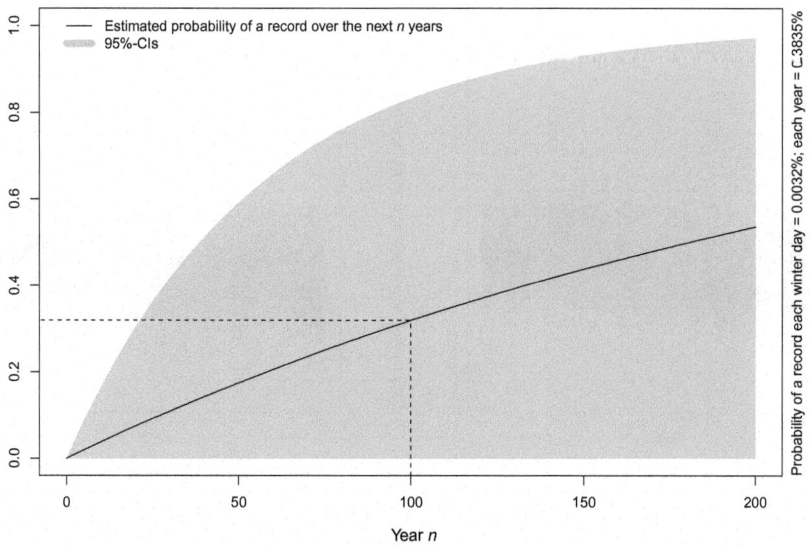

Figure 9.18 Q2: Estimated probability of the exceedance of 3.85 m above NAP (so of a record sea-level height) in at least one of the next n years, together with bootstrap 95%-CIs constructed as described in Remark 9.12. Even though the probability of such an event on a particular winter day is rather small, at 0.0032% with 95%-CI [0.0000, 0.0002], we see that over the next $n = 100$ years (starting in 1957) there is a probability of 32% (with 95%-CI [0.0004, 0.8329]) of such a record happening (indicated by the dashed lines). Source: Authors

of Figure 9.19 visualizes the data together with two fitted lines that are estimated from the data to the left, respectively to the right, of the change point. Whether the change point reflects a smooth change, due to climate change, say, or a more sudden regime switch or a superposition of both needs to be investigated in more detail. For example, a fairly sudden, though small, jump in the mean sea level may be due to a change in the coastal structure (anthropological or natural). We contacted Rijkswaterstaat on this issue and obtained the information that "Around 1965 there was an increase in tidal range (mainly a high water increase) at Hoek van Holland caused by the construction of the (first) Maasvlakte. By the construction of the second Maasvlakte, in the years 2009, ..., 2012, the tidal range at Hoek van Holland was slightly diminished." The Maasvlakte is a massive manmade westward extension of the Port of Rotterdam. As a consequence, a full analysis of the data has to take a combination of these human interventions with climate-change influence into account. Such non-stationarities, when proven statistically significant (which is the case for the two trends shown on the right-hand side of Figure 9.19), will have an effect on the return probability. The return probability will be higher than when estimated under the assumption of no trend. Moreover, for a fixed return probability, the return level will be underestimated. If you are further interested in these important aspects of the data, see Becker et al. (2009).

Although not part of Q3, what can we do here? We can apply an additional modeling layer before applying the POTM; see Remark 9.12. This additional modeling layer takes care of the

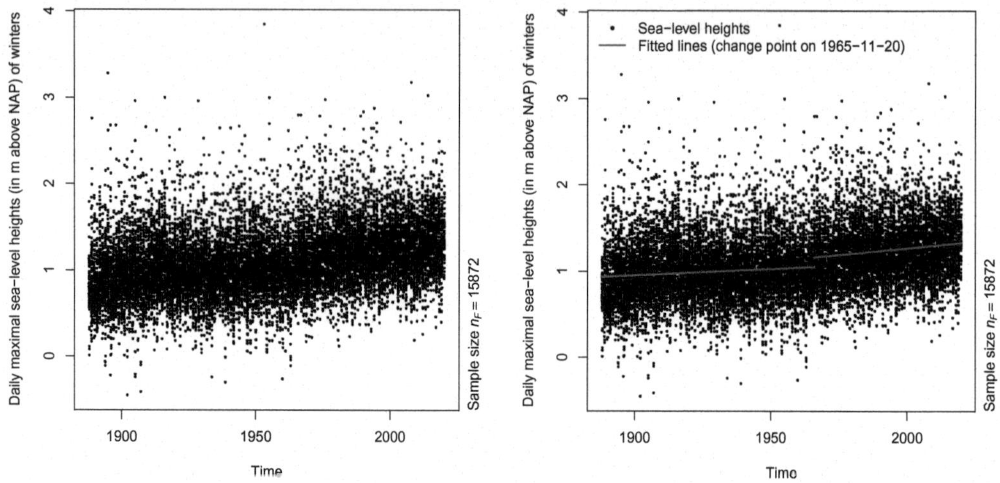

Figure 9.19 Left: Plot of the daily maximal sea-level heights (in meters above NAP) for the winter months November to February from January 1, 1888 to December 31, 2019. Right: Same, but with two fitted lines (before and after the change point of the mean, detected at November 21, 1965). Source: Authors

trend in the data. To this end we estimate the 97.5% quantile of the data as a linear function over time. This results in a linear threshold $u = u_t$ as a function of time t, representing an estimate of the 97.5% quantile of the data at t over time. This time-dependent threshold ranges from 1.75 m (when t is January 1, 1888) to 2.10 m (when t is December 31, 2019, denoted by T) above NAP. This increase is in line with a reported sea-level rise of about 2 to 3 mm a year. We then compute the excess E_t of an exceedance X_t at t as $E_t = X_t - u_t$. The left-hand side of Figure 9.20 shows the excesses obtained in this way. The estimates $\hat{\xi}_{n_u}, \hat{\beta}_{n_u}$ (where $n_u = 395$ for the time-dependent threshold u) of the GPD parameters ξ, β and corresponding bootstrap 95%-CIs (as before, computed as described in Remark 9.12) are $\hat{\xi}_{n_u} = -0.0003$ with 95%-CI $[-0.10, 0.16]$ and $\hat{\beta}_{n_u} = 0.2576$ with 95%-CI $[0.22, 0.30]$. The Q–Q plot of the excesses against the fitted GPD $G_{\hat{\xi}_{n_u}, \hat{\beta}_{n_u}}$ is shown on the right-hand side of Figure 9.20 and there is no strong evidence to believe that $G_{\hat{\xi}_{n_u}, \hat{\beta}_{n_u}}$ is not an adequate model for the excesses over the time-dependent threshold. As we did for Q1 and Q2, we fitted a GPD with parameters depending on the day of the excess as covariate, but also for Q3 found no strong evidence against the hypothesis that the GPD parameters are constant over time, which tells us that the time-dependent threshold effectively has taken care of the time dependence (the "additional modeling layer" we mentioned) and so the corresponding excesses could be modeled with the POTM, as we did.

Figure 9.21 shows the return level plot. The return level $\hat{r}_{u,p}$ is computed as in (9.9), with $u_T = 2.10$ (where T is December 31, 2019), $n_{u_T} = 210$ (the number of exceedances over u_T) and with the GPD parameter estimates $\hat{\xi}_{n_u}, \hat{\beta}_{n_u}$ obtained from the excesses over the time-dependent threshold $u = u_t$.

As for Q2, but now we are at the end of 2019, we can also ask for the probability of a record, that is, the probability $\mathbb{P}(X > 3.85)$, where X is a random daily maximum sea level in a winter month. Again we evaluate $\hat{\bar{F}}(3.85)$ based on $\hat{\xi}_{n_u}, \hat{\beta}_{n_u}$. We obtain this probability as 0.0015%

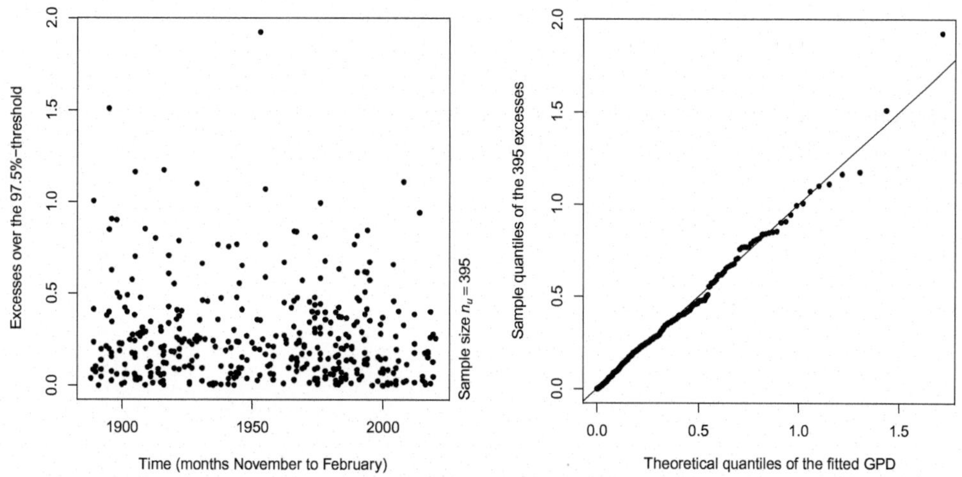

Figure 9.20 Q3: Left: Excesses in meters over a time-dependent 97.5% quantile as threshold (the threshold ranges from 1.75 m to 2.10 m above NAP) for the daily maxima of the winter months from January 1, 1888 to December 31, 2019 at Hoek van Holland. Right: Q–Q plot against the fitted GPD. Source: Authors

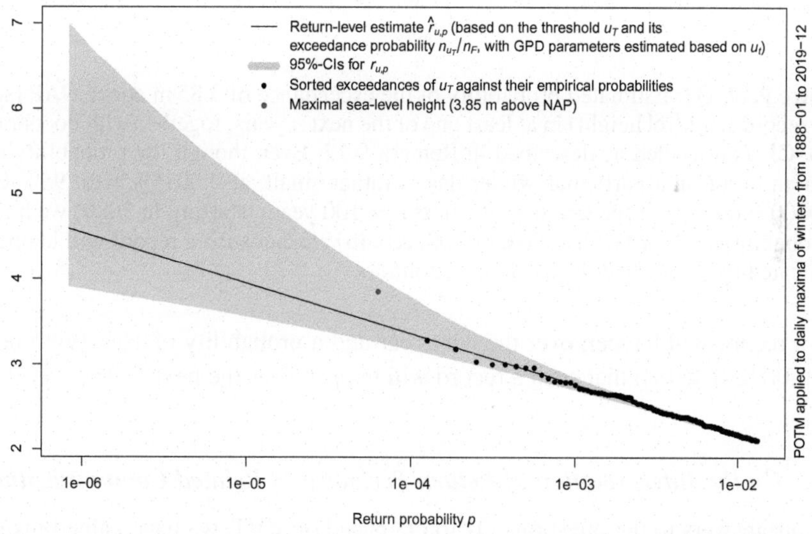

Figure 9.21 Q3: Return level plot of the return level estimate $\hat{r}_{u,p}$ from (9.9) as a function of p, together with 95%-CIs for $r_{u,p}$ based on the (last) threshold $u_T = 2.10$ (where T is December 31, 2019) and corresponding exceedance probability n_{u_T}/n_F, with the GPD parameters fitted using the excesses determined from the time-dependent threshold u_t. Included are the decreasingly sorted exceedances over the threshold u_T (an estimate of the 97.5% quantile of the daily maxima of the winter months at T), plotted against their empirical probabilities. For a 1 in 10 000 years event (so that $p = 1/1\,200\,000$, the smallest p shown), we have $\hat{r}_{u,p} = 4.5903$ with 95%-CI [3.91, 7.00]. Source: Authors

with bootstrap 95%-CI [0.0000, 0.0001]. Similarly to Figure 9.18, Figure 9.22 shows what this small probability translates into over the next n years. It shows the estimated probability of seeing at least one exceedance event of 3.85 m above NAP (so the probability of seeing at least one record) over the next n years as a function of n including bootstrap 95%-CIs. As an example, there is a probability of 0.1783% (with bootstrap 95%-CI [0.0000, 0.0062]) that such a record will happen over the next year and a probability of 16% (with bootstrap 95%-CI [0.0022, 0.4619]) that such a record will happen over the next 100 years.

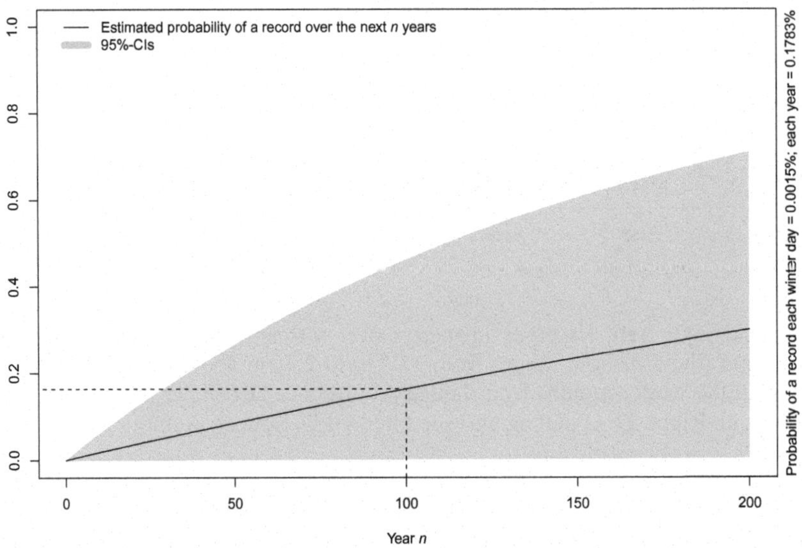

Figure 9.22 Q3: Estimated probability of the exceedance of 3.85 m above NAP (so, of a record sea-level height) in at least one of the next n years, together with bootstrap 95%-CIs constructed as described in Remark 9.12. Even though the probability of such an event on a particular winter day is rather small, at 0.0015% with 95%-CI [0.0000, 0.0001], we see that over the next $n = 100$ years (starting in 2020) we have a probability of 16% (with 95%-CI [0.0022, 0.4619]) that such a record will happen (indicated by the dashed lines). Source: Authors

9.5.3 On the Interpretation of Return Periods and Related Communication

Through our answers to the questions Q1 to Q3 based on EVT, we have gained insight into possible high sea-water levels at Hoek van Holland. Recall from Chapter 1 that the van Dantzig Report, as part of the Delta Report (Deltacommissie, 1961), suggested a value of 6 m above NAP as the dike height needed to withstand a 1 in 10 000 years event. Before we recall and discuss the numbers obtained, we will revisit the interpretation of a return period as a risk measure used for the communication of rare-event risk. We have all experienced, or heard of, a once in a lifetime event or a once in a century event. In particular, flood events due to rain often fall into the latter category. In the case of an engineering construction such as a dike, one should consider the risk of flooding over the dike's design-life period. As part of the Delta Works, the design-life period for the Oosterschedekering (Figure 1.6) stands at

200 years (starting in 1986). For several key components of the Maeslantkering (Figure 1.5), the design-life period is 100 years (starting in 1997). By way of example, we focus on a 1 in 200 years event in what follows; this is also the longest time horizon we have chosen for Figures 9.18 and 9.22. In this context, please take a moment to consider about what a 1 in 200 years event for a dike means. That we expect the dike to break and thus see a flood in about 200 years? That we are safe until then? That we are safe thereafter once it happened? None of these. The correct interpretation of a 1 in 200 years event is that *every year* the *probability* of the event happening is $1/200 = 0.005$. If you still want to know what this implies for the probability that the event will happen over the course of the following 200 years, the derivation in Section 9.5.2 gives you the tools to derive the answer to this question. We will repeat the main steps from that subsection. Assuming that risk exposures between successive years are iid (admittedly a strong assumption, and to be discussed), we get for the probability of the event happening at least once in the next 200 years

$$\begin{aligned}
&\mathbb{P}(\text{“the event happens at least once in the next 200 years”})\\
&= 1 - \mathbb{P}(\text{“the event does not occur in any of the next 200 years”})\\
&= 1 - \prod_{i=1}^{200} \mathbb{P}(\text{“the event does not occur in year } i \text{ from now”})\\
&= 1 - (1 - 0.005)^{200} \approx 0.6330 = 63.30\%;
\end{aligned}$$

again see Section 9.5.2, where we have already presented such a calculation. As such, the probability of a 1 in 200 years event happening at least once over a period of 200 years is 63%. More generally, the probability that a 1 in n years event will happen at least once over the next n years is $1-(1-1/n)^n$, which converges, for $n \to \infty$, to $1-1/e \approx 0.6321 = 63.21\%$; already, for $n \geq 24$, the first two decimals are fixed (0.63 . . .). Again we stress the underlying assumptions. This calculation assumes that the probability, 0.005, of the event happening in year i is the same for all $i = 1, 2, \ldots, 200$. It is perhaps interesting to know what would happen if this probability were to increase linearly from 0.005 to 0.01 over the course of the next 200 years (while still assuming independence across the years). In this case we obtain $1 - \prod_{i=1}^{200}\big(1 - (0.005 + 0.005 \times (i-1)/199)\big) \approx 0.7782 \approx 77.82\%$, which is an increase of almost 23% in comparison with the aforementioned 63.30%.

From an academic point of view, it is too easy to shrug off wrong interpretations of the concept of the return period. Such interpretations occur too often in day-to-day public communication. For a clear communication of rare-event risk, we should be able to do better. An interesting newspaper article clearly voicing this point of view in the context of floods in Australia is Deacon (2021). In order to make our discussion concrete and relevant within the context of sea-level data, let us return to the example of the 1 in 10 000 years safety measure at Hoek van Holland. The conclusions given apply to any return period.

First of all, a literal interpretation of the 10 000 years may distract attention from the urgency of the necessary protective measures to be taken, such as the strengthening of existing dikes as well as the construction of new ones. It is therefore important to shift the risk communication away from return periods towards a formulation of exceedance probabilities. For instance, determine a dike height such that with probability $1/10\,000 = 0.0001$ the dike will not be toppled during a sea surge over the next year. So, at the beginning of each year we face the above exceedance probability formulation; the risk perception then becomes much more imminent. Next, one faces the question why the probability is 0.0001 and not 0.001 or 0.00001, say. There is no easy solution for determining the probability level of this safety measure, nor was there around the discussions of the Delta Report. In any case, based on several, including economic, considerations, 0.0001 was chosen; here Jan Tinbergen (Figure 1.7) played a crucial role. In communicating this number to the public, it is always good to make some comparison. For instance, suppose that at the beginning of the year you toss a fair coin. For heads to come up 13 times in a row, the probability is $2^{-13} = 1/8192 \approx 0.000122 \approx 0.0001$, so the 1 in 10 000 years safety standard corresponds to that rare coin-toss experiment of independently throwing heads 13 times in a row at the beginning of each year going forward. Or, if you prefer, the probability of throwing five dice and obtaining a 6 on each die is also of the same order, since $6^{-5} = 1/7776 \approx 0.000129 \approx 0.0001$; see Figure 9.23. Another way of representing the probability 0.0001 is via distances. Imagine we go

Figure 9.23 My lucky day! Source: Enrico Chavez

on a road trip across the United States from Central Park (in Manhattan, New York) to see the Hollywood Sign (in Hollywood, Los Angeles). The trip is 4500 km long, so a 10 000th of that distance is only 450 m, a distance that does not even get you from Central Park to the Hudson River, not to mention Manhattan. Entering a game with winning chance 0.0001 is equivalent to correctly guessing down to 450 m accuracy where we are on that road trip without knowing when we started.

A somewhat more vague example based on field experiments is the frequency of four-leaf clovers appearing in a field full of clover, which is about 1 in 5000, though one often finds it reported as 1 in 10 000. If, however, you speak to friends and family, they all are surprised to hear of a 1 in 5000 event, and certainly of a 1 in 10 000 event; see Figure 9.23. Why? Well, it may be that we all have found at some time in our lives a four-leaf clover, so that we have moved the four-leafed-clover finding event from that part of our brain storing information on rare events to a part concerned with more common ones. This goes to show how difficult it is to communicate this topic and how much our personal, often fragmented, experience still plays a dominant role. Anyhow, the Delta Report settled on the 1 in 10 000 years events, so let us stick to that.

This may still be such a remote probability that it "surely" will not affect us, one might think, but with a design-life period of 200 years in mind (starting in 1986), we obtain the probability $1-(1-1/10\,000)^{200} \approx 0.0198 = 1.98\%$ for such an event to happen at least once over the next 200 years (assuming that from year to year, the risk of such breaches for the next 200 years can be viewed as iid). So the real question, "What is the probability that the new raised dike height will be breached at least once over the next 200 years?" that needs to be communicated has the answer: "about 2%". In other words, there is about a 2% chance of seeing at least one breach over the design-life period of the dam. This brings the risk within the relevant time range of a couple of generations and hence does concern us. Of course, we made the very important iid risk assumption over the 200 consecutive years. Under a climate-change scenario, it is fair to say that the 2% must be considered a conservative estimate and definitely warrants a much more in-depth analysis. Arguing as above, if that probability increases linearly to twice its value over the next 200 years, the 1.98% increases to 2.96%, so to almost, 3% (while still assuming independence across the years). This is what the Second Delta Report aims for; see Veerman and Stive (2008). Rootzén and Katz (2013) highlighted non-iid aspects of this design-life-period interpretation of risk. We stress once more that the communication of such rare events is better redirected from "1 in 10 000 years" (stating a return period) to "yearly exceedance probability of 0.0001" (stating a return probability). The latter should be presented together with a mental calibration through examples of small probabilities (such as with coins, dice or distances, as we used before) and the incorporation of the notion of the design-life period, if applicable.

9.5.4 Netflix Data

This section covers a financial risk management application of EVT and is rather technical. A basic interest in, and some earlier contact with, econometric modeling of financial data would be helpful here; see for instance McNeil et al. (2015, Chapter 4). We decided to leave this section in the book as it further highlights the versatility and usefulness of the methods introduced so far. For the less technically versed reader, a cursory reading will contain interesting information on the world of econometrics. When we talk about risk, financial risk is no doubt of prime importance; just check your household finances.

We discuss a time series of financial data. As in Example 8.56, we consider the negative log-returns (−log-returns) of the price of one Netflix share in USD and compute EVT-based predictions of $\mathrm{VaR}_{0.99}$ and $\mathrm{ES}_{0.99}$ including CIs, over some time period ahead as required, for example, within the Basel regulatory guidelines for larger banks; see Sections 8.4.2 and 8.4.5 for the necessary background on these risk measures. Although their specific values are not of further interest to us, we once more show how EVT can be used to solve important practical problems.

The choice of the Netflix share price was made mainly for two reasons. First, from a pedagogical point of view, the Netflix data show all the features we are interested in conveying while also satisfying the required modeling assumptions (both those we show and those we have not formally introduced in this book but have checked nonetheless). And second, the name Netflix is certainly one you have heard about; you might even have a Netflix account or have looked into getting one during the coronavirus pandemic. Netflix has become a synonym for personalized streaming services, offering films and television series in your browser (or app) with considerable convenience. If you are further interested in a more detailed analysis of the Netflix share price you can look into the very strong increase until about the end of October of 2021, followed by a sharp decline afterwards, and wonder about the economic reasons.

About the Data

In contrast with the Bear Stearns stock data we mentioned before, it is comparatively easy to obtain financial data about actively traded stocks. The historical price of one Netflix share in USD can be obtained from `finance.yahoo.com`, for example: search for the ticker symbol "NFLX" and once on the corresponding webpage, navigate to "Historical Data". The downloaded daily data for the years 2018 to 2021 are realizations of the prices S_t, $t = 0, 1, \dots, n$, (one typically considers *closing* prices, the prices at the end of each trading day and then builds the −log-returns X_t, $t = 1, 2, \dots, n$, as in (8.54)). One reason for modeling X_t rather than S_t is that the former data show no trend and, at the early stage of econometric modeling, were often even considered iid from some distribution. This distribution was conveniently taken as the normal; that the latter is inadequate from a risk management point of view we have already learned in Chapter 3.

DeGARCHing

We start from the −log-returns X_t, $t = 1, 2, \dots, n$, shown in the right-hand side plot of Figure 8.36. There is obviously structure in the data going beyond iid-ness. For instance, we see periods of comparably large variability (for example, in the second half of 2018) and periods of comparably small variability (for example, in the second half of 2021). This behavior is called *volatility clustering* and is typically found in stock-price return data. As explained in Section 9.5.1, to get iid data we thus need to apply another modeling layer first, one that takes care of the volatility clustering and leaves us with iid data for modeling purposes.

A classical model often applied to address volatility clustering is that of a GARCH model, given by

$$X_t = \sigma_t Z_t, \tag{9.10}$$

$$\sigma_t^2 = \alpha_0 + \alpha_1 X_{t-1}^2 + \beta_1 \sigma_{t-1}^2. \tag{9.11}$$

The quantities α_0, α_1 and β_1 are parameters of the model, and we assume them to fulfill certain conditions (to be more precise, $\alpha_0 > 0$, $\alpha_1 \geq 0$, $\beta_1 \geq 0$, $\alpha_1 + \beta_1 < 1$). To learn more about such models, see McNeil et al. (2015, Chapter 4).

At a high level, (9.10) describes the (random) X_t as a linear transformation, with location term 0 and scale term σ_t, of the (random) Z_t. As we see from (9.11), the scale σ_t (the volatility) depends on time t through its past value σ_{t-1} and the past observation X_{t-1}, both of which are observed and thus known at time t. The Z_t are assumed to be iid with distribution function F_Z known as *innovation distribution*. The iid assumption implies that all time-relevant information is driven by the deterministic σ_t (again, known at time t), whereas the randomness of X_t is driven by the randomness of Z_t. In Section 10.2.2 we will see X_t expressed as another linear transformation of Z_t (there, with a non-zero location μ_t depending on time t and a constant $\sigma_t = \sigma$), leading to a different time series model that is also based on a linear transformation with time-dependent terms.

The innovation distribution F_Z must have mean 0 and variance 1 (any other finite variance could be modeled by a different σ_t^2). Typical innovation distributions are the standard normal $\mathrm{N}(0, 1)$ or a standardized t distribution; we will use the latter. If F_{t_ν} denotes the distribution function of the $t_\nu = t_\nu(0, 1)$ distribution, then we can express the distribution function of the standardized t distribution as $F_Z(x) = F_{t_\nu}(\sqrt{\nu/(\nu-2)}x)$, $x \in \mathbb{R}$, which is the distribution function of the $t_\nu(0, (\nu-2)/\nu)$ distribution; see also Example 8.54. The degrees of freedom parameter ν (which has to satisfy $\nu > 2$ in order for the resulting distribution to have finite variance) is used to control the heavy-tailedness of the innovation distribution.

The process of modeling X_t through Z_t via a GARCH-type model to obtain iid data is known as *deGARCHing*. We cannot go into detail here how such models are fitted, but this can be achieved with software based on a chosen innovation distribution F_Z such as the standard normal or the standardized t distribution. After fitting the model's parameters (including those of F_Z, that is, ν in our case here) from the realizations of the X_t, the realizations of the Z_t in (9.10) are available to us for further modeling. We can extract them from the X_t via $Z_t = X_t/\sigma_t$ (see (9.10)). They are known as *standardized residuals*; "standardized" comes from the fact that F_Z has mean 0 and variance 1. If the model fits well, the standardized residuals should be iid from the (fitted) F_Z and we can check that, as a way to verify the correctness of the GARCH-type model in modeling the volatility clustering. Extreme value theory modeling (we use the POTM) is then performed using the standardized residuals, and the results obtained are transformed back to the original time series.

Figure 9.24 shows the extracted Z_t on the left-hand side and the Q–Q plot against the fitted innovation distribution F_Z on the right-hand side (based on the sample size $n = 1006$ here, the estimated degrees of freedom parameter is $\hat{\nu}_n = 4.39$, so the innovation distribution is fairly heavy-tailed). Comparing the left-hand side of Figure 9.24 with the right-hand side of Figure 8.36 indicates that deGARCHing has taken care of the volatility clustering.

Remark (On the choice of innovation distribution)**.** We gave a classical statistical argument to justify our use of the standardized t distribution as innovation distribution. Recent research (Thomas Mikosch and Olivier Wintenberger, private communication) warns about the use of this and other heavy-tailed innovation distributions without further questioning, as such innovation distributions may imply GARCH models with a heavy-tailedness that is not in agreement with the heavy-tailedness of the data; this happens less for standard normal innovations for which GARCH-type models can still produce heavy-tailed data. A careful analysis of the heavy-tailedness of the data and the statistically most appropriate choice of innovation distribution based on the standardized residuals is required. Whereas we have done such analyses, their ingredients go beyond what we have introduced in this book and we thus do not report these results here; the interested reader is advised to follow up this research more closely.

We now turn to the idea behind (9.11). Since Z_t has mean $\mathbb{E}(Z_t) = 0$, one also has $\mathbb{E}(X_t) = 0$, so the mean remains unchanged. Since Z_t has variance $\mathrm{var}(Z_t) = 1$, one has $\mathrm{var}(X_t) = \sigma_t^2$. Therefore, (9.11) describes the variance of X_t. As we see from (9.11), the variance σ_t^2 of X_t

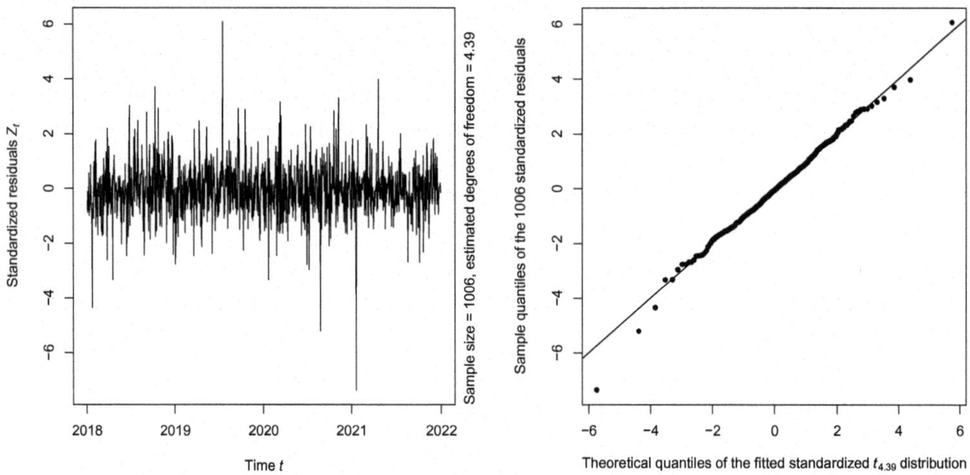

Figure 9.24 Left: Plot of the standardized residuals $Z_1, Z_2, \ldots, Z_n$ for $n = 1006$ after deGARCHing the Netflix return data (compare this picture with the right-hand side picture in Figure 8.36). Right: Q–Q plot of the standardized residuals versus the fitted standardized Student t innovation distribution F_Z. Source: Authors

is determined by two main ingredients, the squared previous values X_{t-1}^2 and the previous variance σ_{t-1}^2. We see that if the previous values of the process fluctuate more (so that the squared values are large) or if the previous standard deviations are large (because already we are in a time period of high volatility), then the standard deviation of X_t tends to be large, too. This creates the effect of volatility clustering on top of the iid Z_t and leads to more realistic models for X_t, as are often seen in financial data. We can now also clarify where the acronym GARCH comes from: it stands for generalized autoregressive conditional heteroskedasticity. Autoregressive conditional heteroskedasticity (ARCH) models were introduced by Robert Fry Engle III in 1982. He, jointly with Clive Granger (1934–2009), received the 2003 Nobel Memorial Prize in Economic Sciences "for methods of analyzing economic time series with time-varying volatility (ARCH)". The term "autoregressive" (meaning "on itself") refers to the dependence of X_t on X_{t-1} and "conditional heteroskedasticity" refers to the fact that the variance $\mathrm{var}(X_t)$, conditional on its past values, is not constant over time. The "not constant" is the important part here and hints at volatility clustering. The word "generalized" in the acronym GARCH refers to the dependence of σ_t^2 also on σ_{t-1}^2; this step was introduced in Bollerslev (1986).

During the week of September 5–9, 2005, the Swiss Association of Actuaries celebrated its centenary. As part of the celebrations, both ASTIN (Actuarial Studies in Non-Life Insurance) as well as AFIR (Actuarial Approach for Financial Risks) international conferences were organized at ETH Zurich, including a common ASTIN–AFIR day on Wednesday, September 7, 2005. On that day, Robert Engle gave a talk with the title "Downside risk – econometric models and financial implications". On this occasion, he told the story that once a journalist asked him why he had received the Nobel Prize. His answer involved dynamic volatility models of the GARCH-type, to which the journalist countered that this answer was too technical for the newspaper readers. A second attempt to answer the question involved a changing variance in the returns of financial data, but was still considered too technical by the journalist. The final answer came almost down to "stock prices go up and down", after which the by then satisfied journalist said: "Did you get the Nobel Prize for just that?" We have come a long way since the first Sveriges Riksbank Prize in Economic Sciences in Memory of Alfred Nobel in 1969, given to Jan Tinbergen, the economist we met in Section 1.3 when discussing the van Dantzig report. Since those early days, econometrics has grown into a vast area of research with important applications in finance and economics.

Applying the Peaks Over Threshold Method

We now apply the POTM to the deGARCHed −log-returns, the standardized residuals $Z_1, Z_2, \ldots, Z_n$. In Section 9.4.3 we briefly mentioned tools such as the mean excess plot to determine a suitable threshold or the plot of the estimated GPD shape parameter ξ as a function of the choice of threshold u. We also consulted a third tool, not discussed here. It also considers an estimated quantity across different threshold values u, and one can trade off bias and variance to determine a suitable threshold. In the end, we chose as threshold $u = 1.0755$, the empirical 0.9-quantile of the standardized residuals $Z_1, Z_2, \ldots, Z_n$. The resulting parameter estimates with bootstrap 95%-CIs are $\hat{\xi}_{n_u} = 0.0597$ with 95%-CI $[-0.14, 0.26]$ and $\hat{\beta}_{n_u} = 0.6538$ with 95%-CI $[0.47, 0.84]$. As before, these CIs are constructed as in Remark 9.12, in particular, without taking into account the statistical uncertainty pertaining to the estimation of the GARCH(1, 1) model.

The left-hand side of Figure 9.25 shows the 101 excesses of the standardized residuals over our chosen threshold u, and the right-hand side of this figure shows the Q–Q plot of these excesses against the fitted GPD, which confirms the quality of the fit.

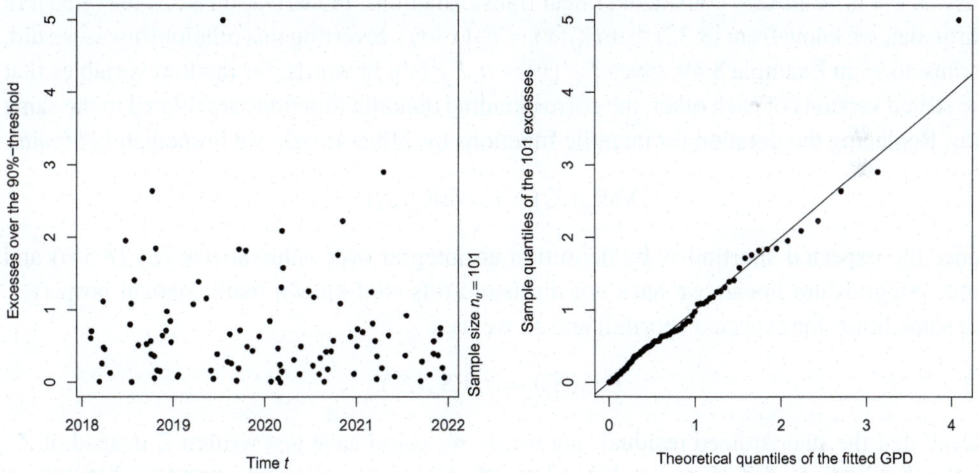

Figure 9.25 Left: Excesses over the 90%-quantile of the standardized residuals $Z_1, Z_2, \ldots, Z_n$ as threshold for the daily −log-returns of the price of one Netflix share in USD. Right: Q–Q plot against the fitted GPD. Source: Authors

Value-at-Risk and Expected Shortfall

In Sections 8.4.2 and 8.4.5 we introduced the quantile-based risk measures VaR and ES. These risk measures are calculated daily throughout the financial industry. In this section, we exemplify how EVT can be used towards their statistical estimation. These estimates may enter as input for a calculation of regulatory capital within the Basel Accords; see McNeil et al. (2015) and the references therein. Based on the chosen threshold u and the fitted GPD parameters $\hat{\xi}_{n_u}$ and $\hat{\beta}_{n_u}$, we obtain estimates of the risk measures VaR and ES with the POTM. For $Z_t \sim F_Z$, they are given by

$$\widehat{\mathrm{VaR}}_\alpha(Z_t) = u + \frac{\hat{\beta}_{n_u}}{\hat{\xi}_{n_u}}\left(\left(\frac{1-\alpha}{n_u/n}\right)^{-\hat{\xi}_{n_u}} - 1\right), \quad \alpha \in [1 - n_u/n, 1), \tag{9.12}$$

and, for $\hat{\xi}_{n_u} < 1$,

$$\widehat{\mathrm{ES}}_\alpha(Z_t) = \frac{\widehat{\mathrm{VaR}}_\alpha(Z_t) + \hat{\beta}_{n_u} - \hat{\xi}_{n_u} u}{1 - \hat{\xi}_{n_u}}, \quad \alpha \in [1 - n_u/n, 1). \tag{9.13}$$

The estimator $\widehat{\mathrm{VaR}}_\alpha(Z_t)$ in (9.12) is the POT quantile estimator (9.8) evaluated at the confidence level α. We have already seen $\widehat{\mathrm{VaR}}_\alpha$ in Section 9.4.5 (note that n_F there is the same as n here). The estimator $\widehat{\mathrm{ES}}_\alpha(Z_t)$ in (9.13) can be derived from (9.12) by calculating (8.55). As mentioned earlier, we focus on the case $\alpha = 0.99$. With $n = 1006$, $n_u = 101$, $u = 1.0755$ (90%-quantile), $\hat{\xi}_{n_u} = 0.0597$ and $\hat{\beta}_{n_u} = 0.6538$, we can compute (9.12) and (9.13). However, this is not what we are primarily interested in. The question is how $\widehat{\mathrm{VaR}}_\alpha(Z_t)$ and $\widehat{\mathrm{ES}}_\alpha(Z_t)$ translate to our original modeling layer, namely $\mathrm{VaR}_\alpha(X_t)$ and $\mathrm{ES}_\alpha(X_t)$ for future time points t, the quantities in which we are actually interested.

Because of the simple linear connection between Z_t and X_t in (9.10), this question is easy to answer. We have already considered linear transformations, in Derivation 8.35; see (8.31). In particular, we know from (8.32) that $F_{X_t}(x) = F_Z(x/\sigma_t)$. Inverting this relationship as we did, for instance, in Example 8.49, gives $F_{X_t}^{-1}(y) = \sigma_t F_Z^{-1}(y)$. In words, for random variables that are scaled versions of each other, the corresponding quantile functions are related in the same way. Replacing the notation for quantile functions by value-at-risk, we immediately obtain

$$\mathrm{VaR}_\alpha(X_t) = \sigma_t \, \mathrm{VaR}_\alpha(Z_t).$$

Since the expected shortfall is by definition an integral over value-at-risk (by (8.55)) and since integrals are linear (we have not discussed this well-known mathematical property), the same holds for expected shortfall and so we have

$$\mathrm{ES}_\alpha(X_t) = \sigma_t \, \mathrm{ES}_\alpha(Z_t).$$

Recall that the standardized residuals are iid, so we could have just written Z instead of Z_t, as the distribution of Z_t does not depend on t (which is also why we wrote its distribution function simply as F_Z). We thus see that the "decomposition" (9.10) (a linear transformation) translates in the same way into the risk measures VaR and ES, in that the dependence on time t appears only in the scaling factor σ_t of the linear transformation.

Figure 9.26 shows the realizations of X_t from July 1, 2021 to December 30, 2021, together with the 30 one-day-ahead predictions of $\mathrm{VaR}_{0.99}(X_t)$ (left) and $\mathrm{ES}_{0.99}(X_t)$ (right), as discussed, including bootstrap 95%-CIs (constructed as in Remark 9.12, in particular; we have not taken into account the statistical uncertainty pertaining to the estimation of the GARCH(1, 1) model). If we were to overlay the red $\mathrm{VaR}_{0.99}(X_t)$ and $\mathrm{ES}_{0.99}(X_t)$ curves, we

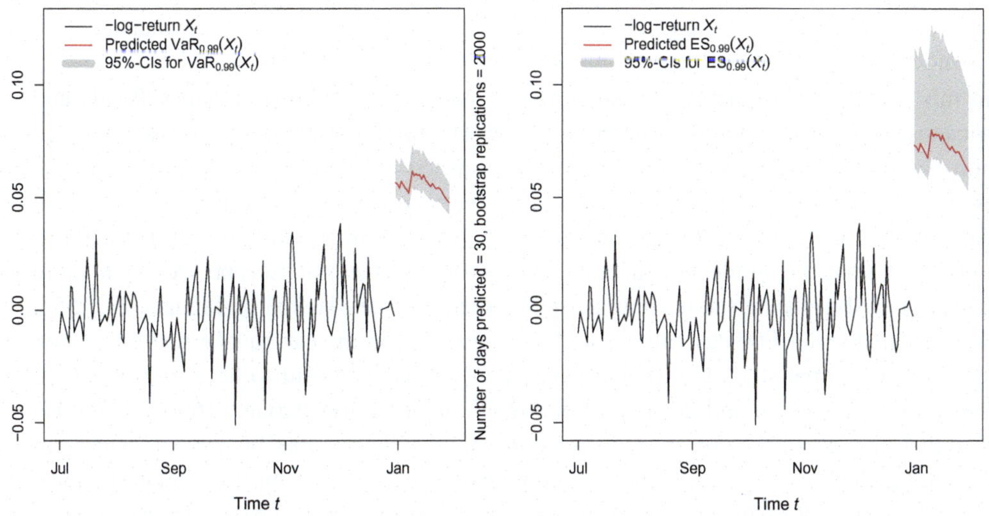

Figure 9.26 Negative log-returns X_t of the Netflix data from July 1, 2021 to December 30, 2021, together with 30 one-day-ahead predictions of $\mathrm{VaR}_{0.99}(X_t)$ (left) and $\mathrm{ES}_{0.99}(X_t)$ (right) and corresponding bootstrap 95%-CIs. Source: Authors

would always have that $\mathrm{ES}_{0.99}(X_t) > \mathrm{VaR}_{0.99}(X_t)$, reflecting the severity (*what if*) interpretation of ES versus the frequency (*if*) interpretation of VaR. We also see that the 95%-CIs for $\mathrm{ES}_{0.99}(X_t)$ are wider than those for $\mathrm{VaR}_{0.99}(X_t)$, which, as we addressed in Example 8.4.5, comes from the fact that $\mathrm{ES}_{0.99}(X_t)$ is more sensitive to changes in the tail of the distribution than $\mathrm{VaR}_{0.99}(X_t)$. Over a longer time horizon than just 30 days, we could also compare our risk measure predictions with the (by then) realized (and thus known) −log-returns in order to assess how accurate our risk measure predictions are. This is known as *backtesting*. A simple procedure for backtesting $\mathrm{VaR}_{0.99}(X_t)$ predictions is based on the relative frequency of times t (typically days) for which X_t exceeds $\mathrm{VaR}_{0.99}(X_t)$. For example, with two years of daily data (about 500 trading days) we expect to see $500 \times (1 - 0.99) = 5$ such exceedances, so if we see them much more rarely or much more frequently, our $\mathrm{VaR}_{0.99}(X_t)$ prediction is not accurate. An appropriate backtesting procedure can also be worked out for $\mathrm{ES}_\alpha(X_t)$; see Acerbi and Székely (2019).

9.6 A Basic Model for Earthquakes

From our earlier discussions in this chapter, it should be clear that EVT yields reliable tools that could, for example, be applied for the estimation of earthquake return periods and endpoints of magnitude distributions. This has indeed been done. The estimation of the endpoint of the magnitude distribution is of particular interest when it concerns earthquakes induced by human intervention, in which case it is not necessarily the laws of physics that play the key role but equally importantly, or even more so, the specific type of human intervention. An excellent example of this is discussed in Beirlant et al. (2019). This paper concludes that, on the basis of the application of several different EVT-related techniques to the Groningen gas field in The Netherlands, the maximum possible anthropogenic origin seismic event magnitude is estimated to be in the range from 3.61 to 3.80. The upper bounds of 90%-confidence intervals vary from 3.85 to 4.50. These estimates are useful when it comes to a resilience analysis of buildings.

In this section, we focus on the modeling of natural earthquakes. Besides floods and hurricanes, (natural) earthquakes belong to the most feared natural disasters. Each affected region has its own memory of human suffering caused by these disasters throughout the ages. For instance, the October 18, 1356 Basel earthquake in Switzerland destroyed all major churches and castles within a 30 km radius of Basel, with an estimated 300 casualties. An event of this magnitude was a rather rare event for Central Europe. A much more devastating earthquake hit the northwestern Chinese province Shaanxi on January 23, 1556, with an estimated 830 000 casualties. On November 1, 1755 the earthquake that hit Lisbon came with an estimated 12 000 to 50 000 casualties. It is even said to have reinforced the philosophical pessimism and deism of Voltaire (François-Marie Arouet (1694–1778)). He argued that, in view of the presence of such evil, there could not possibly exist a benevolent and loving god who rewards the virtuous and punishes the guilty. A more recent earthquake was the 2004 Indian Ocean earthquake and tsunami of December 26, 2004, with an estimated 280 000 casualties; see also Chapter 4. Over the period 1997–2017, nearly 750 000 people died globally due to earthquakes. Hence it does not come as a surprise that man has always searched for a cause. Stories are to be found in mythology all over the world. Known to many, no doubt is Poseidon, the Greek god of the sea, storms, earthquakes and horses. When he is in a bad mood, he strikes the ground with his trident causing severe trembling. Perhaps lesser known examples include the giant catfish Namazu in Japan and the god Loki from Norwegian mythology. If you are further interested in these stories, you may also look for the snake, turtle and elephant myth in Hindu culture. Science has evolved a long way from these mythological views. Our present day understanding is that earthquakes mainly result from geological faults in Earth's crust; see our discussion in Section 4.2. The search for cause(s), predictions and consequences constitutes a major task for modern science. This quest combines a macro view of Earth's geological structure within an evolving universe with that of a micro analysis, for example, of how to achieve societal resilience at the level of engineering and economics. The latter very much includes insurance.

Earthquake risk is known to be highly unpredictable; see our discussion of the L'Aquila trial in Chapter 5. It is perhaps the risk category *par excellence*, to which the famous quote "Those who have knowledge don't predict, those who predict don't have knowledge" applies. In some form or another, this quote is attributed to Lao Tzu, a Chinese philosopher who lived around 550 BCE. The father of the Richter scale, also known as the Gutenberg–Richter scale for earthquake magnitude, Charles Francis Richter (see also Sections 4.1 and 6.6) once said "Only fools, liars, and charlatans predict earthquakes." By now, there exists a considerable literature on earthquake prediction. For an early review of published material, see Geller (1997). From the latter reference we quote:

> Earthquake prediction research has been conducted for over 100 years with no obvious successes. Claims of breakthroughs have failed to withstand scrutiny. Extensive searches have failed to find reliable precursors. Theoretical work suggests that faulting [the movement of tectonic plates] is a non-linear process, which is highly sensitive to unmeasurably fine details of the state of the Earth in a large volume, not just in the immediate vicinity of the hypocentre. Any small earthquake thus has some probability of cascading into a large event. Reliable issuing of alarms of imminent large earthquakes appears to be effectively impossible.

Not all scientists, however, are ready to give up the search. One example is Lerner-Lam (1997) who take a somewhat less defeatist view:

> Can earthquakes be predicted? This is what society wants to know, yet most of us think this question is not properly posed. But declaring earthquakes to be *unpredictable* does a disservice to the state of our science and, quite frankly, doesn't do the public any good. [. . .] The falsification of a particular methodology does not indict an entire field, particularly one in which the set of observations is incomplete. Personally, I am not ready to declare earthquakes inherently unpredictable.

In view of the massive human and economic loss caused by earthquakes, it is no doubt highly desirable to have mathematical models that allow one to quantify extreme earthquakes and thus the associated risk. In this section we briefly present a basic model for earthquakes. It seems obvious to model large shocks with the POTM. However, earthquakes typically occur in clusters: foreshocks occur before mainshocks and mainshocks trigger aftershocks that, in turn, can trigger further aftershocks, etc. Clusters typically occur both in time and in space; in our book we only consider the time component. We start with typical features of earthquake magnitude data.

9.6.1 Typical Features of Earthquake Shock Data

History has come up with numerous popular beliefs that were and to some extent are still deemed to be relevant when it comes to earthquake prediction. Below we offer some of these popular beliefs as reported by the Swiss Seismological Service of ETH Zurich:

(B1) Unusual *animal behavior* just prior to an earthquake has often been linked to the foretelling of an earthquake. These include: catfish, snakes, rats, sheep, dogs and pheasants. For instance, dogs are believed to run away from an imminent earthquake. A similar behavior has been reported for snakes, leaving their protective hideouts or winter shelters.

(B2) Various studies and experiences (also in the case of L'Aquila) have indicated a higher concentration of radon just prior to an earthquake.

(B3) There are a few – in some cases controversial – reports of unusual light effects presaging seismic activity.

(B4) There has been discussion about whether the increased gravitational pull during so-called *supermoons* can be associated with earthquakes.

Multiple scientific studies have shown that none of (B1)–(B4) have any association with earthquake occurrence.

(B5) Foreshocks, the change in electromagnetic signals and the appearance of a period just before a larger earthquake where the usual seismic activity quietens down have all been studied as possible indicators.

Whereas the physical processes underlying (B5) sound relevant, when it comes to prediction so far one has not been able to build a reliable forecasting model based on them. In order to be of any real practical use, the proper prediction of an earthquake will necessarily have to include the date and time, the location and the magnitude. Whereas this may seem obvious, considering the details soon highlights the enormous complexity of earthquake prediction.

There is not only the magnitude as a measure of the total energy released at the earthquake's hypocenter but also its intensity (the amount of shaking). This intensity of an earthquake is a function of magnitude and depth, but also of distance along Earth's surface from the epicenter as well as of the soil and rock composition. Through a combination of such factors, the number of victims and the amount of damage can become more extensive at longer distances from the epicenter's precise location. This is just one example where the modeling and prediction of vulnerabilities as consequences of a natural disaster add a considerable amount of complexity and hence uncertainty to disaster-risk management. Science spends a lot of energy and time on understanding the natural processes leading to extreme events, and rightly so. Such events however only become a catastrophe once the loss of human life and damage to infrastructure are taken into account.

In the case of L'Aquila, it was well known that the greater area around the town is prone to earthquakes. In the past it had been severely struck on several occasions, as in 1315 (magnitude 6.7), 1349 (magnitude 6.5), 1461 (magnitude 6.5) or in 1703 (magnitude 6.7). In 1915, Avezzano, roughly 40 km south of L'Aquila, was hit by a 7.0 earthquake on the moment magnitude scale; see Section 4.1 on how the magnitude of an earthquake is measured. With more than 30 000 victims, it is the largest and deadliest documented earthquake in the region; see Swiss Re Institute (2019). Earthquakes are very much natural phenomena in need of analysis and recording at a global level. For this purpose, worldwide earthquake catalogs offer a wealth of information; see for instance USGS (2021b) from the United States Geological Survey's Earthquake Hazards Program.

From the above website we learn that, during the period 2000–2020, the average number of earthquakes worldwide with a magnitude above 5.0 fluctuated in a relatively narrow band from a minimum of 1341 in 2002 to a maximum of 2481 in 2011. If we were to consider Earth in a first approximation as an isolated system, the law of the conservation of energy would put a limit to the number and magnitudes of earthquakes worldwide. The number of people killed, however, fluctuates much more strongly (as it depends, for example, on how densely populated the location is and on the construction of the houses there). The highly variable number of people killed highlights once more the difficult transition from an extreme natural event to a human catastrophe. As we saw in Chapter 5, for the population of L'Aquila the main worry was whether or not a severe earthquake was deemed imminent in late March and the first days of April, 2009. Here the seismic observations over that period were the only available scientific data on which to base a decision about issuing a warning.

9.6.2 About the Data

Initially, we did not intend to present a mathematical model for earthquakes. However, when we started looking for data related to the L'Aquila earthquake, it became obvious to us that a short discussion of quantitative modeling would do justice to the "understanding" in the title of our book. The technical level of the following subsections is more taxing (we are still on a hike). As before, even a cursory reading will hopefully offer the less technically versed reader with an appreciation of science's quest towards understanding. Before we start on our excursion of earthquake model building, we return to a common theme of our book, "getting the data".

There are some graphs in this book (for example, Figure 3.2 or the right-hand side of Figure 13.1) that can be found online but in such low quality that we wanted to replicate these graphs for this book. One such graph is Figure 5.2 in Chapter 5, showing earthquake shock data; compare the plot found on Wikipedia (2021a) with our Figure 5.2 to know why we were eager to have an improved version of this graph. But for that we needed the data. On June 15, 2021, we wrote to the Istituto Nazionale di Geofisica e Vulcanologia (INGV), the National Institute of Geophysics and Volcanology of Italy, to inquire about the data, since this Institute is mentioned as a source for the graph on Wikipedia (2021a). Six days later came the answer that our request was being forwarded to a team of experts. At this point we had already found other databases, or so we thought. The first one was that of the European–Mediterranean Seismological Centre (EMSC); see EMSC (2021). However, this database provides data only on earthquakes with magnitude 2 or more (mostly reported in the local magnitude scale commonly known as the Richter scale), which provides only a limited picture of the data; compare with Figure 5.2. We then found the database of the United States Geological Survey (USGS); see USGS (2021b). It provides a myriad of options but contains only a small amount of data points in the region around L'Aquila.

We then found the database Istituto Nazionale di Geofisica e Vulcanologia (2021), which is actually the official database of the Istituto Nazionale di Geofisica e Vulcanologia. Via the button "Custom Search" at the top right of the webpage, one can choose "Start date" (here we chose 1985-01-01; the earliest observations in this region are from 1985) and "End date" (here we ultimately chose the last time we downloaded the data, which was 2021-12-25). As "Magnitude" we selected "Minimum magnitude: −1" and "Maximum magnitude: 10". Under "Where" we then choose "Circle" as "Area type". And under "Get coordinates from Italian Municipality name" we typed in "L'Aquila" to obtain the latitude and longitude of the region around L'Aquila (to be more precise, latitude 42.3522 and longitude 13.3994 were selected). As "Radius" we used 50 (in kilometers); because of the low frequency of earthquakes at a single location, consolidating observations from a wider geographical region is crucial. The rest of the default settings were fine, so we hit "Search"; note that we also chose "Timezone: UTC" on the main webpage for better compatibility with our software, but that is not relevant for our story here. Finally, we saw 81 331 hits; it looks good. Now we 'only' have to download that data, right? The button "Export list (UTC)" provides "Text" as an option, and choosing that displays a text-only webpage; all still looks good. Scrolling all the way down, however, reveals that only data until April 16, 2009 are displayed (the data for that day may not even be fully displayed). It turns out that at most 10 000 events can be obtained at a time. Of course one does not know what that means in terms of the considered time period since there are times with only few events and times with lots of measurements because of an active earthquake, for example. So one needs to manually adjust the start and end dates to display the data several blocks at a time while making sure no block exceeds more than 10 000 events (of course this could also be automated by software, but that is still quite involved). Not so great.

Thankfully, one can speed up the manual work a little bit by realizing that the URL (the address of the result webpage) reveals the above choices made. In particular, it contains the strings "starttime=" and "endtime=" and so by just modifying the URL we can directly obtain the result page for the chosen start and end dates. Partitioning the years into 10 blocks, we then wrote a program to iterate over the blocks, download the data of each block (which, once run, took only about 40 s in total) and then read and merged the 10 datasets into a single time series. Saved to a file including all the information about the earthquake events (although we just needed the times and magnitudes), the file size turned out to be of an order acceptable as a standard email attachment nowadays. You might wonder why such file sizes are not allowed to be downloaded all at once (perhaps with a warning or confirmation first), we certainly do, too.

Another point one might wonder about is what we actually downloaded. Similarly to the EMSC database, the INGV database contains magnitudes measured on different scales, predominantly the local magnitude scale (below 3.5) with the moment magnitude scale for larger magnitudes. The site guide of Istituto Nazionale di Geofisica e Vulcanologia (2021) provides more information.

Figure 9.27 shows all the magnitudes of seismic observations we had at our disposal; Figure 5.2 showed a subset of these data. We also see the magnitude 6.0 earthquake of August 24, 2016 with epicenter near the town of Accumoli, 45 km north of L'Aquila, which killed 299 people. Just examining the figure, some features jump out. First, before spring

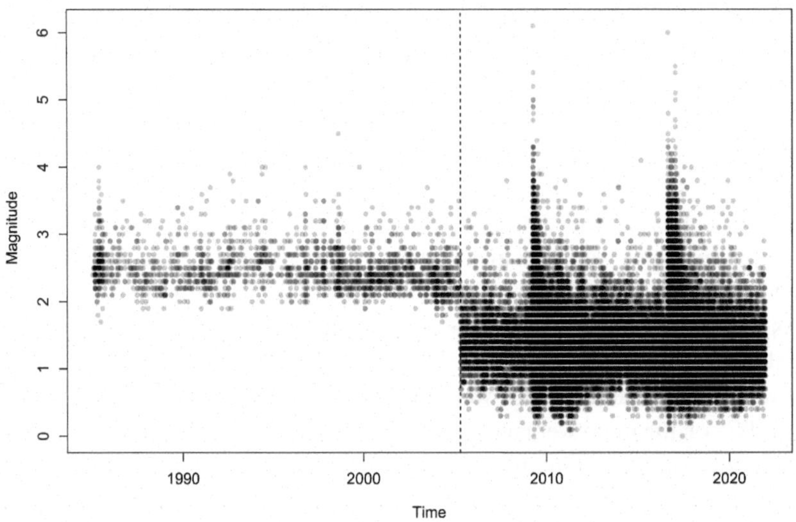

Figure 9.27 L'Aquila shock data from January 1, 1985 to December 25, 2021. The vertical dashed line indicates April 16, 2005 when the earthquake monitoring and analysis system was updated. Source: Authors

2005, we see that there were no smaller magnitudes reported. As Alessandro Amato from INGV explained, their monitoring and analysis system was changed on April 16, 2005 to use data recorded by their new digital seismic network, and this date is indicated by a vertical dashed line in Figure 9.27. The former system was not able to accurately compute lower magnitudes pre-2005, which is why we only see them thereafter. Second, as is natural for that earthquake-prone part of Italy, several thousands of smaller shocks have been recorded since spring 2005, of which several dozen have magnitude above 3. Third, from Figure 5.2 no immediate special behavior of seismic activity can be discerned over the beginning period of 2006 until the middle of March 2009, say. If one zooms in close to April 2009, however, Figure 9.27 reveals six foreshocks of magnitude greater than 3, four of which were on March 30, 2009. The main earthquake on April 6, 2009 (at 01:32:40 UTC to be precise) reached a value of 6.1 on the moment magnitude scale; note that 01:32:40 UTC corresponds to 03:32:40 local time. In various sources online one finds different values of this largest shock, for example 6.3 on Wikipedia. The same is true for the shock on August 24, 2016 of magnitude 6.0 (also reported as 6.2). We let Alessandro Amato from INGV explain such discrepancies in his own words: "There are different estimates for the moment magnitude (Mw), as for all the earthquakes worldwide. The variability depends on the data and on the technique adopted by monitoring institutions. There is not one single 'right' magnitude, therefore." As such, all reported numbers in what follows are based on the INGV data we have. Last but not least, note that the white horizontal gaps visible in Figure 9.27 come from the standard rounding of magnitudes to one decimal place.

Zooming in on the time period from January 1, 2006 to December 25, 2021, and plotting only the daily maxima, we obtain Figure 9.28. It reveals a clear structure of the aftershocks.

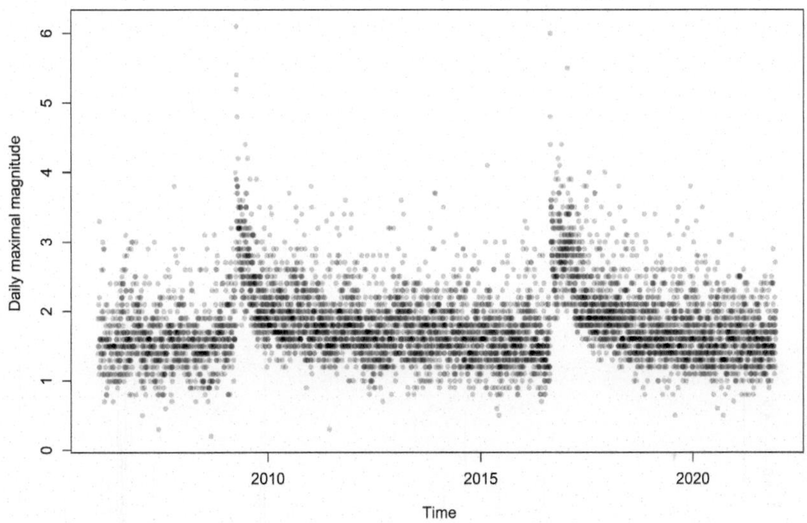

Figure 9.28 Daily maxima of the L'Aquila seismic data for the time period from January 1, 2006 to December 25, 2021.

The peak observations of 6.1 (L'Aquila) and 6.0 (Accumoli) for when the earthquakes struck are followed by aftershocks rapidly decaying in size. Harder to spot is that these mainshocks are followed by smaller mainshocks that themselves are followed by several smaller aftershocks decaying in size. This cascading behavior is known as *self-excitement*. Gradually, more and more low-magnitude shocks mask this *self-exciting* behavior of peaks followed by a rapid decay.

9.6.3 The ETAS Model

Several stochastic models have been proposed to try to model the aforementioned characteristics of earthquake data. We would like to single out two names here, Yosihiko (Yosi) Ogata and Alan G. Hawkes (1938–2023). Ogata has made path-breaking contributions to our understanding of seismic data. His most cited work is Ogata (1988). In this paper, Ogata introduces the *Epidemic Type Aftershock Sequence (ETAS)* model, which we present below. Inspired by Alan G. Hawkes' work, the ETAS model allows one to model the aforementioned self-excitement behavior of earthquakes. Hawkes' name will forever be linked to an important class of stochastic models known as *Hawkes processes*, which capture self-excitement behavior and find applications well beyond earthquake modeling, for example in finance, insurance, cyber risk, epidemiology and neuroscience. In the pages that follow, we give a gentle introduction to the ETAS model with the main aim of showing that it captures important characteristics of earthquake data. A more detailed discussion would need a considerable amount of mathematical background well beyond the level encountered in our book.

We proceed by steps, first introducing how aftershocks of a single mainshock can be modeled. This gives rise to so-called *trigger models*, where mainshocks are understood as completely random events, each triggering a sequence of aftershocks. Ogata (1988) introduced a model where not only the mainshocks but also each aftershock triggers another aftershock sequence, the same again and so on, the ETAS model. We will fit this model to those magnitudes exceeding the threshold $M_r = 3$ (which, because of the aforementioned missing small magnitudes before spring 2005, allows us to utilize the whole dataset we have); see Figure 9.29.

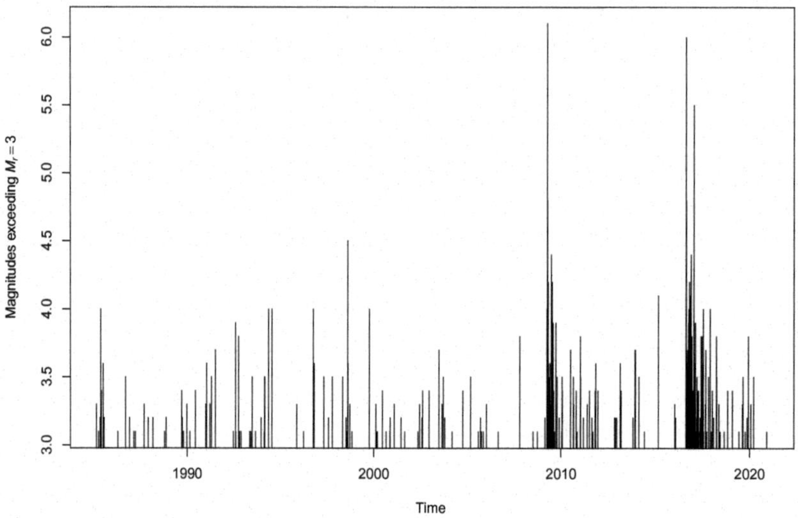

Figure 9.29 Shocks exceeding magnitude $M_r = 3$ from January 1, 1985 to December 25, 2021. Each vertical bar indicates one such exceedance. We also see from this plot that the two mainshocks on April 6, 2009 and August 24, 2016 were followed by smaller shocks. Source: Authors

Mathematically, an earthquake can be modeled as an event that happens at a time point t_i and of a certain magnitude m_i; to incorporate a spatial component, one would include the latitude and longitude of the earthquake as well. An important object to model is the so-called *conditional intensity* $\lambda(t)$, which specifies at each time point t the expected number of events per unit of time (in our case, per day); the word "conditional" reflects that, at time t, all information about earthquakes before t enters the intensity. For example, if there is, on average, one event every 10 days, we have $\lambda(t) = 0.1, t \geq 0$. This constant intensity obviously neglects the clustering behavior of earthquake events, at which point aftershock modeling comes into play.

Conditional Intensity of Aftershocks of a Single Mainshock

At the end of the nineteenth century, Japanese seismologist Fusakichi Omori (1868–1923) had already found that the conditional intensity of aftershocks after a mainshock decreases roughly as the reciprocal of the time passed after the mainshock. Utsu (1970) adapted the original formula to allow for a more general decay of the form $k/(t+c)^p$ for parameters $k, c, p > 0$, and provided the formula $10^{\alpha m - d}$ for the number of aftershocks after a mainshock of magnitude m had occurred in terms of parameters $\alpha > 0$ and $d \in \mathbb{R}$. Combining the two formulas for the number of aftershocks and their decay into the conditional intensity of an aftershock of magnitude m leads to $\lambda_{\mathrm{a}}(t) = 10^{\alpha m - d} k/(t+c)^p = 10^{\alpha m} K/(t+c)^p$ for a parameter $K > 0$. Utsu (1970) used this conditional intensity with $m = M_0 - M_r > 0$, where M_0 is the (large) magnitude of an initial earthquake and, as before, $M_r (< M_0)$ is a threshold above which modeling the magnitudes is of interest. We thus have the aftershock model

$$\lambda_{\mathrm{a}}(t) = 10^{\alpha(M_0 - M_r)} \frac{K}{(t+c)^p}, \tag{9.14}$$

which can be interpreted as the expected number of aftershocks of magnitude at least M_r on day t after an earthquake of magnitude M_0. The integral $\Lambda_{\mathrm{a}}(t) = \int_0^t \lambda_{\mathrm{a}}(s)\,\mathrm{d}s$ can then be used to compute the expected total number of aftershocks of magnitude exceeding M_r over the t days after an earthquake of magnitude M_0. And, for days $0 < t_1 < t_2$, $\Lambda_{\mathrm{a}}(t_2) - \Lambda_{\mathrm{a}}(t_1)$ provides us with the expected number of such aftershocks in the time interval $(t_1, t_2]$.

A calibration to 13 datasets of aftershock sequences led Utsu (1970) to the median estimates $\hat{\alpha} = 0.85$, $\hat{K} = 10^{-1.83} \approx 0.0148$, $\hat{c} = 0.30$ and $\hat{p} = 1.30$. With these choices of parameters, we obtain the estimate $\hat{\lambda}_{\mathrm{a}}(t) = 10^{\hat{\alpha}(M_0 - M_r)} \hat{K}/(t+\hat{c})^{\hat{p}}$ of (9.14). For example, with $M_r = 3$ and $M_0 = 6.1$, we have $\hat{\lambda}_{\mathrm{a}}(t) = 10^{0.85(6.1-3)} \times 0.0148(t+0.3)^{-1.3}$ and thus

$$\hat{\Lambda}_{\mathrm{a}}(t) = \left[10^{0.85(6.1-3)} \frac{0.0148}{-0.3}(s+0.3)^{-0.3}\right]_{s=0}^{t} \approx \frac{-21.2883}{(t+0.3)^{0.3}} + 30.5495.$$

Within the first week ($t = 7$), we thus expect to see about 19 aftershocks of magnitude exceeding 3 after an earthquake of magnitude 6.1. In total ($t \to \infty$), we expect to see about 31 such aftershocks. And, after 70 years, we still expect to see $\int_{70\times 365}^{\infty} \hat{\lambda}_{\mathrm{a}}(s)\,\mathrm{d}s = \hat{\Lambda}_{\mathrm{a}}(\infty) - \hat{\Lambda}_{\mathrm{a}}(70 \times 365) \approx 1.0137 \approx 1$ aftershock in this setting. Note that $\hat{\alpha}, \hat{K}, \hat{c}, \hat{p}$ are not representative of our earthquake data and we will thus have to estimate these parameters later on the basis of the data we have.

Trigger Models

From the aftershock model described above, a straightforward way to model earthquakes in some region is to model the mainshocks of magnitude $m_{\mathrm{m},i}$ (M_0 in (9.14)) exceeding M_r as completely random events with constant conditional intensity $\mu > 0$ and then, at time t

(given the information about each past mainshock at time point $t_{m,i} < t$ with magnitude $m_{m,i}$) to model the conditional intensity of the aftershocks by $\sum_{t_{m,i}<t} \lambda_a(t - t_{m,i})$, the sum of all conditional intensities related to each mainshock. The conditional intensity of the resulting trigger model at time t is thus

$$\lambda(t) = \mu + \sum_{t_{m,i}<t} \lambda_a(t - t_{m,i}) = \mu + \sum_{t_{m,i}<t} 10^{\alpha(m_{m,i}-M_r)} \frac{K}{(t - t_{m,i} + c)^p}. \tag{9.15}$$

A drawback of this model is that we are required to know (or at least to be able to estimate) which time points correspond to mainshocks and which to aftershocks.

Ogata's ETAS Model

The ETAS model of Ogata (1988) also considers mainshocks happening as completely random events with constant conditional intensity $\mu > 0$. Following ideas of Hawkes (1971), the ETAS model not only considers mainshocks and their aftershocks, but lets those aftershocks again have aftershocks, which again trigger aftershocks, etc. The conditional intensity at time point t, given past main- or aftershocks happening at time point t_i of magnitude m_i ($= M_0$ in (9.14)), is

$$\lambda(t) = \mu + \sum_{t_i<t} \lambda_a(t - t_i) = \mu + \sum_{t_i<t} 10^{\alpha(m_i-M_r)} \frac{K}{(t - t_i + c)^p}. \tag{9.16}$$

As both types of shocks can trigger new aftershocks, we see from (9.16) that the potentially complicated classification between mainshocks and aftershocks in the trigger model (9.15) can be avoided altogether.

Model Fitting and Assessment

Using the maximum likelihood estimator, we obtain the (rounded) estimates $\hat{\mu} = 0.0079$, $\hat{\alpha} = 0.0102$, $\hat{K} = 0.0156$, $\hat{c} = 0.8444$, $\hat{p} = 1.1431$ of the ETAS parameters μ, α, K, c, p relating to the L'Aquila magnitudes exceeding $M_r = 3$ in the years from 1985 to 2021. The left-hand side of Figure 9.30 shows the corresponding implied estimate $\hat{\Lambda}(t)$ of $\Lambda(t)$ as a function of $t \geq 0$, together with the actual number of earthquakes exceeding $M_r = 3$; ideally, a complete analysis would include CIs. The right-hand side of Figure 9.30 shows a Q–Q plot of the *transformed interarrival times* or *residuals* $\hat{\Lambda}(t_i) - \hat{\Lambda}(t_{i-1})$ (that is, the change in the conditional intensity between adjacent earthquake events) against Exp(1)-quantiles. Under the true conditional intensity λ, the residuals can be shown to be an iid sample from an Exp(1) distribution, and the Q–Q plot does not hint at any significant departure of $\hat{\Lambda}$ from the true Λ in this sense.

Historical and Simulated Shocks

The left-hand side of Figure 9.31 shows the historical shocks exceeding magnitude $M_r = 3$ as in Figure 9.29 (now in logarithmic scale), with corresponding estimated conditional intensity $\hat{\lambda}(t)$ displayed below the shocks. The fitted ETAS model can also be used for simulation purposes. From the past conditional intensity, one can generate the time of the next earthquake shock (its magnitude distribution is not specified by the ETAS model but can be generated from a fitted distribution or drawn at random from past magnitudes; we do

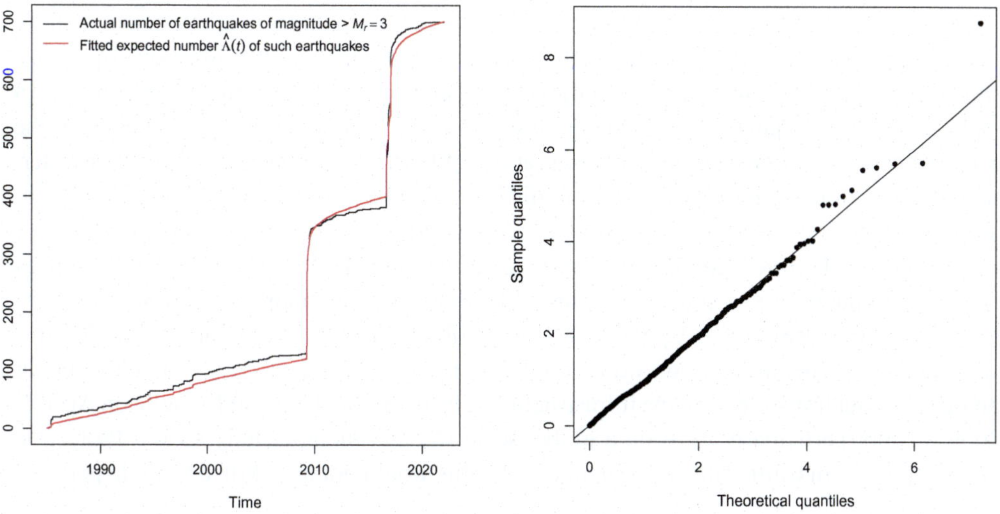

Figure 9.30 Left: The estimate $\hat{\Lambda}(t)$ as a function of t together with the actual number of earthquakes exceeding $M_r = 3$. Right: A Q–Q plot of differences of the form $\hat{\Lambda}(t_i) - \hat{\Lambda}(t_{i-1})$ versus Exp(1)-quantiles. Source: Authors

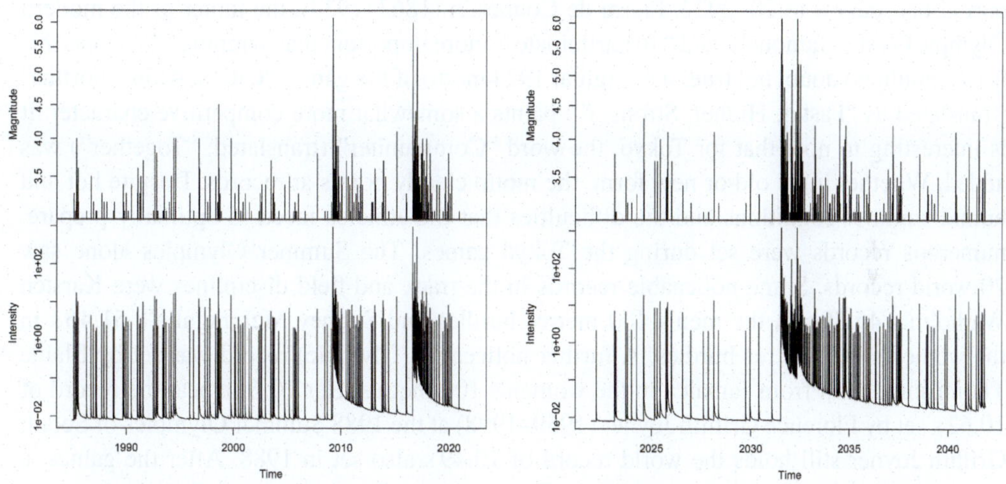

Figure 9.31 Left: Shocks exceeding $M_r = 3$ and corresponding estimated conditional intensity $\hat{\lambda}(t)$. Right: One set of simulated shocks and their conditional intensity over 20 years. Source: Authors

the latter), which then gives an updated conditional intensity for the next shock, and so on. We omit further details here and instead refer to the right-hand side of Figure 9.31, which shows a single set of shocks and their conditional intensity obtained from such a simulation on the basis of the fitted ETAS model; repeated use of such simulations would allow one to make (probabilistic) predictions about future earthquakes.

Ogata (1998) contains a further generalization of the ETAS model in the form in which we have presented it. Besides the propagation of shocks over time, this generalization also allows shocks to propagate through space. Needless to say, appropriately modeling earthquakes with many parameters requires large amounts of data. A ray of hope, no doubt, comes from the availability of ever more fine satellite observations of Earth. This, coupled with a better understanding of tectonic dynamics, may eventually lead to usable prediction technology for earthquakes. Further, remote sensing devices will produce large amounts of data and open the door to applications of artificial intelligence and machine learning. Even with the possible advances hoped for in the not too distant future, understanding and communication problems, as encountered in the case of the L'Aquila earthquake, will remain with us for a longer time. For this specific catastrophe, the solution would not have required more data, better models or better predictions, but investments in the infrastructure. The main shock that hit L'Aquila on April 6, 2009 was not particularly large on a global scale. As an official at Italy's Civil Protection Agency put it: "In California, an earthquake like this would not have killed a single person."

9.7 Records

Owing to the coronavirus pandemic, the 2020 Tokyo Summer Olympic Games took place with a one-year delay from July 23 to August 8, 2021. They were followed by the Paralympic Games, from August 24 to September 5, 2021 and the Winter Olympics in Bejing over the period February 4 to 20, 2022. Pierre de Coubertin (1863–1937), the father of the modern Olympic Games, famously said "To participate is more important than winning." Whereas this is (or should no doubt be) true, the original 1924 motto of the games "Citius, Altius, Fortius" (translated as "Faster, Higher, Stronger") paints a somewhat more competitive character. It is interesting to note that for Tokyo, the word "Communiter" (translated: "Together") was added. Whether in its old or new form, the motto clearly points at records. Despite hot and humid summer conditions and the difficulties that the athletes faced to optimally prepare, numerous records were set during the Tokyo games. The Summer Olympics alone saw 20 world records. Some noticeable records in the track and field disciplines were Karsten Warholm's 45.59 s in the men's 400 metres hurdles and Sydney McLaughlin's 51.46 s in the women's 400 metres hurdles. A further noticeable (Olympic) record was set by Elaine Thompson-Herah from Jamaica in the women's 100 metres. Her 10.61 s beat the record of 10.62 s set by Florence Griffith Joyner (1959–1998) at the 1988 Summer Olympics of Seoul; Griffith Joyner still holds the world record of 10.49 s also set in 1988. After the games, a lot has been written on why so many records were set. Concerning athletics, the question quickly became whether athletes simply had become better or were there other factors at play. A full answer will no doubt involve a combination of both. However, concerning the "other factors", we learned about foam-and-plate super spikes and a novel, fast, springy track design. For the super shoes, the key experience was one of propulsive sensation, so much so that some used the terminology of mechanical doping to describe what was happening on the track. In this regard, a similar story is that of marathon world record holder Eliud Kipchoge, who on October 12, 2019, became the first athlete to break the magical mark of two hours (the final time was 1:59:40) at a special event in Vienna, Austria. He wore a prototype of the Nike Air Zoom Alphafly Next%, which many people believe played an important role in this (unofficial) record; as this was not an open event, the run did not count as a new marathon record. Regarding the Tokyo track, its designer Andrea Vallauri was quoted by BBC Sport on

August 4, 2021 as "it provides shock absorption and some energy return, like a trampoline, and in the end gives a 1%–2% performance advantage to the athletes." Sydney McLaughlin said, after her record-breaking run, that the track "gave you energy right back and pushes you and propels you forward", adding, "It was really cool [. . .] to push the boundaries of what is possible." In the past, we have experienced certain athletic events leading to amazing jumps in performance. Particular examples were seen in the 1968 Mexico Summer Olympic Games, which were held in Mexico City at an altitude of 2240 m. Here the high altitude effect became clearly noticeable. Some of us may still remember the long jump event where Bob Beamon shattered the world record by 55 cm putting it at an almost unbelievable 8.90 m; see Figure 9.32 for a cartoonish interpretation. It was again broken by Mike Powell in 1991, who,

Figure 9.32 World record long jump broken at the Kangaroolympics. Source: Enrico Chavez

in Tokyo, jumped 8.95 m during a most amazing competition with Carl Lewis. Though the wind speed played a role during the event, the numbers (in meters) for the different rounds speak for themselves: Mike Powell: 7.85, 8.54, 8.29, X, 8.95, X; and Carl Lewis: 8.68, X, 8.83w, 8.91w, 8.87, 8.84 ("X" indicates an invalid attempt, "w" a wind-assisted one not counted as a valid record). Another sporting event that will always be linked to Mexico City is the introduction by Dick Fosbury of the Fosbury Flop in the high jump, by jumping over the bar backwards whereas the prevailing methods involved jumping forwards or sideways. With this technique (new at the time but the universally preferred method since) it is possible for the high jumper's body to fly over the bar while his or her center of gravity passes below it. With a lower center of gravity, less energy is needed to pass over the bar. Particularly in the high jump, techniques have evolved dramatically since the early days of the recording of results in athletics.

A record is defined as an event or a measurement that surpasses previous similar events or measurements. It is by thinking more carefully about the meaning of the word 'similar' that the problems start. In athletics, this may be interpreted as under similar or even equal circumstances. Of course, as human and technical progress advances, similar and equal defy a proper, workable definition. Especially in swimming events, records have been shattered at a stunning rate. This is no doubt due to improved training and techniques. You may however recall the appearance of technology-driven knee length swimsuits, referred to as tech suits. Their surface design was deemed to give an unfair advantage and were hence banned. The race towards ever "Faster, Higher, Stronger" will however go on; we can only hope that fairness through a broadly supported "Together" will prevail in the end.

There exists a huge literature on the topic of records. At some time, you may have owned a copy of the *Guinness Book of Records*. The birth of the idea for writing such a summary text on records can be found on `guinnessworldrecords.com`. We found the story so lovely that we decided to include part of it here:

> The idea came about in the early 1950s when Sir Hugh Beaver (1890–1967), Managing Director of the Guinness Brewery, attended a shooting party in County Wexford. There, he and his hosts argued about the fastest game bird in Europe, and failed to find an answer in any reference book. In 1954, recalling his shooting party argument, Sir Hugh had the idea for a Guinness promotion based on the idea of settling pub arguments and invited the twins Norris (1925–2004) and Ross McWhirter (1925–1975) who were fact-finding researchers from Fleet Street to compile a book of facts and figures. Guinness Superlatives was incorporated on 30 November and the office opened in two rooms in a converted gymnasium on the top floor of Ludgate House, 107 Fleet Street in London.

The *Guinness Book of World Records 2021* lists over 40 000 records. Since 1955, over 140 million copies in over 21 languages and in more than 100 countries have been sold. We can just imagine how many pub brawls have been averted!

We have already made the link between EVT and the analysis of records; see, for instance, Example 9.6. We also discussed several examples within the context of the Dutch sea-level data in Section 9.5.2. We stressed that EVT started from the basic assumption that successive observations are realizations from an iid model. In answering the questions Q1 to Q3 in Section 9.5.2, we addressed the importance of non-stationary behavior such as trends and change points, which often appear due to climate change. As we briefly discussed in the examples above, the topic of non-stationarity is also of immediate relevance to sports data. An interesting mathematical question now concerns the growth of records in truly iid data. An answer to this question could serve as a yardstick against which record growth in practice can be gauged. Embrechts et al. (1997, Lemma 6.2.21) yielded such a result. When one discusses extremes, and especially records, from a model point of view, it is standard to assume that the underlying model distribution function F is continuous. This avoids the possibility of ties. In sports events, especially involving the measurement of time, technology has evolved to such a level of precision that we have come very close to that ideal model set-up.

Suppose now that we have a sequence $X_1, X_2, \ldots$ of iid random variables with a continuous distribution function F. The number of records in the first n random variables from this sequence can be defined as

$$N_1 = 1, \quad \text{and} \quad N_n = 1 + \sum_{k=2}^{n} I_{\{X_k > M_{k-1}\}}, \quad n \geq 2,$$

where we recall that the indicator function $I_{\{X_k > M_{k-1}\}}$ is 1 if $X_k > M_{k-1}$ and is 0 otherwise. By definition, the number N_1 of records after the first observation X_1 is 1. Then, the number

N_n of records after the first n observations $X_1, X_2, \ldots, X_n$ is obtained by counting, in addition to the first record, how many X_k exceed the maximum $M_{k-1} = \max\{X_1, X_2, \ldots, X_{k-1}\}$, so how many of the observations $X_2, \ldots, X_n$ are records. As we have done multiple times in Chapter 8, it helps to look at the extreme cases. If $X_1 > X_2 > \cdots > X_n$, then $M_{k-1} = X_1$, which is greater than X_k, so all the indicators $I_{\{X_k > M_{k-1}\}}$ are 0 and we obtain $N_n = 1$ for all $n \geq 1$, and indeed we only have the one record X_1 in this case. And if $X_1 < X_2 < \cdots < X_n$, then $X_k > M_{k-1}$ for all $k = 2, 3, \ldots, n$ and so we have $N_n = 1 + \sum_{k=2}^{n} 1 = 1 + (n-1) = n$, and indeed every one of the n observations is a record in this case. Embrechts et al. (1997, Lemma 6.2.21) provide the mean and variance of the random variable N_n; for the proof, see this reference.

Lemma 9.14 (Mean and variance of N_n). *Suppose $X_1, X_2, \ldots \overset{\text{iid}}{\sim} F$ for a continuous distribution function F. Then*

$$\mathbb{E}(N_n) = \sum_{k=1}^{n} \frac{1}{k} \quad \textit{and} \quad \operatorname{var}(N_n) = \sum_{k=1}^{n} \left(\frac{1}{k} - \frac{1}{k^2} \right).$$

Note that $\mathbb{E}(N_n) - \log(n)$ converges, for $n \to \infty$, to the Euler–Mascheroni constant $\gamma = 0.5772\ldots$, so $\mathbb{E}(N_n) \approx \log(n) + \gamma$ for n large. The left-hand side of Figure 9.33 shows $\mathbb{E}(N_n)$ as a function of n, and the right-hand side of this figure shows the relative error $|(\mathbb{E}(N_n) - (\log(n) + \gamma))/\mathbb{E}(N_n)|$ as a function of n. The left-hand side of Figure 9.34 shows

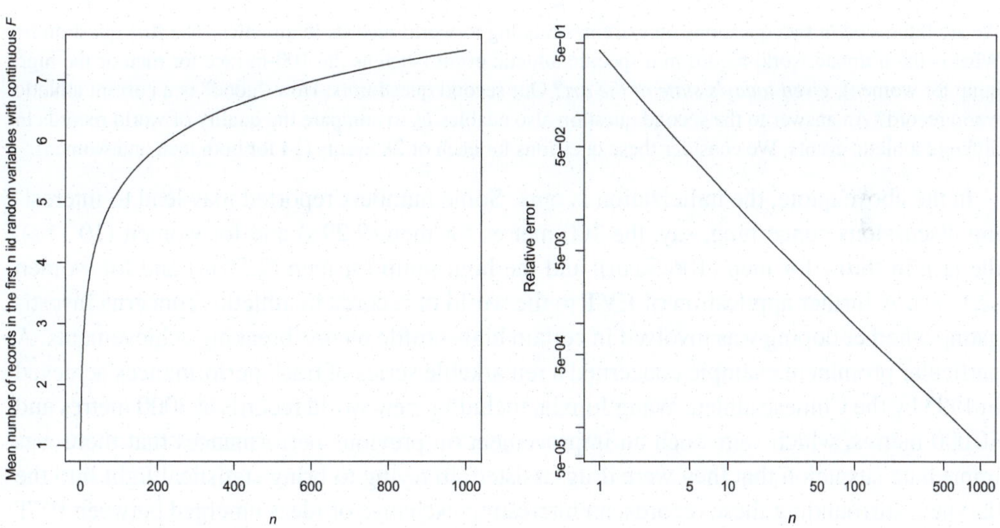

Figure 9.33 Left: Mean number of records in the first n iid random variables $X_1, X_2, \ldots$ from a continuous F. Right: Relative error with respect to the approximation $\log(n) + \gamma$. Source: Authors

both the expected and actual numbers of records for the Dutch data, as used in Q1 in Section 9.5.2. The right-hand side of this figure shows the same plot with a logarithmic x-axis, so that the mean number of records is roughly linear, which simplifies a comparison with the actual number of records.

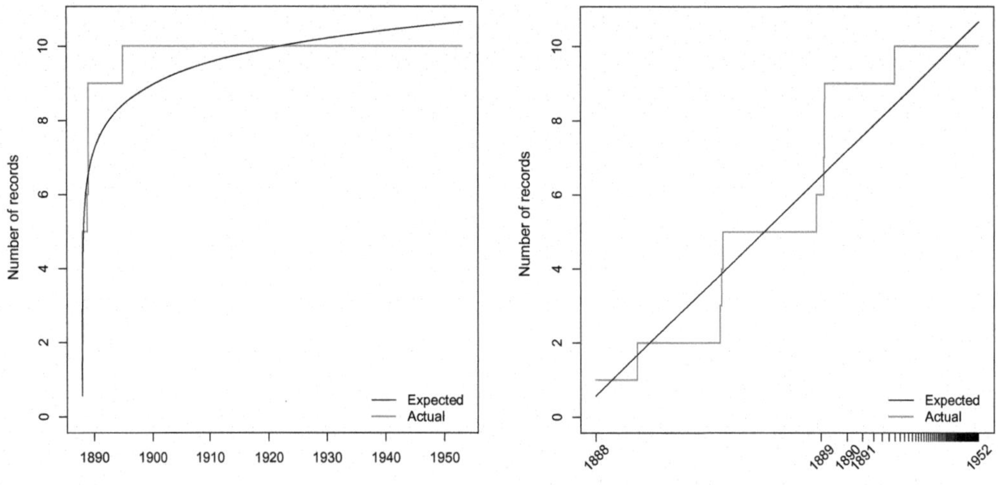

Figure 9.34 Left: Expected and actual number of records for the Dutch data as used in Q1 in Section 9.5.2. Right: The same plot with logarithmic x-axis. Source: Authors

For an application of EVT to records in athletics, see Einmahl and Magnus (2008). These records have a clear finite upper bound, so the Weibull model ($\xi < 0$ in (9.3)) becomes relevant. In this case, the finite right endpoint $\mu - \sigma/\xi$ needs to be estimated. From the abstract of Einmahl and Magnus (2008), we quote:

We are interested in two questions on extremes relating to world records in athletics. The first question is: What is the ultimate world record in a specific athletic event (such as the 100-m race for men or the high jump for women), *given today's state of the art*? Our second question is: How "good" is a current athletic world record? An answer to the second question also enables us to compare the quality of world records in different athletic events. We consider these questions for each of 28 events (14 for both men and women).

In the above quote, the italicization is ours. Some numbers reported may lead to interesting discussions concerning, say, the 100 metres for men (9.29 s) and for women (10.11 s), the javelin throw for men (106.50 m), and the high jump for men (2.50 m) and for women (2.11 m). A further application of EVT to the world of records in athletics concerns investigating whether doping was involved in certain high-profile record-breaking achievements. A particular prominent example concerned a remarkable series of track performances achieved in 1993 by the Chinese athlete Wang Junxia, including new world records at 3000 metres and 10 000 metres, which were such an improvement on previous performances that there was immediate suspicion that they were drug-assisted. In trying to bring statistical light into the darkness surrounding these records, an interesting exchange of ideas emerged between EVT specialists, in particular Richard Smith and Jonathan Tawn; we have already encountered both in Section 9.3.1. Robinson and Tawn (1995) concluded that "Whereas Wang's time is largely inconsistent with past data, our final conclusion is that at present, no legal case can be made that her time is from a different population [. . .]." On the basis of a so-called Bayesian approach to EVT, Smith (1997) concluded however that "This seems to give [. . .] convincing evidence that Wang's record was indeed a very unusual outlier, whether drug-assisted or not." Currently, Wang Junxia still officially holds the world record for the 3000 metres. In Stephenson and Tawn (2013), the authors proposed an EVT-based global model to fit athletics data over all Olympic distances and all times, allowing them to identify and rank the best athletics track performances in history. When applied to the male and female running distances between 100 and 10 000 metres, the best performance scores are for Lee Evans (USA) with a 1968 world record run in the 400 metres and the 1988 100 metres of Florence Griffith Joyner, also from the USA: we mentioned her before. It is perhaps interesting to learn that the exceptional Jamaican sprinter Usain Bolt appears third and eighth on the list for his world records in the 100 and 200 metres events, both achieved in 2009 and still current at the time of publication.

9.8 The Sinking of the MV Derbyshire

Before we discuss the tragic fate of the MV Derbyshire, we would briefly like to comment on why ships in the English language have a female gender. Presumably, the best explanation is "tradition". The custom may indeed be based on the fact that, historically, vessels often supported a bow-statue of a goddess believed to offer protection at sea for the crew that sailed on board. Captain Ernst Lindemann of the German battleship Bismarck in World War II on the other hand referred to his ship as 'he'. It is perhaps interesting to note that the official maritime and shipping industry newspaper, Lloyd's Register of Shipping, now calls ships 'it'. Keeping with tradition, we will refer to a ship as 'she'.

For the historic facts, we used several sources that we will refer to in due course. For the early stage of the story we have relied on documents from UK historical naval museums, in particular from the Merseyside Maritime Museum (2021). The MV Derbyshire was a massive ore-bulk-oil (OBO) combination carrier built in 1976, belonging to the so-called Bridge Class of bulk carriers. Her dimensions were indeed massive: length 294 m, width (also referred to as the beam in nautical terms) 44 m, she could carry about 160 000 (metric) tonnes of cargo. She had a double hull design and, like the Titanic as well as the Bismarck, was considered unsinkable. Her original name, before a later name change, was the Liverpool Bridge. The MV Derbyshire started her final voyage on July 11, 1980, when she left the port of Sept-Îles in Quebec, Canada, for Kawasaki in Tokyo Bay, carrying 157 446 tonnes of iron ore. A few days before her planned arrival in Japan she ran into Typhoon Orchid, a storm that reached wind speeds of up to 150 km/h. This typhoon was extreme but not particularly unusual. A characteristic of Typhoon Orchid, however, was that its wind field caused abnormally high swell. The MV Derbyshire's last known message, on September 9, 1980, reported a severe tropical storm. When she did not arrive at her port of destination, air and sea rescue efforts were launched. These initiatives resulted in finding a large oil spill close to the coordinates of her last reported position. Later, also a torn-off rescue boat was discovered. It became clear that the MV Derbyshire had foundered. Unusually, no mayday distress message was received (the internationally agreed "mayday" call for help at sea was inspired by the French "m'aidez", indeed meaning "help me"). It became clear that the ship must have sunk very suddenly. This tragic event caused the loss of lives of 42 crew members and two of their wives. There were no survivors. The MV Derbyshire is the biggest British-registered ship ever lost at sea. The question early on became "Why?" Was human error involved, a faulty design or was it a tragic Act of God as we already encountered at the beginning of Section 9.2?

The UK Government decided that, on the basis of the little information available at the time, a full and formal investigation was not warranted. However, eager to have these questions answered, the families of the crew formed the Derbyshire Family Association (DFA). Family members definitely wanted the truth on whether human error was to blame. The DFA secured financial support for a continuation of the investigation. As a consequence and after a series of further, less tragic, accidents with sister ships from the Bridge Class, an inquiry was held over the period October 1987 to March 1988. The formal investigation, presented on January 18, 1989 concluded as follows:

For the reasons stated in this Report the Court finds that the Derbyshire was probably overwhelmed by the forces of nature [an Act of God] in typhoon Orchid, possibly after getting beam on to wind and sea, off Okinawa in darkness on the night of 9/10 September 1980 with the loss of 44 lives. The evidence available does not support any firmer conclusion.

The nautical expression "beam on to wind and sea" refers to having full wind and wave forces battering a ship at right angles to a ship's course (so, on her side). This formulation has a hint of wrong maneuvering (human error?) hidden in its wording. Indeed in stormy weather, ships are advised to steer right into the crest of the waves with only a small angle of deviation. Needless to say that this conclusion did not satisfy the members of the DFA who initiated further efforts in order to try to locate the sunken ship.

Figure 9.35 Left: The MV Derbyshire with the nine hatch covers; the number 1 hatch cover as well as the mushroom-type ventilation covers on the front deck played a crucial role in the sinking of the ship. Right: Jonathan Tawn from Lancaster University who was appointed sole statistical expert for the 2000 court case on the sinking of the MV Derbyshire. Sources: National Museums Liverpool (left) and Lancaster University (right)

For the next stage of the story, we quote from Vishal (2019):

In June 1994, the wreck of Derbyshire was found at a depth of 4 km, spread over 1.3 km. An additional expedition spent over 40 days photographing and examining the debris field looking for evidence of what sank the ship [200 hours of video, 137 000 photographs]. Ultimately it was determined that waves crashing over the front of the ship had sheared off the covers of small [mushroom shaped] ventilation pipes near the bow [see Figure 9.35]. Over the next two days, seawater had entered through the exposed pipes into the forward section of the ship, causing the bow to slowly ride lower and lower in the water. Eventually, the bow was completely exposed to the full force of the rough waves which caused the massive hatch on the first cargo hold to buckle inward allowing hundreds of tons of water to enter in moments [see Figure 9.35]. As the ship started to sink, the second, then third hatches also failed, dragging the ship underwater. As the ship sank, the water pressure caused the ship to be twisted and torn apart by implosion.

The crucial question implied by the above quote remained whether the force from the incoming waves, also referred to as *green seas*, could result in the initial collapse of hatch cover 1. Hatch cover 1 can be seen covering the first hold counting from the bow on the left-hand side of Figure 9.35. Typhoon Orchid was reported to produce waves up to 11 m in height, possibly with an occasional higher, so-called *rogue*, wave. For a long time, maritime myth has referred to the existence of wave heights of 100 ft (30 m). So far, and this mainly in the Northeast Atlantic Ocean between Scotland and Norway, solitary rogue waves with heights between 20 m and 30 m have been recorded. The measurement of rogue waves should be calibrated by the height of the surrounding waves. A rogue wave really sticks out like a tower block. In 1995, the Draupner Wave off Norway was measured as 25.6 m with a surrounding wave height of about 12 m, hence with a factor of 2.13. In November 2020, off the coast of Vancouver Island, a rogue wave was measured standing at 17.6 m with surrounding waves at about 6 m, resulting in an extreme, record-breaking, factor close to 3. If you are interested in the topic, consult Gemmrich and Cicon (2022). Whether such a massive solitary wave hit the MV Derbyshire we do not know. High waves, as in Typhoon Orchid, tend to have very steep fronts, which allow little time for a ship to rise sufficiently high into the successive incoming waves. As a consequence, over a two-day period, the first hatch cover was persistently battered by very heavy waves, until it eventually gave way, sealing the tragic fate of the MV Derbyshire.

At this point of the investigation, an EVT-based analysis was called for, in order to compare and contrast its independent findings with the previous contradictory statistical work that had been done. The EVT specialist Jonathan Tawn (see the right-hand side of Figure 9.35) from Lancaster University provided this analysis during a reopened court case presided over by Mr. Justice Colman (1938–2017). The Court sat for a total of 54 days from April 5, 2000, and concluded its oral hearings on July 26, 2000. The final report, prepared by Jonathan Tawn and Janet Heffernan, stood at 125 pages. It no doubt constitutes one of the prime achievements of EVT-based statistical analysis and was acknowledged as such by the prestigious science journal *Nature* on March 25, 2015. Below we will briefly sketch the link between the questions asked in Court and EVT. If you are interested in further details, the following papers are recommended. Heffernan and Tawn (2004) provides a summary aimed at a more general audience. For a more technical discussion, see Heffernan and Tawn (2001, 2003). See Tawn (2017) for an excellent talk given to the Royal Statistical Society's Local Group of Northern Ireland. It refers to the 2014 Research Excellence Framework for UK Universities and highlights how academic research may have considerable impact on important societal questions.

As we have already learned, the critical event was the collapse of the hatch cover on hold 1. These covers were constructed to withstand a force of 42 kPa (kilopascal), one kPa being 0.01 kilogram-force per square cm. For the statistical analysis, data were obtained from several sources. Satellite information from the day of the sinking gave precise meteorological information together with hindcast (as different from forecast) wave data. Laboratory measurements were based on a 1:65 scale replica of the MV Derbyshire in a wave tank provided by the Maritime Research Institute (MARIN), located in Wageningen in The Netherlands. The statistical design for the latter was a so-called $2 \times 4 \times 3$ factorial design accounting for an intact versus a damaged ship (hence the factor 2), 4 wave conditions and 3 speeds. From these combined measurements, several so-called covariates were derived. Advanced EVT became essential for deriving reliable estimates, given that the wave impact on the ship's hatch cover required to have led to the sinking was larger than any derived in the tank studies;

here the time for running these experiments played an important role. Hence, one needed to go beyond the range of the data provided. The essential statistical task became the estimation of the conditional probability

$$\mathbb{P}(\text{“wave impact} > 42\,\text{kPa on hatch cover 1”} \mid \text{“various ship and sea conditions”})$$

for the full extent of the MV Derbyshire's path (around two full days) through the typhoon. This dynamic time component was crucial for the statistical modeling. Recall that "wave impact > 42 kPa" corresponds to "hatch cover 1 failure". From the data available, EVT provided the best-fitting model for the relevant conditional distribution function

$$F_u(x-u; z_1, z_2, z_3) = \mathbb{P}(\text{“wave impact”} \le x \mid \text{“wave impact”} > u; z_1, z_2, z_3)$$

for some high threshold u (10 kPa, say), z_1, z_2, z_3 being the covariates, that is, the variables encoding the covariate information (z_1 encoding the intact/damaged state, z_2 the wave condition and z_3 the speed in the aforementioned MARIN set-up). Whereas this is exactly the set-up for the application of the POTM approach to EVT, as discussed in Section 9.4, Tawn and Heffernan had to extend existing EVT theory in order to take account of the specific, especially dynamic, aspects of the various covariates.

This study became the first to use EVT in the context of naval architecture. It provided a review of the adequacy of the current international design standards for shipping in relation to the hatch strength of ocean-going carriers. As a result of a further, more involved, EVT-based statistical analysis, Lloyd's Register of Shipping eventually proposed a 35% strengthening of the hatch covers. For all 1720 bulk carriers built between 2008 and 2012 the strength of the hatch covers was increased by 35% from the previous design standards, and, for the 5830 previously built bulk carriers, hatches were strengthened and new inspection and maintenance procedures were required. The hatch strengthening was one of several recommendations that were eventually implemented. There have been no sinkings of ocean-going bulk carriers since the new design standards were introduced in 2004, whereas on past evidence over 100 such sinkings of ocean-going bulk carriers would have been expected over a ten-year period, resulting in numerous lives of seamen lost. The relevance of EVT, through the work of Jonathan Tawn and Janet Heffernan, becomes clear from the following conclusion by Lord Justice Colman (see Colman, 2000, Sections 6.13, 6.14, 11, 14 and Appendix 17):

> The contribution that the extreme values group at Lancaster made to identify the cause of the sinking was of absolutely fundamental importance to the outcome of this Investigation.

Very much in line with our discussion of the 1953 Great Flood from Chapter 1, for the investigation into the sinking of the MV Derbyshire, interdisciplinary collaboration also played a very important role.

Every disaster discussed so far hides untold human suffering. In the case of the MV Derbyshire, this suffering became a face through the DFA. Wednesday, September 9, 2020, saw the 40th anniversary of the sinking of the MV Derbyshire. In Lambert (2020), Paul Lambert (MBE, Chairman of the DFA, and brother of one of the seamen who died) gave a moving account of the struggle towards finding the truth:

> You know, I honestly believe that the MV Derbyshire tried her best to protect her crew. Every time her head would go down into the waves, she would fight to raise her head to try and stay afloat. She would do this

time and time again until she was too tired, her bow and Hold No 1 finally so heavy with the sea filling them up that she could not fight anymore and would start to slowly sink beneath the waves, taking all hands with her.

We offer this section on the MV Derbyshire as a tribute to all who died on its final voyage and to those who contributed persistently towards the ultimate pursuit of truth.

9.9 Lessons Learned

The risks we discuss in this book all have to do with rare events for which the probability or frequency of occurring (*if*) is small but, given that they occur, the consequences or severities (*what if*) can be catastrophic. In this chapter we introduced the modeling of extreme events such as maxima, records and return periods or high quantiles. After a brief historical excursion, we introduced the two main probabilistic approaches from EVT and their statistical counterparts, the BMM and the POTM. The two approaches were compared and contrasted. We applied EVT to numerous examples, the Dutch sea-level data from Chapter 1, the L'Aquila earthquake data from Chapter 5 and the Netflix data as an example from the realm of finance. An important lesson learned was how to obtain the relevant data, how to clean them and convert them to a format ready to be analyzed with the relevant statistical tools. For the earthquake data we went a step further by introducing the ETAS model from the realm of "stochastic processes". These models are also relevant for the modeling of pandemic data. Some further examples finally led to an important EVT application, the sinking of the MV Derbyshire. We learned in the latter example that an EVT-based analysis of the data turned out to be crucial for reaching the final court decision on the cause underlying the sinking of the ship.

10

On Climate Change and Related Risk

We do not inherit the earth from our ancestors,
we borrow it from our children.

Native American proverb

10.1 From IPCC to COP26

The quote "We have only one planet. If we screw it up, we have no place else to go." sounds as though it might have been taken from today's newspaper, or from a banner of the environmental movement Extinction Rebellion or from social platforms like X (formerly Twitter). However, these warning words were made by US senator John Bennett Johnston from Louisiana on June 23, 1988. That day, US senators met in the Dirksen Senate Office Building in Washington DC to discuss issues related to global warming. During this meeting, three key conclusions were formulated. We quote from Sullivan (2018):

(1) The Earth is warmer than at any other time in recent history.
(2) This global warming can be attributed, with 99% certainty, to a man-made increase in the greenhouse effect, primarily from the burning of fossil fuels and changes in the way we use land.
(3) This greenhouse effect is making extreme weather events like heat waves, storms, and droughts more frequent and intense.

Following this event, in the same year, a major United Nations climate body was established by the United Nations Environment Programme and the World Meteorological Organization: the *Intergovernmental Panel on Climate Change (IPCC)*. The latter's role is defined in IPCC (1998):

The role of the IPCC is to assess on a comprehensive, objective, open and transparent basis the scientific, technical and socio-economic information relevant to understanding the scientific basis of risk of human-induced climate change, its potential impacts and options for adaptation and mitigation. IPCC reports should be neutral with respect to policy, although they may need to deal objectively with scientific, technical and socio-economic factors relevant to the application of particular policies.

The IPCC currently has 195 member countries with thousands of people from different fields contributing to its findings. Assessment reports summarize numerous scientific publications (assessed by IPCC scientists) to disseminate state-of-the-art knowledge on the drivers of climate change, its impact and projections, but also on the possible strategies of adaptation and mitigation. In 2007, the IPCC and former US Vice President Al Gore were jointly awarded the Nobel Peace Prize "for their efforts to build up and disseminate greater knowledge about man-made climate change, and to lay the foundations for the measures that are needed to counteract such change".

In IPCC (2021), the IPCC observes changes in the Earth's climate in every region and across the whole climate system. Many of these changes are not only unprecedented across thousands of years, but, like the continued sea-level rise in Holland, are irreversible over a very long future time period. On an almost daily basis we are confronted with the social and economic consequences of extreme climate events. These events, and the way in which they are reported in the media, highlight the world's vulnerability to the consequences of climate change.

The Glasgow International Conference on Climate Change, COP26, took place from October 31 to November 12, 2021; COP stands for Conference of the Parties. The United Nations established this series of conferences with the inaugural one, COP1, in Berlin in 1995. At that conference, the first joint measures in international climate action were agreed upon. For its 26th version in Glasgow, almost 40 000 delegates from 200 countries registered. Despite travel restrictions due to the coronavirus pandemic, this made COP26 comfortably the best attended COP in history. The various IPCC reports yield chilling reading about the state of our global climate. Science provides politicians worldwide with a blueprint of "what is to come" and "which actions are needed". It is up to world leaders to take these actions; in Glasgow, over 100 politicians were present. Numerous actions were taken, many more were promised. During the conference, several highly attended protest marches also took place. Greta Thunberg, a Swedish climate activist, told young protesters at a march on Friday November 5 that the UN talks were "now a global north greenwash festival, a two-week long celebration of business as usual and blah blah blah" and branded Glasgow a failure. The future will tell how important this summit was. One concrete deal has become *The Glasgow Climate Pact*. Indeed, at the end of the conference, it was officially announced that more than 40 countries – which include major coal-users like Poland, Chile and Vietnam – agreed to shift away from coal. Unfortunately, some of the world's most coal-dependent countries, including Australia, India, China and the United States, have not yet signed up. Moreover, the agreement does not cover other fossil fuels such as oil or gas. For next steps in the direction of moving away from fossil fuels we certainly have to wait for COPn with $n > 26$.

The full IPCC-AR6 scientific report (IPCC, 2021) amounts to a staggering 4000 pages. The "AR6" in IPCC-AR6 stands for the Sixth Assessment Report. At a summary level, its key findings are:

A. The Current State of the Climate

A.1. It is unequivocal that human influence has warmed the atmosphere, ocean and land. Widespread and rapid changes in the atmosphere, ocean, cryosphere [the frozen water part of the Earth system] and biosphere have occurred.

A.2. The scale of recent changes across the climate system as a whole – and the present state of many aspects of the climate system – are unprecedented over many centuries to many thousands of years.

A.3. Human-induced climate change is already affecting many weather and climate extremes in every region across the globe. Evidence of observed changes in extremes such as heatwaves, heavy precipitation, droughts, and tropical cyclones, and, in particular, their attribution to human influence, has strengthened since AR5.

A.4. Improved knowledge of climate processes, paleoclimate evidence [based on ancient records] and the response of the climate system to increasing radioactive forcing [the change in energy flux in the atmosphere caused by natural or anthropogenic factors of climate change] gives a best estimate of equilibrium climate sensitivity of 3 °C, with a narrower range compared to AR5.

B. Possible Climate Futures

B.1. Global surface temperature will continue to increase until at least mid-century under all emissions scenarios considered. Global warming of 1.5 °C and 2 °C will be exceeded during the 21st century unless deep reductions in CO_2 and other greenhouse gas emissions occur in the coming decades.

B.2. Many changes in the climate system become larger in direct relation to increasing global warming. They include increases in the frequency and intensity of hot extremes, marine heatwaves, heavy precipitation, and, in some regions, agricultural and ecological droughts; an increase in the proportion of intense tropical cyclones; and reductions in Arctic sea ice, snow cover and permafrost.

B.3. Continued global warming is projected to further intensify the global water cycle [the continuous movement of water on, above and below the surface of Earth], including its variability, global monsoon precipitation and the severity of wet and dry events.

B.4. Under scenarios with increasing CO_2 emissions, the ocean and land carbon sinks [natural systems that suck up and store carbon dioxide from the atmosphere] are projected to be less effective at slowing the accumulation of CO_2 in the atmosphere.

B.5. Many changes due to past and future greenhouse gas emissions are irreversible for centuries to millennia, especially changes in the ocean, ice sheets and global sea level.

The report then continues with a global risk assessment (**C**) and possible mitigation measures in order to limit future climate change (**D**). The bulk of the report gives scientific evidence underlying the claims made in **A** to **D**.

It is clear that in the context of our book there is only very little we can contribute to the overall wealth of scientific material on climate change. On the other hand, referring to the title of our book, we cannot bypass this discussion that is so fundamental to current and future generations. First of all, through a very brief summary of the material produced in the various IPCC reports, we hope to convince you as reader to spend some time going over parts of these reports and indeed get involved in the quest for solutions, possibly as scientists, certainly as citizens wherever you are, and as inhabitants of our blue planet. In any case, recall how we started this section: "We have only one planet. If we screw it up, we have no place else to go." Concerning the final part of the above quote, entrepreneurs like Elon Musk, the CEO of Tesla, and Richard Branson, the boss of Virgin Atlantic have set their mind, and money, on space travel, the former through SpaceX, the latter though Virgin Galactic. In the case of Elon Musk "SpaceX wants to reach Mars so that humanity is not a single planet species." At the same time, the discovery of ever-new exoplanets (planets outside our solar system; there are already more than 4500 of them), and even the first planet(s) outside our galaxy, have further raised (some of) humanity's hope of eventually being able to "pack one's bags and move to another place". Who knows what future civilizations will be able to achieve. At present, it is highly doubtful, though, that such a life would be more pleasant than on Earth. As such, we very much stress that we have immediate problems to solve, with our feet (at least for a longer while) still firmly resting on Mother Earth.

Throughout our book, we have looked at rather isolated examples of risks such as floods, earthquakes, economic events, and technological disasters, mostly without considering a more holistic view of the ambient environment, that is our global planet, where these events take place. This restriction is especially felt when we discuss natural disasters. It is impossible to gain a full understanding of, for example, the flood risk in a particular country without getting involved with more global atmospheric modeling. This more global approach stands at the center of the various IPCC reports. We hope, however, that through giving some pointers to the wider IPCC issues, while at the same time adding some concrete examples, we can achieve an increased awareness of the urgency and importance of the climate debate. Here the "Think globally, act locally" comes very much to the forefront. Our more strenuous hike in Chapters 8 and 9 was there to help give a better understanding and appreciation of the statistical work that enters into the discussions and conclusions of the IPCC reports.

In the media, climate change is often equated to global warming. An increase in global temperature underlies many of the extreme weather events we observe worldwide. In the sections to follow, we first discuss a global-mean-temperature anomaly ("think globally") before we present two examples where warming at a more local level poses important challenges ("act locally"). We chose these examples from our own geographic environment. Similar stories can be told for many other regions of the world. The examples presented concern the melting of alpine glaciers and the harvesting of grapes for wine making in France. We see these examples as representative mental images on climate change. Each one can be elaborated in much more detail, and certainly many more examples can be added. For the former example, we refer to some of the literature mentioned; for the latter example, the IPCC reports offer ample extra reading material.

In our discussion on exponential growth in Section 6.3, we mentioned the important 1972 Club of Rome Report Meadows et al. (1972). The authors warned in great detail about the danger that unbridled exponential growth poses for finite resources. One obvious finite resource that as humans we have is our planet and hence it should not come as a surprise that the Club of Rome also raised a warning finger with respect to our climate in general, and more specifically how industrial pollutants may have a harmful effect. The following sentence in the report should have been an early wake-up call for politicians worldwide: "It is not known how much CO_2 or thermal pollution can be released without causing irreversible changes in the earth's climate [...]." This statement (about the so-called *tipping point*) was further supported by numerous projections; most of them, unfortunately, turned out to be true. Today "It is not known" has changed into "We know very well". The date of publication is 1972, hence 50 years ago!

10.2 Mean Temperature Anomaly

The *Paris Agreement* is an international treaty on climate change negotiated in 2015 on the occasion of COP21 in Paris. The following statement is a goal frequently cited in the media coverage of climate change:

> Holding the increase in the global average temperature to well below 2 °C above pre-industrial levels and pursuing efforts to limit the temperature increase to 1.5 °C above pre-industrial levels, recognizing that this would significantly reduce the risks and impacts of climate change;

see United Nations (2015, Article 2, 1. (a)). There is no deadline given for this goal, but United Nations (2015, Article 4, 1.) mentions "as soon as possible". Another frustrating aspect of the vagueness of this goal is that the term "pre-industrial levels" is not further specified. Hawkins et al. (2017) concludes that the 1720 to 1800 period is the most appropriate one, but mentions the problem of incomplete information and missing data during this period. The paper suggests the 1850 to 1900 period as a reasonable surrogate.

In this section, we investigate Earth's mean surface temperature to provide us with an oversimplified, but at least quantitatively supported, global assessment of climate change.

Other indicators of climate change not further discussed here are for instance the Earth's *albedo* (the solar radiation reflected back to space), the solar intensity and greenhouse insulation.

10.2.1 About the Data

We start by briefly considering the dataset with the longest available history of temperature measurements in the world. This is the Hadley Centre Central England Temperature (HadCET) dataset available on the website Met Office (2021). On this website, hit "download page" and right-click "Monthly_HadCET_mean.txt" to download these monthly mean temperatures, which are representative of a roughly triangular area of the United Kingdom enclosed by Lancashire, London and Bristol. At the time of writing, measurements are available from January, 1659 to October, 2021. Entries of −99.9 are already listed for the upcoming November and December, 2021, which appear nonsensical. Such entries are not unknown to data scientists; they are placeholders for missing values. Needless to say, encoding missing values (often symbolized by "NA" for "not available") by numbers bears the risk of including such values when computing averages and other values, at which point they can largely distort results, or – even worse – only mildly but nonetheless significantly distort the results and remain undetected. For more information about this HadCET dataset, especially the types of manipulations applied to the data, see Parker et al. (1992).

Plotting the monthly mean temperature, the HadCET data reveal a very slight trend in the data over time. To see trends more clearly, we will consider mean annual temperatures (averaged over all months of a year). Furthermore, instead of temperature itself, we consider its *anomaly*, the temperature location-transformed by an average over multiple years. To this end we compute the average yearly temperature over the 30 years from 1951 to 1980 as a baseline of 0 °C and subtract this average from the data; one also often finds in the literature an average taken over the time period from 1961 to 1990. Temperature anomalies allow for a comparison of how much warmer or colder it is than "normal". Figure 10.1 shows the mean annual temperature anomaly A_t in Central England as a function of year t. The plot speaks for itself, it does not require any modeling to see that, more recently, A_t lies mostly above 0 °C.

If you (still) think that this observation only concerns Central England, let's zoom out to the global scale. Global-mean surface temperature anomaly data can be obtained from the Goddard Institute for Space Studies (GISS), a laboratory in the Earth Sciences Division of NASA's Goddard Space Flight Center. Goddard Institute for Space Studies (2021) provides global-mean monthly surface temperature anomaly data, also with respect to the time period from 1951 to 1980. These data are combined land-surface air and sea-surface water temperatures averaged across all days of each month and across all measurement stations around the globe. The frequently-asked-questions section referred to on Goddard Institute for Space Studies (2021) provides insight into this combination of land-surface air and sea-surface water temperatures. The former are measured at weather stations which only cover about a third of our planet, the rest refer to oceans. On oceans, sea-surface water temperatures are available from ships and buoy reports (also satellite data). Although the two types of temperature are different (as air warms and cools faster over land than over water), their anomalies are very similar and can thus be combined to obtain a more global perspective

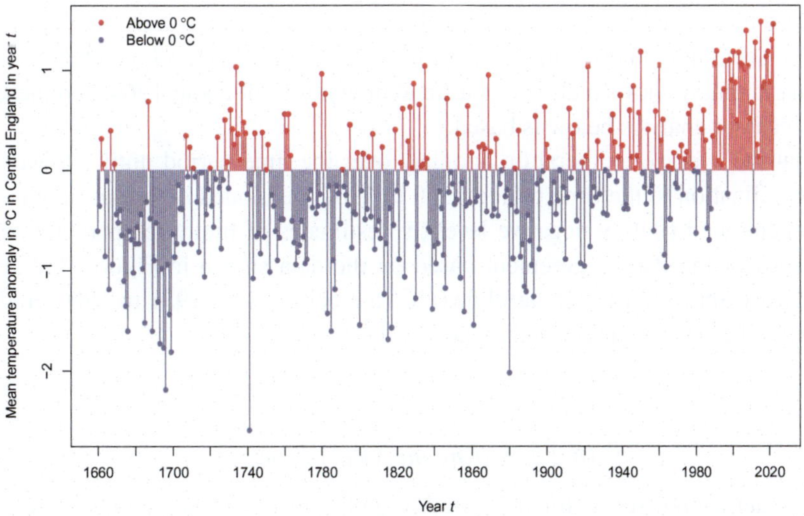

Figure 10.1 Mean annual temperature anomaly in Central England based on the HadCET dataset; the baseline 0 °C is the average over the 30 years from 1951 to 1980. Source: Authors

on temperature anomaly, which is clearly of interest. On the downside, the data are only available from January, 1880, so essentially we zoom out in the spatial dimension and zoom in on the time dimension in comparison with the HadCET dataset. Due to the latter and the fact that we want to model the data, we focus on monthly data here, so do not further aggregate the data.

Figure 10.2 shows the GISS mean monthly surface temperature anomaly data. We see fluctuations around a slightly negative level until about 1930. Over the course of the next 20

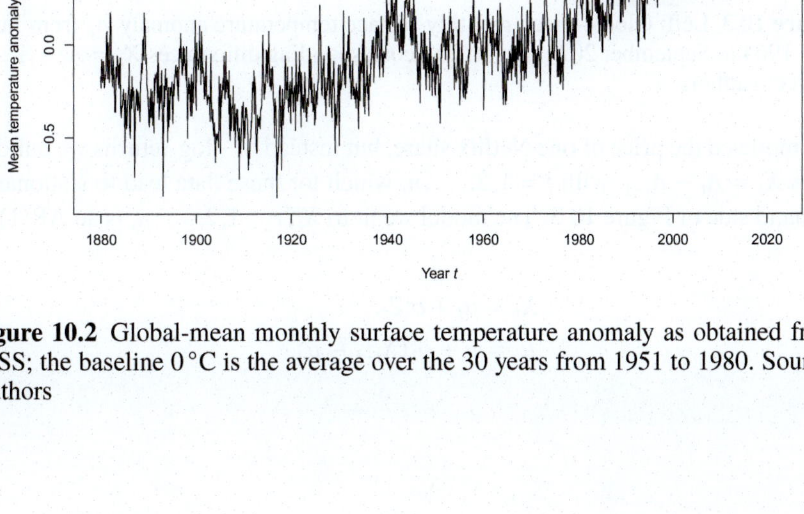

Figure 10.2 Global-mean monthly surface temperature anomaly as obtained from GISS; the baseline 0 °C is the average over the 30 years from 1951 to 1980. Source: Authors

years, we see first an increasing trend until about 1945, followed by a decreasing trend until about 1950. Seemingly fluctuating around a new, slightly higher level than before 1930, the temperature anomaly shows a clear linear trend upwards as of about 1964, and these are the data on which our analysis below is based.

Note that the average temperature anomaly over the time period until the end of 1900 (as an approximation to the time period 1850 to 1900 mentioned before) is about −0.21 °C (the aforementioned slightly negative level), so estimates of future temperature anomalies in the context of the Paris Agreement based on the data shown in Figure 10.2 are rather conservative, as they refer to the baseline 0 °C (the period from 1951 to 1980), rather than the baseline −0.21 °C, as the pre-industrial level.

10.2.2 Fitting an AR(1) Process

We now model a time series for the monthly global-mean surface temperature anomaly, say A_t, $t = 0, 1, \ldots, n$, where $t = 0$ refers to January 1964 and $t = n = 692$ refers to September 2021; see the left-hand side of Figure 10.3. We proceed as in Section 9.5.4,

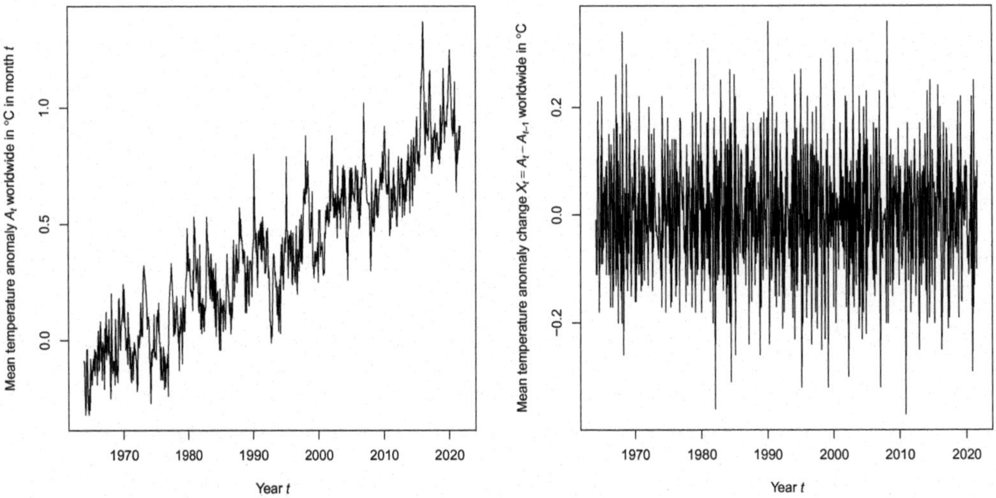

Figure 10.3 Left: Global-mean monthly surface temperature anomaly A_t from January 1964 to September 2021. Right: The corresponding differences $X_t = A_t - A_{t-1}$. Source: Authors

where we modeled the price of one Netflix share, but instead of −log-returns we build simple differences $X_t = A_t - A_{t-1}$, with $t = 1, 2, \ldots, n$, which for these data lead to stationarity; see the right-hand side of Figure 10.3. The model we fit to X_t, $t = 1, 2, \ldots, n$, is an AR(1) model, given by

$$X_t = \mu_t + \sigma Z_t,$$
$$\mu_t = \mu + \phi_1(X_{t-1} - \mu).$$

The model parameter ϕ_1 is assumed to satisfy $|\phi_1| < 1$, and, as innovation distribution F_Z, we use the standard normal.

Figure 10.4 shows the standardized residuals $Z_1, Z_2, \ldots, Z_n$ (with $n = 692$) on the left-hand side and the corresponding Q–Q plot against the standard normal distribution on the right-hand side. We also applied two other statistical procedures to assess whether the standardized residuals can be considered iid and there was no strong evidence against this hypothesis.

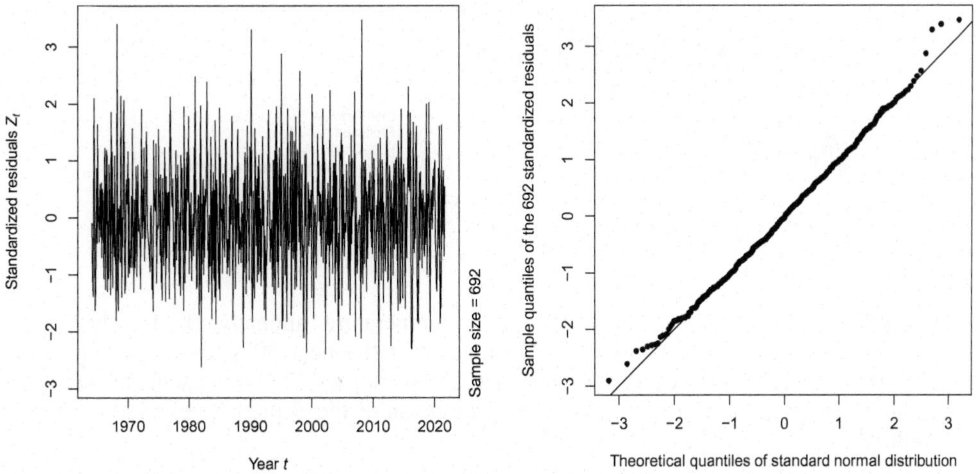

Figure 10.4 Left: Plot of the standardized residuals $Z_1, Z_2, \ldots, Z_n$ for $n = 692$ after fitting an AR(1) model. Right: Q–Q plot of the standardized residuals against the standard normal innovation distribution F_Z. Source: Authors

10.2.3 Prediction

On the basis of the fitted AR(1) model, we can now predict monthly values of X_t going forward. Not taking into account covariates other than time, we predict the global-mean monthly surface temperature anomaly A_t from October 2021 to September 2051 on the basis of the last available temperature anomaly A_t. Figure 10.5 shows the past values of A_t (the solid black line) up until and including September, 2021, the prediction of A_t (the solid red line) and corresponding bootstrap 95%-CIs (gray region) based on $B = 2000$ replications, constructed as in Remark 9.12, in particular, not taking into account the statistical uncertainty pertaining to the estimation of the AR(1) model. The mean over the corresponding B predictions is also shown in Figure 10.5 (the purple line) and we clearly see an upward trend here. The figure also contains the 1.5 °C and 2 °C goals of the Paris Agreement, with −0.21 °C taken as "pre-industrial levels" as mentioned before. As our basic statistical analysis of these temperature anomaly data shows, we urgently need significant changes to address climate change. You can read, for example, on Wikipedia (2021e) and in more recent documents of the IPCC about the (expected and likely consequential) effects of breaching the Paris Agreement goals. It does not look good.

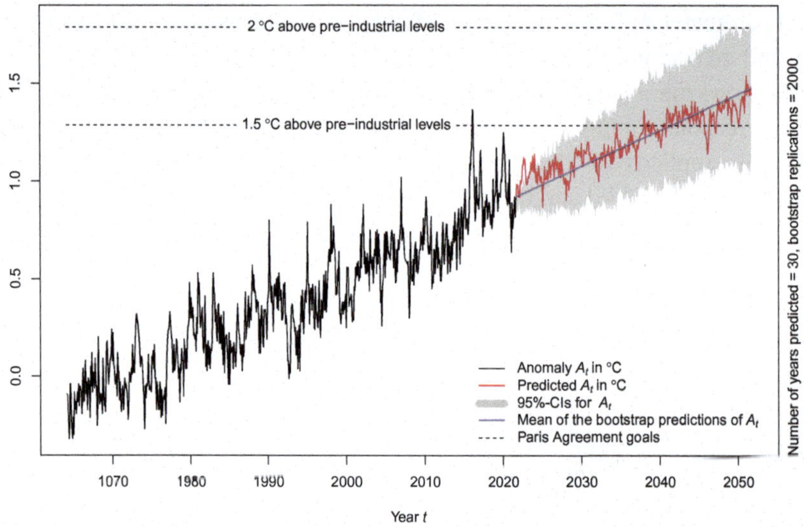

Figure 10.5 Global-mean monthly surface temperature anomaly A_t together with its prediction from a fitted AR(1) model over the next 30 years, including bootstrap 95%-CIs based on $B = 2000$ bootstrap replications. We have also included a solid line (purple), indicating, for each t, the mean of the B predictions. Source: Authors

10.3 Alpine Glaciers

Arolla is a village in the municipality of Evolène, in the canton of Valais, Switzerland. It is located south of Sion at the end of the alpine valley Val d'Hérens at an altitude of about 2000 m, offering a high-altitude ski resort equipped with five ski lifts on natural snow. The village was built 4000 years ago on the road to the Collon Pass (3130 m), an ancient route linking the Valais to the Aosta Valley (Northwest Italy). This road became impossible to use later on with the glacier advances of the seventeenth and nineteenth centuries. Between 1860 and 1910 guides and hotels appeared in Arolla, essentially to serve mountaineers who sought to climb the surrounding mountains such as the Pigne of Arolla, the Evêque or the Mont Collon (3637 m). The access road to Arolla became passable in 1948 and it was paved in 1965. Nowadays, Arolla is the starting point of the small course of the ski mountaineering race Patrouille des Glaciers and the obligatory passage of routes such as the Haute Route Chamonix-Zermatt or the Matterhorn tour.

The effect of (rising) temperature on glaciers is certainly one of the most visible signs of the effect of climate change. The left-hand side of Figure 10.6 shows a well-known lithograph of Mont Collon and the Lower Arolla Glacier, situated below it, by James D. Forbes from about the mid nineteenth century. One can clearly see the majestic glacier, with a cave at the bottom where water flows down from the glacier in the form of a creek. The situation is very different in the right-hand side picture, which shows Mont Collon and the Lower Arolla Glacier in 2011. The Lower Arolla Glacier currently extends over 1.5 km, at the foot of the Glacier of Mont Collon, and its tongue stops at about 2150 m of altitude. Figure 10.7 shows the tongue and its extension at three different points in time, which documents the recession

Figure 10.6 Left: The lithograph "Mont Collon, and the Lower Arolla Glacier; Vallée d'Erin" by James D. Forbes (1809–1868), as it appeared in Forbes (1843). Right: Mont Collon and the Lower Arolla Glacier in 2011. Sources: Viatimages/Bibliothèque de Genève, Fb 325 (left) and Stuart Lane, University of Lausanne (right)

Figure 10.7 The Lower Arolla Glacier in 2005 with curves indicating the ice tongue in 1856 (light blue), in 1895 (red) and in 1986 (purple). Source: Stuart Lane

of the glacier. Below we provide the results of a brief analysis of historical temperature data from Switzerland and the evolution of the Lower Arolla Glacier.

10.3.1 About the Data

We work with two datasets, the annual mean temperature, in degrees Celsius, for the regions above 1000 m in Switzerland from 1864 to 2021 and the changes in length of the Lower Arolla Glacier in meters from 1886 to 2019.

The first dataset can be obtained from MeteoSwiss (2021). It currently requires one to navigate to the tab "Climate", then choose "Swiss temperature mean" and select one of the two links "Data on the Swiss temperature mean" on this webpage. On the latter page, there are datasets in two versions listed. We worked with "Northern Switzerland high", version 1.1, and the text file can be downloaded via the link with this name. The dataset has its own *digital object identifier (DOI)* `10.18751/Climate/Timeseries/CHTM/1.1`, a unique way of identifying objects that we also occasionally use in the list of references in this book, when available. More about how the data were obtained can be learned from Begert and Frei (2018). The downloaded dataset contains as its last column the mean annual temperature above 1000 m in Switzerland, with which we are working. As we did, you might wonder what "Northern" means here and why, given that Arolla is situated in the south of Switzerland we did not use the "Southern Switzerland high" data. As expert Stuart Lane clarified, the cardinal directions are to be interpreted with respect to the principal barrier of the Alps, of which Arolla lies to the north (again an easy trapdoor, but our findings would not be affected much in this case).

The second dataset was obtained from GLAMOS (2021); GLAMOS stands for Glacier Monitoring Switzerland. After choosing the English option ("EN") on the website, navigate to the tab "Downloads", then choose "Length change" and click the link "DOWNLOAD". The dataset can also be referred to by its DOI; see GLAMOS (2020). It contains the changes in length of several glaciers, so we extracted the rows starting with "Bas Glacier d'Arolla", the Lower Arolla Glacier. As columns, we extracted the end date of the measurement and the change of length in meters. The first length change is reported to be −600 m, much larger in absolute value than the others. This comes from the fact that the first length change is reported for the years 1856 to 1886; thereafter the length changes are reported much more frequently, every couple of years, yet the times in between are not equal, a fact important for plotting the data properly as time series at their correct measurement dates instead of on an equidistant grid (another trapdoor wide open for you to step on and fall into).

10.3.2 Annual Mean Temperature and Glacier Length

Since the end of the Little Ice Age in the second half of the nineteenth century, a gradual rise of temperatures has been recorded on a global scale. In Switzerland particularly, a period of climate stabilization was observed between the late 1950s and early 1980s, followed by a period of rapid warming since the late 1980s; before you continue reading, first cover the right-hand side picture in Figure 10.8. The left-hand side of Figure 10.8 shows the mean annual temperature in degrees Celsius above 1000 m in Switzerland from 1864 to 2021; can you summarize noticeable features in the plot? The right-hand side of Figure 10.8 shows the same data together with two fitted lines that are estimated from the data to the left, respectively to the right, of a detected significant change point of the mean in the year 1987. It also includes a linear prediction for 2022 to 2050 (based on the data starting in year 1987) including asymptotic 95%-CIs constructed using a limiting normal distribution (similarly to how we constructed CIs in Section 8.6.3). The rather basic linear prediction provides us with an answer to the grim *what if* question "What if the mean annual temperature continues to increase like this?" In our preliminary analysis of the temperature data we note the 1987 change point of the mean. A much more detailed statistical as well as climatological analysis is called for; see for instance Sippel et al. (2020) for an in-depth discussion. In this section and also in Section 10.4.3 we use the statistical term "change point" within the broader context of the above paper.

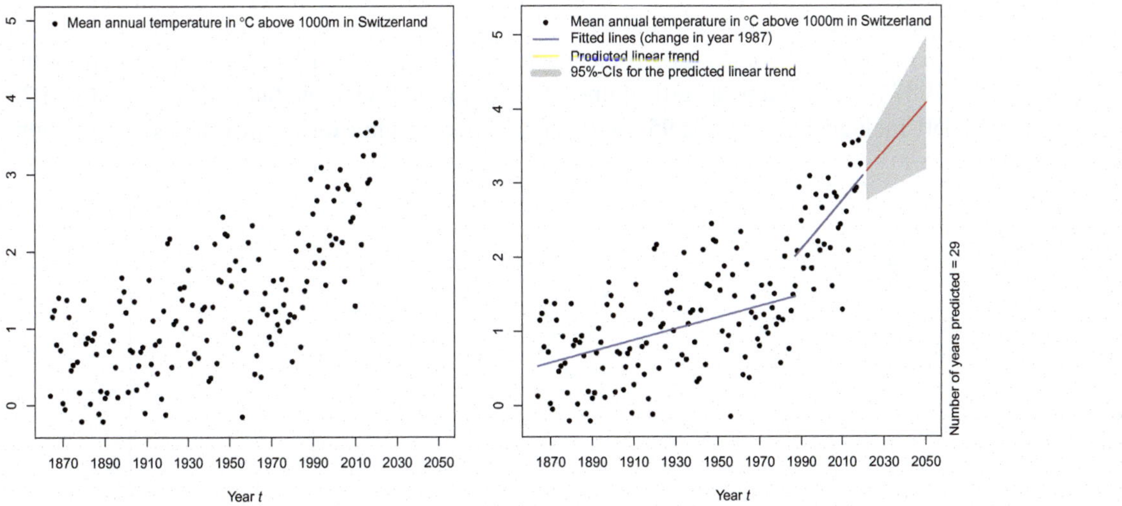

Figure 10.8 Left: Plot of the mean annual temperature in degrees Celsius in Switzerland from 1864 to 2021 at altitudes above 1000 m. Right: Two fitted lines (before and after a change point of the mean statistically detected for these data in the year 1987) and a linear prediction with 95%-CIs for the years 2022 to 2050. Source: Authors

Figure 10.9 shows the annual temperature anomaly (with respect to the average over the years from 1951 to 1980) above 1000 m in Switzerland. You can see that the 1987 change

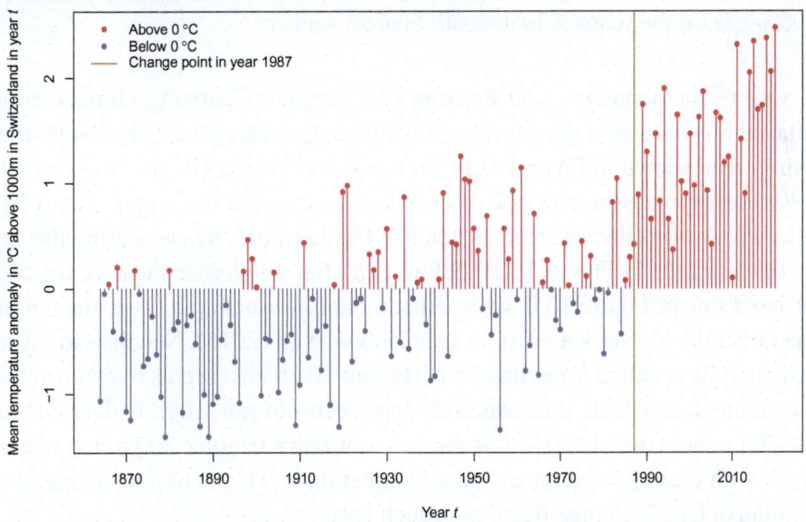

Figure 10.9 Mean annual temperature anomaly above 1000 m in Switzerland with baseline 0 °C, which is the average over the 30 years from 1951 to 1980. The change point of the mean in year 1987 is indicated by a vertical line. Source: Authors

point of the mean (indicated by the vertical line) separates a period of positive and negative variations (1960–1980) from one with only positive mean temperature anomalies, starting around the end of the 1980s and recently exceeding 2 °C. It should thus not come as a surprise that glaciers react accordingly. Figure 10.10 shows the cumulative (this is, summed) length change in meters of the Lower Arolla Glacier from 1886 to 2019. A local maximum for the year 1987 is indicated by a vertical line. We also provide a simple linear prediction for 2020 to 2050, including asymptotic 95%-CIs. The glacier's decline results in a change of its length,

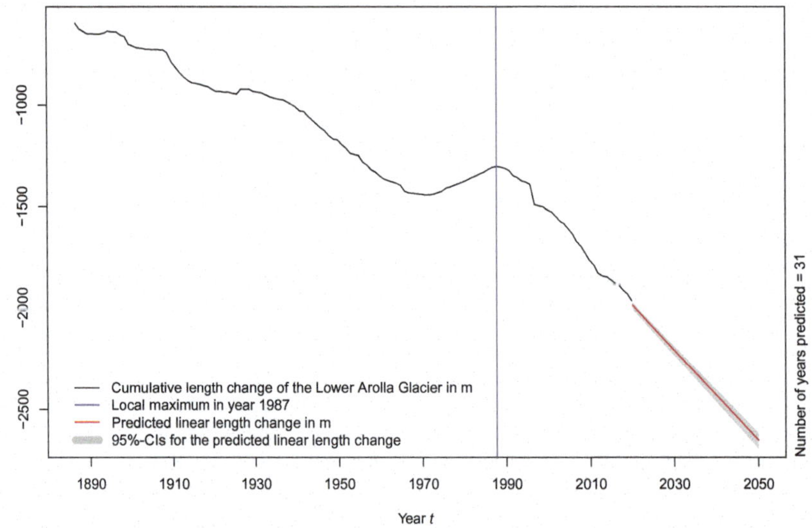

Figure 10.10 Cumulative length change in meters of the Lower Arolla Glacier from 1886 to 2019 with a local maximum at 1987 (vertical line) and a linear prediction with 95%-CIs for the years 2020 to 2050. Source: Authors

its mass as well as the geometry near its front (the tongue). Changing climatic conditions induce a delayed response on a glacier: typically, the bigger the glacier, the longer the delay. The increasing trend shown in Figure 10.10 for the Lower Arolla Glacier between 1970 and the mid 1980s corresponded to a period of exceptional floods in the region. Apart from this period, the Lower Arolla Glacier receded steadily. Furthermore, by comparing the slopes of the roughly linear trends in Figure 10.10 before and after 1987 with those of the linear fits on the right-hand side of Figure 10.8, we see that there is a match: the larger the temperature increase, the larger the glacier's recession; see Gabbud et al. (2016). Needless to say, a much more refined analysis is called for at this point. Figure 10.10 also seems to confirm the delay of the glacier's length response; the change of slope seems to start slightly delayed at around 1990. Finally, let us note that the 95%-CIs for the cumulative length change are rather narrow in comparison with those for the mean annual temperature. This is of course due to the fact that the cumulative length change fluctuates much less.

There are several risks associated with the recession of glaciers. For instance, the melting of the permafrost (the permanently frozen ground in the Alps above 2500 m), together with the formation of underground lakes formed under glaciers, lead to an increase in the risk of avalanches and the inundation of villages in the valleys. An increase in temperature also has consequences for the fragile ecosystems at high altitude. One such example relates to Arolla, which obtained its name from the arolla forests that surround(ed) it. The arolla (*Pinus cembra*, in French l'arol(l)e) is a majestic fir tree, known as the Queen of the Alps. The smell of its wood enters several poetic texts and travelers' stories. The increase in temperature forms a threat to its existence. Future generations may eventually have some difficulty in telling their children where the name Arolla comes from.

10.4 Measuring Wine Barrels and Harvesting Grapes

When it comes to risks and extremes, environmental risk poses a serious challenge to agricultural production worldwide. One such example concerns the production of wine, a topic we turn to in the subsequent sections. As we gradually move from our more strenuous hike of the previous chapters to a more gentle stroll, we start with another campfire story from the realm of mathematics. The story links the second wedding of Johannes Kepler and the measurement of geometric volumes. The particular application of Kepler's work concerns wine barrels. We will also include a related optimal stopping problem from probability theory. After these introductory stories, we will turn to the analysis of wine harvesting data under climate change scenarios.

10.4.1 Johannes Kepler and the Volume of a Wine Barrel

Johannes Kepler (1571–1630) is universally known for his three fundamental laws of planetary motion. Together, these laws describe the orbits of planets around the Sun. Perhaps less known is his work on the calculation of surfaces and volumes. In these calculations, he developed an early version of infinitesimal calculus through slicing a rotational volume in ever-smaller (hence eventually infinitesimal) cylinder-like volumes. Kepler lost his first wife to cholera and remarried in 1613 in Linz, Austria. For this wedding, he bought a barrel of wine. In order to determine the price of a barrel, at the time, wine merchants used a special method for determining a barrel's volume. Kepler wanted to better understand this method and, in particular, became interested in how to maximize the volume of a wine barrel under given geometric and measurement constraints. His solution formed the basis of the publication *Nova Stereometria Doliorum Vinariorum* (translated: "New Solid Geometry of Wine Barrels"); see Kepler (1615).

As a warm-up towards the solution of Kepler's wine barrel problem, we consider the following problem. Consider all rectangles with circumference L and sides a and b. Show that the rectangle with maximal area is a square, that is, $a = b = L/4$. This problem clearly shows some affinity to Kepler's in going from area to volume and from rectangle to "a barrel". An easy, one-line solution to the rectangle problem uses derivatives (see also later), a concept not yet known to Kepler. We will explain next how Kepler circumvents this lack of knowledge and yet arrives at the correct solution.

The left-hand side of Figure 10.11 illustrates how the merchant determined the volume and thus the price of a barrel of wine. With the barrel lying on its side, the merchant would insert a stick through the tap hole, also known as a bunghole, at the center top (the middle side if the barrel was standing) to the opposite edge at the lower left or the lower right of the barrel. The length of the stick would determine the volume of the barrel and therefore its price; see Cardil (2019).

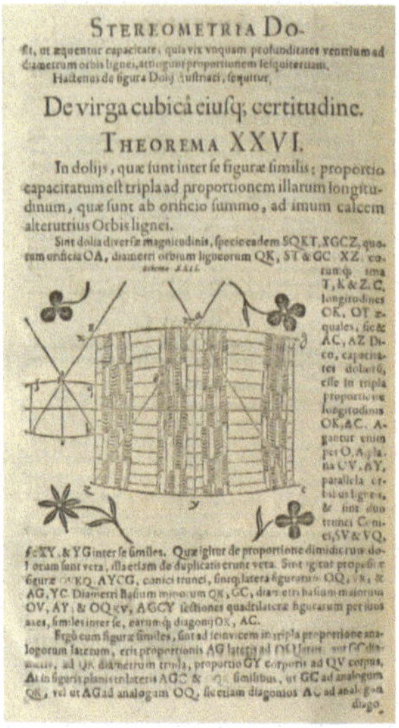

STEREOMETRIA DO-

Haltenus de figura Dolij Austriaci, sequitur,

De virga cubicâ ciusq; certitudine.

THEOREMA XXVI.

In dolijs, quæ sunt inter se figuræ similis; proportio capacitatum est tripla ad proportionem illarum longitudinum, quæ sunt ab orificio summo, ad imum calcem alterutrius Orbis lignei.

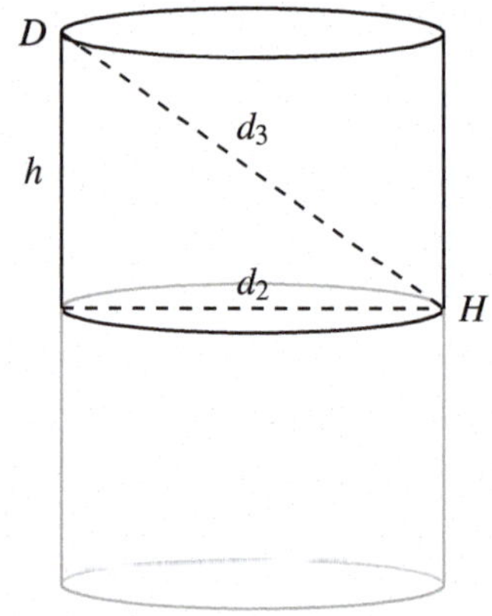

Figure 10.11 Left: Page 98 of Kepler (1615), illustrating how the wine merchant determined the price of a barrel of wine by inserting a stick through the barrel's tap hole to one of the sides. Right: A cylinder approximating a wine barrel standing upright with the quantities Kepler used to determine the approximate volume of a barrel. Sources: Kepler (1615) (left) and authors (right)

Derivation 10.1 (Optimal base-diameter to height ratio of a cylinder with given diagonal). We start by reproducing Kepler's calculations. You may find this derivation a bit cumbersome notation-wise; however, let's try to follow his seventeenth-century thinking. First, Kepler notices that the volume V_{barrel} of a barrel can be approximated by that of a cylinder; see the right-hand side of Figure 10.11, which shows the cylinder standing upright, with tap hole H at middle right and the point D denoting the location at which the merchant points his stick. The volume V_{cylinder} of the cylinder is simply twice the volume V of the top half of the cylinder with diagonal d_3 (the distance between H and D) and diagonal d_2 of the cross section of the cylinder; the indices 2 and 3 refer to the dimensions in which the diagonals are measured. With h the height of the top half of the cylinder, as in Figure 10.11, Pythagoras' theorem implies that $d_2^2 = d_3^2 - h^2$. Since the volume of a cylinder is the area of its base times its height, we obtain that the volume of the top half of the cylinder in Figure 10.11 is

$$V = (d_2/2)^2 \pi h = \frac{\pi}{4} d_2^2 h = \frac{\pi}{4}(d_3^2 - h^2)h.$$

The volume of the whole cylinder (approximating the volume V_{barrel} of the barrel) is thus

$$V_{\text{cylinder}} = 2V = \frac{\pi}{2}(d_3^2 - h^2)h. \tag{10.1}$$

The left-hand side of Figure 10.12 shows p. 66 of Kepler (1615). In the middle of the page (above the table) we can find that Kepler was interested in the case $d_3 = 20$. We could not find which unit of length was used in Kepler's case; this however is not relevant for the solution presented but was surely highly relevant for the wedding guests! Here we see a further example of the importance of using precise units of measurement, as discussed in Section 6.6. Recall that the meter was only defined as a standard unit of length in 1793. Its

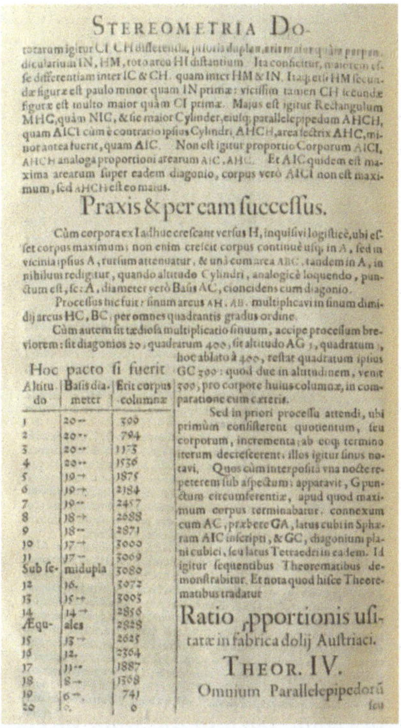

STEREOMETRIA DO-

rotarum igitur CI CH differentia, prioris dupla, erit maior quàm perpendicularium IN, HM, toto arcu HI distantium. Ita conficitur, maiorem esse differentiam inter IC & CH, quam inter HM & IN. Itaq; etsi HM secundæ figuræ est paulo minor quam IN primæ: vicissim tamen CH secundæ figuræ est multo maior quàm CI primæ. Majus est igitur Rectangulum MHC, quàm NIC, & sic maior Cylinder, eiusq; parallelepipedum AHCH, quam AICI cùm è contrario ipsius Cylindri AHCH, area sectrix AHC, minor antea fuerit, quam AIC. Non est igitur proportio Corporum AICI, AHCH analoga proportioni arearum AIC, AHC. Et AIC quidem est maxima arearum super eadem diagonio, corpus verò AICI non est maximum, sed AHCH est eo maius.

Praxis & per eam successus.

Cùm corpora ex I adhuc crescant versus H, inquisivi logisticè, ubi esset corpus maximum; non enim crescit corpus continuè usq; in A, sed in vicinia ipsius A, rursum attenuatur, & unà cum area ABC, tandem in A, in nihilum redigitur, quando altitudo Cylindri, analogicè loquendo, punctum est, sc: A, diameter verò Basis AC, cioncidens cum diagonio.

Processus hic fuit: sinum arcus AH, AB, multiplicavi in sinum dimidij arcus HC, BC, per omnes quadrantis gradus ordine.

Cùm autem sit tædiosa multiplicatio sinuum, accipe processum breviorem: sit diagonios 20, quadratum 400, sit altitudo AG 1, quadratum 1, hoc ablato à 400, restat quadratum ipsius GC 399: quod duc in altitudinem, venit 399, pro corpore huius columnæ, in comparatione cum cæteris.

Sed in priori processu attendi, ubi primùm consisterent quotientum, seu corporum, incrementa; ab eoq; termino iterum decrescerent: illos igitur sinus notavi, Quos cùm interposita una nocte repeterem sub aspectum: apparavit, G punctum circumferentiæ, apud quod maximum corpus terminabatur, connexum cum AC, præbere GA, latus cubi in Sphæram AIC inscripti, & GC, diagonium plani cubici, seu latus Tetraedri in eadem. Id igitur sequentibus Theorematibus demonstrabitur. Et nota quod hisce Theorematibus tradatur

Hoc pacto si fuerit

Altitudo	Basis diameter	Erit corpus columnæ
1	20--	399
2	20--	794
3	20--	1173
4	20--	1536
5	19--	1875
6	19--	2184
7	19--	2457
8	18--	2688
9	18--	2871
10	17--	3000
11	17--	3069
Sub se-	midupla	3080
12	16.	3072
13	15--	3003
14	14--	2856
Æqu-	ales	2828
15	13--	2625
16	12.	2304
17	11--	1887
18	8--	1368
19	6--	741
20	0.	0

Ratio proportionis usitatæ in fabrica dolij Austriaci.

THEOR. IV.

Omnium Parallelepipedorũ seu

h	d_2	V_{box}
1	19.9750	399
2	19.8997	**792**
3	19.7737	1173
4	19.5959	1536
5	19.3649	1875
6	19.0788	2184
7	18.7350	2457
8	18.3303	2688
9	17.8606	2871
10	17.3205	3000
11	16.7033	3069
12	16.0000	3072
13	15.1987	3003
14	14.2829	2856
15	13.2288	2625
16	12.0000	**2304**
17	10.5357	1887
18	**8.7178**	1368
19	6.2450	741
20	0.0000	0

Figure 10.12 Left: Kepler (1615, p. 66) showing the table Kepler made (with heights h, diameters d_2 and corresponding volumes V_{box} as columns). Right: A table with exact values (rounded data for d_2; boldface if deviating from Kepler's numbers). Sources: Kepler (1615) (left) and authors (right)

current definition is the length of the path traveled by light in vacuum during a time interval of 1/299 792 458 of a second. But let us return to Kepler. As the page in Figure 10.12 shows, he created a table in which the first column is the height h, the second column is the diagonal d_2 of the cross sectional circle and the third column contains the respective volumes. The volumes Kepler calculated are those in (10.1) but without the factor $\pi/2$. You can verify that such volumes $(d_3^2 - h^2)h$ are precisely those of a box with square base inscribed in the cylinder with volume V_{cylinder}. Let us denote these volumes by $V_{\text{box}} = (d_3^2 - h^2)h$. Although the approximation of V_{cylinder} by V_{box} is wrong by a factor $\pi/2 \approx 1.5708$, the idea is that if the volume of the box is maximal, so is the volume of the cylinder and thus the barrel's volume. The right-hand side of Figure 10.12 shows a table that reproduces the numbers from Kepler's table for $h = 1, 2, \ldots, 20$ (first column). The third column in the table on the right shows V_{box} for the corresponding values of h (and $d_3 = 20$). Although we do not need them explicitly, the second column in the table shows the exact values of $d_2 = \sqrt{d_3^2 - h^2}$ for $d_3 = 20$ and corresponding h. Values in bold are those deviating from Kepler's calculations; note that for d_2, Kepler rounded to integer values. So this reproduces the results of Kepler (and more).

Kepler came to the conclusion that among all cylinders with the same d_2, the one with largest volume satisfies $d_2/h = \sqrt{2}$; see (Kepler, 1615, Theorem V on pp. 69). We now derive

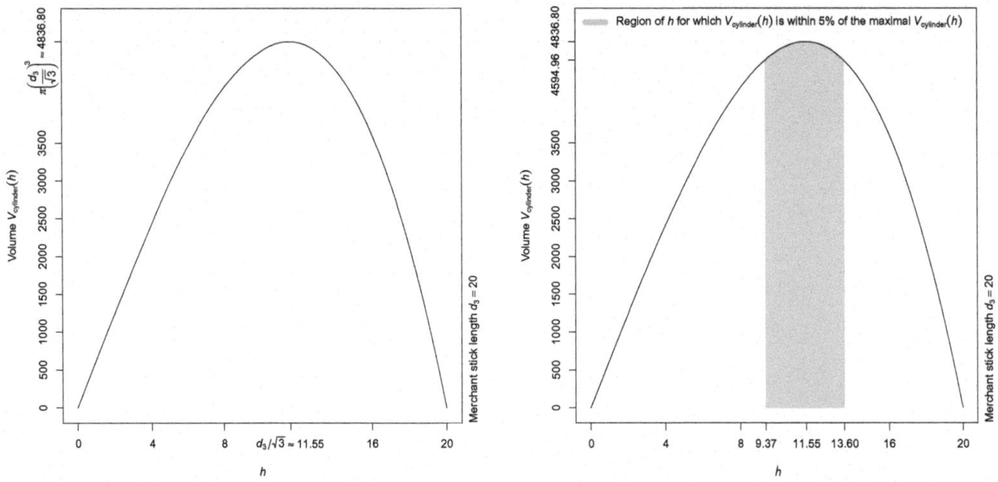

Figure 10.13 Left: The volume $V_{\text{cylinder}}(h)$ as a function of $h \in [0, d_3]$ for $d_3 = 20$, with maximum at $h = d_3/\sqrt{3}$. Right: The same function with an interval of h-values highlighted (gray band) for which $V_{\text{cylinder}}(h)$ is within 5% of the maximum $V_{\text{cylinder}}(d_3/\sqrt{3}) = \pi(d_3/\sqrt{3})^3$. Source: Authors

this result in the usual way we learned in high school (going back to the work of Fermat, Newton and Leibniz). Formula (10.1) provides us with the volume of the whole cylinder in terms of the height h of the top half of the cylinder for a given length d_3 that the merchant measured, so $V_{\text{cylinder}} = V_{\text{cylinder}}(h)$. One can easily see from this formula that $V_{\text{cylinder}}(h) = 0$ for $h \in \{0, d_3\}$, and that $V_{\text{cylinder}}(h) > 0$ only for $h \in (0, d_3)$. A local optimum of $V_{\text{cylinder}}(h)$ for $h \in (0, d_3)$ must satisfy the necessary condition $V'_{\text{cylinder}}(h) = 0$. With

$$V'_{\text{cylinder}}(h) = \frac{\pi}{2}(d_3^2 - 3h^2)$$

we see that $V'_{\text{cylinder}}(h) = 0$ for $h \in (0, d_3)$ if and only if $h = d_3/\sqrt{3}$. Since $V''_{\text{cylinder}}(h) = -3\pi h < 0$ for all $h \in (0, d_3)$, we know that $h = d_3/\sqrt{3}$ is the height that maximizes $V_{\text{cylinder}}(h)$, with corresponding maximal volume $V_{\text{cylinder}}(d_3/\sqrt{3}) = \pi(d_3/\sqrt{3})^3$. For the optimal $h = d_3/\sqrt{3}$, we obtain that $h^2 = d_3^2/3 = (d_2^2 + h^2)/3$, so $2h^2 = d_2^2$ and thus $d_2/h = \sqrt{2}$, as reported by Kepler. Note that this ratio is independent of d_3, and so applies to any size of barrel!

The left hand-side of Figure 10.13 shows the volume $V_{\text{cylinder}}(h)$ as a function of $h \in [0, d_3]$ for $d_3 = 20$, with maximum at $h = d_3/\sqrt{3}$. The right-hand side of this figure shows the same function but with an interval of h-values highlighted for which $V_{\text{cylinder}}(h) \geq 0.95\pi(d_3/\sqrt{3})^3$, that is, for which $V_{\text{cylinder}}(h)$ is within 5% of the maximum, $V_{\text{cylinder}}(d_3/\sqrt{3})$. As we can see, $V_{\text{cylinder}}(h)$ is fairly flat for h in this interval, so even if h deviates from its optimum $d_3/\sqrt{3}$ a little, $V_{\text{cylinder}}(h)$ is still rather large.

As we have just seen at the end of Derivation 10.1, the function representing the volume of a cylinder with a given diagonal is rather flat near its optimal height. Kepler realized this. Although, if applied to other barrels, the price determination method of the merchant would be considered fraudulent, Kepler came to the conclusion that the proportions of the wine merchant's barrels in Linz, Austria, were such that the merchant's method to calculate the volume was sufficiently accurate despite some minor differences. Cheers!

As we are still sitting around the campfire, a further story involving Kepler can be told. The story concerns a well-known problem in probability theory. As such, it would have also fit nicely as an example on probability calculations in Chapter 8. So the analysis of grape harvesting under climate risk still has to wait a little bit. For now there is more to Kepler's second marriage.

The story goes that over two years, Kepler had considered 11 possible candidates in order to find a new life companion with whom to be married. In probability theory, the problem of having to make a choice in life in a sequential fashion is known as the *secretary problem*. From today's perspective, this name is no longer acceptable and it is gradually being replaced by the more neutral and indeed more correct *best choice problem*. Also, in the various interpretations of the problem the word "secretary" loses its original meaning; we can for instance think about job applicants, or even tasks. The problem asks one to maximize the probability of selecting the best among n applicants for a position if the applicants are rankable, interviewed one at a time and in random order, and a decision about an applicant is to be made immediately after the interview; once rejected, an applicant cannot be recalled. This is an *optimal stopping* problem as an interviewer needs to find the optimal time point when to stop the interviewing process and thus select the best applicant even without having seen the quality of the remaining candidates. The following derivation briefly considers a basic version of the problem.

Derivation 10.2 (The $1/e$ stopping rule)**.** The optimal stopping rule prescribes interviewing and rejecting the first r candidates, and then selecting the first of the remaining $n-r$ applicants who is better than all other applicants so far. The probability that the best applicant is indeed selected in this way is then given as follows. Recall that "|" means "given" and "∩" means "and". By using conditional probabilities, see Section 8.1.4, we obtain that

$$
\begin{aligned}
&\mathbb{P}(\text{“best applicant is selected”}) \\
&= \sum_{k=1}^{n} \mathbb{P}(\text{“applicant } k \text{ is selected”} \cap \text{“applicant } k \text{ is the best”}) \\
&= \sum_{k=1}^{n} \mathbb{P}(\text{“applicant } k \text{ is selected”} \,|\, \text{“applicant } k \text{ is the best”}) \\
&\qquad \times \mathbb{P}(\text{“applicant } k \text{ is the best”}) \\
&= \frac{1}{n} \sum_{k=1}^{n} \mathbb{P}(\text{“applicant } k \text{ is selected”} \,|\, \text{“applicant } k \text{ is the best”}) \\
&= \frac{1}{n} \sum_{k=r+1}^{n} \mathbb{P}(\text{“applicant } k \text{ is selected”} \,|\, \text{“applicant } k \text{ is the best”}) \qquad (10.2) \\
&= \frac{1}{n} \sum_{k=r+1}^{n} \mathbb{P}\left(\begin{matrix} \text{“the best of the first } k-1 \text{ applicants} \\ \text{is among the first } r \text{ applicants”} \end{matrix} \,\middle|\, \text{“applicant } k \text{ is the best”} \right) \qquad (10.3)
\end{aligned}
$$

$$= \frac{1}{n} \sum_{k=r+1}^{n} \frac{r}{k-1} = \frac{r}{n} \sum_{k=r}^{n-1} \frac{1}{k}. \tag{10.4}$$

Some explanations are in order. Equation (10.2) holds since none of the first r applicants is selected. For understanding (10.3), Figure 10.14 provides a visual explanation. Finally, (10.4) holds since the probability of the best among the first $k - 1$ applicants be-

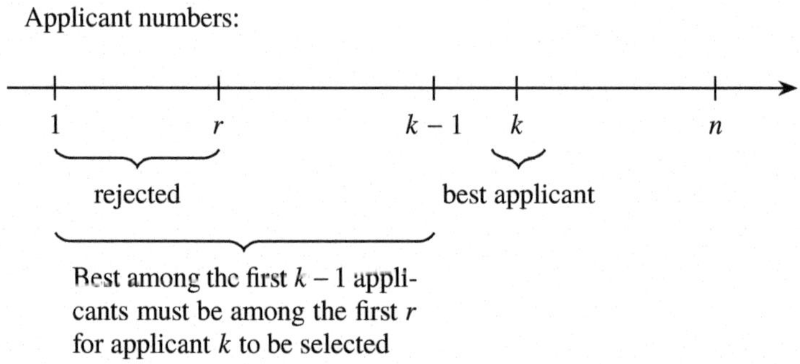

Figure 10.14 Sketch explaining (10.3). Source: Authors

ing among the first r applicants is $r/(k-1)$. For large n, the sum $\sum_{k=r}^{n-1} 1/k$ is approximately equal to the integral $\int_r^n (1/x)\,\mathrm{d}x = \left[\log(x)\right]_r^n = \log(n) - \log(r) = -\log(r/n)$ and so $\mathbb{P}$("best applicant is selected") $\approx (-r/n)\log(r/n) = f(r/n)$ for $f(x) = -x\log(x)$, $x = r/n > 0$. Check that the derivatives of f are $f'(x) = -\log(x) - 1$ and $f''(x) = -1/x$. As in Derivation 10.1, we can conclude that f attains its maximum at $x = 1/e \approx 0.37$ (recall that $e = 2.7182\ldots$); indeed, we meet Leonhard Euler again! Given that $x = r/n$ we see that, for large n, the optimal stopping point (of rejections) for n applicants is $r = n/e$. The corresponding maximal value of f is $1/e$, so the probability of selecting the best applicant is (roughly) $1/e$ with this strategy. This probability is independent of n; this strategy has the name *$1/e$ stopping rule* or *37% rule*.

Much more on this (and related) problem(s) can be found in Ferguson (1989). According to this publication, which refers to a letter of Kepler to a baron on October 23, 1613, as mentioned above Kepler considered 11 candidates for his hand. He was strongly attracted to the fifth, but he listened to the advice of friends who persuaded him to propose to the fourth candidate. But he had waited too long; she turned him down. As such, Kepler decided on candidate 5, coincidentally exactly according to the $1/e$ stopping rule; the latter suggests rejecting the first $11/e \approx 4$ candidates.

10.4.2 Harvesting Time and Risks to Viticulture

As in the case of Kepler's wedding, wine has been, and still is, present at many ceremonies as a symbol of happiness and friendship. It is considered the drink of the gods, Bacchus being the Roman God of wine and Dionysius the Greek God of wine. France is one of the largest wine producers in the world with its winemaking history dating back to Roman times. France is the origin of many grape varieties such as Chardonnay, Pinot Noir, Cabernet Sauvignon, Syrah, to name a few. All can now be found around most of the world's wine regions. The quality of wine is intimately related to its *terroir* (from the French "terre", meaning "land" or "soil"). Terroir describes the environmental factors that affect a grape's phenotype (its observable characteristics) including its unique environmental context, farming practices and a specific growth habitat. Collectively, these contextual characteristics are said to give a particular wine its *character*. If you are a wine producer, the weather is not only your worst enemy but also acts as your best friend when in need. In this section we will very briefly highlight the influence of temperature on the complicated, though exciting, path from the grapes to the wine in your glass. In this process, determining the ideal time to harvest is crucial. Before the French Revolution, harvesting time was dictated by a public order, namely the "ban des vendanges" following a seigniorial decision. After the Revolution, even though winegrowers were officially free to start the harvest whenever they wanted, it was still the parishes who set a compulsory minimum date for the harvest. After 1791, harvesting times were under the control of the municipality's mayor and the dates that were decided on for the most important varieties were the earliest possible among all the varieties grown in the municipality. The law of July 9, 1889 abandoned the "ban des vendanges" in almost all of France, letting the winegrowers freely decide on the harvesting time. Ideally, harvesting time is determined by the ripeness of the grapes, measured by their sugar content, tannin and acidity, which are all related to weather conditions. When climatic conditions are good, the vintage is expected to be great. On the other hand, severe weather conditions such as heavy precipitation, heat waves, droughts, hail, strong winds or frost can damage grapes and cause various vine diseases or even destroy whole vineyards.

As projected by the IPCC, both the rise of the average annual temperature and the increase in frequency of climate extremes will negatively impact viticulture (the cultivation and harvesting of grapes). The former presents a long-term risk for the suitability of certain varieties (especially in Mediterranean climate regions) and for the sustainability of viticulture in traditional wine regions, while the latter has short-term effects on both quality and volume.

10.4.3 About the Data

In this section, we briefly present two datasets that both have been provided to us by Cornelis (Kees) van Leeuwen, professor of viticulture at Bordeaux Sciences Agro and Bordeaux University's Institut des Sciences de la Vigne et du Vin. Similar data can be found on the website MTECT (2020). The first dataset contains the average growing-season temperature in the Bordeaux region from 1951 to 2020 as measured by the Bordeaux Mérignac weather station. Figure 10.15 shows a plot of these data. The second dataset contains the grape-harvesting dates (GHDs) of a Saint-Émilion property in the Bordeaux region from 1951 to 2019, which kindly asked to remain anonymous. Figure 10.16 shows a plot of these data. The missing GHD value for 1956 is due to the fact that in that year the harvest was destroyed by frost.

For each dataset in Figures 10.15 and 10.16 we fitted a smooth curve through the data (solid black lines). In Figure 10.15, we clearly notice an increasing trend in the average growing-season temperature for the Bordeaux region, with a change point of the mean around 1986 (indicated in the plot by a vertical line). In Figure 10.16 from about 1980 onwards, we observe an overall reversed effect, a decreasing trend in the GHDs for the specific Saint-Émilion property. This time, a change point of the mean is detected two years

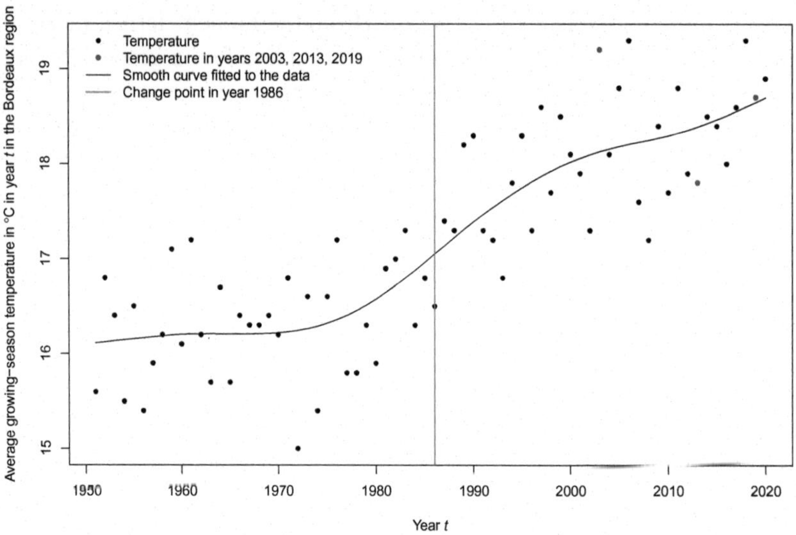

Figure 10.15 Average growing-season temperature in the Bordeaux region from 1951 to 2020. Source: Authors

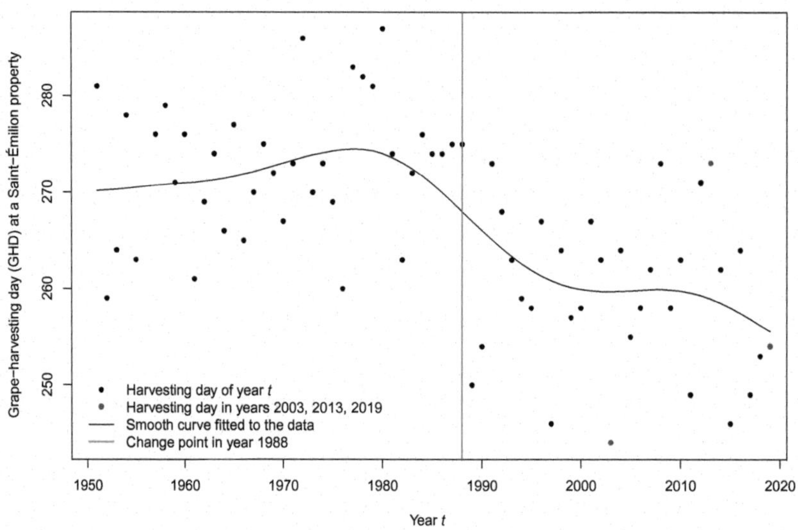

Figure 10.16 Grape-harvesting day per year at a Saint-Émilion property in the Bordeaux region from 1951 to 2019. Source: Authors

later, in 1988. This very preliminary analysis clearly shows an inverse relationship between GHD and growing-season temperature.

10.4.4 Climate Change and the Quality of Wine

The 1988 change point referred to in Figure 10.16 coincides with the year in which US senator Johnston made his famous statement referred to in the beginning of this chapter.

Whereas since that year the harvest of grapes has occurred on average ten days earlier, the effect of climate change on the quality of the wine is still unclear. It has become increasingly difficult to predict whether an upcoming vintage is of good quality. We will exemplify this point in the case of the Saint-Émilion winery data. We do stress "exemplify", as a detailed analysis is beyond the scope of our book. Also note that grapes and winemaking are just one agricultural sector; the implications of climate change on the global agricultural sector is no doubt one of the main areas of worry for humanity.

We have already noted that high mean seasonal temperatures lead to early GHDs. This was for instance the case for the years 2003 and 2019. Were wines from these vintage years exceptionally good? Days of heat at the end of June can work wonders in this direction. At that point into the season, grapes are still forming and could be susceptible to mildew, a fungus that loves humidity and can destroy the grapes. A heat spike considerably reduces that risk. During heat spikes, vines protect their grapes against a lack of water rather than using extra energy for growing leaves, which has a positive effect on the quality of the wine. This was the case in the Bordeaux region in 2003 and 2019, where ideal GHDs were chosen and excellent vintages resulted. Other wine growing regions were less fortunate. Therefore, local weather forecasting with sufficient pre-warning time of extreme weather conditions is of the utmost importance. Long-term predictions based on climate change are crucial when it comes to the replanting of vines or the consideration of new varieties. Ideally, an optimal wine production portfolio needs to be found through geographic spreading and variety diversification. Of course, such a statement is easily made; turning it into practical decisions is quite another matter.

In Section 9.3.1 we met the old proverb "The proof of the pudding is in the eating"; in this case "The proof of the wine is in the drinking" is more appropriate. When it comes to the quality appraisal of wine, technical terminology exposes every possible molecule of wine texture. But when it comes to the actual tasting and trying to put the various sensations experienced into words, the simple "Wine is bottled poetry" by the Victorian poet and novelist Robert Louis Stevenson says it all. So, in a world increasingly experiencing the effects of global climate change, how does this bottled poetry fare?

France is not the only country where wine production goes back to Roman times; the same is true for England, for example, for Biddenden in Kent. In part due to climate change, this town offers excellent wine, including some of the best sparkling wines around. Note that one cannot refer to the latter as Champagne, as the name is restricted to the Champagne region of northeast France. Not without some pride, the *Mail Online* of November 25, 2020 ran an article with as title "What a corking result! British sparkling wines beat champagne in blind taste test". In its May 16, 2018 edition of the *Sunday Times Wine Club*, the heading points in the same direction "Why English sparkling wine is the new Champagne"; see Sunday Times Wine Club (2018). Stripping away the chauvinistic overtone, what results are English wines that do not have to shy international competition. From a risk management point of view, French producers of sparkling wines have started to invest in wine-growing properties in the south of England (an example of proper diversification), and this is all due to an excellent local terroir and climate change. This kind of story can be multiplied all over the world. We quote from McAllister (2021):

> Global warming is already extending the boundaries of viable grape growing. Northern Germany, Belgium, England and even Scandinavia have entered the modern fine wine scene. Some wine producers have already relocated to cooler climes. In Australia, for instance, the southern island state of Tasmania is rising in

popularity because its lower temperatures are suitable for Chardonnay and Pinot Noir. [. . .] Another option is to grow different grapes, introducing varieties from southern Italy, Sicily and Greece that are more heat tolerant. Alternatively, wine producers can just accept [that] their wine will have a different flavor profile. [. . .] Climate change will impact the costs of production, revenues and profits of wine producers. They are already, or will be, adjusting their practices and adapting their winemaking business for a warmer world. But whether this adaption is successful, may come down to consumers. Wine drinkers are creatures of habit. Will they happily quaff wines from grape varieties suitable for hotter climes, such as Nero d'Avola, Vermentino, Fiano, Vranec and Xinomavro? Will they accept a different flavor profile from their Pinot Noir grown in Burgundy? Or can they bring themselves to drink a Pinot Noir from England? You decide.

In an excellent article, Asimov (2019) summarizes the situation as "The accelerating effects of climate change are forcing the wine industry, especially those who see wine as an agricultural product rather than an industrial beverage, to take decisive steps to counter or adapt to the shifts. So far, these efforts are focused on five factors that are inherently crucial to growing and producing wine." These five factors are: (1) the wine map is expanding; (2) wine makers are seeking higher grounds; (3) growers are curtailing sunlight; (4) regions are considering different grapes; and (5) the weather is no longer as predictable. We completely concur with the final conclusion that "Viticulture by its nature is complicated. As the world's climates are transformed, it is only becoming more so."

The Navarre-based winemaker Luis Fernando Olaverri once said "Wine is the only artwork one can drink." Whether we consider wine as an artwork or as an agricultural crop, the risk related to its production can be insured. Some readers may even be lucky enough to have an extensive collection of expensive wines, so that a household insurance plan for your wine collection may even be under consideration. Whereas we three authors love a good glass of wine, the value of the content of our wine cellars combined lies well below the necessary threshold, also referred to as *deductible* in insurance jargon, to make an insurance cover viable. On the issue of wine-crop insurance, the publication AXA Insurance Group (2019) highlights "A parametric yield protection coverage can help protect clients all the way through the wine industry supply chain." We quote:

A parametric, or index-based, insurance coverage is based on objective and transparent indices such as temperature, total rainfall or number of burnt hectares. The payout is triggered as soon as the agreed-upon threshold is met. This means that no costly farm visits are required and payouts can swiftly reach the insureds allowing for a quicker recovery process.

A parametric yield protection coverage can help protect clients all the way through the wine industry supply chain. This coverage offers protection against a drop in crop production based on the chosen historical yield data for a particular Appellation – the protected geographic area in which a certain wine is grown. This coverage also provides protection for crops against disease, which typically is excluded from traditional insurance policies. Certain diseases that affect vines can be caused by humidity or too much sun, for example. Clients do not need to buy the coverage for their entire crop value, but just the portion that is vulnerable to a particular weather event, such as frost.

For the wine industry, this can be an efficient solution to ensure that the revenues are protected, and payouts are swiftly transferred, should a severe weather event threaten a harvest.

It should gradually be becoming clear that the successful vintner of the future not only needs an excellent wine palate, (s)he also needs to understand the basics of risk management. Wouldn't it be exciting to find, one day, our book on the library shelves of some famous château?

Figure 10.17 The future of afternoon tea? Source: Enrico Chavez

10.5 Lessons Learned

We discussed data examples where some methodology introduced in Chapters 8 and 9 was applied. With climate change being of major concern, we paid attention to the 2021 Glasgow COP26 climate conference. Whereas the agreements made were not overwhelming, the Glasgow Climate Pact on the reduction of the use of coal as fossil fuel raises hope that the world is willing to tackle the global warming problem. At the moment, however, we could only write "The future will tell how important this Summit has been." As a feasible data analysis in the spirit of the COP26 conference, we considered temperature data. As in the previous examples in the book, we have taken the time to explain where the data can be found and how an initial analysis can be performed. As before, we stress the quantification of statistical uncertainty when presenting forecasts. The effect of a changing climate is exemplified in terms of the melting of glaciers, including the Lower Arolla Glacier in Switzerland, and in terms of wine (as part of agricultural) production, with data from the Bordeaux area in France. We started the wine example with the story of the marriage of Johannes Kepler and the measurement of the volume of wine barrels. The omnipresence of mathematical ideas is further linked to "marriage" through the famous secretary problem, the solution of which brought us back to Leonhard Euler. As Jacob Bernoulli used to say *Eadem mutate resurgo*; see Figures 7.3 and 7.4.

11

Further Examples from the World of Extremes

Share your knowledge. It is a way to achieve immortality.
Tenzin Gyatso, fourteenth Dalai Lama

11.1 Longevity

The words *Omnium versatur urna serius, ocius sors exitura* are attributed to Quintus Horatius Flaccus (Horace, 65 to 8 BCE). They appear as the subtitle of Section 6.3.5 on "Oldest Ages" in Gumbel (1958). In that section, Gumbel discusses how EVT can be applied to answer questions related to the life expectancy of humans. The translation of the quote by Horace is "From the urn of death, shaken for all, sooner or later the lot must come forth." Gumbel uses the somewhat more modern translation "Age at death is a chance [random] variable." Indeed the combination of *urna* and *sors* (urn and lot) immediately leads to the notion of randomness (chance) underlying an individual person's time of death. Here we do stress "individual", but more on this as well as on relevant EVT-based research later in this section. You may recall from Section 9.2 that we started our excursion within the realm of EVT through the actuarial problem of calculating the mean duration of life of the *last survivor* among n men of equal age who all die within t years, as reported by Nicolaus I Bernoulli in 1709.

We first recall some cultural, historical, comments related to longevity. "Old as Methuselah" is an idiom for being extremely old. Methuselah, "the man of the javelin", was a biblical patriarch who is supposed to have lived for 969 years. As such, Methuselah's name has become synonymous with longevity; in 2011 a giant tortoise named Methuselah died at age 130. In Japanese culture, Jurōjin, one of the Seven Gods of Fortune, symbolizes longevity. In Chinese culture, this role is taken over by Shóu, The Old Man of the South Pole. Here is the wonderful legend of Shóu from Wikipedia:

> According to legend, the Old Man of the South Pole was once a sickly boy named Zhao Yen who had been predicted to die when he was 19 years old. He was therefore advised to visit a certain field and to bring with him a jar of wine and dried meat. In that field, he would find two men intent on playing checkers under a tree. He should offer them wine and meat, but should avoid answering their questions. Zhao Yen followed the advice and when the two men had consumed the meat and the wine, they decided to thank him by exchanging the figures of his life expectancy from 19 to 91 years. Later he was told that one of the two men was the star of the North Pole, which fixes the date of birth of the men, and the other the star of the South Pole, which fixes the date of death.

Today's medical advances at the level of the body's metabolism, which are chemical reactions in the cell turning food into energy, for some foster hope for eternal life. In a 2008 ARTE television documentary by Gérald Caillat, with as its title "Immortality", Aubrey de Grey, an English author and biomedical gerontologist (a person specializing in the study of aging), claimed that the first human to live 1000 years was probably already alive, and might even be between 50 and 60 years old. De Grey introduced the concept of "longevity escape velocity (LEV)", which models the extension of life expectancy as gradual incremental steps (of the order of some decennia) that are often a consequence of scientific or medical advances. If you are interested in the debate around the possibility of achieving human immortality, you may read the following BBC News items from December 3, 2004. One side of the immortality coin is presented by de Grey's "We will be able to live to 1,000"; see de Grey (2004). In "Don't fall for the cult of immortality", S. J. Olshansky (2004) presents an alternative view. By now, a whole industry is building up around the extension of human lifespan as we know it. For instance, on October 16, 2021, the New York Times ran a DealBook Dialogue podcast event on the topic of "What's next for the business of longevity?"; see Hirsch (2021). In this podcast David Sinclair, the author of *Lifespan*, made a statement on life expectancy similar to the one above by Aubrey de Grey: "The first person to live to 150 has already been born." Though from 150 to 1000 is quite a step, from there to immortality is much more than just speculation. For humans to be able to live to the age of Methuselah is still a somewhat distant dream. Within the animal kingdom, however, some species have achieved near immortality. Two well-known examples are *turritopsis dohrnii* (from the broader family of jellyfish) and *hydra* of the zoological taxonomy of *cnidarians*. The "multi-headed" shape of cnidaria led to the Greek mythological serpent Learnean Hydra, which Hercules had to slay in his second task. These species achieve their longevity either by being able to turn back time, by reverting to an earlier stage of their life cycle, as in the case of *turritopsis dohrnii*, or, in the case of hydra, through a special type of genes leading to an infinite self-renewal of stem cells. Lobsters also have the capacity of living for a very long time. In this case, a special enzyme seems to be able to endlessly repair their DNA and hence keep them forever young. Unfortunately, the energy consumed by the lobster needed for the regrowth of its ever-bigger shells finally prevents it from reaching immortality. So far, science has not been able to turn these *elixirs of life*, also referred to as the *philosopher's stone*, into ready-to-use biological processes to achieve the ultimate dream of mankind. We will therefore concentrate on the life expectancy of humans as we observe it statistically, or indeed actuarially.

For the actuarial profession, the construction of accurate life tables is of the utmost importance. The life and pensions industry very much depends on its findings. According to the World Health Organization, on the basis of data published in December 2020, life expectancy at birth (female, male) ranges from a high of $(86.9, 81.5)$ for Japan to a low $(54.2, 47.7)$ for Lesotho. Overall, the values are, for Africa $(66.6, 62.4)$, the Americas $(79.8, 74.5)$, Europe $(81.3, 75.1)$ and the World $(75.9, 70.8)$. These numbers reflect existing worldwide social and economic inequalities, as well as the well-known gender gap.

On September 25, 2021, the Swiss newspaper NZZ ran an article under the (translated) title of "In the age of centenarians, we have to rethink our view on the course of life." Part of the article concerns Jeanne Calment, a French woman who was born in Arles on February 21, 1875 and died there on August 4, 1997. This makes her, at 122 years and 164 days, the oldest human whose age is well documented. As a child, she occasionally saw Vincent van Gogh, who stayed in Arles from 1888 to 1889. Once asked by a journalist what impression she has of her own future, she answered "Short, very short." In whichever way one addresses the question of human longevity, the key input are data followed by a careful statistical analysis. Lifetime data are available in various formats, referred to by actuaries as *life tables*. Life tables typically split into cohort versus period tables. A *cohort* is a group of subjects who share a defining characteristic, for example year of birth. A *cohort life table* shows the probability of a person from a given cohort dying at each age over the course of their lifetime and hence yields age-specific probabilities of death. *Period* or *static life tables* show the current probability of death (for people of different ages, in the current year). Further, there can be numerous stratifications into underlying factors; obvious ones are gender and nationality. For research in biostatistics, one typically needs much more detailed factor information, such as for instance medical preconditions, profession, lifestyle, etc. Survival data typically involve various types of censoring, as statistical information on individuals may not be fully available at the time of analysis. Right censoring, for example, occurs when an individual was included in the database at onset but was later removed, for any reason other than death. As a consequence, we have only a lower bound on that person's age at death. This is of course information that should not be disregarded but has to be treated with proper statistical techniques. Finally, before a statistical analysis can be performed, data has to be validated as much as possible. It therefore should not come as a surprise that answering the question "How long can humans live?" has no easy answer.

With an increase in life expectancy, insurance companies and pension funds face so-called *longevity risk*. For the pensions industry, we can make this type of risk clear by looking at life expectancy at age 65, say. This has increased significantly among both women and men over the past several decades in all countries of the Organisation for Economic Co-operation and Development (OECD). According to the OECD:

> In 2007, life expectancy at age 65 in OECD countries stood, on average, at over 20 years for women and close to 17 years for men. This represents a gain of almost five years for women and four years for men on average across OECD countries since 1970.

At age 80, these numbers become 9.2 years for women (up from 6.5 years in 1970) versus 7.6 years for men (up from 5.6 years in 1970). Life expectancy has increased throughout the ages, in lock-step with improved sanitation, medical care and nutrition. From an historical point of view, there are some interesting observations concerning life expectancy improvements. Some are obvious, others are less so. For instance in Caspari and Lee (2012) the authors analyze the relative ages of skeletons and conclude that longevity only began to significantly increase – that is past the age of 30 or so – about 30 000 years ago, which is quite late in the span of human evolution. They call the shift the "evolution of grandparents", as it marks the first time in human history that three generations might have co-existed. We have indeed come a long way since then.

Let us denote by T a positive random variable representing the length of life under study. In our case, this will typically be the lifespan of a human. In reliability engineering, T may correspond to the time to default of a mechanical component. Denote the distribution function of T by F and its survival function by $\bar{F}(x) = 1 - F(x)$, and suppose that F has density f. In lifetime analysis, the concept of *hazard* or *failure rate* (in engineering) or *force of mortality* (in actuarial science and biostatistics) plays a crucial role. It is defined as follows:

$$\lambda(t) = \frac{f(t)}{\bar{F}(t)} = -\frac{\mathrm{d}}{\mathrm{d}t}\log(\bar{F}(t));$$

calculate the right-hand side to verify that it is indeed equal to $f(t)/\bar{F}(t)$. Integrating both sides from 0 to $x > 0$, we obtain

$$\int_0^x \lambda(t)\,\mathrm{d}t = -\log(\bar{F}(x))$$

and thus

$$\bar{F}(x) = \exp\left(-\int_0^x \lambda(t)\,\mathrm{d}t\right), \quad x > 0,$$

or $F(x) = 1 - \exp(-\int_0^x \lambda(t)\,\mathrm{d}t)$, $x > 0$. With a slight misuse of notation (as in the last sentence of Remark 8.32 (6)), we can make the following interpretation, for small $\varepsilon > 0$:

$$\lambda(t) \approx \frac{\mathbb{P}(t < T \leq t + \varepsilon)}{\varepsilon \bar{F}(t)} = \frac{\mathbb{P}(t < T \leq t + \varepsilon)}{\varepsilon \mathbb{P}(T > t)} = \frac{\mathbb{P}(t < T \leq t + \varepsilon \mid T > t)}{\varepsilon};$$

hence $\lambda(t)$ can indeed be interpreted as the rate of dying (or defaulting) over the next ε units of time beyond t, given that the person (or component) of interest has survived (or functioned) until time t. The value of ε very much depends on the type of application. Survival functions $\bar{F}$ can then be modeled via the corresponding hazard rate function $\lambda(t)$. Recall from Example 8.40 the memoryless property of the exponential distribution $T \sim \mathrm{Exp}(\lambda)$; hence $F(t) = 1 - \exp(-\lambda t)$, $t > 0$, for $\lambda > 0$. As a consequence, the hazard rate is constant, that is $\lambda(t) = \lambda$ in this case, an idealized one with no aging effect. Both in reliability theory as well as mortality modeling, the more realistic, but still somewhat crude, *bathtub* (or *U-shaped*) *curve* model for the hazard rate plays a fundamental role; see the left-hand side of Figure 11.1. Using the terminology of mortality modeling, after an initial phase of infant death and improving mortality, a fairly stable, normal or mid-life phase takes over with constant force of mortality, which is then followed by an end-of-life phase where the force of

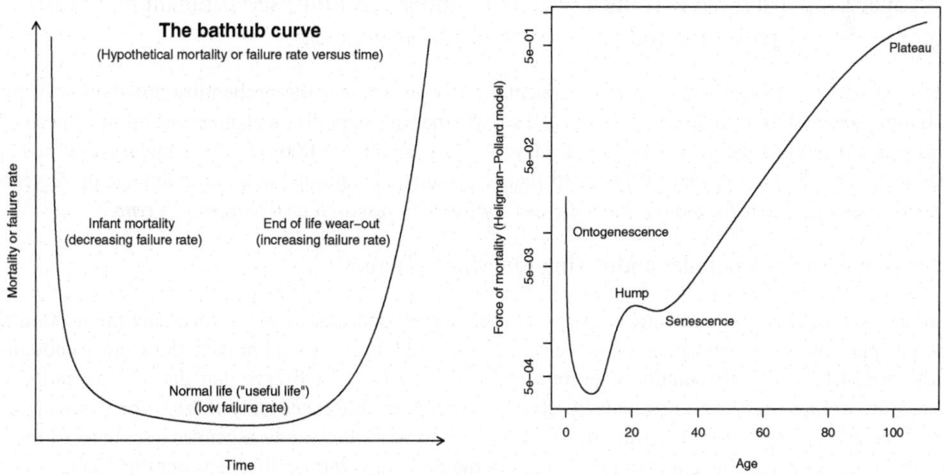

Figure 11.1 Left: The bathtub curve for hazard rate modeling in survival analysis. Right: The more realistic S-shaped force of mortality curve for humans on a log scale. Source: Authors (based on figures by Dennis J. Wilkins and Remund et al. (2018))

mortality increases. The period of advanced age is referred to as *senescence* (from the Latin, senēscere, to grow old), whereas the early infant ages are called *ontosenescence* ("onto" meaning "before"). As stated in Levitis and Martinez (2017) regarding the two halves of the U-shaped curve:

> Things tend to get better before they get worse. Given this pattern, it may be unsurprising that age-specific mortality is U-shaped for a wide range of organisms. Risk of death is high early in life, declines rapidly during ontogenesis, bottoms out around the age of maturity and then increases exponentially through advanced age. The details of the decline [. . .] and increase [. . .] of mortality risk are variable, but both are very widely observed, to the point that U-shaped mortality is a reasonable default hypothesis for most biological populations.

For the modeling of the force of human mortality, the S-shaped curve on the right-hand side of Figure 11.1 represents observed reality better; note that mortality curves often use a logarithmic scale. Besides infant mortality, we observe an unfortunate but well-known bump around early adolescence before aging starts to take its toll. Extreme value theory plays an important role in describing the upper end of the mortality curve. Obvious questions are:

(1) Is there an upper bound on the lifespan of humans?
(2) If so, what is it?
(3) Is there a plateauing (leveling off) effect setting in at high age?
(4) If yes, at what age and how can we interpret this?
(5) And what are the social and economic consequences of the findings coming out of (1)–(4)?

Question (1) has an obvious EVT character: given (assumed iid) mortality data $T_1, T_2, \ldots, T_n$ (ages at death), model the random variable $\max\{T_1, T_2, \ldots, T_n\}$ and test whether the underlying distribution function has a finite right endpoint. In Section 9.3, we have seen that this corresponds to a negative shape parameter ($\xi < 0$) for the resulting GEV or GPD distribution. For a paper supporting the existence of a finite upper age limit, see Einmahl et al. (2019). We quote from the paper's introduction (with our italicization):

> There is no scientific consensus on the fundamental question whether the probability distribution of the human life span has a finite endpoint or not and, if so, whether this upper limit changes over time. Our study uses a unique dataset of the ages at death – in days – of all (about 285,000) Dutch residents, born in The Netherlands, who died in the years 1986–2015 [hence 30 separate datasets] at a minimum age of 92 years and is *based on extreme value theory, the coherent approach to research problems of this type*.

Further, the authors conclude, addressing Question (2) above:

> For women, the average estimated upper endpoint [over the 30 datasets] is 115.7 years and the maximum estimated upper endpoint, corresponding to the year 2001, is 123.7 years [. . .] For men, the average endpoint estimate is 114.1 and the maximum endpoint estimate is 124.7 [. . .] Observe that the average endpoint estimates for women and men are relatively close (1.6 years in difference only [higher for women]). In contrast, the average age of death during the 30 years that we consider shows a gender difference of more than 5 years. [. . .] there are no clear global changes in the maximum human lifespan over the years.

Turning to Question (3), an EVT-based paper that addresses the possibility of the plateauing of the force of mortality is Belzile et al. (2021); see also Belzile et al. (2022) for a statistical review on a longevity cap. The authors conclude as follows:

Once the effects of the sampling frame are taken into account by allowing for truncation and censoring of the ages at death, a model with constant hazard after age 108 [see Question (4) above] fits all three datasets well; it corresponds to a constant probability of 0.49 that a living person will survive for one further year, with 95% confidence interval (0.48, 0.51). The power calculations make it implausible that there is an upper limit to the human lifespan of 130 years or below.

Clearly, the two papers are somewhat at odds when it comes to the existence or non-existence of an upper limit of the human lifespan. Of course, as statistical analyses, the differences in data used, as well as a comparison of the estimation uncertainties (clearly reported in both studies) have to be taken into account. There is indeed "no scientific consensus" on the issue, even among EVT specialists. Both analyses however add value to the broader discussion of longevity. As we have already stated, for the life and pension insurance industry the issue is highly relevant. Gbari et al. (2017) combine insurance thinking with an EVT-based approach. In some ways it bridges the gap between the two papers above. We quote:

In this paper, the force of mortality at the oldest ages is studied using the statistical tools from extreme value theory. A unique data basis recording all individual ages at death above 95 for cohorts born in Belgium between 1886 and 1904 is used to illustrate the relevance of the proposed approach. No levelling-off in the force of mortality at the oldest ages is found and the analysis supports the existence of an upper limit to human lifetime for these cohorts. Therefore, assuming that the force of mortality becomes ultimately constant, that is, that the remaining lifetime tends to the Negative Exponential distribution as the attained age grows is a *conservative strategy for managing life annuities*. [...] As a byproduct of our analysis, we also [draw conclusions] about the existence of an ultimate age. However, we do not claim that this limit must be interpreted in the biological or demographic sense, induced by genes or other natural mechanism. Rather, the ultimate age that has been obtained in the present paper serves as a working upper bound on policyholder's lifetime when actuarial calculations need to be performed. From a practical point of view, closing the life table by assuming that the missing [excess survival probabilities] beyond the last available age [x], that is, *assuming that the remaining lifetimes ultimately conform to the Negative Exponential distribution, is equally efficient and even conservative for products exposed to longevity risk.*

Again, the italicization is ours. The authors report as ultimate ages for males 114.82 and for females 122.73, thus 122 years and 267 days for the latter. With her 122 years and 164 days, Jeanne Calment came very close! An in-depth discussion of Question (5), that is, what are the socio-economic consequences of a non-trivial increase in human lifespan, would necessitate multiple studies on its own. A start is Scott et al. (2021); for a more actuarial point of view, see Gorvett (2014). If you want to read a short paper giving a statistical analysis of human longevity, written in a very accessible way, go for Pearce and Raftery (2022). We quote its final conclusion as it nicely summarizes our discussions above (the italicization is ours):

Predicting the extremes of humanity is a challenging task filled with unknowns. Just as it is conceivable that a medical breakthrough could let humans live indefinitely, every individual [reaching] age 123 could simply die the next day. Instead, our study has taken a statistical, data-driven approach without delving into untestable hypotheticals. The results are promising, but incremental: *humans will most likely break Ms. Calment's record of 122 this century, but likely not by more than a decade.*

For a cartoonist's interpretation of longevity, see Figure 11.2.

Figure 11.2 Supercentenarians (aged 110 or older) as future rebels. Source: Enrico Chavez

11.2 From Katrina to Ida

On August 29, 2005, the category-5 hurricane Katrina made landfall close to New Orleans in Louisiana. It caused over 1800 deaths and 125 billion USD in damages. The catastrophic consequences of severe tropical storms are typically a consequence of three main weather-related factors: wind, rain and sea surge. In the case of Katrina, the available system of levees, floodgates, pumps and canals failed to protect the city and its surroundings. The word "levee" in American English comes from the French verb "lever" (translated: "to lift"), to heighten, and more particularly from "levée", the feminine past participle of "lever". In Commonwealth English one uses "dyke"; both "dike" and "dyke" stem from the Dutch word "dijk" for "embankment" or "floodbank". In the aftermath of Katrina, the US Army Corps of Engineers were interested in discussing how EVT could be used in helping to set appropriate heights of levees in order to protect metropolitan New Orleans. The measures taken since have combined a raising and strengthening of existing levees, as well as the construction of new ones, the addition of moveable floodgates, the further improvement of the network of canals as well as the installation of extra, massive, pump systems to drain eventual floodwater. New Orleans is not only known as The Big Easy, but also locally as the Fishbowl. As can be seen from the left-hand side of Figure 11.3, New Orleans is very much surrounded by water, with more than 50% of the city lying at or below sea level. Built on marshland, it is surrounded by the Mississippi River, Lake Pontchartrain and the Gulf of Mexico. So pumping systems become crucial; the current (2021) system in place has a capacity of about 1400 m^3 (somewhat over half an Olympic-size swimming pool) per second. These protective works were finished in 2017 at a cost of about 14.5 billion USD. The hurricane protection levees for metropolitan New Orleans were built to a rather minimal 1-in-100-years standard, hence according to a 1% probability of overtopping each single year. In order to reflect the dramatic effect of the storm Ida, we present the events as they unfolded. We are writing these lines on Sunday, August 29, 2021, 16 years to the day after Katrina hit the city. For today, the category-4

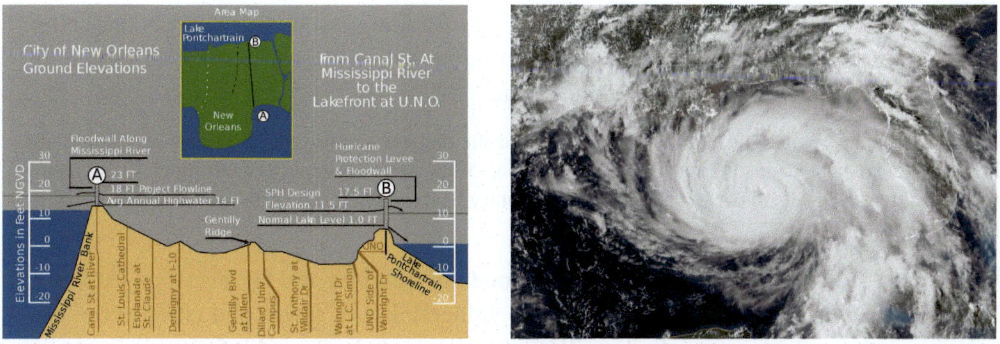

Figure 11.3 Left: Vertical cross-section of New Orleans, showing a maximum levee height of 23 feet. Right: A satellite image of hurricane Ida over the Gulf of Mexico before making landfall near New Orleans. Sources: Wikimedia Commons (left) and National Oceanic and Atmospheric Administration (NOAA) (right)

hurricane Ida is to make landfall just south of New Orleans. In the week before, citizens have asked themselves whether the levees would hold. On Sunday, Governor John Bel Edwards of Louisiana expressed optimism that the city would handle the storm on CNN's *State of the Union*: "The new storm system surrounding New Orleans, with its 350 miles of levees, flood walls, gates and pumps, will withstand the storm surge. There's been tremendous investment in this system since Hurricane Katrina. This will be the most severe test of that system."

We have since moved on, to Monday, August 31; so far, the levees have kept their protective promise. With an electricity blackout affecting more than one million people, the consequences of this storm will still be considerable. The remnants of hurricane Ida moved north, bringing record rainfall and flash floods to the northeastern states of the USA. On September 2, a state of emergency was declared in New York City, New York state and New Jersey. New York City experienced the wettest hour on record. The National Weather Service recorded 3.15 inches (8.00 cm) of rain in Central Park in New York City. In one hour on Wednesday night, September 1, the amount of rain surpassed by far the previous record of 1.94 inches (4.93 cm) set only 11 days earlier on August 21 by tropical storm Henri. On August 23, the New York Times reported the following sobering fact:

> In 36 hours from Saturday night to Monday morning alone, 8.05 inches of rain fell in Central Park. [...] Maybe eight inches doesn't seem like that much. But it's actually quite a lot. An inch of rain falling on an acre of land weighs about 113 tons. Eight inches of rain on the 302.6 square miles of New York City weighs about 175 million tons and comprises more than 42 billion gallons of water. This is where you run into problems. The city's ancient combined storm-sewer system was not built to handle that. [...] What it was built to handle is only 3.8 billion gallons a day. [...] With a warming climate you're going to see more water vapor in the air, and more water vapor in the air leads to more rain [this effect is known as the *Clausius–Clapeyron relation* and it states that the water-holding capacity of the atmosphere increases by about 7% for every 1 °C rise in temperature].

You may want to convert the US measurements in the above statement to their metric equivalents. Whichever unit you take, the facts clearly point to major problems faced by heavily built up areas when extremely heavy rain occurs over a very short period of time. The death toll of Ida in the northeast may run up to a factor beyond ten when compared with South Louisiana, where Ida made landfall. Up to a week ahead, science is good at predicting where and when a hurricane will make landfall. The prediction of the time and place of major flash rains with any accuracy remains difficult. Warning and preparing the population for such events in a timely manner is even harder. Building levees and dikes is one thing; upgrading or replacing older sewer systems in bigger cities is quite another matter. In its September 3 edition, the New York Times wrote:

> Speaking from the White House, President Biden said the damage showed that "extreme storms and the climate crisis are here", constituting what he called "one of the great challenges of our time". Overlapping disasters in the US have laid bare a stark truth: the country is not ready for the extreme weather brought by climate change.

Which country is ready? The 2021 summer floods in Belgium and Germany, as well as in Zhengzhou, China, stand sad testimony to the vulnerability of towns and cities in the face of these weather events, especially for cities with a large underground infrastructure; on the latter topic, see, for example, Normile (2021).

Remark (From historical floods to sponge cities)**.** So far, we have discussed several examples of floods, each one telling its own story of tragedy and human suffering. In Chinese history, we encounter several examples of catastrophic floods. China has two main rivers, the Yellow River in the north and the Yangtze River further south. A considerable part of China lies in the drainage basin of these two rivers. In more recent history, three calamitous events stand out. These are the Yellow River floods of 1887 and 1938, and the Yangtze River flood of 1931. Each of these floods claimed several million deaths. Throughout history, China has battled against the natural forces of water. An early story relates to emperor Yu the Great, the founder of the Xia Dynasty some 4000 years ago. Yu the Great was a legendary ruler of ancient China who is heralded in mythical stories as having battled against the water-forces of nature by reinforcing dikes and constructing drainage canals to tame the Yellow River. Let us fast forward to the twenty-first century; like any country in the world, China still battles against flooding, for example due to overflowing rivers or typhoon- and monsoon-induced heavy rainfall. An interesting component of this never-ending battle is the development of so-called *sponge cities*.

As we learn from World Future Council (2016):

> The sponge city indicates a particular type of city that does not act like an impermeable system not allowing any water to filter through the ground, but, more like a sponge, actually absorbs the rain water, which is then naturally filtered by the soil and allowed to reach into the urban aquifers. This allows for the extraction of water from the ground through urban or peri-urban wells. This water can be easily treated and used for the city water supply.

And further:

> A sponge city needs to be abundant with spaces that allow water to seep through them. Instead of only impermeable concrete and asphalt, the city needs more:
>
> - **Contiguous open green spaces**, interconnected waterways, channels and ponds across neighbourhoods that can naturally detain and filter water as well as foster urban ecosystems, boost bio-diversity and create cultural and recreational opportunities.
> - **Green roofs** that can retain rainwater and naturally filter it before it is recycled or released into the ground.
> - **Porous design** interventions across the city, including construction of bio-swales and bio-retention systems to detain run-off and allow for groundwater infiltration; porous roads and pavements that can safely accommodate car and pedestrian traffic while allowing water to be absorbed, permeate and recharge groundwater; drainage systems that allow trickling of water into the ground or that direct storm water run-off into green spaces for natural absorption.

- **Water saving and recycling**, including extending water recycling particularly of grey water at the building block level, incentivizing consumers to save water through increased tariffs for increase in consumption, raising awareness campaigns, and improved smart monitoring systems to identify leakages and inefficient use of water.

We have indeed come a long way since emperor Yu the Great. It is telling that one of the main proponents of sponge cities is Yu Kongjian, a Chinese ecological urbanist, urban planner and landscape architect, professor of landscape architecture at Peking University; see, for instance, Yu (2016). He formulates the following basic equation of the movement from gray (concrete) to green cities: Nature's economy = ecology.

11.3 Lessons Learned

The two examples treated in this chapter are only pointers to a much wider realm of potential manifestations of extreme events and their influence on life as we know it. In the case of longevity, we touched a very personal, individual, level. Science finds itself on the verge of basic developments that bring the Elixir for Eternal Life a little closer to the fulfillment of one of humanity's dreams. The potential benefits of such a genetic elixir to humanity may, most likely, only become available in the rather distant future. We gave some examples of possible interpretations of this distant future. More concretely, EVT plays an important role in studying the possible human life span as of today. An important lesson learned so far is the plateauing of the force of mortality at high age, that of supercentenarians, people of age 110 and beyond. At the other end of the risk spectrum, we exemplified environmental risk as manifested through hurricanes. Through the particular examples of Katrina and Ida we highlighted both society's vulnerability to such catastrophic events, as well as, for instance, in the case of New Orleans, how resilience against the consequences of such extreme weather events can be provided. The development of sponge cities offers an interesting example of this resilience building. Finally, several examples stressed the need for clear, evidence-based risk communication.

12

Networks

Now we live in fairyland. The only slightly disappointing thing about this land is that it is smaller than the real world has ever been.

Frigyes Karinthy (1887–1938) in *Chains*

12.1 A Small World

A nice party game goes as follows. Suppose you are together with a group of n people and you make the following statement: "I bet that there are at least two people among us who have their birthday on the same day." We assume that there are 365 days in a year, that all days are equally probable for births to occur and that the participants' birth events are independent. Of course, this is not fully true but making the assumptions more realistic does not change the outcome in any significant way. We make the bet more specific by assuming that there are $n = 30$ people in the group. Would you accept the bet? Most people will accept, reasoning from their own birthday and that the probability of a specific person having the same birthday is small, indeed it is $1/365$. Let us do the calculation, also known as the *birthday problem*. Denote by B_n the event of having at least two of n people with the same birthday. Recall the Pavlovian reaction from Section 8.1.1. When you read "at least two", go for the complement B_n^c, meaning "no two people have the same birthday", hence

$$\mathbb{P}(B_n) = 1 - \mathbb{P}(B_n^c) = 1 - \frac{365}{365} \times \frac{364}{365} \times \cdots \times \frac{365 - n + 1}{365};$$

the product of terms of the form $(365 - k)/365$, $k = 0, 1, \ldots, n - 1$ (it is a product because of independence), comes from the fact that the first person can have any birthday (probability $365/365 = 1$), the second person can have any birthday except the same as the first person (probability $364/365$), etc. Figure 12.1 shows the probability $\mathbb{P}(B_n)$ as a function of n. As you can see from the specific values of $\mathbb{P}(B_n)$ printed in the lower right corner of the plot, the somewhat surprising result is that the changeover for the above bet takes place from $n = 22$ to $n = 23$ ($\mathbb{P}(B_n)$ switches from < 0.5 to > 0.5, indicated by a dashed line). Hence as soon as 23 (or more) people are in a group, the probability of having a least two people with the same birthday is greater than 50%. For $n = 30$, as in the above bet, it is 70.63%, and for $n \geq 57$, it is already above 99%. You can check this experiment next time you go to a party. But why do most of us get the solution intuitively wrong and expect the changeover value for n to be much larger than 23? The answer is *networks*, although here the word *graph* would be more standard; we will come back to this terminology in the next section. As we already said before, every individual reasons from her/his own perspective: my birthday versus the

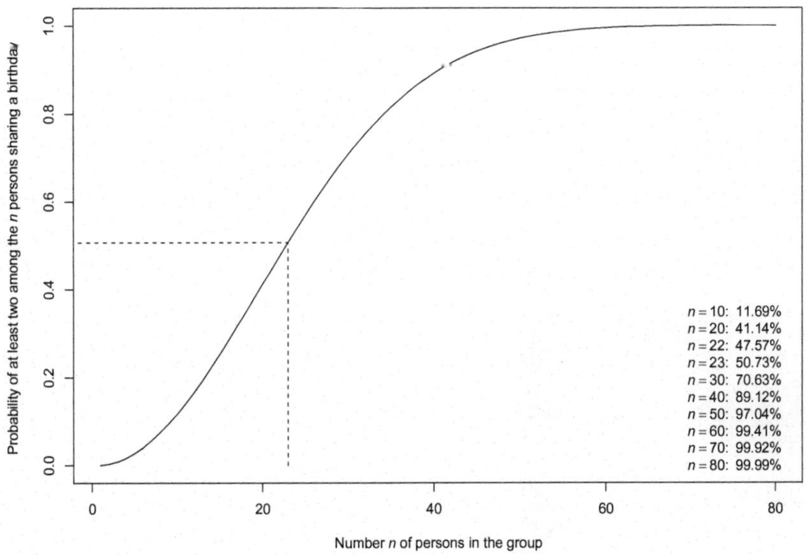

Figure 12.1 Probability of at least two among n people sharing a birthday as a function of n ($n = 23$ is indicated by a dashed line). Source: Authors

$n - 1$ others. This reasoning links that person to the $n - 1$ others and hence results in an easy graph with $n - 1$ links. We have to do this however for every participant to the party, which leads to a more involved graph linking each of the $\binom{n}{2} = n(n-1)/2$ pairs. Once you start thinking about the problem, the added complexity becomes obvious; our intuition however does not follow this more analytical approach.

We want to add a little story at this point. Suppose that at the same party, you are sitting around a (large, round) dinner table, and, as is the custom in Switzerland, for example, at the start of the dinner you raise your glasses and clink the glass of all the other participants; clinking glasses became a custom, many centuries ago, in order to avoid being poisoned (a heavy clink of two full medieval tin beakers would spill some fluid from one beaker into the other). How often do you hear the sound of glasses clinking? Using the same logic as in the birthday problem (can you see why?), the solution is

$$(n-1) + (n-2) + \cdots + 1 = \frac{n(n-1)}{2};$$

formally, this equality is easily proven by induction. Concerning adding successive numbers, as on the left-hand side of the above equation, there is a nice story about eight year old Carl Friedrich Gauss, whom we met in Chapter 8. The teacher of his class wanted some time to snooze and gave the pupils the task of adding up the first 100 numbers. Great was his surprise when young Carl Friedrich gave the right answer, namely 5050, almost instantly. Gauss realized that $1 + 100 = 101, 2 + 99 = 101, \ldots, 50 + 51 = 101$, makes in total 50 times the number 101 and thus 5050. Cheers!

Let us do a final party game where a network structure plays an important role. Take any person at the party (Marius, say) and ask, starting from Marius, in how many steps you can reach, for example, the Dalai Lama (not of course present at the party). A step in the chain means "has shaken hands with" or "has spoken to"; in pre-COVID times, this game was also referred to as "how many handshakes are you away from". In this case, an upper bound surely is 4. From Marius to Valérie is 1, from Valérie to the economist Robert Engle is 2; after having received the Nobel Prize in Economics, Robert Engle was invited to the White House by former US President George W. Bush (so 3) who met the Dalai Lama (so indeed 4). Of course, there may well be a shorter path. The amazing thing is that it is hardly possible to find a path longer than 6 between any two people on this planet; this is referred to as *six degrees of separation*. The idea goes back to the Hungarian writer Frigyes Karinthy (1887–1938); see Figure 12.2. In 1929, Karinthy wrote the wonderful short story *Láncszemek* ("Chains"),

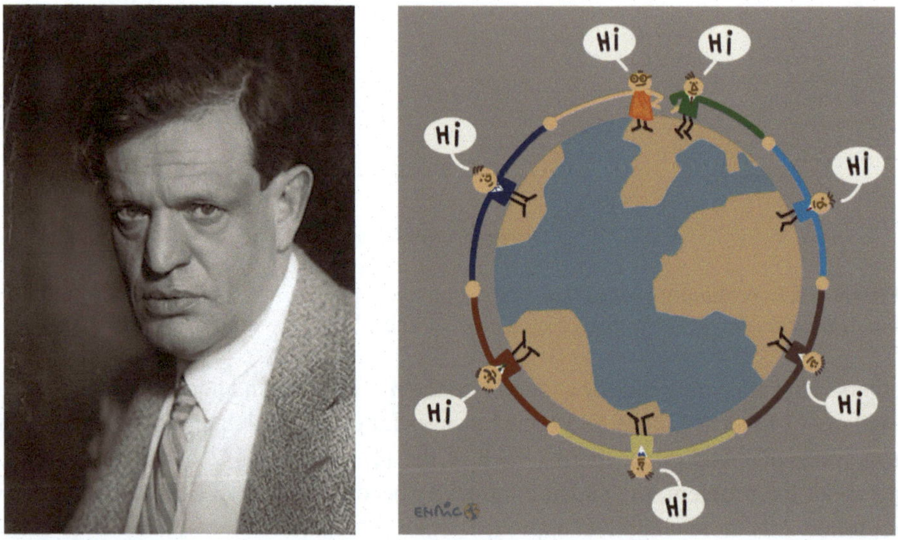

Figure 12.2 Left: Frigyes Karinthy (1887–1938). Right: Six degrees of separation. Sources: Wikimedia Commons (left) and Enrico Chavez (right)

in which we can read:

> Let me put it this way: Planet Earth has never been as tiny as it is now. It shrunk – relatively speaking of course – due to the quickening pulse of both physical and verbal communication. This topic has come up before, but we had never framed it quite this way. We never talked about the fact that anyone on Earth, at my or anyone's will, can now learn in just a few minutes what I think or do, and what I want or what I would like to do. If I wanted to convince myself of the above fact: in couple of days I could be – Hocus pocus! – where I want to be.

Karinthy then continues with the above game, reaching the conclusion that, whichever two names the participants chose, the chain never needed more than five links; try it yourself! The wonderful story can be read on the website Karinthy (1929). It is truly amazing that Karinthy wrote it as early as 1929, well before computers took over the world of communication. We have not been able to find out why he did so beyond the fact that the late 1920s heralded an increase of technological and industrial awareness and international traveling. Six degrees was rediscovered almost four decades later, in 1967, by Stanley Milgram, a Harvard professor who turned the concept into a much celebrated groundbreaking study on our interconnectivity; see Milgram (1967). Through the 1990 play "Six Degrees of Separation" by John Guare, the wider public became very well aware of the concept as well as the game. Now almost 100 years after the publication of Karinthy's prophetic story, the world has shrunk even more, in particular through social networks. In Chapter 6 on the coronavirus pandemic we saw what this shrinking has for consequences. In a research publication (Edunov et al., 2016), the authors write:

> How connected is the world? Playwrights, poets, and scientists have proposed that everyone on the planet is connected to everyone else by six other people. In honor of Friends Day, we've crunched the Facebook friend graph and determined that the number is 3.57. Each person in the world (at least among the 1.59 billion people active on Facebook) is connected to every other person by an average of three and a half other people. The average distance we observe is 4.57, corresponding to 3.57 intermediaries or "degrees of separation." Within the US, people are connected to each other by an average of 3.46 degrees. [For Mark Zuckerberg it is 3.17 and for Facebook's COO Sheryl Sandberg it is 2.92] Our collective "degrees of separation" have shrunk over the past five years. In 2011, researchers at Cornell [University], the Università degli Studi di Milano, and Facebook computed the average across the 721 million people using the site then, and found that it was 3.74. Now, with twice as many people using the site, we've grown more interconnected, thus shortening the distance between any two people in the world.

12.2 A Walk Through Königsberg

Königsberg used to be a German town at the east coast of the Baltic Sea. It was badly damaged towards the end of World War II and is now part of Russia and known as Kaliningrad. We have already met the Swiss mathematician Leonhard Euler on several occasions throughout the book. During his time at the Saint Petersburg Academy of Sciences, he was confronted with an amusing pastime problem from the burghers of Königsberg. As shown in Figure 12.3, the river Pregel divides the city into four separate parts, which are connected by seven bridges. The question posed was whether it is possible to walk around the city crossing

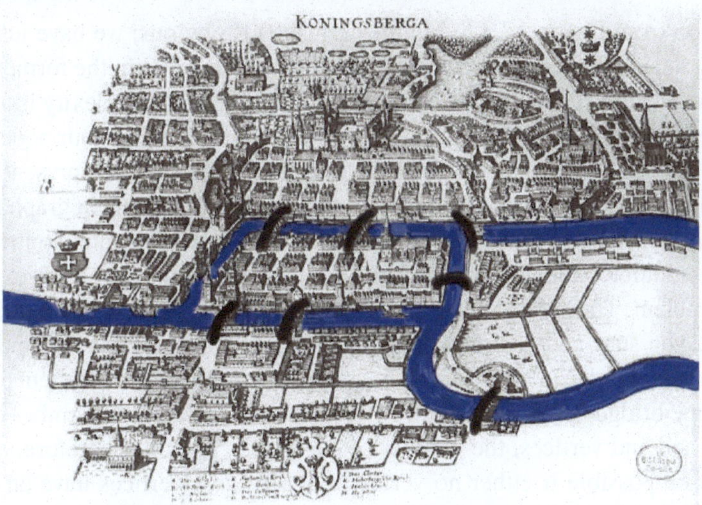

Figure 12.3 The map of Königsberg at the time of Leonhard Euler, showing the river Pregel, the seven bridges and the four separate parts of the city. Source: Authors (based on a drawing from Merian Erben in 1652, Wikimedia Commons)

each bridge, but only once. One can start and finish at different locations; you may try a couple of routes yourself. Euler gave a wonderful solution reducing the problem to its bare essentials and in doing so provided the foundations of the highly relevant fields of mathematics known as *graph theory* and *network theory*. The idea of the solution is sketched graphically in Figure 12.4. We will use both the words "graph" and "network" for two very

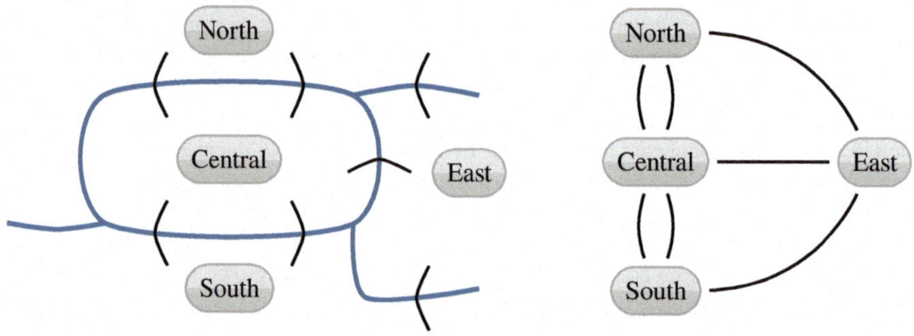

Figure 12.4 Königsberg's topography first reduced by stripping out the detailed layout of houses and squares (left) and then further reduced to its bare essentials (right). Source: Authors

much related mathematical structures; their usage depends on the context. Both consist of points and lines between some of the points; for a graph these are called *vertices* and *edges* and for a network they are called *nodes* and *links*. There is a lot more structure that can be added, but let us return to Euler, walking in Königsberg around 1736, the year he solved the problem.

The transition from Figure 12.3 to Figure 12.4 (left) is obvious; we have just left out the city's planning details that are not relevant (except for the fact that in the former we may see where to go for a drink). The crucial step in this cutting through complexity is to be found in going from (left) to (right) in Figure 12.4, where we have just kept the four nodes for the four parts of Königsberg that are connected by bridges and the seven edges representing the seven bridges. In doing so, Euler reduced the problem to a mathematical one in graph theory, at the time an entirely new field of research to which Euler made fundamental contributions. The minimal representation on the right-hand side of Figure 12.4 contains all the essentials from the original problem. The solution now stares you in the face! Leave out for the moment the vertices where you start and/or finish the walk. Whenever you enter a remaining vertex (a part of the city) via an edge (a bridge), you must leave that vertex by a different edge so as not to walk the same bridge twice. Hence, all those vertices need an even number of edges. For the beginning and end vertices, the number of edges may be odd. Therefore, a Königsberg walk can only be possible if either no vertex or exactly two vertices have an odd number of edges. As all the vertices have an odd number of edges (either five or three) there is no solution. The answer to this famous *Königsberg bridge problem* is hence negative. For those for whom this problem and its solutions are new, take a pause and go over the arguments again (and possibly again). Let the beauty of Euler's thinking sink in. In the above solution by Euler we see that the number of edges (the *degree*) of a vertex plays a crucial role. And since Euler was the first to discuss such paths through graphs that visit every edge exactly once, they are called *Eulerian paths* in graph theory.

Networks and graphs are now ubiquitous throughout the modeling landscape of risk. Just think of computer networks (cyber risk), social networks (fake news risk), professional networks, electricity grid networks (black-out risk), financial networks (systemic risk), road and traffic networks (obvious risk), sea and air transportation networks (supply chain risk and the risk of pandemic spreading), water canals and sewer network systems (flood risk), etc. In the next section, we pay a short visit to this landscape, starting from a historical perspective before coming back to Frigyes Karinthy.

12.3 On Email, WWW and Social Networks

In Chapter 10 we discussed how society is increasingly worried about global risks stemming from climate change. A different type of risk concerns our technologically networked society, which creates not only mental, but also physical, spider webs around us from which escape can be difficult, if not impossible. Email came about in the early seventies through its inventor Raymond Samuel ("Ray") Tomlison (1941–2016) and the Advanced Research Projects Agency Network (ARPANET). Tomlison is also responsible for the now ubiquitous @ (read: "at") symbol.

The *World Wide Web* came much later. On March 12, 1989, Timothy Berners-Lee, an independent contractor at the European Organization for Nuclear Research (CERN, Geneva), implemented the first workable version of what later became to be known as the WWW. At the top of a CERN internal document that spells out the structure of this novel computer-based information system appear the words, by now famous, scribbled by a superior: "Vague but exciting...". No doubt vague at the time, exciting for sure, but, as we know by now, world-changing. Let us listen to Berners-Lee himself highlighting these exciting moments in history (from Wikipedia): "I just had to take the hypertext idea and connect it to the TCP [Transmission Control Protocol] and DNS [Domain Name System] ideas and – ta-da! – the World Wide Web." And:

> Creating the web was really an act of desperation, because the situation without it was very difficult when I was working at CERN later. Most of the technology involved in the web, like the hypertext, like the Internet, multifont text objects, had all been designed already. I just had to put them together. It was a step of generalising, going to a higher level of abstraction, thinking about all the documentation systems out there as being possibly part of a larger imaginary documentation system.

The email system as well as the web came about in order to facilitate research communication within universities or research institutions like CERN. Search engines like Yahoo! Search, AltaVista and Google Search saw the light in the mid 1990s. At the same time, the first social networks appeared. An interesting one, at least by name (as it refers to the six degrees of separation), is `sixdegrees.com`. The deluge of such platforms then came early in the twenty-first century with Skype, Facebook, Twitter (now X) and the like. The resulting social network forest now caters for all specific tastes and desires. Interesting, reflective, questions for all of us surely are:

- What was the first email you ever sent?
- What was the first WWW search you conducted?
- What was your first "like" on a social network?

Few of us will remember these early actions though they stood at the user's entrance of a world highly determined by computers. This computerized world no doubt offers a multitude of benefits, but it has also led to a world where risk is omnipresent. You do not have to cross the street to encounter risk; it is there, right in front of you, at the touch of a (computer) button (think of cyber crime, for example). Through all these developments, risk goes well beyond its purely technological or physical interpretation; it increasingly relates to a much wider spectrum of societal, political and psychological platforms and channels. An obvious benefit is the democratization of knowledge. With rare exceptions such as North Korea, all over the world a simple smartphone gives the user the key to a treasure trove of unbounded knowledge. Think, for example, of the online encyclopedia *Wikipedia* (or, in China, *Baidu Baike*). On the other side of the coin however, we find more modern terms like fake news, influencers, bots, trolls, microtargetting and the like. Often news is packaged into very short time spans of breaking or bombshell news as if the human mind could not hold an attention span beyond that short period of time or is not able itself to judge relevance and impact. And if this 30-second news coverage is not sufficient, we have X (formerly Twitter), a so-called microblogging environment through which communication was originally restricted to 140 characters. It became 280 in November 2017; audio and video tweets are typically limited to 140 seconds, hence are "micro". During the coronavirus pandemic, science communication through X became a preferred medium. Modern society, through its democratic structures, will have to find a careful balance between the two sides of the technological coin unleashed by these developments. Likewise, the science community cannot disregard this development; scientist will have to become aware and indeed able to reach out to society through the social media which are consulted (especially) by the younger generation. An excellent example of how this can work well is due to David Spiegelhalter from the Winton Centre for Risk and Evidence Communication of the University of Cambridge; see `wintoncentre.maths.cam.ac.uk`. Especially during the coronavirus pandemic, his Twitter feeds contained a lot of interesting comments (indeed tweets) and highly informative links to related work.

If the above network stories have whetted your appetite for more, then we can recommend Barabasi and Frangos (2013), in which the chapters are numbered as "The first link: Introduction" up to "The fourteenth link: Network economy". In "The third link: Six degrees of separation" we find an excellent, partly historical, discussion on Karinthy's work. The last paragraph of this 'link' is worth quoting:

> "Small worlds" are a generic property of networks in general. Short separation is not a mystery of our society or something peculiar about the Web: Most networks around us obey it. It is rooted in their structure – it simply doesn't take many links for me to reach a huge number of Webpages or friends. The resulting small worlds are rather different from the Euclidean world to which we are accustomed and in which distances are measured in miles. Our ability to reach people has less and less to do with the physical distance between us. Discovering common acquaintance[s] with perfect strangers on worldwide trips repeatedly reminds us that some people on the other side of the planet are often closer along the social network than people living next door. Navigating this non-Euclidean world repeatedly tricks our intuition and reminds us that there is a new geometry out there that we need to master in order to make sense of the complex world around us.

Throughout the ages, instruments have been created to help humans in their daily lives, from the first hand axe to the steam engine to the computer and the smartphone. Now, as we stand on the verge of the AI (artificial intelligence) revolution, perhaps for the first time our actions are reduced to merely helping these instruments, and do so perhaps only as long as it remains necessary or even possible.

In all these developments, risk swims along, surfacing occasionally to raise a warning finger. It took more than two million years to evolve from the first stone tools to a modern day screwdriver. On the other hand, from Guglielmo Marconi's invention of the radio telegraph system in 1901 to IBM's first smartphone (1992) was less than 100 years. Darwinian evolution and human instinctual behavior definitely function at the time scale of the former but surely not at that of the latter. A consequence of this is that, from a risk perspective, our brain has not evolved at the same rate. The incredibly fast pace of technological change, often referred to as an example of disruptive change, poses considerable challenges to society. A typical example is our vulnerability to cyber risk in a highly computerized world. Children at school are taught from very early on how to safely cross a road where the obvious and visible risks are cars and other vehicles. The three words "Stop. Look. Listen" are used across the globe to teach young children the importance of road safety. And yet, when it comes to safety guidelines concerning the ever-present, though invisible, risks whenever we use a smartphone or a computer, our society takes a much more distant and relaxed approach. This leads to all of us crossing blindfolded a dangerous IT-road dozens of times a day. Even at a legal level, hardly any liability law protects the user from flaws or security weaknesses in the software installed on all these devices or in the cloud (where not infrequently personal data end up and are thus under constant potential attack). Society as well as many individuals have witnessed devastating effects of the many risk-Achilles-heels embedded in this technological evolution. The speed with which these technological changes are taking place implies that Darwinian adaption to the risks posed will not help here; IT-mentally we are still looking for danger lurking in the trees while walking in the savanna. Achieving risk awareness as well as protection in this environment of a highly networked society poses no doubt great challenges. Let us start by teaching children from an early age on to stop and think before they click a link. Even better, let us make sure they do not get trapped in cyberspace and that they spend sufficient time outside in "naturespace".

In the near future, AI will keep surprising us through its range of applications. For statisticians, AI is not really "a brave new world" but more the next phase in the joint development of statistical methodology together with algorithms coming from computer science. No doubt, a key push has come from the incredible growth of computational power embedded in modern chip technology, together with large companies knowing how to leverage this power; recall Moore's law of the doubling about every two years of the number of transistors in a dense integrated circuit (a further example of exponential growth). Despite the physical limits affecting Moore's law, we may only dream of what quantum computing may bring us in the, perhaps not so distant, future. In its section on science and technology, on Friday, December 31, 2021, the Swiss newspaper NZZ ran an article with an AI-like title "Das Smartphone als Sommelier" (Das = The, als = as). The article reported on the use of chemistry and AI in determining which wine should be offered to a client on the basis of his/her preferred taste descriptions. Few of us can come up with a wording like "Delicate but concentrated with silky texture and tannins, fine acidity, and a kaleidoscope of complex flavors ranging from red and black berries, rose, black tea and allspice." Or for another bottle "A more pronounced nose of berries, violets and spice, fresh apple pie and butter, turning into a more yellow apple with well-integrated tasty oak notes. And not to miss the crisp acidity and electric core of limestone minerality." Most of us have a slightly more restricted vocabulary like "dry, heavy, fruity" or more audaciously "with a slightly sweet touch, in search for a longer finish". The blind-tasting experiment of one particular company that combines machine learning/AI with chemistry in order to match consumers to products scored in the same way as a random choice (50% correct). The AI field is learning day by day, however. Anyhow, it may be a while before a song is written to praise an AI algorithm. There is a song that praises the possibly most famous wine in the world, La Romanée Conti. In this 1975 love song to the quality of a wine, Anne Sylvestre (1934–2020) sings the words "Before I die, I would like to have tasted La Romanée Conti!" (Sylvestre, 1975). Machine learning and AI still have a long way to go. We wonder how Johannes Kepler, whom we met in Section 10.4.1 measuring the content of a wine barrel for his wedding, would have reacted to these developments.

12.4 Coincidences

We started this chapter with the famous birthday problem that involves calculating that, as soon as there are 23 or more people in a group, the probability that at least two have their birthday on the same day is more than 50%. If 23 sounded a bit low, then your intuition indeed was fooled through our person-centered thinking: "my birthday compared to the other $n-1$ birthdays". In thinking about the birthday problem, however, we have to be aware that within a group of n people, there are $\binom{n}{2} = n(n-1)/2$ pairs of people and hence we are looking for a common birthday pair among these. At first, finding such a pair might be greeted with "What a coincidence!", but, as we explained, it really is not so. In day-to-day life, rare coincidences do occur much more often than we care to admit. Below we discuss a rather representative example from the realm of lotteries.

On February 14, 1986, the New York Times ran an article with the heading "Odds-defying Jersey woman hits lottery jackpot second time":

> Defying odds in the realm of the preposterous – 1 in 17 trillion – a woman who won $3.9 million in the New Jersey state lottery last October has hit the jackpot again and yesterday laid claim with her fiancé to an additional $1.4 million prize. "Shocking – definitely shocking," said 32-year-old Evelyn Marie Adams, the manager of a 7-Eleven convenience store in Point Pleasant Beach, after she redeemed her latest winning ticket in last Monday's Pick-6 Lotto game. [. . .] For aficionados of miraculous odds, the numbers were mind-boggling: In winning her first prize last Oct. 24, Mrs. Adams was up against odds of 1 in 3.2 million. The odds of winning last Monday, when numbers were drawn in a somewhat modified game, were 1 in 5.2 million. And after consultation with a professor of statistics at Rutgers University, lottery officials concluded that the odds of one person winning the top lottery prize twice in a lifetime were 1 in about 17.3 [US] trillion – that is, 17,300,000,000,000 [see also Section 6.6].

The choice of words like "preposterous", "shocking" and "mind-boggling" indeed comes to mind. But does the reporting of these odds do full justice to the communication of such an incredible coincidence? The 1 in 17 trillion, as stated in the case of Mrs. Adams, is correct in the context of the lottery reported. Hence these odds surely reflect well on the part of the story that concerns this one specific person, Mrs. Adams. It is however the correct answer to a not-very-relevant question. For the general public, the truly interesting and relevant question is: What are the odds of some non-specific person out there winning the jackpot of this lottery twice over a well-defined period, four months in this case? Answering this question, we have to consider the many millions of people that play the lottery over that period. Can mathematics help us here? Diaconis and Mosteller (1989) studied the statistical theory underlying coincidences defined as follows (the italicization is theirs):

> A coincidence is a surprising concurrence of events, perceived as meaningfully related, with no apparent causal connection. [. . .] The definition aims at capturing the common language meaning of coincidence. The observer's psychology enters at [the words] *surprising*, *perceived*, *meaningful*, and *apparent*. A more liberal definition is possible: a coincidence is a rare event; but this includes too much to permit careful study.

The paper's Section 7.1 contains an interesting discussion on the wider class of birthday problems. For instance, in order to have a greater than 50% chance of having a triple occurrence of the same birthday, one needs a group of at least $n = 88$ people, for a quadruple occurrence, $n = 188$. We learn that the standard birthday problem that we discussed goes back to Richard von Mises in 1939. This is indeed the same Richard von Mises we encountered in our discussion of extreme value theory in Section 9.3. But let us return to the lottery tickets of Mrs. Adams. We quote once more from the above paper by Persi Diaconis and Frederick Mosteller (1916–2006):

> Stephen Samuels and George McCabe of the Department of Statistics at Purdue University arrived at some relevant calculations. They called the event "practically a sure thing", calculating that it is better than even odds to have a double winner in seven years someplace in the United States. It is better than 1 in 30 that there is a double winner in a four-month period – the time between winnings of the New Jersey woman [Mrs. Adams].

And indeed, if you scan the web, you will find several double-jackpot winners.

In the previous chapters we have encountered two versions of the law of large numbers, in both their weak and strong versions. Diaconis and Mosteller (1989) formulated yet another, admittedly less mathematical, law of large numbers (the italicization is ours):

> **The Law of Truly Large Numbers.** Succinctly put, the law of truly large numbers states: *With a large enough sample, any outrageous thing is likely to happen.* The point is that truly rare events, say events that occur only once in a million [as the mathematician Littlewood (1953) required for an event to be surprising] are bound to be plentiful in a population of 250 million people [the US population in 1989]. If a coincidence occurs to one person in a million each day, then we expect 250 occurrences a day and close to 100,000 such occurrences a year. Going from a year to a lifetime and from the population of the United States to that of the world (5 billion at this writing), we can be absolutely sure that we will see incredibly remarkable events. *When such events occur, they are often noted and recorded.* If they happen to us or someone we know, it is hard to escape that spooky feeling. [. . .] Events [that are] rare per person occur with high frequency in the presence of large numbers of people; therefore, even larger numbers of interactions occur between groups of people and between people and objects.

The last sentence stresses the importance of interconnectivity through networks. Networks without any doubt act as facilitators of rare coincidence events. If you want to read more on this topic at a very introductory level, see Hand (2014). The title of this book can be viewed as a one-sentence summary of this section: *The Improbability Principle: Why Coincidences, Miracles and Rare Events Happen Every Day.*

12.5 Lessons Learned

We are aware of the omnipresence of social media and the influence these media have on society's response to risk. As individuals, we typically observe and try to understand risk from an egocentric position, which is natural. Networks of all types have become ubiquitous and form so much an integral part of society's fabric that they have achieved the level of "not being noticed any more", almost like the air we breathe. This implies that we have to train our rational thinking in order to take network phenomena into account. We have to move from an egocentric contemplation of the world around us to a network-driven one, where the number of contacts and the flow of information grows exponentially. The important consequences of the latter were discussed in Section 6.3. A central example was provided by the famous six degrees of separation idiom. Once more, on our stroll, Leonhard Euler crossed our path. Through the historic example of the Königsberg Walk, Euler initiated the currently highly important fields of network and graph theory. Of course, the Internet had to be mentioned, and indeed we did so but, unfortunately, only in passing. We reflected on our first encounters with these media. It is always difficult to gauge the importance of developments while one is part of them; the emergence of applications of AI (together with technological advances such as automation, data exchange, the internet of things and cloud computing, referred to collectively as the Fourth Industrial Revolution) have the hallmark of being such a pivotal moment in time.

13

The Black Tulip and February 3, 1637

Il est certaines catastrophes que la plume d'un pauvre écrivain ne peut décrire, et qu'il est obligé de livrer à l'imagination de ses lecteurs dans toute la simplicité du fait.
(There are some catastrophes that a poor writer's pen cannot describe and which he is obliged to leave to the imagination of his readers with a bald statement of the facts.)

Alexandre Dumas (1802–1870) in *La Tulipe Noire*

13.1 On Alexandre Dumas and Johan de Witt

We have reached the final chapter of our book. As you are about to discover, the title hides some links to topics already treated. The *Black Swan* of Nassim Taleb (2007) has reached considerable prominence in the world of risk where it stands for an event so rare that even the possibility that it might occur is unknown; it has a catastrophic impact when it does occur, though, and it often is explained in hindsight as if it were actually predictable. A typical example is 9/11 but not the coronavirus pandemic as the latter event clearly fails the first criterion of predictability *ex ante*. In the Introduction to our book, we mentioned some other species from the animal kingdom of risk; you may indeed recall the gray rhino and the dragon king. There is however one further "black something" to which we want to bring your attention. It concerns a species from botany, the *Black Tulip*. From Wikipedia we learn:

> The name "tulip" is thought to be derived from a Persian word for turban, which it may have been thought to resemble. Tulips originally were found in a band stretching from Southern Europe to Central Asia [...] Originally growing wild in the valleys of the Tian Shan Mountains [in Central Asia], tulips were cultivated in Constantinople as early as 1055. By the 15th century, tulips were among the most prized flowers; becoming the symbol of the Ottomans.

In the sixteenth to seventeenth centuries, western diplomats at the Ottoman court brought tulip bulbs home, in particular to the Low Countries (that is, Flanders and the Netherlands), where the more spectacular species soon caused a frenzy. Most books on financial bubbles and crashes contain a version of *tulip mania*, telling the story how ever increasing prices for rare bulbs reached levels where one bulb could fetch the equivalent of a nice house in the center of Amsterdam. A particularly prized tulip was *Semper Augustus*, occasionally referred to as the most beautiful tulip in the world, now unfortunately extinct; see Figure 13.1. As we learn from the right-hand side of Figure 13.1, the price index for tulips collapsed in the early days of February 1637, with a possible turning point on February 3, and hence this chapter's title. We will shortly come back to whether or not this was indeed a bubble driven by irrational behavior. But first we would like to take up a recurring theme from our book, our efforts to acquire quality data.

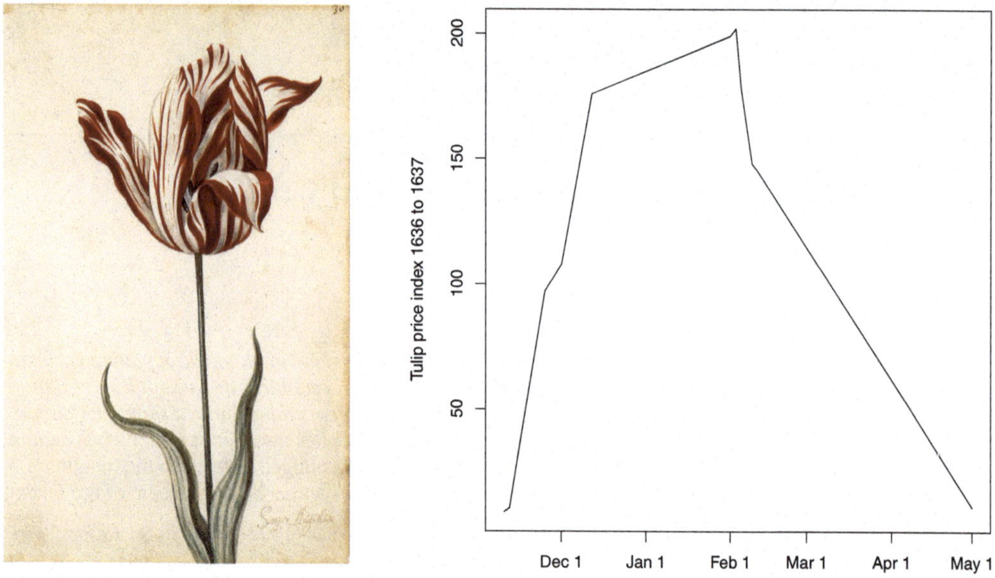

Figure 13.1 Left: Semper Augustus. Right: The evolution of the price index for tulip bulbs (peak on February 3, 1637, followed by a crash) according to data from Thompson (2007). Sources: Wikimedia Commons (left) and authors (right)

Remark 13.1 (About the data)**.** The right-hand side of Figure 13.1 is a graph we produced. How did we get these data? Searching for it led us to a Bachelor's thesis (van Doorn, 2010). Figure 6 in Chapter 3 of that thesis shows the price index for tulip bulbs, but not as a proper time series (so without correct distances along the x-axis). Table 1 in Chapter 3 of the thesis lists the price index data, but not all days are provided and months appear as numbers from 00 to 12 (thus as 13 different numbers). The caption of Table 1 refers to the paper of Thompson (2007), but at first glance, this publication does not provide the data. However, a careful reading of it reveals that, in its Appendix I, the author describes the dates and index values, in words. The reason is that the data were actually extracted from a conversation between two symbolical persons named Waermondt and Gaergoedt, published by an unknown author in 1637; see Posthumus (1929). The various dialogues between Waermondt (who speaks the truth) and Gaergoedt (who is a money grubber) satirized the folly of tulip mania. They were written as pamphlets aiming to educate the general public on this folly. The title of the 1937 text, in old Dutch, reads as "Samen-Spraeck tusschen Waermondt Ende Gaergoedt, Nopende de opkomste ende ondergang van Flora", which can be translated as "Dialogue between Waermondt and Gaermoedt concerning the rise and fall of flora." This text serves as a wonderful example of risk communication in the seventeenth century, very much like Galileo's *Dialogue* from Section 5.3. By going through Thompson (2007, Appendix I) word for word we were able to create the time series of the tulip bulb price index shown on the right-hand side of Figure 13.1. We also posted the data in the Quantitative Finance Stack Exchange forum, where users had already asked for these data; see Quantitative Finance Stack Exchange (2017).

At the time, it was totally unclear what made tulips break color, thus producing alternating color stripes radiating upwards like the red and white in a Semper Augustus. As a consequence, collectors used all sorts of unsuccessful tricks to achieve these much desired effects through breeding, eventually leading to a pure game of chance. In a way, the famous *A Random Walk Down Wall Street* (Malkiel, 2003) became a kind of early version, as *A Random Walk Down Tulip Land*. By the 1920s, the mystery of the color breaking was conjectured to be a consequence of a virus; a final confirmation for this only came in the 1960s. This viral disease also caused the plants' degeneration and eventually its extinction, as in the case of Semper Augustus. As such, the source of the tulip's beauty turned out to be its curse. Today, broken tulip varieties do exist, but none come close to the superb beauty of Semper Augustus and some of its seventeenth-century competitors on the pageant stage of "Who is the most beautiful tulip in The (Nether)land(s)?" For examples of recent broken varieties, you may for instance look for the *Tulipa Absalon* (a special type of "Rembrandt tulip") which genetically goes back to 1780.

The above story stresses the omnipresence of viral diseases and the difficulty of reigning them in. In the world of tulip breeding, working with strains of broken varieties poses the hazard of viral spreading; this risk should be treated with great care. Some form of "garden plot distancing" rather than "social distancing" is surely called for.

Tulip mania, a market bubble or not, that's the question! Most economists, including Thompson, agree on the fact that the tulip mania in the seventeenth-century The Netherlands was not a market bubble, as we currently define market bubbles and crashes as, for example, in Reinhart and Rogoff (2009). The episode surely was a folly for a small group of investors and cultivators. Prices however can be explained as either corresponding to a small supply versus an excessive demand (the kind we no doubt witness in modern markets, as for instance for art works) or coming from an imperfect functioning of the financial market at the time in relation to more complicated (derivative) products. Whichever explanation is favored, irrational behavior is less of an issue. If you are further interested in the topic, you can consult Pavord (1999) or Garber (2000).

A 1841 publication that, more than any other, contributed to the fame of the tulip mania, and indeed proclaimed it as an example of a market waiting to implode, was by the Scottish journalist Charles Mackay. Its original title was *Memoirs of Extraordinary Popular Delusions*. For a re-edited version, see Mackay and De La Vega (1995). The latter also contains Joseph de la Vega's *Confusión de Confusiones*, written in 1688, in which the author describes the (first) Stock Exchange in Amsterdam in the seventeenth century; you are encouraged to read this early observation on the functioning of a stock market. You may be somewhat surprised that, even then, options (puts and calls), short selling ("windhandel", literally "wind trade", the Dutch word for "trading in hot air") and their potential uses and, more importantly, misuses were intensely debated. The participants taking part in these debates are a merchant, a philosopher and a trader. The last acts as a teacher for the former two novices, on the topic of trading in stocks. You may recall the discussion between Salviati, Sagredo and Simplico from Galileo's *Dialogue* in Section 5.3. One surely could have wished for a similar, widely read, debate around the (mis)use of derivatives markets leading up to the 2007–2008 financial crisis, discussed in Chapter 3. The book was published shortly after the stock exchange crisis of 1688; see Petram (2014). Interestingly, several points of discussion in de la Vega's debates very much reflect the situation in today's markets; perhaps none more so than the topic of *windhandel* (short selling), which in seventeenth-century markets was on-and-off forbidden. A short seller bets on the market value of a stock, say, falling and hence borrows, for a fee, that stock from a broker, sells it at today's price and buys it back on the market at a future time when that same stock, as speculated upon, has fallen in value. The short seller nets the difference in value. All this of course assumes that the value of the stock will fall below its initial value; if not, one could face an unlimited loss potential. Just as credit default swaps (CDS), which we encountered in Chapter 3, take the temperature of credit risk, the short selling of stocks takes the temperature of market risk. A problem may occur if one starts piling up these bets using borrowed money (leverage); then seventeenth century meets twenty-first century. In the case of the 2007–2008 financial crisis, we saw whence this speculation can lead. Short sellers who bet on the fact that a stock is irrationally overvalued in a relentless bull market should always keep the famous adage of John Maynard Keynes in mind: "The market can stay irrational longer than you can stay solvent."

But let us return to the title of this chapter. The brochure *The Black Tulip* of the Amsterdam Tulip Museum (very much worth a visit) contains the wonderful story of the quest for the perfectly black tulip; see `amsterdamtulipmuseum.com`. As the story is so beautiful, and perhaps less known to you, we decided to include it in full (with thanks to the museum). Plan a future visit to the museum, and for now, enjoy reading!

The Black Tulip

Myth became reality in the story of the black tulip we know as "Paul Scherer." For centuries, tulip lovers dreamed of such a flower.

The myth of the existence of a black tulip inspired the novel of Dumas (1850), a story that influenced generations. It is a powerful tale about love, jealousy, and obsession. However, it also includes in the story a magnificent prize, offered to the first man or woman to produce a pure, black tulip. Dutch growers worked for years to create a black tulip cultivar in real life.

A few tulip breeders came close. In 1891, [a] well-known grower [by the] name E. H. Krelage declared victory in creating the fictional flower, going so far as to name his new breed "La Tulipe Noire" after the book. But, while none can doubt the marketing genius of tying his new breed to the story, those who saw it would note its color as being not black, but dark purple. Breeder [J. J.] Grullemans introduced the extremely popular "Queen of Night" in 1944, and in 1955 "Black Beauty" was introduced by M. van Waveren. All were undoubtedly dark, but [there] were also undoubtedly deep shades of purple. So the search continued, with many breeders enthralled.

The darkest tulip now in existence bloomed for the first time thirty years later on a cold winter night in a greenhouse in the tiny village of Oude Niedorp, where residents still went about in wooden shoes.

A 29-year-old horticulturist Geert Hageman had been obsessed with creating a truly black tulip from the time he first read Dumas' novel as a boy. In 1979 as a young hybridizer funded by a consortium of growers, he crossed a handful of promising varieties, hoping for just the right combination and a bit of luck, too. And then, like all tulip makers, he could do nothing but wait.

Just after midnight on February 18, 1986, Geert decided to check his greenhouse one last time before going home. It was well below freezing outside, but a pleasant 68 degrees Fahrenheit [20 degrees Celcius] inside.

He'd planted thousands of tulip seeds seven years earlier, and now these plants were about to bloom for the first time. Each planted in its own pot, sported a green bud, just beginning to show color. As he scanned the greenhouse, a small dark flower caught his eye. Was this the black tulip he had hoped and worked for?

Geert had no one with whom to share his excitement that night. His wife was sleeping soundly; his colleagues had all gone home. Instead he wandered among the plants, quietly drinking a celebratory beer. Though he later told the Chicago Tribune he had expected to find such a flower, its appearance was a near miracle. Among his myriad plants, only one, just one, produced that black flower.

The next day, he took his single tulip to the West Frisian Flora show at Bovenkarspel where flower and hybridizer created quite a stir. Suddenly, Geert was a celebrity, interviewed on television and by many international news organizations. When the furor subsided, Geert returned to his patient growing operation. It took 11 more years to build sufficient stock to bring the new tulip to market. In that time, Geert considered many names, Winnie Mandela and Dr. Martin Luther King Jr. among them, but in the end, Geert settled on "Paul Scherer". Geert's flower is widely available today, the blackest tulip on the market, a credit to the patience and persistence that leads to alchemy.

But is it a true Black Tulip? Both yes and no. These tulips are darker than any that came before, and are widely considered to be the darkest breed of tulips today. However, the breed still maintains a faint purple hue, and so is still not truly black.

Today, the search continues.

The French novelist Alexandre Dumas (1802–1870) is well known, in particular for his historic novels, including *The Count of Monte Cristo* and *The Three Musketeers*. With *The Black Tulip* he wrote a novel combining the historical and political developments in the seventeenth-century The Netherlands and the quest for a black tulip; see Figure 13.2.

Figure 13.2 Left: Alexandre Dumas (1802–1870). Right: A "Paul Scherer" black tulip. Sources: Wikimedia Commons (left) and `amsterdamtulipmuseum.com` (right)

This brings us to a human drama hidden in the title of this chapter, as well as to the final story of this book. An important link to actuarial risk will emerge. Alexandre Dumas starts his novel with the lynching by an Orangist mob of the brothers Cornelis and Johan de Witt on August 20, 1672, in Den Haag (The Hague), The Netherlands. In 1653 Johan de Witt was elected the Grand Pensionary (like a Prime Minister) of the States of Holland. In the midst of the various wars with France and England, the brothers de Witt got caught between their more republican convictions for the future state as opposed to the more royalist, centralist orientation of the Orangists, supporting William III from the House of Orange–Nassau. The latter would become the King of England, Ireland and Scotland from 1689 till his death in 1702. It is unproven, but at least suggested that William had a hand in the lynching of Cornelis and Johan in 1672; he definitely did not try to prevent it. Because of the various wars, in Dutch history 1672 is referred to as "Rampjaar" (translated: "Disaster Year"). In his novel, Alexandre Dumas used some artistic freedom by merging the dates of the Tulip Mania (the 1630s) with the events involving the brothers de Witt (the 1670s). In doing so, he linked both of these through the quest for the magic black tulip; for the details, we leave you to read Dumas' book. It is to Johan de Witt and his important publication *Waerdye*, a series of letters containing actuarial calculations, that we want to turn our attention; see Figure 13.3.

Figure 13.3 Left: Johan de Witt (1625–1672). Right: The cover of his *Letters to the States-General* on the topic of annuities. Sources: Wikimedia Commons (left) and Koninklijke Bibliotheek (right)

While in Leiden, where he studied law, Johan de Witt stayed in the house of Frans van Schooten (1615–1660), who was a brilliant mathematician. From van Schooten, Johan de Witt learned a fair amount of mathematics, which served him very well in his later life as a statesman. He developed a strong interest in actuarial questions, especially related to life insurance, and provided the first mathematically correct formula for life annuity valuation; see Ciecka (2008). A life annuity is a kind of longevity insurance, where for a premium paid at onset of the contract, the buyer (the annuitant), receives from the seller (typically a life insurance company, or as in de Witt's case, the government) a series of future payments at fixed time intervals, and this until the death of the buyer. Of course, for modern annuities, there exist many different variants of such products. At the time of de Witt, life annuities constituted an important source of income to the government. Recall that the Dutch were at war with England and France. Reading the *Waerdye* in detail reveals that the author tries to straddle his convictions as a mathematician, politician and lawyer. The lawyer enters in the definition of what constitutes a fair price, fair to both parties involved. The mathematician explains, through propositions, how such a fair price can be derived. Finally, the politician has the funding of the state in mind in these difficult times. A real tour de force is the language used by de Witt, aimed at a broad audience with no mathematical background. No doubt the *Waerdye* can join the likes of Galileo's *Dialogue*, de la Vega's *Confusión de Confusiones* and the discussion between *Waermondt and Gaergoedt*, by an unknown writer. All these early publications shine as communication gems. De Witt was not satisfied with the convention at the time of selling life annuities at a fixed price without reference to the age of the annuitant and proposed a method of calculating the price of life annuities that would vary with age. This remarkable contribution to actuarial science and mathematical finance was published in 1671, a year before his violent death, in a series of letters to the

States-General. The original version is written in Dutch, with title *Waerdye van Lyf-renten naer Proportie van Los-renten*. You can find an excellent, abridged, English translation by Frederick Hendriks under the title *De Witt's Treatise on Life Annuities in a Series of Letters to the States-General* (referred to as *Letters*) in de Witt (1671). This 15-page document is well worth reading.

In Johan de Witt's *Letters* several stories told in our book come together. We finish our historical discourse in this chapter by revisiting Jacob Bernoulli's *Ars Conjectandi*; see the left-hand side of Figure 7.3. In doing so, Johan de Witt's name reappears. One of the main reasons that the *Ars Conjectandi* was not published during Jacob Bernoulli's lifetime (recall that he died in 1705) was that, whereas he was convinced of its mathematical importance, he lacked a telling practical application for his theory. We take up the story as told in Mattmüller (2014):

> In the 1666 *Journal des Savants* [Jacob] Bernoulli had read a review of John Graunt's *Natural and Political Observations made upon the Bills of Mortality*, the first survey of population dynamics ever published; however, he seems never to have seen the book itself. To underpin his study of life annuities, Bernoulli urgently hunted for a printed memorandum by the Dutch civil servant Johan de Witt that contained an extensive mortality table, but his appeals to Leibniz to borrow his copy came to nought. And Halley's paper on the demography of Breslau in the *Philosophical Transactions* does not even seem to have come to his notice. So Bernoulli's intent to prop his favourite brainchild up at least by one meticulously elaborated application failed by an almost complete lack of data. We can only speculate what prospects Jacob Bernoulli could have opened up, and whether his view of stochastics would have been taken up sooner and more broadly, with greater enthusiasm and readier success, if he had lived to perfect his plans.

This is indeed a story from well before the email, the web, and Google; how much easier it is for us today to find and retrieve material at the push of a (computer) button (if publicly available). How much more difficult all this was for the scientific giants and great personalities from the realm of risk that came well before us.

13.2 Lessons Learned

On our journey through the hilly landscape of risk, we met several interesting personalities. In a certain way, Figure 13.3 of Johan de Witt and his *Waerdye* captures very well an important message we have tried to convey in our book. Risk is a topic that permeates all aspects of daily life, and this for all people on our planet. Its scientific understanding is no doubt challenging. Science on its own, however, does not suffice. We need people like Johan de Witt who are not only able to advance science, but at the same time, are willing to communicate scientific findings to a broader public, and, when necessary, take up related societal (even political) responsibilities. The discussion in Chapter 6 on the coronavirus pandemic brought this necessity acutely to the foreground. Reaching the end of our journey, the word "humbleness" comes to mind. Indeed, as authors of this book, we feel humbled thinking about the numerous giants of the field of risk that we have brought to your attention. As multifaceted as risks and risk modeling are, some of these giants of the field of risk, such as Jacob Bernoulli, stayed firmly on the path of science; others, such as Johan de Witt and Emil Julius Gumbel, used their scientific knowledge very much in the political arena. In their own ways, these personalities have contributed greatly to a better understanding of the world around us. To our students, and also to you, dear reader, we want to convey that, even with all the scientific knowledge we may have learned, we always have to stay humble in the face of real applications. The latter definitely applies when a considerable aspect of risk is involved. As a final message to you, we recall, from the Introduction of our book, Hamlet's advice to Horatio:

There are more things in heaven and earth, Horatio,
Than are dreamt of in your philosophy.

A Note About the References

Throughout the book we have tried hard to give you all the relevant references concerning quotes and further reading. In the current technological environment, references not only concern published books, papers and reports available in paper format; an increasing amount of material is stored solely electronically in repositories such as GitHub, cloud environments for videos, or an online encyclopedia like Wikipedia (often, neither the author(s) nor the publication year is provided). Beyond these, scientists have grown accustomed to put early, not-yet-refereed, versions of their work on electronic preprint archives like arXiv or SSRN. Finding these papers can normally be done easily with a search engine in the browser. For references not otherwise available, we have provided web addresses with date stamps of when we accessed them. We very much hope that these will continue to work in the future, but bear in mind that books can go out of print, websites can be closed or relocated, videos may be taken offline, etc. In the latter cases, `archive.org/web` may be able to take you back in time and find the material but, if not, please be assured that we have checked all the material and that it indeed existed at the time we referred to it. In fact, as a final About the Data story, Marius Hofert has written a program to extract all web addresses listed below and open them in the browser as a proof of their existence (in April 2022). With these comments added, you should be able to find your way through the library; see Figure R.

Figure R Where do I find. . . ? Source: Enrico Chavez

References

Acerbi, C., and Székely, B. 2019. The minimally biased backtest for ES. *Risk*. September 2019.

Acharya, V. V., Cooley, T., Richardson, M., and Walter, I. 2010. Manufacturing tail-risk: A perspective on the Financial Crisis of 2007–2009. *Foundations and Trends® in Finance*, **4**(4), 247–325.

Andrews, D. F., and Herzberg, A. M. 1985. *Data: A Collection of Problems from Many Fields for the Student and Research Worker*. New York: Springer-Verlag.

Andrews, R. G. 2022. Tonga shock wave created tsunamis in two different oceans. January 25, 2022. `science.org/content/article/tonga-shock-wave-created-tsunamis-two-different-oceans`. Last accessed April 24, 2022.

Andriano, P., and McKelvey, B. 2007. Beyond Gaussian averages: Redirecting international business and management research toward extreme events and power laws. *Journal of International Business Studies*, **38**, 1212–1230.

Artzner, P., Delbaen, F., Eber, J. M., and Heath, D. 1999. Coherent measures of risk. *Mathematical Finance*, **9**, 203–228.

Asimov, E. 2019. How climate change impacts wine. *The New York Times*, October 14. `nytimes.com/interactive/2019/10/14/dining/drinks/climate-change-wine.html`. Last accessed April 24, 2022.

AXA Insurance Group. 2019. From grape to glass: Managing wine risk. `axaxl.com/fast-fast-forward/articles/from-grape-to-glass-managing-wine-risk`. Last accessed April 24, 2022.

Barabasi, A.-L., and Frangos, J. 2013. *Linked: The New Science of Networks*. New York: Perseus Books.

BBC. 1996. Fermat's last theorem. `bbc.co.uk/programmes/b0074rxx`. Last accessed April 24, 2022.

Becker, M., Karpytcheva, M., Davy, M., and Doekes, K. 2009. Impact of a shift in mean on the sea level rise: Application to the tide gauges in the Southern Netherlands. *Continental Shelf Research*, **29**(4), 741–749.

Begert, M., and Frei, C. 2018. Long-term area-mean temperature series for Switzerland – Combining homogenized station data and high resolution grid data. *International Journal of Climatology*, **38**(6), 2792–2807.

Beirlant, J., Keiko, A., Reynkens, T., and Einmahl, J. H. J. 2019. Estimating the maximum possible earthquake magnitude using extreme value methodology: The Groningen case. *Natural Hazards*, **98**(3), 1091–1113.

Belzile, L. R., Davison, A. C., Rootzén, H., and Zholud, D. 2021. Human mortality at extreme age. *Royal Society Open Science*, **8**(9), 1–15.

Belzile, L. R., Davison, A. C., Gampe, J., Rootzén, H., and Zholud, D. 2022. Is there a cap on longevity? A statistical review. *Annual Review of Statistics and Its Application*, **9**, 21–45.

Bernoulli, J. 1713. *Ars Conjectandi*. Basel: Thurneysen Brothers.

Bernstein, P. L. 1996. *Against the Gods: The Remarkable Story of Risk*. New York: Wiley.

Bingham, N. H. 2013. The worldwide influence of the work of B. V. Gnedenko. *Teoriya Veroyatnostei i ee Primeneniya*, **58**(1), 27–36.

Bingham, N. H., Goldie, C. M., and Teugels, J. L. 1987. *Regular Variation*. Cambridge: Cambridge University Press.

Binswanger, K., and Embrechts, P. 1994. Longest runs in coin tossing. *Insurance: Mathematics and Economics*, **15**, 139–149.

Bird, J., and Fortune, J. 2007. Bird & Fortune – silly money – investment bankers. youtube.com/watch?v=9z70BKwfSUA. Last accessed April 24, 2022.

BIS. 2020. Statistical release: OTC derivatives statistics at end-June 2020. bis.org/publ/otc_hy2011.pdf. Last accessed April 24, 2022.

Blastland, M., Freeman, A. L. J., van der Linden, S., Marteau, D. M., and Spiegelhalter, D. 2020. Five rules for evidence communication. *Nature*, **587**(19), 362–364.

Blum, P. 2019. *The Weather Machine: How We See Into the Future*. London: The Bodley Head.

Bollerslev, T. 1986. Generalized autoregressive conditional heteroskedasticity. *Journal of Econometrics*, **31**(3), 307–327.

Bortkiewicz, L. J. 1898. *Das Gesetz der Kleinen Zahlen*. Leipzig: B. G. Teubner.

Brenner, A. D. 1990. *A Guide to the Microfilm Edition of the Emil J. Gumbel Collection: Political Papers of an Anti-Nazi Scholar in Weimar and Exile, 1914–1966*. New York: Leo Baeck Institute.

Brenner, A. D. 2002. *Emil J. Gumbel: Weimar German Pacifist and Professor*. Brill Academic Publishers.

Brown, P. 2015. How math's most famous proof nearly broke. *Nautilus Newsletter*. nautil.us/how-maths-most-famous-proof-nearly-broke-rp-235446/. Last accessed April 24, 2022.

Bui, H. M. 2010. Mathematicians and physicists on banknotes. personalpages.manchester.ac.uk/staff/hung.bui/Collection/Notes/banknotes.htm. Last accessed April 24, 2022.

Camus, A. 1947. *La Peste*. Gallimard.

Cantor, G. 1891. Ueber eine elementare Frage der Mannigfaltigkeitslehre. *Jahresbericht der Deutschen Mathematiker-Vereinigung*, **1**, 75–78.

Cardil, R. 2019. Kepler: The volume of a barrel, another look. www.matematicasvisuales.com/english/html/history/kepler/keplerbarrel2.html. Last accessed April 24, 2022.

Carroll, L. 1865. *Alice's Adventures in Wonderland*. London: Macmillan and Co.

Cartlidge, E. 2015. Italy's supreme court clears L'Aquila earthquake scientists for good. *Science*. doi:10.1126/science.aad7473.

Cartlidge, E. 2016. Seven-year legal saga ends as Italian official is cleared of manslaughter in earthquake trial. *Science*. doi:10.1126/science.aah7374.

Caspari, R., and Lee, S.-H. 2012. The evolution of grandparents: The rise of senior citizens may have played a big role in the success of our species. *Scientific American*, **22**(1S), 38–43. doi:10.1038/scientificamericanhuman1112-38.

CERN. 2013. New results indicate that particle discovered at CERN is a Higgs boson. *CERN Press Release*, 14 March. home.cern/news/press-release/cern/new-results-indicate-particle-discovered-cern-higgs-boson. Last accessed April 24, 2022.

Ciecka, J. E. 2008. The first mathematically correct life annuity. *Journal of Legal Economics*, **15**(1), 59–63.

Cohen, A. N. 2020. False positives in PCR tests for COVID-19. icd10monitor.medlearn.com/false-positives-in-pcr-tests-for-covid-19/. Last accessed April 24, 2022.

Colman, A.D. 2000. *Report of the Re-opened Formal Investigation into the Loss of the MV Derbyshire*. London: The Stationery Office Books.

Congressional Oversight Panel. 2010. June Oversight Report. The AIG rescue, its impact on markets, and the government's exit strategy. govinfo.gov/content/pkg/CPRT-111JPRT56698/pdf/CPRT-111JPRT56698.pdf. Last accessed April 24, 2022.

Conway, J. H., and Kochen, S. 2006. The free will theorem. *Foundations of Physics*, **36**(10), 1441–1473.

CRED (Centre for Research on the Epidemiology of Disasters). 2021. EM-DAT – The International Disaster Database. emdat.be. Last accessed April 24, 2022.

Dalal, S. R., Fowlkes, E. B., and Hoadley, B. 1989. Risk analysis of the Space Shuttle: pre-Challenger prediction of failure. *Journal of the American Statistical Association*, **84**(408), 945–957.

d'Alembert, J. L. R. 1754. Croix ou pile. Pages 512–513 of: Diderot, D., and d'Alembert, J. L. R. (eds.), *Encyclopédie ou Dictionnaire Raisonné des Sciences, des Arts et des Métiers*, vol. 4.

Das, B., Embrechts, P., and Fasen, V. 2013. Four theorems and a financial crisis. *International Journal of Approximate Reasoning*, **54**, 701–716.

Davison, A. C., and Hinkley, D. V. 1997. *Bootstrap Methods and Their Application*. Cambridge: Cambridge University Press.

Daw, R. H., and Pearson, E. S. 1972. Studies in the history of probability and statistics. XXX. Abraham De Moivre's 1733 derivation of the normal curve: A bibliographical note. *Biometrika*, **59**(3), 677–680.

de Grey, A. 2004. We will be able to live to 1,000. news.bbc.co.uk/2/hi/uk_news/4003063.stm. Last accessed April 24, 2022.

de Witt, J. 1671. *Treatise on Life Annuities*. English translation by Frederick Hendriks: Contributions to the history of insurance, and of the theory of life contingencies, with a restoration of the Grand Pensionary De Wit's treatise on Life annuities. *Journal of the Institute of Actuaries*, **2**(3), 222–258 (1852).

Deacon, B. 2021. 'One-in-100-years' flood talk disastrously misleading and should change, risk experts say. ABC Weather, March 28. abc.net.au/news/2021-03-28/one-in-100-years-flood-talk-misleading/100030144. Last accessed April 24, 2022.

Delta Committee. 1962. *Report of the Delta Committee: Final Report. Part 1.* The Hague: State Printing and Publishing Office.

Deltacommissie. 1961. *Rapport Deltacommissie*. 's-Gravenhage: Staatsdrukkerij en Uitgeverijbedrijf.

Devlin, K. 2002. The most beautiful equation in mathematics. *Wabash Magazine*, Winter/Spring. wabash.edu/magazine/2002/WinterSpring2002/mostbeautiful.html. Last accessed April 24, 2022.

Diaconis, P., and Mosteller, F. 1989. Methods for studying coincidences. *Journal of the American Statistical Association*, **84**(408), 853–861.

Dodge, M. M. 1866. *Hans Brinker, or The Silver Skates: A Story of Life in Holland.* New York: James O'Kane.

Drake, S. 1967. *Dialogue Concerning the Two Chief World Systems – Ptolemaic and Copernican.* Berkeley and Los Angeles: University of California Press.

Dumas, A. 1850. *La Tulipe Noire*. Paris: Baudry.

Edunov, S., Bhagat, S., Burke, M., Diuk, C., and Filiz, I. O. 2016. Three and a half degrees of separation. research.facebook.com/blog/2016/02/three-and-a-half-degrees-of-separation/. Last accessed April 24, 2022.

Efron, B. 1979. Bootstrap methods: Another look at the jackknife. *Annals of Statistics*, **7**(1), 1–26.

Einmahl, J. H. J., and Magnus, J. R. 2008. Records in athletics through extreme-value theory. *Journal of the American Statistical Association*, **103**(484), 1382–1391.

Einmahl, J. J., Einmahl, J. H. J., and Haan, L. de. 2019. Limits to human life span through extreme value theory. *Journal of the American Statistical Association*, **114**(527), 1075–1080.

Embrechts, P. 1999. Die Zauberformel gibt es nicht. Pages 28–29 of: Wüstholz, G. (ed.), *Mathematik – Das Geistige Auge*. Departement Mathematik, ETH-Zurich.

Embrechts, P. 2018. January 31, 1953, and September 11, 2001: Living with risk. Farewell Lecture. video.ethz.ch/speakers/lecture/32c992d0-4586-45de-98ea-dea16af0c154.html. Last accessed April 24, 2022.

Embrechts, P., Klüppelberg, C., and Mikosch, T. 1997. *Modelling Extremal Events for Insurance and Finance*. Berlin: Springer.

Engle, R. F. 1982. Autoregressive conditional heteroscedasticity with estimates of the variance of United Kingdom inflation. *Econometrica*, **50**(4), 987–1007.

EMSC (European–Mediterranean Seismological Centre). 2021. Search for earthquakes. emsc-csem.org/Earthquake/?filter=yes. Last accessed April 24, 2022.

Euclid. 300 BCE. *Elements*. Vol. 5. Edited and translated into English by Richard Fitzpatrick. farside.ph.utexas.edu/books/Euclid/Euclid.html. Last accessed April 24, 2022.

FBIIC (Financial and Banking Information Infrastructure Committee). 2008. Market wide pandemic exercise 2006 progress report. https://www.fbiic.gov/public/2008/june/Market%20Wide%20Pandemic%20Exercise%202008%20Progress%20Update%20May%202008.pdf. Last accessed April 24, 2022.

Ferguson, T. S. 1989. Who solved the secretary problem? *Statistical Science*, **4**(3), 282–289.

Ferreira, A., and Haan, L. de. 2015. On the block maxima method in extreme value theory: PWM estimators. *Annals of Statistics*, **43**(1), 276–298.

Feynman, R. P., and Leighton, R. 1985. *"Surely You're Joking, Mr. Feynman!": Adventures of a Curious Character*. New York: W. W. Norton & Company.

Feynman, R. P., and Leighton, R. 1988. *"What Do You Care What Other People Think?": Further Adventures of a Curious Character*. New York: W. W. Norton & Company.

Flusfeder, D. 2018. *Luck*. London: Fourth Estate Ltd.

Forbes, J. D. 1843. *Travels through the Alps of Savoy and Other Parts of the Pennine Chain: With Observations on the Phenomena of Glaciers*. Bibliothèque de Genève, FB 325.

Frei, G., and Stammbach, U. 2007. *Mathematicians and Mathematics in Zürich, at the University and the ETH*. ETH Zürich Bibliothek.

Gabbud, C., Micheletti, N., and Lane, S. N. 2016. Response of a temperate alpine valley glacier to climate change at the decadal scale. *Geografiska Annaler: Series A, Physical Geography*, **98**(1), 81–95.

Gage, J., and Spiegelhalter, D. 2018. *Teaching Probability*. Cambridge: Cambridge University Press.

Gaivoronski, A. A., and Pflug, G. 2005. Value-at-risk in portfolio optimization: Properties and computational approach. *Journal of Risk*, **7**(2), 1–31.

Galeano, E. 2001. *Upside Down: A Primer for the Looking-Glass World*. London: Picador.

Galileo, G. 1623. *The Assayer*. Translated from the Italian "Il Saggiatore" by S. Drake and C. D. O'Malley, 1960. Philadelphia: University of Pennsylvania Press.

Galileo, G. 1632. *Dialogue Concerning the Two Chief World Systems – Ptolemaic and Copernican*. English Translation by Stillman Drake, 1953. Second edition 1967. Berkeley and Los Angeles: University of California Press.

Garber, P. M. 2000. *Famous First Bubbles: The Fundamentals of Early Manias*. Cambridge, MA: The MIT Press.

Gardner, M. 1970a. Mathematical games: The fantastic combinations of John Conway's new solitary game "life". *Scientific American*, **223**(4), 120–123.

Gardner, M. 1970b. The paradox of the nontransitive dice. *Scientific American*, **223**(4), 110–111.

Gascoigne, T., Schiele, B., Leach, J., Riedlinger, M., Lewenstein, B. V., Massarani, L., and Broks, P. 2020. *Communicating Science: A Global Perspective*. ANU Press. `press.anu.edu.au/publications/communicating-science`. Last accessed April 24, 2022.

Gates, W. H. 2015. The next outbreak? We're not ready. `ted.com/talks/bill_gates_the_next_outbreak_we_re_not_ready`. Last accessed April 24, 2022.

Gbari, K. Y. A. S., Poulain, M., Dal, L., and Denuit, M. 2017. Extreme value analysis of mortality at the oldest ages: A case study based on individual ages at death. *North American Actuarial Journal*, **21**(3), 397–416.

Geller, R. J. 1997. Earthquake prediction: A critical review. *Geophysical Journal International*, **131**(3), 425–450.

Gemmrich, J., and Cicon, L. 2022. Generation mechanism and prediction of an observed extreme rogue wave. *Scientific Reports*, **12**(1718).

Gerritsen, H. 2005. What happened in 1953? The big flood in the Netherlands in retrospect. *Philosophical Transactions of the Royal Society A*, **363**(1831), 1271–1291.

GLAMOS. 2020. Swiss glacier length change. Release 2020, Glacier Monitoring Switzerland. `doi:10.18750/lengthchange.2020.r2020`.

GLAMOS. 2021. Swiss glaciers. `glamos.ch`. Last accessed April 24, 2022.

Goddard Institute for Space Studies. 2021. Combined land–surface air and sea–surface water temperature anomalies (Land–Ocean Temperature Index, L-OTI). Global-mean monthly, seasonal, and annual means, 1880–present, updated through most recent month. `data.giss.nasa.gov/gistemp`. Last accessed April 24, 2022.

Goodell, J. 2017. *The Water Will Come. Rising Seas, Sinking Cities, and the Remaking of the Civilized World*. New York: Little, Brown and Company.

Gorroochurn, P. 2011. Errors of probability in historical context. *The American Statistician*, **65**(4), 246–254.

Gorvett, R. 2014. Living to 100: Socioeconomic implications of increased longevity. Living to 100 Symposium in Orlando, Florida. `soa.org/globalassets/assets/files/resources/essays-monographs/2014-living-to-100/mono-li14-4b-gorvett.pdf`. Last accessed April 24, 2022.

Gould, S. J. 1992. *Bully for Brontosaurus: Reflections in Natural History*. New York: W. W. Norton & Company.

Graham, D. W. 2021. Heraclitus. In: *The Stanford Encyclopedia of Philosophy*, Zalta, E. N. (ed.).

Guldimann, T. M. 2000. The story of RiskMetrics. *Risk*, **13**(1), 56–58.

Gumbel, E. J. 1958. *Statistics of Extremes*. New York: Columbia University Press.

Haan, L. de. 1970. On regular variation and its application to the weak convergence of sample extremes. Thesis, Mathematical Centre Tract, vol. 32, University of Amsterdam.

Haan, L. de. 1990. Fighting the arch-enemy with mathematics. *Statistica Nederlandica*, **44**(2), 45–68.

Hand, D. J. 2014. *The Improbability Principle: Why Coincidences, Miracles and Rare Events Happen Every Day*. London: Transworld-Penguin.

Hand, D. J. 2020a. Dark data, on being misled by what isn't there and what to do about it. *Significance*, **17**(3), 42–44.

Hand, D. J. 2020b. *Dark Data: Why What You Don't Know Matters*. Princeton, NJ: Princeton University Press.

Harris, W. 1994. Heraclitus: The complete fragments: Translation and commentary, and the Greek text. Middlebury College. wayback.archive-it.org/6670/20161201173100/http:/community.middlebury.edu/~harris/. Last accessed April 24, 2022.

Hasian, Jr., M., Paliewicz, N. S., and Gehl, R. W. 2014. Earthquake controversies, the L'Aquila trials, and the argumentative struggles for both cultural and scientific power. *Canadian Journal of Communication*, **39**(4), 557–576.

Hawkes, A. G. 1971. Spectra of some self-exciting and mutually exciting point processes. *Biometrika*, **58**, 83–90.

Hawkins, E., Orgega, P., Suckling, E., Schurer, A., Hegerl, G., Jones, P., Joshi, M., Osborn, T. J., Masson-Delmotte, V., Mignot, J., Thorne, P., and van Oldenborgh, G. J. 2017. Estimating changes in global temperature since the preindustrial period. *Bulletin of the American Meteorological Society*, **98**(9), 1841–1856.

Heffernan, J. E., and Tawn, J. A. 2001. Extreme value analysis of a large designed experiment: A case study in bulk carrier safety. *Extremes*, **4**, 359–378.

Heffernan, J. E., and Tawn, J. A. 2003. An extreme value analysis for the investigation into the sinking of the M. V. Derbyshire. *Journal of the Royal Statistical Society: Series C (Applied Statistics)*, **52**(3), 337–354.

Heffernan, J. E., and Tawn, J. A. 2004. Extreme values in the dock. *Significance*, **1**(1), 13–17.

Heilprin, J. 2013. Now confident: CERN physicists say new particle is Higgs boson (Update 3). phys.org/news/2013-03-confident-cern-physicists-higgs-boson.html. Last accessed April 24, 2022.

Herzberg, A. M., and Krupka, I. (eds.). 1997. *Statistics, Science and Public Policy II: Hazards and Risks*. Proceedings of the Conference on Statistics, Science and Public Policy, Queen's University, Kingston, Ontario, Canada, April 23–25, 1997.

Hesterberg, T. C. 2015. What teachers should know about the bootstrap: Resampling in the undergraduate statistics curriculum. *The American Statistician*, **69**(4), 371–386.

Higgs, P. W. 1964. Broken symmetries and the masses of gauge bosons. *Physical Review Letters*, **13**(16), 508–509.

Hirsch, L. 2021. What's next for the business of longevity? *New York Times*, October 16. www.nytimes.com/2021/10/16/business/dealbook/the-business-of-longevity.html. Last accessed April 24, 2022.

Hoehn, A., and Huber, M. 2005. *Pythagoras. Erinnern Sie Sich?* Zürich: Orell Füssli.

Hofert, M., and Scherer, M. 2011. CDO pricing with nested Archimedean copulas. *Quantitative Finance*, **11**(5), 775–787.

Hofert, M., Frey, R., and McNeil, A. J. 2020a. *The Quantitative Risk Management Exercise Book*. Princeton, NJ: Princeton University Press.

Hofert, M., Frey, R., and McNeil, A. J. 2020b. *The Quantitative Risk Management Exercise Book Solution Manual*. github.com/qrmtutorial/qrm/releases/download/TQRMEB/The_QRM_Exercise_Book.pdf. Last accessed April 24, 2022.

Hofstadter, D. 1979. *Gödel, Escher, Bach: An Eternal Golden Braid*. New York: Basic Books.

Holton, G. A. 2014. *Value-at-Risk: Theory and Practice*. Second edition. www.value-at-risk.net/.

Ibragimov, R., and Walden, J. 2007. The limits of diversification when losses may be large. *Journal of Banking and Finance*, **31**(8), 2551–2569.

Ibrion, M., Paltrinieri, N., and Nejad, A. R. 2020. Learning from non-failure of Onagawa nuclear power station: An accident investigation over its life cycle. *Results in Engineering*, **8**, 100185.

Imperiale, A. J., and Vanclay, F. 2019. Reflections on the L'Aquila trial and the social dimensions of disaster risk. *Disaster Prevention and Management*, **28**(4), 434–445.

Iovenko, C. 2018. Dutch masters: The Netherlands exports flood-control expertise. *Earth Magazine*, August 31. `https://www.earthmagazine.org/article/dutch-masters-netherlands-exports-flood-control-expertise`. Last accessed April 24, 2022.

IPCC. 1998. Principles governing IPCC work. `ipcc.ch/site/assets/uploads/2018/09/ipcc-principles.pdf`. Last accessed April 24, 2022.

IPCC. 2021. Climate Change 2021: The physical science basis. Sixth Assessment Report. `ipcc.ch/report/sixth-assessment-report-working-group-i/`. Last accessed April 24, 2022.

Istituto Nazionale di Geofisica e Vulcanologia. 2021. Earthquake list with real-time updates. `http://cnt.rm.ingv.it/en`. Last accessed April 24, 2022.

Jones, S. 2009. On couples and copulas: The formula that felled Wall St. *Financial Times*, 14 January. `www.ft.com/content/912d85e8-2d75-11de-9eba-00144feabdc0`. Last accessed April 24, 2022.

Jorion, P. 2006. *Value at Risk: The New Benchmark for Managing Financial Risk*. Third edition. New York: McGraw–Hill.

Jorion, P. 2008. Risk management lessons from Long-Term Capital Management. *European Financial Management*, **6**(3), 277–300.

Kahane, J.-P. 2018. Les mathématiques responsables de la crise? *Progressistes*. N° 19, Où va la Finance? `revue-progressistes.org/2018/05/10/les-mathematiques-responsables-de-la-crise-jean-pierre-kahane/`. Last accessed April 24, 2022.

Karinthy, F. 1929. Chain links. `vadeker.net/articles/Karinthy-Chain-Links_1929.pdf`. Last accessed April 24, 2022.

Kepler, J. 1615. *Nova Stereometria Doliorum Vinariorum*. Posner Memorial Collection, Carnegie Mellon University Libraries. `http://posner.library.cmu.edu/Posner/books/book.cgi?call=520_K38PN`. Last accessed April 24, 2022.

Kind, J. M. 2014. Economically efficient flood protection standards for the Netherlands. *Journal of Flood Risk Management*, **7**(2), 103–117.

Kolmogorov, A. N. 1933. *Grundbegriffe der Wahrscheinlichkeitsrechung*. Berlin: Springer-Verlag.

Kruizinga, S., and Lewis, P. 2018. How high is high enough? Dutch flood defenses and the politics of security. *BMGN – Low Countries Historical Review*, **133**(4), 4–27.

Kurokawa, K., Ishibashi, K., Tanaka, K., Oshima, K., Tanaka, M., Sakiyama, H., Hachisuka, R., Sakurai, M., and Yokoyama, Y. 2012. The Official Report of The Fukushima Nuclear Accident Independent Investigation Commission. *The National Diet of Japan*. `nirs.org/wp-content/uploads/fukushima/naiic_report.pdf`. Last accessed April 24, 2022.

Lagarias, J. C., Rains, E., and Vanderbei, R. J. 2009. The Kruskal count. Pages 371–391 of: *The Mathematics of Preference, Choice and Order: Essays in Honor of Peter J. Fishburn*, Brams, S., Gehrlein, W. V., and Roberts, F. S. (eds.). Berlin: Springer-Verlag.

Lambert, P. 2020. MV Derbyshire – 40 years on. National Museums Liverpool. `liverpoolmuseums.org.uk/stories/mv-derbyshire-40-years`. Last accessed April 24, 2022.

Leadbetter, M. R. 1991. On a basis for "peaks over threshold" modeling. *Statistics & Probability Letters*, **12**, 357–362.

Lecat, M. 1935. *Erreurs de Mathématiciens des Origines à nos Jours*. Bruxelles: Librairie Castaigne.

L'Ecuyer, P. 1999. Good parameters and implementations for combined multiple recursive random number generators. *Operations Research*, **47**(1), 159–164.

Leemis, L. M., and McQueston, J. T. 2008. Univariate distribution relationships. *The American Statistician*, **62**(1), 45–53.

Leeson, N. 2015. *Rogue Trader*. London: Sphere.

Leibniz, G. W. 1768. *Opera Omnia*. Vol. 6. Geneva: Fratres de Tournes.

LePan, N. 2020. Visualizing the history of pandemics. *Visual Capitalist*, March 14. `www.visualcapitalist.com/history-of-pandemics-deadliest/`. Last accessed April 24, 2022.

Lerner-Lam, A. L. 1997. Predictable debate. *Seismological Research Letters*, **68**(3), 381–382.

Levitis, D. A., and Martinez, D. E. 2017. The two halves of U-shaped mortality. *Frontiers in Genetics of Aging*, **4**(31), 1–6.

Li, D. X. 2000. On default correlation: A copula function approach. *Journal of Fixed Income*, **9**(4), 43–54.

Littlewood, J. E. 1953. *Miscellany*. London: Methuen.

Livio, M. 2020. Did Galileo truly say, "And yet it moves"? A modern detective story. *Scientific American*, May.

Mace, R. 2021. What's a 100-year flood? A hydrologist explains. *The Conversation*, June. theconversation.com/whats-a-100-year-flood-a-hydrologist-explains-162827. Last accessed April 24, 2022.

Mackay, C., and De La Vega, J. 1995. *Extraordinary Popular Delusions and the Madness of Crowds & Confusión de Confusiones*. New York: Wiley.

Malkiel, B. G. 2003. *A Random Walk Down Wall Street*. New York: W. W. Norton & Company.

Margaritondo, G. 2005. Explaining the physics of tsunamis to undergraduate and non-physics students. *European Journal of Physics*, **26**(3), 401–407.

Maslow, A. H. 1966. *The Psychology of Science: A Reconnaissance*. New York: Harper & Row.

Masters, A.. 2022. Behind the numbers: The RSS puts the statistical skills of MPs to the test. *Statistics News*, February 11. rss.org.uk/news-publication/news-publications/2022/general-news/behind-the-numbers-the-rss-puts-the-statistical-sk/. Last accessed April 24, 2022.

Matsuzawa, T. 2014. The largest earthquakes we should prepare for. *Journal of Disaster Research*, **9**(3), 248–251.

Mattmüller, M. 2014. The difficult birth of stochastics: Jacob Bernoulli's *Ars Conjectandi* (1713). *Historia Mathematica*, **41**(3), 277–290.

McAllister, S. 2021. How does climate change affect wine? *Zurich Magazine*, July. zurich.com/en/media/magazine/2021/how-does-climate-change-affect-wine. Last accessed April 24, 2022.

McConway, K., and Spiegelhalter, D. 2021. Communicating statistics through the media in the time of COVID-19. blogs.lse.ac.uk/impactofsocialsciences/2021/02/03/communicating-statistics-through-the-media-in-the-time-of-covid-19/. Last accessed April 24, 2022.

McLean, B., and Nocera, J. 2010. *All the Devils are Here: The Hidden History of the Financial Crisis*. New York: Portfolio/Penguin Press.

McNeil, A. J., Frey, R., and Embrechts, P. 2015. *Quantitative Risk Management: Concepts, Techniques and Tools*. Second edition. Princeton NJ: Princeton University Press.

Meadows, D. H., Meadows, D. L., Randers, J., and Behrens III, W. W. 1972. *The Limits to Growth*. New York: Universe Books.

Merseyside Maritime Museum. 2021. Life on Board: The sinking of MV Derbyshire. liverpoolmuseums.org.uk/merseyside-maritime-museum/mv-derbyshire#section-derbyshire-family-association. Last accessed April 24, 2022.

Met Office. 2021. Hadley Centre Central England Temperature (HadCET) dataset. metoffice.gov.uk/hadobs/hadcet/. Last accessed April 24, 2022.

MeteoSwiss. 2021. Federal Office of Meteorology and Climatology MeteoSwiss. meteoswiss.admin.ch. Last accessed April 24, 2022.

Milgram, S. 1967. The small-world problem. *Psychology Today*, **1**, 61–67.

Modena, C., Valluzzi, M. R., da Porto, F., and Casarin, F. 2011. Structural aspects of the conservation of historic masonry constructions in seismic areas: Remedial measures and emergency actions. *International Journal of Architectural Heritage*, **5**(4-5), 539–558.

Moore, H. 2011. The story behind the financial crisis tattoo. *Marketplace* May 13, 2011. marketplace.org/2011/05/13/story-behind-financial-crisis-tattoo/. Last accessed April 24, 2022.

Morgan, J. P. and Reuters. 1996. RiskMetrics™ – Technical Document. Fourth Edition. https://www.msci.com/documents/10199/5915b101-4206-4ba0-aee2-3449d5c7e95a. Last accessed April 24, 2022.

Morris, R. L. 2016. Lazy Mathematicians? rebeccaleamorris.com/2016/09/30/lazy-mathematicians/. Last accessed April 24, 2022.

MTECT (Ministère de la Transition Écologique et de la Cohésion des Territoires). 2020. Impacts du changement climatique: Agriculture et forêet. ecologie.gouv.fr/impacts-du-changement-climatique-agriculture-et-foret. Last accessed April 24, 2022.

National Research Council. 2014. *Lessons Learned from the Fukushima Nuclear Accident for Improving Safety of U.S. Nuclear Plants.* Washington, D.C.: The National Academies Press.

Nordhaus, W. 2009. An analysis of the Dismal Theorem. Yale University: Cowles Foundation Discussion Paper 1686. papers.ssrn.com/sol3/papers.cfm?abstract_id=1330454. Last accessed April 24, 2022.

Normile, D. 2021. Zhengzhou subway flooding a warning for other major cities: The unprecedented rainfall showed the vulnerability of underground infrastructure. *Science*, July 29.

Ogata, Y. 1988. Statistical models for earthquake occurrences and residual analysis for point processes. *Journal of the American Statistical Association*, **83**(401), 9–27.

Ogata, Y. 1998. Space-time point-process models for earthquake occurrences. *Journal of the American Statistical Association*, **50**(2), 379–402.

Ogawa, Y. 2009. *The Housekeeper and the Professor.* London: Picador.

O'Harrow, Jr., R., and Dennis, B. 2008a. The beautiful machine. *Washington Post*, December 29. washingtonpost.com/wp-dyn/content/article/2008/12/28/AR2008122801916.html. Last accessed April 24, 2022.

O'Harrow, Jr., R., and Dennis, B. 2008b. A crack in the system. *Washington Post*, December 30. washingtonpost.com/wp-dyn/content/article/2008/12/29/AR2008122902670.html. Last accessed April 24, 2022.

O'Harrow, Jr., R., and Dennis, B. 2008c. Downgrades and downfall. *Washington Post*, December 31. washingtonpost.com/wp-dyn/content/article/2008/12/30/AR2008123003431.html. Last accessed April 24, 2022.

Olshansky, S. J. 2004. Don't fall for the cult of immortality. BBC News December 3. news.bbc.co.uk/1/hi/uk/4059549.stm. Last accessed April 24, 2022.

Osinski, M. 2018. The oyster farmer of Wall Street. youtube.com/watch?v=031HpyOMAU4. Last accessed April 24, 2022.

Pais, A. 1982. How Einstein got the Nobel Prize. *American Scientist*, **70**(4), 358–365.

Parker, D. E., Legg, T. P., and Folland, C. K. 1992. A new daily Central England temperature series, 1772–1991. *International Journal of Climatology*, **12**, 317–342.

Pavord, A. 1999. *The Tulip: The Story of a Flower That Has Made Men Mad.* London: Bloomsbury Publishing Ltd.

Peacock, E. 2013. When the gambler's fallacy comes true: Beating the online casino. *Significance*, **10**(6), 40–42.

Pearce, M., and Raftery, A. E. 2022. Will this be a record-breaking century for human longevity? *Significance*, **18**(6), 6–7.

Persily, N., and Tucker, J. A. 2020. *Social Media and Democracy. The State of the Field and Prospects for Reform.* Cambridge: Cambridge University Press.

Petram, L. 2014. *The World's First Stock Exchange.* New York: Columbia Business School Publishing.

Pollack, L. 2012. The formula that Wall Street never believed in. *Financial Times*, June 15. ft.com/content/a19cdaf1-f5db-3abc-823a-8671e8169c5a. Last accessed April 24, 2022.

Pollard, J. M. 1978. Monte Carlo methods for index computation (mod p). *Mathematics of Computation*, **32**(143), 918–924.

Posthumus, N. W. 1929. The tulip mania in Holland in the years 1636 and 1637. *Journal of Economic and Business History*, **1**(3), 434–466.

Preece, D. A., Ross, G. J. S., and Kirby, S. P. J. 1988. Bortkewitsch's horse-kicks and the generalised linear model. **37**(3), 313–318.

Puccetti, G., and Scherer, M. 2018. Copulas, credit portfolios, and the broken heart syndrome: An interview with David X. Li. *Dependence Modeling*, **6**(1), 114–130.

Quantitative Finance Stack Exchange. 2017. Data for the tulip mania. quant.stackexchange.com/questions/36804/data-for-the-tulip-mania. Last accessed April 24, 2022.

Rajan, R. G. 2005a. The Greenspan era: Lessons for the future. International Monetary Fund, August 27. imf.org/en/News/Articles/2015/09/28/04/53/sp082705. Last accessed April 24, 2022.

Rajan, R. G. 2005b. Has financial development made the world riskier? National Bureau for Economic Research, NBER Working Paper 11728. nber.org/papers/w11728. Last accessed April 24, 2022.

Rajan, R. G. 2010. *Fault Lines: How Hidden Fractures Still Threaten the World Economy*. Princeton, NJ: Princeton University Press.

Raspe, R. E. 1785. *Baron Munchausen's Narrative of His Marvellous Travels und Campaigns in Russia*. Oxford: Smith.

Reinhart, C. M., and Rogoff, K. S. 2009. *This Time is Different: Eight Centuries of Financial Folly*. Princeton, NJ: Princeton University Press.

Remund, A., Camarda, C. G., and Riffe, T. 2018. A cause-of-death decomposition of young adult excess mortality. *Demography*, **55**, 957–978.

Rijkswaterstaat. 2021. Ministerie van Infrastructuur en Waterstaat. waterinfo.rws.nl/#!/kaart/Waterbeheer/. Last accessed April 24, 2022.

Robinson, M. E., and Tawn, J. A. 1995. Statistics for exceptional athletics records. *Journal of the Royal Statistical Society: Series C (Applied Statistics)*, **44**(4), 499–511.

Rocard, M. 2008. La crise sonne le glas de l'ultralibéralisme. *Le Monde*, November 1, 2008. lemonde.fr/la-crise-financiere/article/2008/11/01/michel-rocard-la-crise-sonne-le-glas-de-l-ultraliberalisme_1113586_1101386.html. Last accessed April 24, 2022.

Rogers, W. P. 1986. Report to the President on the Space Shuttle Challenger Accident. sma.nasa.gov/SignificantIncidents/assets/rogers_commission_report.pdf. Last accessed April 24, 2022.

Rootzén, H., and Katz, R. W. 2013. Design life level: Quantifying risk in a changing climate. *Water Resources Research*, **49**, 5964–5972.

Rose, T., and Sweeting, T. 2016. How safe is nuclear power? A statistical study suggests less than expected. *Bulletin of the Atomic Scientists*, **72**(2), 112–115.

Ruf, D. 2019. Statistik des Verbrechens – Ein Mathematiker gegen das Naziregime. Sudwestrundfunk, September 19. swr.de/unternehmen/kommunikation/pressedossiers/swrfernsehen-junger-dokumentarfilm-2019-statistik-des-verbrechens-100.html. Last accessed April 24, 2022.

Sagan, C. 1997. *Billions and Billions: Thoughts on Life and Death at the Brink of the Millennium*. New York: Ballantine Books.

Salmon, F. 2009. Recipe for disaster: The formula that killed Wall Street. *Wired*, February 23. wired.com/2009/02/wp-quant/?currentPage=all. Last accessed April 24, 2022.

Salmon, F. 2012. The formula that killed Wall Street. *Significance*, **9**(1), 16–20.

Schiff, D. 2005. AIG replaces Hank Greenberg as CEO: The inevitable hour. *Schiff's Insurance Observer*, **17**(5). insuranceobserver.com/PDF/2005/031505.pdf. Last accessed April 24, 2022.

Schmitt, R. 2009. Prophet and loss. *Stanford Magazine*, March/April. stanfordmag.org/contents/prophet-and-loss. Last accessed April 24, 2022.

Schneider, I. 1984. The role of Leibniz and Jakob Bernoulli for the development of probability theory. *LLULL, Boletin de la Sociedad Española de Historia de las Ciencias*, **7**, 69–89.

Scott, A. J., Martin, E., and Sinclair, D. A. 2021. The economic value of targeting aging. *Nature Aging*, **1**, 616–623.

Shafer, G., and Vovk, V. 2018. The origins and legacy of Kolmogorov's *Grundbegriffe*. Available at arxiv.org/abs/1802.01219. Last accessed April 24, 2022.

Shreve, S. E. 2008. Don't blame the quants. *Forbes*, October 8. forbes.com/2008/10/07/securities-quants-models-oped-cx_ss_1008shreve.html. Last accessed April 24, 2022.

Sigmund, K. 2017. *Exact Thinking in Demented Times: The Vienna Circle and the Epic Quest for the Foundations of Science*. New York: Basic Books.

Sinclair, D. A. 2019. *Lifespan*. New York: HarperCollins.

Singh, S. 1997. *Fermat's Last Theorem*. London: Fourth Estate Ltd.

Sippel, S., Fischer, E. M., Scherrer, S. C., Meinshausen, N., and Knutti, R. 2020. Late 1980s abrupt cold season temperature change in Europe consistent with circulation variability and long-term warming. *Environmental Research Letters*, **15**, 094056.

Slager, K. 2003. *De Ramp*. Amsterdam: Olympus Pockets.

Smith, R. L. 1997. Letter to the editors. *Journal of the Royal Statistical Society: Series C (Applied Statistics)*, **46**(1), 123–128.

Sornette, D. 2009. Dragon-kings, black swans and the prediction of crises. *International Journal of Terraspace Science and Engineering*, **1**(3), 1–17.

Sornette, D., Kröger, W., and Wheatley, S. 2019. *New Ways and Needs for Exploiting Nuclear Energy*. Berlin: Springer.

Spiegelhalter, D. 2011. Fear and numbers in Fukushima. *Significance*, **8**(3), 100–103.

Spiegelhalter, D. 2019. Good luck Professor Spiegelhalter. BBC Radio 4, February 22. bbc.co.uk/programmes/b09kpmys. Last accessed April 24, 2022.

Spiegelhalter, D. 2020. David Spiegelhalter and false positives. MathsWorldUK, October 14. youtube.com/watch?v=XmiEzi541BI. Last accessed April 24, 2022.

Spiegelhalter, D., and Masters, A. 2021. *Covid by Numbers. Making Sense of the Pandemic with Data.* Milton Keynes: Pelican Books, Penguin Random House.

Steinhaus, H., and Trybula, S. 1959. On a paradox in applied probabilities. *Bulletin of the Polish Academy of Sciences, Mathematics*, **7**, 67–69.

Stephenson, A. G., and Tawn, J. A. 2013. Determining the best track performances of all time using a conceptual population model for athletics records. *Journal of Quantitative Analysis in Sports*, **9**(1), 67–76.

Stigler, S. M. 1990. *The History of Statistics: The Measurement of Uncertainty Before 1900.* Cambridge, MA: The Belknap Press of Harvard University Press.

Stiglitz, J. E. 1992. Banks versus markets as mechanisms for allocating and coordinating investment. Pages 15–38 in: *The Economics of Cooperation: East Asian Development and the Case for Pro-Market Intervention*, J. A. Roumasset and S. Barr (eds.). Boulder, CO: Westview Press.

Sullivan, J. 2018. The historic 1988 Senate climate hearing: 30 years later. UN Foundation, June 22. unfoundation.org/blog/post/the-historic-1988-senate-climate-hearing-30-years-later/. Last accessed April 24, 2022.

Sunday Times Wine Club. 2018. Why English sparkling wine is the new Champagne. *The Sunday Times*, May 16. thetimes.co.uk/static/why-english-sparkling-is-the-new-champagne/. Last accessed April 24, 2022.

Swain, F. 2021. The device that reverses CO_2 emissions. BBC Future Planet, March 12. bbc.com/future/article/20210310-the-trillion-dollar-plan-to-capture-co2. Last accessed April 24, 2022.

Swiss Federal Banking Commission. 2008. Subprime crisis: SFBC investigation into the causes of the write-downs of UBS AG. finma.ch/FinmaArchiv/ebk/e/publik/medienmit/20081016/ubs-subprime-bericht-ebk-e.pdf. Last accessed April 24, 2022.

Swiss Re Institute. 2019. L'Aquila, 10 years on. *Sigma*, Special feature. swissre.com/dam/jcr:a1800651-558e-403a-b8df-cb4d8ce9e2ec/Sigma_2_2019_feature_EN.PDF. Last accessed April 24, 2022.

Sylvestre, A. 1975. Une sorcière comme les autres. *EPM Musique*.

Synolakis, C., and Kânoğlu, U. 2015. The Fukushima accident was preventable. *Philosophical Transactions: Mathematical, Physical and Engineering Sciences*, **373**(2053), 1–23.

Tahan, M. 1993. *The Man Who Counted: A Collection of Mathematical Adventures.* New York: W. W. Norton & Company.

Taleb, N. N. 2007. *The Black Swan: The Impact of the Highly Improbable.* New York: Random House.

Taleb, N. N. 2018. *Skin in the Game: Hidden Asymmetrics in Daily Life.* New York: Penguin Random House.

Tawn, J. A. 2017. Extreme value theory: An impact case study for international shipping standards. youtube.com/watch?v=-6NPV5vb4yM. Last accessed April 24, 2022.

Tawn, J. A., and Heffernan, J. E. 2001. Summary of statistical analysis of the seakeeping model tests. Pages 41–54 of: *Proc. Royal Institution of Naval Architects Conference. Design & Operation of Bulk Carriers: Post M.V. Derbyshire*, London.

TEPCO (Tokyo Electric Power Company). 2012a. Fukushima nuclear accident analysis report <summary attachment>. www.tepco.co.jp/en/press/corp-com/release/betu12_e/images/120620e0103.pdf. Last accessed April 24, 2022.

TEPCO (Tokyo Electric Power Company). 2012b. The scale of the tsunami far exceeded all previously held expectations and knowledge. www.tepco.co.jp/en/nu/fukushima-np/info/12042401-e.html. Last accessed April 24, 2022.

Thompson, E. A. 2007. The tulipmania: Fact or artifact? *Public Choice*, **130**, 99–114.

United Nations. 2015. Paris Agreement. unfccc.int/sites/default/files/english_paris_agreement.pdf. Last accessed April 24, 2022.

USGS (United States Geological Survey). 2021a. Cool earthquake facts. usgs.gov/programs/earthquake-hazards/cool-earthquake-facts?qt-science_center_objects=0#qt-science_center_objects. Last accessed April 24, 2022.

USGS (United States Geological Survey). 2021b. Search earthquake catalog. earthquake.usgs.gov/earthquakes/search/. Last accessed April 24, 2022.

Utsu, T. 1970. Aftershocks and earthquake statistics (II): Further investigation of aftershocks and other earthquake sequences based on a new classification of earthquake sequences. *Journal of the Faculty of Science, Hokkaido University, Series 7, Geophysics*, **3**(4), 197–266.

van Dantzig, D. 1956. Economic decision problems for flood prevention. *Econometrica*, **24**(3), 276–287.

van Doorn, L. 2010. Tulip mania. Bachelor Thesis, Tilburg University. arno.uvt.nl/show.cgi?fid=129437. Last accessed April 24, 2022.

van Son, C. 2015. Space is the place. sites.psu.edu/vansonspace/. Last accessed April 24, 2022.

van Staveren, I. 2014. The Lehman sisters hypothesis. *Cambridge Journal of Economics*, **38**(5), 995–1014.

Veerman, C. P. and Stive, M.J. 2008. Working together with water: A living land builds for its future. Deltacommissie. https://repository.tudelft.nl/islandora/object/uuid:af79991f-31e7-47a4-a6ef-bfd54ca59c57?collection=research. Last accessed April 24, 2022.

Vishal, K. 2019. The mystery of MV Derbyshire. *Marine Chronicle*, May 10.

Weitzman, M. 2009. On modeling and interpreting the economics of catastrophic climate change. *Review of Economics and Statistics*, **91**(1), 1–19.

Wells, H. G. 1903. *Mankind in the Making*. Chapman & Hall.

Wigner, E. 1960. The unreasonable effectiveness of mathematics in the natural sciences. *Communications on Pure and Applied Mathematics*, **13**(1), 1–14.

Wikipedia. 2020a. 2009 swine flu pandemic. wikipedia.org/w/index.php?title=2009_swine_flu_pandemic&oldid=952432460. Last accessed April 24, 2022.

Wikipedia. 2020b. Tyche. wikipedia.org/wiki/Tyche. Last accessed April 24, 2022.

Wikipedia. 2021a. 2009 L'Aquila earthquake. wikipedia.org/wiki/2009_L'Aquila_earthquake#Aftershocks. Last accessed April 24, 2022.

Wikipedia. 2021b. Actuarial notation. wikipedia.org/wiki/Actuarial_notation. Last accessed April 24, 2022.

Wikipedia. 2021c. Power law. wikipedia.org/wiki/Power_law#Examples. Last accessed April 24, 2022.

Wikipedia. 2021d. Simpson's paradox. wikipedia.org/wiki/Simpson's_paradox. Last accessed April 24, 2022.

Wikipedia. 2021e. Special report on global warming of 1.5 °C. https://en.wikipedia.org/wiki/Special_Report_on_Global_Warming_of_1.5_%c2%b0C. Last accessed April 24, 2022.

World Future Council. 2016. Sponge cities: What is it all about? January 20. worldfuturecouncil.org/sponge-cities-what-is-it-all-about/. Last accessed April 24, 2022.

Wucker, M. 2016. *The Gray Rhino: How to Recognize and Act on the Obvious Dangers we Ignore*. New York: St Martin's Press.

Yu, K. J. Z. 2016. *Sponge City: Theory and Practice*. China Architecture & Building Press.

Index